ANIMAL LOCOMOTION

ANIMAL LOCOMOTION

PHYSICAL PRINCIPLES AND ADAPTATIONS

MALCOLM S. GORDON

REINHARD BLICKHAN

JOHN O. DABIRI

JOHN J. VIDELER

CRC Press
Taylor & Francis Group
Boca Raton London New York

CRC Press is an imprint of the
Taylor & Francis Group, an **informa** business

CRC Press
Taylor & Francis Group
6000 Broken Sound Parkway NW, Suite 300
Boca Raton, FL 33487-2742

Printed in Canada on acid-free paper

International Standard Book Number-13: 978-1-138-03576-8 (Hardback)

Library of Congress Cataloging-in-Publication Data

Names: Gordon, Malcolm S., author. | Blickhan, Reinhard, author. | Dabiri, John O. (John Oluseun), author. | Videler, John J., author.
Title: Animal locomotion : physical principles and adaptations / Malcolm S. Gordon, Reinhard Blickhan, John O. Dabiri, and John J. Videler.
Description: Boca Raton : Taylor & Francis, 2017. | Includes bibliographical references and indexes.
Identifiers: LCCN 2016052112 | ISBN 9781138035768 (hardback : alk. paper)
Subjects: LCSH: Animal locomotion. | Animal mechanics.
Classification: LCC QP301 .G59 2017 | DDC 591.5/7--dc23
LC record available at https://lccn.loc.gov/2016052112

Visit the Taylor & Francis Web site at
http://www.taylorandfrancis.com

and the CRC Press Web site at
http://www.crcpress.com

CONTENTS

PREFACE

INTRODUCTION

This book is an advanced-level introduction to the current state of knowledge in three active research areas: comparative biophysics, biomechanics, and bioengineering of animals. The general subject is animal locomotion including its evolutionary background and context. The three specific areas are swimming, natural flight, and terrestrial locomotion. This book is published in hardcover and as an online eBook. Both the volume itself and the chapters within it are organized so that they may be downloaded separately.

Chapters 1 and 2 provide contexts for the Chapters 3 through 5 within the frameworks of the scientific method and modern evolutionary biology. The specific discussions focus on the biophysical, biomechanical, and bioengineering bases for how macroscopic animals move around and survive in natural environments. Biomedical perspectives are secondary. The intended primary audience is advanced-level students and researchers mainly interested in and trained in mathematics, physical sciences, and engineering. Students with interests and backgrounds in many areas of biology (e.g., organismic biology, functional morphology, organ system and ecological physiology, physiological ecology, molecular biology, molecular genetics, and systems biology) should also find the volume useful, although some may find the mathematical approaches used occasionally challenging.

Modern biophysics, biomechanics, and bioengineering are actively growing, intellectually challenging, and highly visible interdisciplinary fields of research, development, and application. The largest parts of these fields study living (as opposed to fossil) animals, are biomedically oriented, and are increasingly centered on studies of what are called model organisms (e.g., in addition to the traditional laboratory rats, mice, and guinea pigs, roundworms [*Coenorhabditis*], fruit flies [*Drosophila*], and zebrafish [*Brachydanio*]).

Recently, these fields have undergone processes of internal differentiation, subdivision, and diversification that have broadly followed those existing in the more traditional areas of engineering. An analogy could be that relevant selected aspects of biology and biomedicine have been grafted onto major areas of engineering (e.g., mechanical, aeronautical, civil, marine, electrical, and chemical engineering). Regarding the physical sizes of the systems studied, there has been further differentiation into work on nano-, micro-, meso-, and macroscale systems.

The experience of the authors is that the basic training of many people primarily interested in mathematics, the physical sciences, and engineering often shows substantial gaps and imbalances with respect to some of the most relevant fundamentals of modern biology. This discrepancy is particularly the case for comparative organismic biology, biodiversity, phylogenetic systematic methods, paleobiology, and organic evolution. It is also often true that their course work in biology emphasized molecular genetics and molecular, cellular, and tissue biology, whereas it de-emphasized or omitted organismic, evolutionary, organ system, comparative, ecological, environmental, behavioral, and population genetics and biology.

No single book can remediate all of these gaps and imbalances. Recognizing this fact, this volume tries to provide an essential, foundational level of information in the most relevant of these de-emphasized or omitted areas in ways that are both interesting and challenging.

THIS BOOK

Primary goals of this book are as follows:

1. To introduce interested people to the state of current knowledge and understanding in three active major subareas of comparative animal locomotion: swimming, natural flight, and terrestrial locomotion.
2. To present both the broad subjects and the specific fields as parts of an overall scientific effort to understand the organization, structure, function, and history of the living world in the context of a modern understanding of evolutionary biology.
3. To help people who come to this subject area from nonbiological backgrounds to gain better understanding and perspective with respect to the intellectual position of biomedically oriented studies of humans and of model systems that illuminate important human-related questions. We (humans) are the way we are and do what we do in large part because of our evolutionary history. This book discusses many biomedically important topics, but it approaches them from the perspective of how the biodiversity of animals survive and function in the natural world.
4. To introduce people to some of the biophysical-, biomechanical-, and bioengineering-related aspects of the tree of life, with examples selected from the diversity of organisms that share our planet with us.
5. To provide selected, and hopefully provocative and stimulating, examples of ways in which biophysics, biomechanics, and bioengineering can help to develop new technologies and products based on studies of living organisms viewed from the perspectives of biomimetics and bioinspired design.

The past 20 years has seen remarkable growth in interest and research literature in all major areas of biomechanics, biophysics, and bioengineering; thus, we believe there is a need for more quantitative, physical sciences, and engineering-oriented books like this one.

To achieve that goal, the organization and content of this book reflect several important trade-offs:

1. Although less comprehensive in scope than some current books, subject areas included are discussed in greater depth and at more advanced levels.
2. The evolutionary and biodiversity-oriented (cladistic) framework presented has rarely, if ever, been used in previous textbooks in bioengineering.
3. Students interested in areas not discussed in this volume are best served by referring to more specialized books and review articles on those topics. The reference sections in this book include useful referrals.

This book represents a four-person team effort devoted to providing authoritative and in-depth treatments of the included subject areas. Each discussion is based on the research interests and activities of the respective co-author(s). An important result of this approach is that it exposes students to varied theoretical and stylistic approaches to subject matter. The contents of most chapters are based on courses taught by their author(s).

Recommended background for people using this book includes knowledge of basic organismic (including evolutionary) biology, some organismic or organ system animal physiology, and basic physics and chemistry. Mathematical background should include 2 years of university-level calculus directed toward students in the physical sciences and engineering (as taught at many research universities), including vector calculus, linear algebra, differential equations, and complex analysis.

Major features of this book are as follows:

1. It is selective and not intended to be encyclopedic in its coverage of its subject fields.

2. Treatment of each major subject area is structured around discussions of basic physical principles and then of selected adaptations. It emphasizes the important underlying theoretical and mathematical foundations of the areas covered and illustrates the real-world manifestations of these foundations with selected examples of processes and adaptations occurring primarily in living animals. Discussions of fossil organisms are included where relevant.
3. Primary contexts for all discussions are biodiversity-based, comparative, evolutionary, environmental, ecological, and behavioral.
4. Levels of structural and organizational complexity discussed are mostly at the macroscopic organismic and organ system levels.
5. Human-related materials are included where they best fit these other contexts.
6. Occasional limited mentions of microorganisms and plants are included.

ACKNOWLEDGMENTS

Many people have contributed to the development and writing of this book in a wide range of ways. The project had a longer and more complex developmental process and history than any of us anticipated. The authors heartily thank everyone. Omissions from this list are unintentional, and we apologize for any omissions.

Charles R. ("Chuck") Crumly, Senior Acquisitions Editor/Life Sciences for CRC Press/Taylor & Francis, supported and oversaw the project. Laura K. Jordan acted as coordinator and editorial associate with primary responsibilities for meeting the exacting requirements of CRC Press/Taylor & Francis for the art program and permissions. Michael Davis of the UCLA Life Sciences South Administration information technology staff provided essential computer support.

Other people who contributed in important ways to scientific content include, in alphabetical order, Donald G. Buth, Morteza ("Mory") Gharib, Hao Liu, Paul W. Webb, and Daniel Weihs.

AUTHORS

Malcolm S. Gordon, PhD, is professor of biology in the Department of Ecology and Evolutionary Biology, University of California (UCLA), Los Angeles. He was born in Brooklyn, New York. He earned his undergraduate degree in zoology (with high honors) from Cornell University, Ithaca, New York, in 1954 and his PhD in zoology from Yale University, New Haven, Connecticut, in 1958. He did most of his doctoral research at the Woods Hole Oceanographic Institution, Woods Hole, Massachusetts. He spent a postdoctoral year (1957–1958) in the Department of Zoology, University of Cambridge, United Kingdom. He joined the faculty in the then Department of Zoology (now Department of Ecology and Evolutionary Biology) at UCLA. He has been at UCLA since, except for sabbatical years and leaves of absence at several other places. He has been a visiting professor at various times at the University of Chile (Santiago and Montemar); the Chinese University of Hong Kong; Chulalongkorn University (Bangkok, Thailand); and the University of California, Santa Cruz. He was for several years a visiting associate in aeronautical engineering and bioengineering at the California Institute of Technology, Pasadena. In 1968–1969, he spent a year as assistant director for research for the then proposed U.S. National Fisheries Center and Aquarium in the Department of the Interior, Washington, DC.

Prof. Gordon has been at various times a National Science Foundation Predoctoral Fellow, a Fulbright Postdoctoral Fellow (UK), a Guggenheim Fellow, and a Senior Queen's Fellow in Marine Science (Australia). On several occasions, he was a U.S. National Academy of Sciences lecturer visiting laboratories of the then Academy of Sciences of the USSR. In 2000, he was the Irving-Scholander Memorial Lecturer, University of Alaska, Fairbanks.

Prof. Gordon's primary basic research interests are in several areas of comparative ecological animal physiology. Much of his work has been at the organismic and organ system levels, on marine animals, especially fishes, which has had a strong evolutionary component. Major research areas have included osmotic and ionic regulation in fishes and amphibians, metabolic adaptations of fishes to life in the open sea and the deep ocean, adaptations of amphibious fishes to life out of water, ecophysiology of environmental factors having adverse effects on amphibian survival, and (since the late 1990s) biomechanics and hydrodynamics of fish locomotion. He has done applied research relating to recycling of dissolved nutrients in treated wastewaters, to aspects of fish aquaculture, and to fish and crustacean behavior in relation to fishing nets. He has published several papers on the theoretical aspects of evolutionary biology and also conducts reviews on interactions between comparative physiological knowledge and public policy related to animal health and welfare.

Prof. Gordon has published more than 140 research papers in refereed scientific journals.

He has co-written and edited eight books, five of which are university-level textbooks. Four of the textbooks are editions of one of the (at the time) most widely used texts in comparative and evolutionary physiology. This volume is his ninth book.

Aspects of the basic research Prof. Gordon and his colleagues did on fish locomotion have had two unexpected practical applications. One application has been the development and use by the U.S. Navy of a new type of remotely operated underwater vehicle (ROV). The other was the development and sale by several major automobile manufacturers of multiple lines of passenger vehicles, the shapes of which were influenced by the shapes of the bodies of some of the fishes studied.

A major theme in Prof. Gordon's research has been to combine field studies of the biology of animals studied with laboratory studies of physiological processes. In this context, he has organized and participated in research expeditions to many parts of the world.

Prof. Gordon is a long-time member of four scientific societies: the American Association for the Advancement of Science (AAAS), American Physiological Society (APS), Society for Integrative and Comparative Biology (SICB), and Society for Experimental Biology (SEB). He is a Fellow of AAAS.

Prof. Gordon was treasurer of the International Union of Physiological Sciences (IUPS) from 2005 to 2013. In that role, he was an ex-officio member of the Joint Managing Board of the review journal *Physiology*. That journal is a joint publication of IUPS and APS.

Prof. Gordon was the major professor for 35 doctoral students (PhD and DEnv) and 30 master's students. He has also mentored nine postdoctoral researchers.

Reinhard Blickhan, PhD, was professor of motion science at the Friedrich Schiller University, Jena, Germany, until September 2016. He studied physics at the Justus-Liebig-University, Gießen, and the Technical University Darmstadt, Darmstadt, Germany. His experimental master's thesis (1976) dealt with "critical slowing down," a cooperative phenomenon observed in the vicinity of phase transitions in materials. Combining his background in physics with a continuing interest in living systems, he joined the group of "Sensory Physiology" around Prof. F.G. Barth at the Johann-Wolfgang-Goethe University, Frankfurt am Main. In his PhD dissertation (1983), he investigated how the movements of spiders generate strains in their exoskeletons that can be detected by their slit sense organs. A postdoctoral stay in the stimulating environment created by C.R. Taylor and T.A. McMahon at the Concord (Massachusetts) Field Station, Harvard University, allowed him to deepen his experimental and theoretical research on the biomechanics of legged locomotion. Returning to Germany, he joined the group of W. Nachtigall. By introducing new flow visualization techniques (particle image velocimetry [PIV]) and combining this with kinematic and electromyography (EMG) studies, he investigated the energetics of fish locomotion (Walther-Arndt Habilitationspreis of the Deutschen Zoologischen Gesellschaft). Equipped with a Heisenberg fellowship, he received a call from Friedrich Schiller University for the chair of biomechanics at the Institute of Sport Science. He served for 7 years as director of the institute and headed the Innovation College "Motion Systems."

With a group of young scientists having backgrounds in biology, sports science, engineering, and physics, he followed up his interests in basic principles of locomotion. Objects of research have been trout, ants, spiders, cockroaches, crabs, lizards, birds, mice, rabbits, macaques, and humans. Major questions are as follows: Can we improve the prediction and classification of muscle properties to understand their composition and use within the motion system? To what extent do external mechanical boundary conditions or the conditions of an ecological niche determine biological solutions? How do the mechanical properties, the composition of the motion system, and its nonlinear dynamics facilitate control?

John O. Dabiri is a full professor of civil and environmental engineering and of mechanical engineering at Stanford University, California. His research focuses on science and technology at the intersection of fluid mechanics, energy and environment, and biology. Honors for this work include a MacArthur Fellowship, an Office of Naval Research Young Investigator Award, and a Presidential Early Career Award for Scientists and Engineers (PECASE). *Popular Science* magazine named him one of its "Brilliant 10" scientists for his research in bioinspired propulsion. For his research in bioinspired wind energy, *Bloomberg Businessweek* magazine listed him among its technology innovators, and *MIT Technology Review* magazine named him one of its 35 innovators under 35. In 2014, he was elected a Fellow of the American Physical Society. He currently serves on the editorial boards of the *Journal of Fluid Mechanics* and the *Journal of the Royal Society Interface*, and he is a member of the U.S. National Committee for Theoretical and Applied Mechanics.

John J. Videler, PhD, (Heerlen, The Netherlands, 1941) is a biologist and freelance scientist, writer, and publisher. After finishing his secondary education in Heerlen in 1960, he was trained as a reserve officer in the Royal Dutch Navy in their minesweeping division. He retired as Captain in 1986.

He received a master's (Doctorandus) degree in biology from the University of Amsterdam in 1969. His main subjects of study were functional morphology and fisheries biology with plant ecology and zoological taxonomy as subsidiaries.

From 1969 to 1976, he was a lecturer at the zoological laboratory of Groningen University (RUG) and was offered the opportunity to study for a doctorate under the supervision of ethologist Prof. Dr. G.P. Baerends. This resulted in his 1975 dissertation entitled, "On the Interrelationships between Morphology and Movement in the Tail of the Cichlid Fish *Tilapia nilotica* (L.)." While attending the Symposium on Swimming and Flying in Nature held at Caltech (Pasadena) in 1974, he decided to devote much of his career to animal locomotion.

As a senior lecturer at the RUG Department of Ethology (1976–1986), he established and headed the Functional Morphology and Marine Zoology Research Group. Research highlights in those days were kinematic studies of fish and dolphin swimming by using high-speed film and the discovery of the functions of the sliding pelvis of the African clawed frog, *Xenopus laevis*.

A close cooperation with Dr. Clement S. Wardle at the Marine Laboratory Aberdeen started several projects dedicated to kinematics and dynamics of fish swimming; initially aided by a North Atlantic Treaty Organization (NATO) grant. A sabbatical year in Aberdeen financed by a Royal Society Fellowship and cooperation with Prof. Sir James Lighthill (University of Cambridge) greatly facilitated these projects.

In 1987, a new RUG Department of Marine Biology appointed him as associate professor in marine zoology and leader of the Marine Zoology Research Group. Group research included kinematics; energetics and hydrodynamics of copepods, shrimps, fishes, and dolphins; interaction between moving animals and water; structure and function of mollusk radula teeth; parrot fish grazing on coral reefs; and reproductive biology of deep-sea squid. His group was the first to apply flow visualization and interpretation using PIV around swimming animals. Underwater field research (in Bonaire, Eritrea, and Corsica) became a major component of the work of this group. From 1995 to 2006, he was appointed as project leader and coordinator of a Dutch Marine Biology and Fisheries project in Eritrea.

Bird flight studies became an important part of his research interest due to cooperation with Profs. Serge Daan (RUG) and Daniel Weihs (Technion, Haifa) on the kinematics, dynamics, and energetics of the flight of the kestrel, *Falco tinnunculus*. These studies led to novel insights in the evolution of bird flight based on *Archaeopteryx* fossils.

He held a special professorship in evolutionary mechanics at the University of Leiden from 2000 to 2006. The discovery of leading edge vortices on a model of a swift wing by using PIV (published in *Science*) dates from that period. The RUG appointed him as Leonardo da Vinci Professor in Marine Zoology in 2004 and as Professor in Bionics in 2006. He retired from the RUG in 2010.

His current research interests are bionics (the application in technology of principles evolved in nature), the biophysics of swimming and flight, and the evolution of complex locomotion systems. An ongoing interest in drag reducing mechanisms led to the publication in June 2016 of the discovery of novel drag reducing adaptations in the head of the swordfish, *Xiphias gladius*, including an oil gland connected to capillaries and oil-excreting pores, based on magnetic resonance imaging (MRI) and scanning electron microscopy. Prof. Videler authored two seminal books on animal locomotion: *Fish Swimming* (1993) and *Avian Flight* (2006).

CHAPTER ONE

Basics

MALCOLM S. GORDON

1.1 INTRODUCTION

This book is about animal locomotion as viewed from the perspectives of comparative animal biophysics, biomechanics, and bioengineering. Animal biophysics may be defined as the study of biological phenomena, structures, and processes in animals, including intact animals, by using the concepts and methods of physics. The field overlaps broadly with and complements comparative animal physiology, much of which emphasizes chemical and biochemical approaches to animal function. Animal biomechanics and bioengineering may be defined as the application of mechanical and other engineering principles to the study of the same animals, phenomena, structures, and processes (Leondes 2007–2009; Chien et al. 2008; Chien 2013).

The word "comparative" here refers to the use in these fields of taxonomic and phylogenetic positions of animals (more broadly their evolutionary lineages) as a major research variable. An array of specific methods for making comparisons has developed in recent years (Felsenstein 1985; Nunn 2011). See Chapter 2 for the evolutionary contexts of this statement. We also interpret comparative to imply emphasizing those features and properties of animals that have the closest relationships to evolutionary, environmental, and ecological conditions and situations in the lives of the forms studied.

The overall organization, structure, content, and approaches to subject matter used in the book are outlined in the Preface. In addition to materials relating directly to the contents of the chapters, the references sections include key references to books, review articles, and research papers that introduce readers to relevant major subject areas not covered by this volume. Relevant, reliable, and useful websites are also listed. More general discussions of animal locomotion include Alexander (2003), Biewener (2003), Vogel (2009, 2013), Hanson and Åkesson (2015), and Irschick and Higham (2016).

Three foundational areas not discussed further here are cosmology, the relationship between physics and biology, and the philosophy of science. Some useful references include Hofstadter (1999), Laughlin and Pines (2000), Nagel and Newman (2001), McCarthy (2004), Berto (2009), Chang (2012), Eigen (2013), Love (2015), Carroll (2016), Otto (2016), and Tyson et al. (2016).

1.2 LIFE ON EARTH

There was no life on Earth for a long time after the origin of the Solar System about 4.5 billion years ago (Gy). Complex processes of both physical and chemical evolution occurred in the prebiotic world. The circumstances under which life first arose, the when, where, and how of the origin(s) of life, remain subjects of active debate. Identifiable living things appeared and biological (organic) evolution began about 3.7 Gy (the age of the oldest known fossils of microorganisms). Multicellular macroscopic animals began to appear about 0.8 Gy, and much of the diversity of major groups appeared during a relatively short time interval that has come to be called the "Cambrian explosion" (0.541–0.515 Gy). Living animals are the current results of ongoing organic evolutionary experiments that nature has been pursuing since then. The most recent best estimates of the numbers of different kinds (species) of extant multicellular organisms (metazoans) included in the kingdom Animalia is 8.7 (±1.0) million (Costello et al. 2013). Evidence from the fossil record indicates that larger numbers of species lived for varying lengths of time in the past.

Prebiotic evolution and the origin(s) of life are large active research areas. References providing entry into the rapidly growing literature include Anderson (1983), Dose (1988), Hogeweg and Takeuchi (2003), Markovitch and Doron (2013), Archibald (2014), Becker et al. (2016), Mesler and Cleaves (2016), and Olson and Straub (2016).

The Cambrian explosion is also an area of active interest (Marshall 2006, Erwin and Valentine 2013, Smith and Harper 2013).

Animals are classified by biologists as belonging to 32 major groups (phyla). Based on evidence from both the fossil record and from molecular phylogenies, many phyla have existed for very long periods (0.8+ Gy). Most of these groups have evolved substantially during these periods. Living representatives of at least some of these groups occur in almost every physical environment on Earth that has conditions compatible with water and carbon-based life forms. These environments range from the aquatic and subaquatic to the terrestrial (including subterranean) and aerial. Individual animals range in size from microscopic to the blue whale (the largest animal that has ever lived). Excellent introductions to this biodiversity are Pechenik (2014) and Pough et al. (2013).

This book is about selected aspects of the science deriving from study of current biodiversity and its evolutionary history. It begins with the recognition that nature is the designer, architect, engineer, builder, and manufacturer of life on Earth. The biological, physical, and chemical aspects of nature have interacted with each other over huge amounts of time and space, using vast amounts of energy and raw materials. During these processes, nature has explored large numbers of possible designs, structures, mechanisms, and modes of production of living organisms. From a utilitarian perspective, we can view nature as an endless store of good, well-tested ideas on how best to achieve many different important goals significant in our lives.

1.3 PERSPECTIVES

This book is about a continuing set of what we regard as mystery stories (many biologists may be viewed as practitioners of evolutionary forensics). The mysteries are both retrospective and prospective. They are retrospective because the molecular, developmental, biological, morphological, and fossil records of the evolutionary past of animal lineages are in many ways fragmentary and incomplete. With respect to the fossil record in particular what we have found is largely the result of both chance and stochastic processes. These features of the fossil record gradually become more and more important the further one goes back in time (Foote and Miller 2006; Benton 2014).

The mysteries are also prospective because the environments of organisms are always changing and varying in unpredictable ways over wide ranges of temporal and spatial scales. Describing and understanding what is happening presently and what may happen even in the near future are both uncertain and speculative.

As scientists, we use two basic approaches to study the parts of the universe of interest to us: analysis and synthesis. Analyses and their results make up the great majority of this book. Syntheses of information are discussed wherever they are justified.

1.4 FRAMEWORKS

Animals differ from nonliving physical–chemical systems in multiple important ways. Most animals are autonomously motile during at least some parts of their life histories, if not throughout their lives. All animals grow and age as individuals. Many, but not all, reproduce, often in complicated ways. As populations and genetically related groups (lineages and clades), they evolve over time, often in complex ways.

1.4.1 Organization and Structure

The bodies of animals are organized and operate as hierarchical systems. Depending upon the definitions used, one can distinguish at least 10 structural and organizational levels of complexity, beginning with the intact animal and going down to the level of atoms.

There are three main levels of concern in this book: intact organisms, organ systems, and organs. Organisms are made up of multiple organ systems (e.g., integumentary system, gastrointestinal system, respiratory system). Each organ system is composed of multiple organs (e.g., stomach, intestine).

A major aspect of the hierarchical framework within which animals live is the fact that, from evolutionary perspectives, multiorganismic groupings of many kinds are often as important as are the properties of individual animals. The hierarchy continues upward to include populations, systematic categories (species and higher taxa), communities, habitats, ecosystems, environments, regions, and the planet as a whole.

1.4.2 Emergent Properties

An important phenomenon becomes apparent if, instead of taking things apart, one tries to synthesize an organism from its component parts. New properties, not predictable from syntheses of the properties of all lower levels, emerge at each successive level of complexity on the way up from atoms. The incompleteness theorem of Gödel is relevant here (Hofstadter 1999; Nagel and Newman 2001; McCarthy 2004; Berto 2009). For present purposes, the best example is that of intact organisms.

1.5 META-ISSUES

Intact animals are not simply organized constructs of their component organ systems. Intact animals do many things that are unpredictable on the basis of as detailed and complete descriptions of the structures and functions of their organ systems as one can produce. The most obvious and important emergent property is that they behave. They can move around in more or less directed and controlled ways, and they can carry out the functions that are essential to their survival (e.g., finding food, avoiding predators, finding mates).

Most of what animals do is directed at surviving. Survival in the natural world means that animals have to work adequately, but not necessarily optimally, as physical, chemical (abiotic), and biological systems (Denny 1993; Denny and Helmuth 2009; Denny and Benedetti-Cecchi 2012). They must cope over time with changing abiotic spatial and temporal variability. They also must cope with a wide range of biological challenges (the biotic environment).

The second most important thing that animals do is reproduce. From the evolutionary perspective reproductive capacities and the differential survival of offspring are central to evolutionary process. The degree to which an organism contributes to the composition of the next generation is called its fitness.

Physiology is a life science that has historically included studies of how animals work as highly organized, integrated collections of physical and chemical processes and systems. Much of physiology is concerned with the internal workings of animals (their internal environments), with humans having been the primary subjects for much of the field. Other animals, however, also have their own physiologies, hence comparative physiology. Physiology in the context of functional interactions of animals with real-world external environments is environmental and ecological physiology. Physiology in the context of long time periods is evolutionary physiology. Behavioral physiology recognizes the inseparable relationships between bodily functions and activities. The often adverse impacts of many human activities on animals have led to the development of conservation physiology. Excellent introductions to many aspects of comparative, ecological, and evolutionary physiology are Martin et al. (2015) and Hill et al. (2016). The most authoritative ongoing online source on all aspects of animal physiology is edited by Pollock et al. (2015).

The materials, designs, and structures that support and organize the physical–chemical processes and systems in animals have historically been included in biophysics, biomaterials, anatomy, and functional morphology (Wainwright and Reilly 1994; Liem et al. 2001; Vincent 2012; Schilling and Long 2014). Animals working as integrated physical systems are the subject matter of biomechanics (statics and kinematics; Currey 2006; Leondes 2007a, 2007b, 2009) and bioengineering (Chien et al. 2008). Relationships between structure and function are central themes of study. Each of these fields can also be approached in multiple ways just outlined for physiology (i.e., comparative, environmental, ecological).

The advent in more recent years of molecular biological and molecular genetic technologies, theories, and methods has added multiple new dimensions to these classical fields. The aspects most relevant to the concerns of these books have become to significant extents parts of what is now called systems biology (Dubitzky et al. 2013). Systems biology includes the various fields of omics such as genomics, epigenetics, proteomics, physiomics, and metabolomics. The evolutionary and phylogenetic histories and relationships of animal groups, and also the processes involved in their morphological development and growth, are two important subject areas that are in process of being radically changed by the advent of systems biological knowledge.

1.6 ADDITIONAL CONSIDERATIONS

1.6.1 More about the Fossil Record

Careful studies of the fossil record permit multiple insights and inferences relating to the events and processes that have produced the designs, organizations, and structures of living animals. Chapter 2 describes in greater detail the fossil record and many of its properties. Some features of the record that strongly influence and often limit understanding were mentioned in Section 1.2. Additional observations relevant here include the following:

1. New lineages based upon previously unknown designs often appear suddenly in the record, without clear antecedents. It is generally apparent from the complexities of these new designs that they had to have had substantial previous evolutionary histories. The vagaries of chance and stochasticity in fossilization, erosional losses of deposits, and stochasticity of fossil discovery seem to be sufficient to account for the great majority of these sudden appearances. Over time later discoveries of additional fossils of previously unknown or unrecognized more basal forms have shown that many of these apparent saltatory jumps were actually more gradual transitions.
2. Broad trends are sometimes discernible within specific evolutionary lineages. Early diversification of new lineages often has been associated with substantial experimentation with varied morphologies. This variation often decreases gradually over evolutionary time. The surviving groups often share a narrower range of basic body plans (bauplans).
3. Survival of animals in a variety of environments often involves such stringent natural selective pressures that, over time, some animals belonging to phylogenetically unrelated lineages that survive and thrive in those environments end up having surprisingly similar morphologies, life histories, and modes of operation. General terms describing such situations are convergences and, in more recent parlance, homoplasies. Homoplasies may be thought of as evolutionary analogies, contrasting with the usual evolutionary homologies that are the subjects of most evolutionary studies. Considering vertebrate animals as examples, there are striking homoplasic convergences, from genetically disparate ancestral origins, within both swimming and flying vertebrates. In the context of active, open water swimming, consider living fishes such as marlins, swordfish, sailfishes, and great white sharks compared with the reptilian mosasaurs, plesiosaurs, ichthyosaurs, and spinosaurs of the Mesozoic Era. In the context of powered flight, consider modern birds, which are derived theropod dinosaurs, and bats, which are mammals, compared with Mesozoic reptilian, nondinosaur pterosaurs and pterodactyls (Gordon and Notar 2015).
4. A second important group of homoplasies is called evolutionary reversals. Reversals occur when a lineage that has been evolving in one direction, such as moving from aquatic to terrestrial environments, turns around and goes back the other way. Reversals have occurred in multiple lineages of both invertebrates and vertebrates. One of the best known is the derivation of the major groups of cetacean marine mammals (dolphins, porpoises, whales) from terrestrial ancestors related to the modern hippopotamus (even-toed ungulates). The derivations from terrestrial basal forms of the four lineages of large Mesozoic Era marine reptiles (listed in the last paragraph) are others.

1.6.2 Scaling for Size

Scaling for size is a general issue in both biology and engineering. How best to find simple principles that permit useful comparisons between animals of different sizes and shapes has concerned many scientists for a long time (Calder 1984; Schmidt-Nielsen 1984). Scaling is important to studies of animals in two major respects: comparisons between adult animals of different sizes and comparisons of growth stages of individual kinds of animals at different stages of their life histories. Each section of this book discusses scaling.

1.6.3 Statistics

Biologists analyzing living groups and systems almost always have to deal with variable data. Individual animals vary in almost all ways all the time. Comparisons within and between groups almost always are complicated by variability in properties. As a result, to a larger extent than is often the case in the physical sciences, biological approaches to the subjects covered in this book are designed around statistics. Analytical, mathematical, and deterministic approaches provide useful foundations and frameworks, but they require modification to take variability into account.

The training of many physical scientists and engineers is sometimes deficient in statistics. References useful in study design and statistical analysis include Denny and Gaines (2002), Quinn and Keough (2002), Holmes et al. (2011), Ruxton and Colgrave (2011), Gelman et al. (2013), Bickel and Doksum (2015), and McElreath (2015).

1.6.4 Multivariate Processes

Many important features of living systems are resultants of complex multivariate phenomena and processes. So many factors are involved that deterministic approaches may be inadequate to contribute to mechanistic understanding. Consequently, mathematical approaches using a variety of models and techniques such as fuzzy logic and chaos theory, especially fractals, can contribute to an understanding of the designs and functions of important animal systems (Adam 2006; Klemens 2008; Zenil 2012; Hector 2015; Physiome Project 2016).

The circulatory and respiratory systems of animals are good examples of this complexity (Noble et al. 2012; Guillot and Lecuit 2013; Keller 2013; Physiome Project 2016). As a model, consider animals that have closed circulatory systems (such as vertebrates). In addition to a central pump (a heart), those systems are made up of arrays of blood vessels that progress from the small number of the largest arteries carrying blood away from the heart to a graded series of progressively smaller and much more numerous smaller arteries, arterioles, and then near microscopic capillaries. For the return trip, the capillaries connect with venules, small veins, and finally the large veins that carry the blood to the heart.

The total blood flow (cardiac output) has to be carried through each of these segments. The system has to be arranged in such a way that all parts of the animal are supplied with sufficient blood flow to provide the nutrients and oxygen needed to support the tissues and cells and to carry away the metabolic products of cellular activities. It turns out that the overall physical distributions and the numbers of the different categories of blood vessels are consistent with fractal distributions (Weibel 2000).

1.6.5 Physical Chemistry and Thermodynamics

The second law of thermodynamics provides evolution with its direction through time. At any given point in time, individual animals may be viewed as analogous to physical black boxes having a range of properties and functions. They are semi-enclosed thermodynamic systems that work hard, for as long as they live, to avoid having the many internal processes contributing to their survival reach equilibrium positions. They often, not always, seem to work toward steady states in their internal compositions and activities. Dynamic nonequilibrium steady states are often essential to survival. Transition states sometimes occur almost continuously. Metastable states of different kinds are not rare. Equilibrium states mostly occur after death (Cheetham 2011).

The application of thermodynamic concepts to living systems provides an effective framework for studying, characterizing, and understanding how they function in the wide variety of abiotic environments in which they live. A major component of this work results from the fact that the great majority of abiotic environments change with time, on many different time scales and simultaneously in many different ways. Measurements of in, out, and net fluxes of both properties and materials; analyses of the structural, biochemical, biophysical, and biomechanical properties of bounding membranes; and applications of chemical kinetic principles to a range of rate processes can be combined to produce quantitative understanding of what happens to and within organisms as they deal with ongoing environmental changes.

The hierarchical organization of animal bodies is important to this discussion. Organisms can be viewed and studied as wholes, but they are also internally subdivided (e.g., the different organs in an organ system; the different cells within tissues). They are complex multicompartment systems with many different exchanges occurring between the various compartments. Each exchange has its own rate constants and temporal properties.

As with all other thermodynamic systems, the entropy of the universe including an animal system continuously increases. At different places within the system, however, for varying lengths of time, system entropy may locally decrease. This is most likely the case for animals that are developing and growing as embryos and juveniles. Anabolic (synthetic) and organizing processes in young animals generally occur at higher rates and throughputs than catabolic (breakdown) processes.

Thermodynamic concepts and principles apply directly to biochemical reactions and processes and also to physiological processes in such areas as metabolism.

1.7 FRONTIERS

Animal locomotion studied from the perspectives of comparative biophysics, biomechanics, and bioengineering is an active and promising field for both basic science and applications. People going into these areas have many opportunities for intellectually and materially satisfying careers. Below are a few observations:

1. Major growth areas in both the near and longer term are likely to be associated with applications of new technologies to many of the established subject areas (Manneville 2010; Ramshaw 2011; Buresti 2012; Smits and Lim 2012; Physiome Project 2016).
2. Areas likely to produce new insights and products are molecular genetics (including bioinformatics), epigenetics, and nanotechnology.
3. Paying serious attention to biodiversity and evolutionary histories, looking carefully and creatively at wide ranges of relatively poorly studied animals living in different poorly understood environments, is likely to produce startling new knowledge and products (Schwenk et al. 2009).

REFERENCES

Adam, J.A. 2006. *Mathematics in Nature: Modeling Patterns in the Natural World*. Princeton University Press, Princeton, NJ, 416 pp.

Alexander, R.M. 2003. *Principles of Animal Locomotion*. Princeton University Press, Princeton, NJ, 371 pp.

Anderson, P.W. 1983. Suggested model for prebiotic evolution: The use of chaos. *Proc Natl Acad Sci USA* 80: 3386–3390.

Archibald, J. 2014. *One Plus One Equals One: Symbiosis and the Evolution of Complex Life*. Oxford University Press, Oxford, UK, 205 pp.

Becker, S., I. Thoma, A. Deutsch, T. Gehrke, P. Mayer, H. Zipse, and T. Carell. 2016. A high-yielding, strictly regioselective prebiotic purine nucleoside formation pathway. *Science* 352: 833–836.

Benton, M. 2014. *Vertebrate Paleontology*, 4th edn. Wiley-Blackwell, Chichester, UK, 480 pp.

Berto, F. 2009. *There's Something about Gödel: The Complete Guide to the Incompleteness Theorem*. Wiley-Blackwell, Malden, MA, 233 pp.

Bickel, P.J., and K.A. Doksum. 2015. *Mathematical Statistics: Basic Ideas and Selected Topics*, 2 vols. Chapman & Hall/CRC, Boca Raton, FL, 1021 pp.

Biewener, A.A. 2003. *Animal Locomotion*. Oxford University Press, Oxford, UK, 281 pp.

Buresti, G. 2012. *Elements of Fluid Dynamics*. Imperial College Press, London, UK, 604 pp.

Calder, W.A., III. 1984. *Size, Function, and Life History*. Harvard University Press, Cambridge, MA, 431 pp.

Carroll, S. 2016. *The Big Picture. On the Origins of Life, Meaning, and the Universe Itself*. Dutton, New York, 480 pp.

Chang, H. 2012. *Is Water H_2O? Evidence, Realism and Pluralism*. Springer, New York, 338 pp.

Cheetham, N.W.H. 2011. *Introducing Biological Energetics: How Energy and Information Control the Living World*. Oxford University Press, Oxford, UK, 334 pp.

Chien, S. (ed.). 2013. *Mechanotransduction in Vascular Physiology and Bioengineering: Selected Works of Shu Chien*. World Scientific, Singapore, 620 pp.

Chien, S., P.C.Y. Chen, and Y.C. Fung (eds.). 2008. *An Introductory Text to Bioengineering*. World Scientific, Singapore, 564 pp.

Costello, M.J., R.M. May, and N.E. Stork. 2013. Can we name Earth's species before they go extinct? *Science* 339: 413–416.

Currey, J.D. 2006. *Bones: Structure and Mechanics*. Princeton University Press, Princeton, NJ, 456 pp.

Denny, M.W. 1993. *Air and Water: The Biology and Physics of Life's Media*. Princeton University Press, Princeton, NJ, 341 pp.

Denny, M.W., and B. Helmuth. 2009. Confronting the physiological bottleneck: A challenge from ecomechanics. *Integr Comp Biol* 49: 197–201.

Denny, M.W., and L. Benedetti-Cecchi. 2012. Scaling up in ecology: Mechanistic approaches. *Annu Rev Ecol Evol Syst* 43: 1–22.

Denny, M.W., and S. Gaines. 2002. *Chance in Biology: Using Probability to Explore Nature*. Princeton University Press, Princeton, NJ, 416 pp.

Dose, K. 1988. Prebiotic evolution and the origin of life: Chemical and biochemical aspects. *Progr Mol Subcell Biol* 10: 97–112.

Dubitzky, W., O. Wolkenhauer, H. Yokota, and K.H. Cho (eds.). 2013. *Encyclopedia of Systems Biology*. Springer, New York, 4 vols, 2367 pp.

Eigen, M. 2013. *From Strange Simplicity to Complex Familiarity: A Treatise on Matter, Information, Life, and Thought*. Oxford University Press, Oxford, UK, 754 pp.

Erwin, D.H., and J.W. Valentine. 2013. *The Cambrian Explosion: The Construction of Animal Diversity*. Roberts, Greenwood Village, CO.

Felsenstein, J. 1985. Phylogenies and the comparative method. *Am Nat* 125: 1–15.

Foote, M., and A.I. Miller. 2006. *Principles of Paleontology*, 3rd edn. WH Freeman, San Francisco, CA, 480 pp.

Gelman, A., J.B. Carlin, H.S. Stern, D.B. Dunson, A. Vehtari, and D.B. Rubin. 2013. *Bayesian Data Analysis*, 3rd edn. Chapman & Hall/CRC, Boca Raton, FL, 675 pp.

Gordon, M.S. and J.C. Notar. 2015. Can systems biology help to separate evolutionary analogies (convergent homoplasies) from homologies? *Progr Biophys Mol Biol* 117: 19–29.

Guillot, C., and T. Lecuit. 2013. Mechanics of epithelial tissue homeostasis and morphogenesis. *Science* 340: 1185–1189. DOI: 10.1126/science.1235249.

Hanson, L.-A., and S. Åkesson (eds.). 2015. *Animal Movement across Scales*. Oxford University Press, Oxford, UK, 279 pp.

Hector, A. 2015. *The New Statistics with R: An Introduction for Biologists*. Oxford University Press, Oxford, UK, 199 pp.

Hill, R.W., G.A. Wyse, and M. Anderson. 2016. *Animal Physiology*, 4th edn. Sinauer Associates, Sunderland, MA, 828 pp.

Hofstadter, D.H. 1999. *Gödel, Escher, Bach: An Eternal Golden Braid*. Basic Books, New York, 777 pp.

Hogeweg, P., and N. Takeuchi. 2003. Multilevel selection in models of prebiotic evolution: Compartments and spatial self-organization. *Orig Life Evol Biosph* 33: 375–403.

Holmes, D., P. Moody, and D. Dine. 2011. *Research Methods for the Biosciences*, 2nd edn. Oxford University Press, Oxford, UK, 457 pp.

Irschick, D., and T. Higham. 2016. *Animal Athletes: An Ecological and Evolutionary Approach*. Oxford University Press, Oxford, UK, 255 pp.

Keller, P.J. 2013. Imaging morphogenesis: Technological advances and biological insights. *Science* 340: 1234168. DOI: 10.1126/science.1234168.

Klemens, B. 2008. *Modeling with Data: Tools and Techniques for Scientific Computing*. Princeton University Press, Princeton, NJ, 472 pp.

Laughlin, R.B., and D. Pines. 2000. The theory of everything. *Proc Natl Acad Sci USA* 97: 28–31. DOI: 10.1073/pnas.97.1.28.

Leondes, C.T. (ed.). 2007–2009. *Biomechanical Systems Technology: Vol. 1: Computational Methods*. World Scientific, Singapore, 328 pp.

Leondes, C.T. (ed.). 2007a. *Biomechanical Systems Technology: Vol 2: Cardiovascular Systems*. World Scientific, Singapore, 280 pp.

Leondes, C.T. (ed.). 2007b. *Biomechanical Systems Technology: Vol 4: General Anatomy*. World Scientific, Singapore, 344 pp.

Leondes, C.T. (ed.). 2009. *Biomechanical Systems Technology: Vol 3: Muscular Skeletal Systems*. World Scientific, Singapore, 316 pp.

Liem, K., W. Bemis, W.F. Walker, and L. Grande. 2001. *Functional Anatomy of the Vertebrates: An Evolutionary Perspective*, 3rd edn. Harcourt, Philadelphia, PA, 784 pp.

Love, A.C. (ed.). 2015. *Conceptual Change in Biology: Scientific and Philosophical Perspectives on Evolution and Development*. Springer, New York, 490 pp.

Manneville, P. 2010. *Instabilities, Chaos and Turbulence*, 2nd edn. Imperial College Press, London, UK. 456 pp.

Markovitch, O., and L. Doron. 2013. Prebiotic evolution of molecular assemblies: From molecules to ecology, Chapter 20. *12th European Conference on Artificial Life: General Track*. DOI: 10.7551/978-0-262-31709-2-ch020.

Marshall, C. 2006. Explaining the Cambrian "explosion" of animals. *Annu Rev Earth Planet Sci* 34: 355–384.

Martin, L.B., C.K. Ghalambor, and H.A. Woods (eds.). 2015. *Integrative Organismal Biology*. Wiley-Blackwell, Hoboken, NJ, 341 pp.

McCarthy, T. 2004. "Gödel's Proof" reviewed. *Not Am Math Soc* 51: 333–337.

McElreath, R. 2015. *Statistical Rethinking: A Bayesian Course with Examples in R and Stan*. Chapman & Hall/CRC, Boca Raton, FL, 469 pp.

Mesler, B. and H.J. Cleaves, II. 2016 *A Brief History of Creation: Science and the Search for the Origin of Life*. WW Norton, New York, 312 pp.

Nagel, E., and J.R. Newman. 2001. *Gödel's Proof*, rev edn. (D.R. Hofstadter, ed.). New York University Press, New York, 125 pp.

Noble, D., Z. Chen, C. Auffray, and E. Werner (eds.). 2012. *The Selected Papers of Denis Noble CBE FRS: A Journey in Physiology towards Enlightenment*. Imperial College Press, London, UK, 625 pp.

Nunn, C.L. 2011. *The Comparative Approach in Evolutionary Anthropology and Biology*. University of Chicago Press, Chicago, IL, 392 pp.

Olson, K.R., and K.D. Straub. 2016. The role of hydrogen sulfide in evolution and the evolution of hydrogen sulfide in metabolism and signaling. *Physiology* 31: 60–72.

Otto, S. 2016. *The War on Science. Who's Waging It, Why It Matters, What We Can Do About It*. Milkweed Editions, Minneapolis, MN, 530 pp.

Pechenik, J.A. 2014. *Biology of the Invertebrates*, 7th edn. McGraw-Hill Education, New York, 624 pp.

Physiome Project. 2016. On-line at physiomeproject.org A group of websites organized and maintained by an international consortium sponsored by the International Union of Physiological Sciences (IUPS). Websites establish mathematical modeling standards, ontologies, tools and software for describing and modeling biological systems of many kinds using ordinary differential and algebraic equations. *Primarily biomedical*.

Pollock, D.M. (ed. in chief). 2015. *Comprehensive Physiology*. Wiley-Blackwell, San Francisco, CA.

Pough, F.H., C.M. Janis, and J.B. Heiser. 2013. *Vertebrate Life*, 9th edn. Pearson, Boston, MA, 720 pp.

Quinn, G.P., and M.J. Keough. 2002. *Experimental Design and Data Analysis for Biologists*. Cambridge University Press, Cambridge, UK, 556 pp.

Ramshaw, J.D. 2011. *Elements of Computational Fluid Dynamics*. Imperial College Press, London, UK, 140 pp.

Ruxton, G.D., and N. Colegrave. 2011. *Experimental Design for the Life Sciences*, 3rd edn. Oxford University Press, Oxford, UK, 178 pp.

Schilling, N., and J.H. Long. 2014. Axial systems and their actuation: New twists on the ancient body of craniates. *Zoology* 117: 1–92.

Schmidt-Nielsen, K. 1984. *Scaling: Why Is Animal Size so Important?* Cambridge University Press, Cambridge, UK, 241 pp.

Schwenk, K., D.K. Padilla, G.S. Bakken, and R.J. Full. 2009. Grand challenges in organismal biology. *Integr Comp Biol* 49: 7–14.

Smith, M.P., and D.A.T. Harper. 2013. Causes of the Cambrian explosion. *Science* 341: 1355–1356.

Smits, A.J. and T.T. Lim. 2012. *Flow Visualization: Techniques and Examples*, 2nd edn. Imperial College Press, London, UK, 444 pp.

Tyson, N.D., M.A. Strauss, and J.R. Gott. 2016. *Welcome to the Universe: An Astrophysical Tour*. Princeton University Press, Princeton, NJ, 470 pp.

Vincent, J. 2012. *Structural Biomaterials*, 3rd edn. Princeton University Press, Princeton, NJ, 240 pp.

Vogel, S. 2009. *Glimpses of Creatures in their Physical Worlds*. Princeton University Press, Princeton, NJ, 328 pp.

Vogel, S. 2013. *Comparative Biomechanics*, 2nd edn. Princeton University Press, Princeton, NJ, 820 pp.

Wainwright, P.C., and S.M. Reilly (eds.). 1994. *Ecological Morphology: Integrative Organismal Biology*. University of Chicago Press, Chicago, IL, 448 pp.

Weibel, E.R. 2000. *Symmorphosis: On Form and Function in Shaping Life*. Harvard University Press, Cambridge, MA, 263 pp.

Zenil, H. (ed.). 2012. *A Computable Universe: Understanding and Exploring Nature as Computation*. Imperial College Press, London, UK, 856 pp.

CHAPTER TWO

Animal Biodiversity

ORIGINS AND EVOLUTION

MALCOLM S. GORDON

2.1 INTRODUCTION

The main subject of this chapter is the animal portion of what biologists call the tree of life (see Section 2.2.4). It provides a summary overview, tailored to the context of this volume, of the who, what, when, where, and how of the processes and results of organic evolution in animals. The possible why(s) of evolution we leave for others elsewhere. Recent overviews of modern evolutionary biology include Levinton (2001), Hedges and Kumar (2009), Ruse and Travis (2009), Bell et al. (2010), Roff (2010), Koonin (2011), Shapiro (2012), Avise and Ayala (2013), Dial et al. (2015), and Eldredge (2015).

This chapter has two main goals:

1. To provide readers having primarily mathematical, physical science, or engineering backgrounds with sufficient information concerning the origins, history, and variety of the animal biodiversity with which we share Earth that they can place in an informed context the descriptions and discussions of the biophysics, biomechanics, and bioengineering of animal locomotion that make up the core of the book.
2. To also provide those readers with a basic understanding of the main principles and features of contemporary evolutionary biology. Our hope is that this will help them understand that, no matter what aspects of these fields they may work in, the organisms they work with are the way they are largely as a result of their evolutionary histories.

A theatrical metaphor may help people unfamiliar with the complexities of evolutionary biology better understand the organization and content of this chapter. Visualize Earth as a giant stage on which the story of organic evolution plays itself out.

Earth itself has a long, complex evolutionary history (see Chapter 1). Prebiotic Earth (from the initiation of the Solar System to the first appearance of life) was the product of processes best described by cosmology, astrophysics, geophysics, geology, meteorology, physical chemistry (thermodynamics), organic chemistry, and biochemistry (Hazen 2012; Pross 2012).

Things became more relevant to us with the appearance of the first living things. The plot soon got complicated. As always, the general time line was dictated by thermodynamics (Hedges and Kumar 2009; Harte 2011). Major events and long-term trends in the conditions on stage resulted from many perturbations, including but not limited to variations in solar radiation, formation of the oceans and atmosphere, plate tectonics, volcanic eruptions, asteroid collisions, changes in atmospheric gas composition, and climate change.

A second long period of time passed during which microorganisms evolved, multiplied, and diversified. This was the period of formation of the roots of the tree of life. Many of the major groups that developed at these early stages are still important in the living world today.

Macroscopic multicellular (metazoan) animals became important actors about 0.8 Gy (Hedges and Kumar 2009; Erwin and Valentine 2013). The earliest known metazoan fossils are about 0.6 Gy old. The best recent estimate is that there are about 8.7 (±1.0) million species of living animals (Costello et al. 2013). This chapter outlines major features of the rest of the play. Three useful active websites for reference are the Map of Life, Tree of Life, and Open Tree of Life.

2.2 UNDERSTANDING ANIMAL BIODIVERSITY

2.2.1 Classification, Taxonomy, and Systematics

Since the origins of language, humans have given names to objects and living things that they encountered. Specific kinds of animals therefore have multiple common names, depending upon the different languages used by the people talking about them. As the science of biology developed, it became apparent that a system of uniform scientific names was necessary to ensure that scientists in different places and times could be sure that they were talking about the same animal. Some general references that provide essential background for this discussion are MacLeod (2008), Schuh and Brower (2009), and Nielsen (2012).

Carl Linnaeus, a Swedish scientist who lived during the eighteenth century CE, developed the basic system of classification and naming that is still used today. Linnaeus was a classically trained European. He adopted Greek and Latin as the primary languages to be used for naming living things. For species of organisms, both plants and animals, he developed a system of binominal nomenclature. Two names were needed because studies of organismic diversity (anatomy and natural history) demonstrated that most species showed clear similarities to some other species and clear differences from others. The first larger grouping he called a genus. Thus the scientific names of particular kinds of animals consist of both the name of a genus and of a species. The common carp, for example, is a Linnaean species of fish (he gave it its scientific name) called *Cyprinus carpio.*

Linnaeus and the many other biologists in multiple countries who developed the foundations of our knowledge of animal biodiversity also recognized that multiple levels of higher categories were needed to fully classify and describe the animals they studied. The hierarchical classification used today has the following seven major levels, starting from the Animal Kingdom at the highest level:

1. Kingdom Animalia
2. Phylum (32 presently recognized)
3. Class (variable numbers within different phyla)
4. Order (variable numbers within classes)
5. Family (variable numbers within orders)
6. Genus (variable numbers within families)
7. Species (variable numbers within genera)

There are also multiple sublevels within this classification. How many sublevels are used varies widely between different groups of animals. In some groups, for example, both superorders and suborders may be recognized.

The system of animal classification and naming is codified in a book and online: the International Code of Zoological Nomenclature. The code is overseen by an International Commission on Zoological Nomenclature appointed by the International Union of Biological Sciences. The commission also acts on occasion as the body that resolves important disputes about the application and interpretation of the rules. The website for the code is www.iczn.org/iczn/index.jsp.

Each taxonomic unit in these classifications is called a taxon (plural taxa). The taxon for the common carp mentioned above is *Cyprinus carpio* (by convention the binominal names of species are italicized). There are other species in the genus *Cyprinus*, the type genus of the multigeneric fish family Cyprinidae. The many families of fishes that are closely related to the carps all belong to the order Cypriniformes. Beyond this, the higher levels of fish classification are complex and, in many respects, still unresolved. For this discussion, we simply say that the common carp belongs to the group of fishes that have jaws (Gnathostomes), bony skeletons (Teleostomes), and fins that include cartilaginous or bony rays (Actinopterygii) (Nelson 2006).

There is no objective, experimentally testable way to define the boundaries of the higher categories. As a result the scientific literature contains multiple definitions of many of these groups. Each definition is based upon an array of characteristics chosen by the authors of the papers discussing them. The bases for the different arrays often vary and may be controversial (Schuh and Brower 2009).

Biologists who work in the ongoing effort to describe, classify, name, and catalog the diversity of organisms are taxonomists. The basic first level of their work is called alpha taxonomy. Recent estimates are that possibly as few as 25% of living species of animals have been described and named. This fraction varies widely between different groups, being low for insects and nematodes and high for vertebrates such as birds and mammals (Costello et al. 2013).

Alpha taxonomy today has become complex and technical, using both classical methods and modern technology. A major effort in modern taxonomy is the development of a system of barcoding of organisms, based both on anatomical features and molecular biological information. There is uncertainty as to the efficacy of this approach (MacLeod 2008).

For hundreds of years, biologists viewed the edifice of named and classified organisms as static. It was not until the publications of Charles Darwin and Alfred Wallace in the mid-nineteenth century CE that it became apparent that the reality was instead one of constant change and dynamism. Modern genetics began with Gregor Mendel's work in the later nineteenth century. Modern molecular biology and genetics began in the early to mid-twentieth century (Rogers 2016).

These historic changes in perception and understanding transformed many aspects of both science and society. Taxonomy was one of the fields transformed. It became apparent that classifications could describe both organisms themselves and the evolutionary relationships between organisms. Such classifications are called natural classifications. Taxonomy evolved and produced phylogenetic systematics. The major paradigm shift was the recognition that the evolutionary process as a whole could be modeled as an ongoing computer (see Section 2.2.3; Kitching et al. 1998; Wiley and Lieberman 2011; Broom and Rychtar 2013).

2.2.2 Databases

The raw materials for all classifications of animals are the observed characteristics of the animals. Because most animals possess large numbers of characteristics that might be used for classification purposes, disputes and disagreements often have occurred with respect to which characteristics, or sets of characteristics, are the most informative and useful. The development of evolutionary theory added to those disputes questions relating to which characteristics are most important for survival and which are most directly subject to natural selection. The advent of high-capacity computers capable of data mining in very large databases along with the development of the mathematics of phylogenetic analysis (cladistics; Section 2.2.3) resulted in the development of quantitative statistical approaches that make more precise and objective both the processes of estimating levels of selection exerted on particular characteristics and the processes of developing phylogenetic trees and networks of evolutionary relationships (Lemey et al. 2009; Huson et al. 2011; Stevens 2013).

The earliest stages of alpha taxonomy were based primarily upon gross external morphology and various internal anatomical features of adult animals belonging to contemporary groups (note that many species that were common in the fifteenth or sixteenth centuries, also some much more recently, have since gone extinct). Early biologists also recognized that there are often similarities between the sequences of events that occur during embryonic development and growth of particular kinds of animals and the inferred evolutionary histories of those animals. These observations led to the development of a theoretical statement (demonstrably correct only in broad outline): "Ontogeny recapitulates phylogeny." A substantial part of modern evolutionary developmental biology (evo-devo) focuses on these aspects (Hall and Olson 2006; Gilbert and Epel 2009; Tollefsbol 2014; Bustin and Misteli 2016).

Paleontology became central to the study of the evolution of life once the fossil record was properly identified and characterized. The fossil record is literally the ground truth for evolutionary biology. It is the most direct evidence we have of the actual events that occurred during the history of life. Section 2.4 describes the major features of the fossil record and the multiple lines of evidence that derive from it.

The recent development of many new technologies and fields of study stimulated development of an array of additional lines of evolutionarily informative evidence (databases). Most of these databases derive from work on organisms alive today. A few derive from studies of DNA (ancient DNA) extracted from fossils of relatively recently extinct species (Orlando and Willerslev 2014). Their use has added a multitude of new approaches and new insights into evolutionary biology. Many of them constitute molecular evolution.

A partial list of other major databases includes updated versions of several more classical approaches.

These include functional morphology (most relevant to this book), biogeography, life history studies, and studies of animal behavior. Advances in microscopy have enabled use of the microscopic and ultramicroscopic anatomy of cells, tissues, and organs.

The mergers of molecular biology and molecular genetics with biochemistry, cell biology, and phylogenetic systematics produced multiple new fields, all now parts of systems biology (Dubitzky et al. 2013). These include protein and enzyme compositions (proteomics, transcriptomics) and metabolic pathways (metabolomics). One of the most visible and active areas studies DNA relationships (genomics, applied to both mitochondrial and nuclear genomes, also studies of gene families [Lynch 2007; Cannarozzi and Schneider 2012; Rogers 2016]). RNA is also central to many aspects of evolution. Evo-devo was mentioned above. Physiomics, the combination of classical physiology with bioengineering and molecular genetics, is discussed in Section 1.6.4.

Epigenetic processes both expand and complicate these approaches. Epigenetic studies are based on methods for studying gene activity and expression in addition to identifying genes (Nelson and Nadeau 2010; Mattick 2012; Ci and Liu 2015). Many new and different insights into evolutionary relationships in many groups are resulting from these newer parts of systems biology. Section 2.5 includes further discussion of the far-reaching implications of epigenetic studies.

The availability of multiple lines of evidence has both clarified and complicated the grand project of developing a system of natural classifications for all animal groups. It has clarified many questions because it made possible empirical testing of phylogenetic hypotheses. It is now possible to quantitatively compare the different patterns of relationships (phylogenetic trees and networks) generated through use of the different databases derived from the same groups of animals. The inference is that higher degrees of congruence imply more natural classifications (Huson et al. 2011; Wiley and Lieberman 2011).

Multiple lines of evidence have also complicated the project because they have led to many ongoing disputes related to substantial differences of opinion as to the reasons for many cases of only fractional or otherwise incomplete congruence between trees and networks of relationships. Important considerations can relate to questions of which data set best demonstrates the selective pressures producing the evolutionary changes observed within given groups. A wide range of technical, methodological, and interpretive differences also may contribute to differences of opinion. These disputes should ultimately be resolved on the basis of additional information obtained from additional research. The long-term result should be a more precise and accurate understanding of the properties of the tree of life (Vinicius 2010; Wagner 2011).

2.2.3 Cladistics

As mentioned above, the development of information theory and computer science had major impacts on evolutionary biology. Beginning during World War II, Willi Hennig, a German entomologist, developed the system of phylogenetic systematics called cladistics that is almost universally accepted and used today. The first major publication was a book that appeared in 1950. The cladistic approach combines the fundamental features of organic evolutionary process with the data processing power of computers (the field is called bioinformatics) to produce what are hoped to be natural classifications using data from any relevant database.

The scientific literature about the cladistic method and its array of applications is huge and growing rapidly. There are thousands of research papers in hundreds of research journals and dozens of book-length discussions, analyses, reviews, and technical manuals. Useful entries to this literature include Kitching et al. (1998), Lemey et al. (2009), and Huson et al. (2011).

Diagrams showing cladistic relationships within animal groups are cladograms. When considering cladograms, remember that there are multiple kinds of trees and networks that can be calculated from any one or more of the databases mentioned. Two broad categories are based respectively on assumptions of maximum parsimony and on maximum likelihood. Multiple algorithms are available for performing these calculations, each based on different sets of mathematical and statistical assumptions. Which choice is optimal for any particular group or data set can be controversial. It is almost always necessary to run any given calculation multiple times, often hundreds or thousands of times. Significant variations usually occur in the details of the shapes of the calculated trees or networks, in the relative lengths of branches, and in the inferred relationships shown within them. Here we present only a brief outline of the central ideas and principles underlying cladistics.

The central concept as applied to animals is that, at specific places and times, all animal lineages began monophyletically (Gordon 1999 discusses some of the issues arising from this formulation). This means that each natural group of animals began from single, probably small, populations of an ancestral (basal) species. Over time and space the descendants of those basal forms multiplied; distributed themselves geographically; diversified genetically, morphologically, and functionally; and changed. All of the descendants of that basal species, no matter how varied they became genetically, morphologically, or in other ways, are members of a clade. The groups within that clade can be named and characterized to form a set of natural taxa (a lineage).

Associated with this monophyletic model, Hennig and his successors developed new terminology useful in describing the changes that occur in animals as they evolve. The characters (traits, features) of the ancestral basal taxon are plesiomorphies. Plesiomorphic characters that persist with little or no change in two or more descendant species are symplesiomorphies. Successor characters derived from ancestral characters are apomorphies. Derived characters shared by two or more taxa of descendants (shared derived characters) are synapomorphies.

The definitional limits of taxa established using cladistic methods frequently do not agree exactly with those limits as established using more traditional classification methods. Cladistic categories generally include significant ranges of geological time (Hedges and Kumar 2009). Discussions continue in the literature about how best to resolve such disputes.

The process of developing cladistic classifications that are widely agreed to and accepted is not always fast or easy. One problem is that some widely recognized taxa turn out not to be clades. In vertebrate biology that has turned out to be the case with what most people call reptiles. Cladistic analyses indicate that turtles may have derived from a different ancestral basal form than the form that was basal to the snakes and lizards. A named taxon that contains representatives of more than one clade is called a paraphyletic group.

Cladistic methods have helped resolve longstanding questions about several common evolutionary phenomena that are exceptions to the basic phenomenon of monophyletic homology. The general term used to describe these phenomena is homoplasy. Homoplasic taxa are evolutionary analogues. They are taxa within lineages that are phylogenetically distant from each other that have evolved similar important morphological or functional features. There are three types of homoplasies: convergences, parallelisms, and reversions. Cladistically generated phylogenetic trees for many lineages often highlight the widespread occurrence of numbers of convergent homoplasies. Parallelisms and reversions are less abundant, but still appear frequently. Animal flight (see Chapter 4) is one of the most striking of many sets of convergent homoplasies considered in this book. The chapters on both aquatic and terrestrial locomotion (Chapters 3 and 5) also discuss multiple sets of convergent homoplasies. Useful discussions of homoplasies include Sanderson and Hufford (1996), McGhee (2011), Wake et al. (2011), and Gordon and Notar (2015).

Most taxa in most clades are evolutionary homologues. Homologous features of animals (most often structures) are defined as features having the same developmental origins even if the features themselves are substantially different. This is the case for the wings of pterosaurs, birds, and bats because they derive from the tetrapod forelimb. The anatomies of the wings in the three groups have important structural and functional differences.

The homologous wings of pterosaurs, birds, and bats and the wings of insects are traditionally considered to be convergently homoplasic because the four groups are in different phyla and their wings have very different embryonic origins and structures. However, molecular genetic evidence uncovered in evo-devo studies indicates that the genes responsible for wing development in all groups have significant similarities. It is possible to make an argument that epigenetic effects on gene expression played important roles in the origins of wings. The application of cladistic principles and systems biological methods to multiple data sets in this case has yielded significant new insights into how flight structures evolved (see Chapter 4; Gilbert and Epel 2009).

Cladistic approaches facilitate analyses and evolutionary modeling efforts that were previously more subjectively done (Vinicius 2010; Huson et al. 2011; Broom and Rychtar 2013). There are many clades in many animal groups for which the fossil records are minimal or nonexistent. Other lines of evidence are often also fragmentary or inconsistent with one another. In those situations it is difficult to work out the possible, hopefully the probable, properties of the undiscovered basal forms from which the clade in question is descended. These efforts at predicting the major properties of the basal forms are a category of biological reverse engineering.

The intersection of cladistic principles and methods with both the mathematical theory of neutral selection and the realization that large amounts of molecular genetic variation in organisms is manifested in the form of single-nucleotide polymorphisms (SNPs) led to the development of molecular clocks. A central assumption is that SNPs are point mutations that gradually accumulate within lineages over evolutionary time. If the rates of accumulation of numbers of SNPs are fairly constant over long periods and in different groups of organisms cladistic methods make it possible to generate a molecular clock to estimate how much time has passed since the existence of the last common ancestor(s) within and between lineages (Hedges and Kumar 2009; Koonin 2012; Eldredge 2015).

This approach is widely used to estimate the ages at which important branch points in cladograms arose. If the lineage studied also has a substantial fossil record, and that record has been physically well dated, it is possible to test the accuracy of the molecular clock calculations against the measured dates of appearances in the fossil record. Comparisons like these have led to many interesting discussions in the literature.

Recent results deriving from cladistic analyses of very large molecular genetic data sets for important groups of animals have produced the recognition that within genomes substantial numbers of regions have persisted almost completely unchanged for extended periods of evolutionary time. Some of these regions are considered to be pseudogenes ("molecular fossils" no longer active in protein synthesis). Others are regulatory factors controlling epigenetic gene expression in diverse parts of the transcriptomes within different genomes. Still others are called evolutionarily ultraconserved elements. It is probable that many of these stable sequences will play significant roles in future efforts analyzing phylogenetic relationships within specific lineages in many different clades (Dunn et al. 2008; Boyle et al. 2014; Gerstein et al. 2014; Sisu et al. 2014).

2.2.4 The Tree of Life

Our understanding of the cladistically derived structure of the tree of life continues to change with time as large amounts of new information develop resulting from widespread applications of high-throughput techniques to genome and gene sequencing in increasingly numerous groups of organisms. A particularly important aspect of that work is the addition of substantial numbers of fully sequenced genomes deriving from groups of organisms previously unstudied. A second aspect is the addition of multiple genomes within relatively unknown groups. Within group variations are sometimes significant. Two of the better sources of reliable, carefully reviewed, and critiqued information on the present state of understanding are the websites for the Tree of Life Web Project and for the Open Tree of Life. References helpful in understanding the following discussion are Pough et al. (2013) and Pechenik (2014).

The published universal tree of life (UTL) summarizes phylogenetic systematic knowledge of the relative positions and evolutionary relationships of all known and described living things. The present consensus structure of the UTL recognizes four major branches: Eubacteria (true bacteria), two groups of what are called Archaea (Euryarchaeota and Crenarchaeota), and Eukarya. There is an ongoing dispute about the most correct cladistic arrangement of these groups. There are two major alternate models (the "Archaea tree" and the "Eocyte tree"). In both models, animals are a group located on a separate branch of the Eukarya (common name eukaryotes) (Figure 2.1).

Eukaryotes are organisms sharing three main characteristics. Most of the DNA in their cells is enclosed in formed nuclei. They produce most of the proteins they use in complex membranous systems called the endoplasmic reticulum. They generate most of the energy they use in mitochondria. The other three groups listed in the last paragraph are prokaryotes and lack these characteristics.

The bases for the differences of opinion concerning the correct relationships between the prokaryote groups are complex. Most are not directly relevant here. One major aspect is worth mentioning, however, because some of the considerations also apply to the eukaryotes. The issue is that, among the prokaryote groups, lateral gene exchanges between even evolutionarily remotely related organisms have happened multiple times

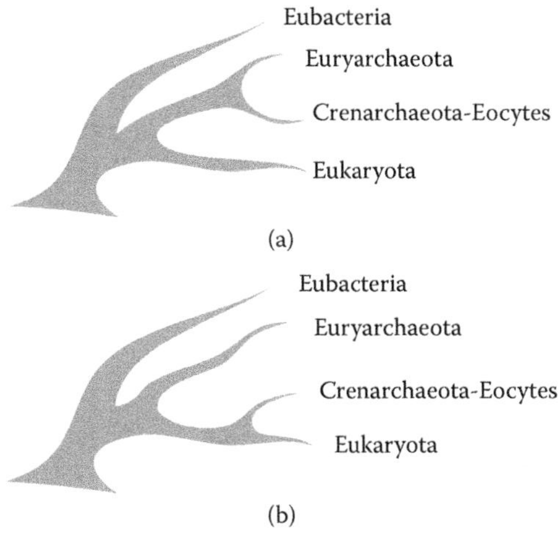

Figure 2.1. Simplified diagrammatic phylogenetic trees showing the two currently widely accepted alternative hypotheses (a, b) for the cladistic relationships between the four major domains of living things. The two trees are equally probable most parsimonious outcomes of large numbers of iterations of phylogenetic tree algorithms applied to very complex large databases. There is presently no clear objective basis for choosing one or the other. Metazoan animals are included in the Eukaryota. (For more information see text, references sections, and Tree of Life Web Project: http://tolweb.org.)

over evolutionary time and apparently occur occasionally today.

A major result of these relatively frequent and easy exchanges of genetic materials is that the genomes of all prokaryotes are actually mosaics of fragments and pieces of the genomes of many different groups of organisms. This makes the cladistic search for what is called the last universal common ancestor of life on Earth complicated, if not impossible. Another way of saying it is that the roots of the UTL are tangled. Because eukaryotes were derived from ancestral prokaryotes, it turns out that eukaryote genomes are also mosaics of multiple parts from multiple sources (Figure 2.2).

There are also multiple complexities and uncertainties with respect to the cladistic relationships between the 32 currently recognized phyla of multicellular (metazoan) animals that concern us here. The broad brush outline of those relationships recognizes three major categories (Table 2.1).

The first category is based upon the number of layers of different tissues that make up the basic structure of their bodies. There are three subcategories: two phyla have poorly defined tissue layers; two phyla seem to have (there is some uncertainty about this) two layers (they are diploblastic). The remaining 28 phyla have three layers (they are triploblastic).

The second category includes all of the triploblastic phyla. Those phyla are subdivided into groups based upon whether or not they have a specific type of body cavity, the coelom. Three phyla are listed as still uncertain. Eight phyla have no coelom (they are acoelomate). Five phyla have another type of body cavity that arises from a different embryonic process (they are pseudocoelomate). The remaining 11 phyla all have coeloms (they are coelomate).

The third category is the coelomates. The relationships of two phyla are still uncertain. The nine remaining phyla are divided into two groups based upon the embryonic origins of their mouth openings. Six are called protostomes and three are called deuterostomes. Our phylum, the Chordata, is considered (by us) the most highly evolved of the deuterostomes.

2.3 WHERE ANIMALS LIVE AND HAVE LIVED

2.3.1 Ecological Niches

Large parts of animal ecology are devoted to studying the factors and mechanisms that determine where animals live (their distributions in both space and time), what determines their numbers, and what determines major features of their life histories. Here we focus on their geographic distributions, both in the short and long term (Harte 2011; Upchurch et al. 2011; Nosil 2012).

Animal distributions are always changing, as are the genetic compositions of their populations.

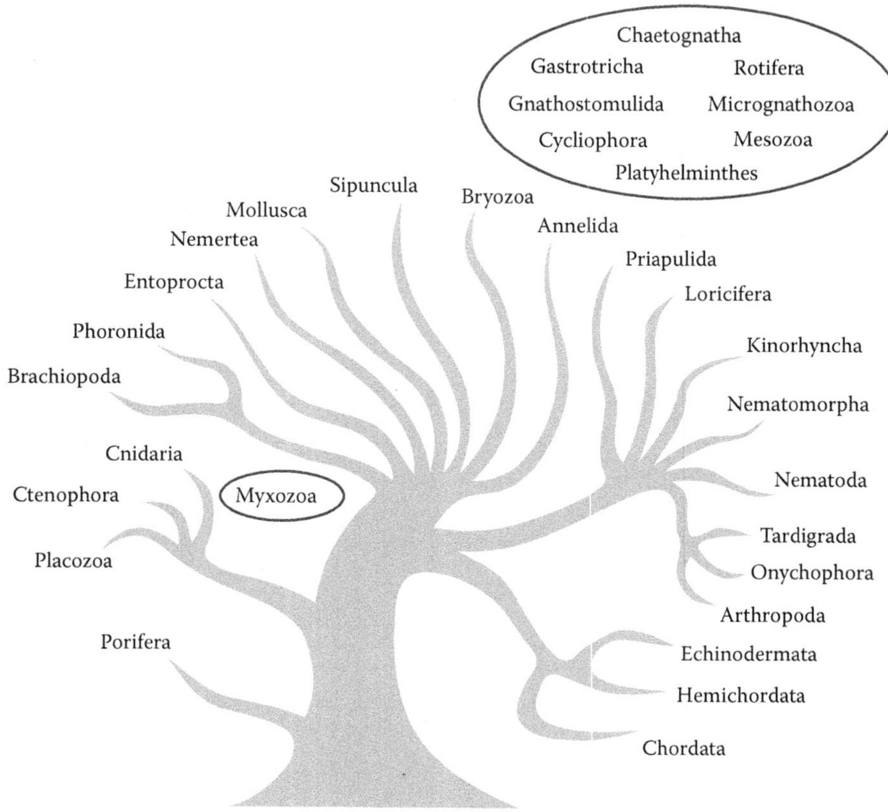

Figure 2.2. Simplified diagrammatic phylogenetic tree showing the currently accepted cladistic relationships among the 32 phyla of living metazoan animals. Major branches represent the larger groupings summarized in the text and in Table 2.1. Phyla named within the oval have uncertain relationships with the phyla listed on the tree itself. For more information, see text and References. (Redrawn and modified from primary sources, Tree of Life Web Project: http://tolweb.org; Pechenik, J.A., *Biology of the Invertebrates*, 7th edn., McGraw-Hill Higher Education, New York, 624 p, 2014.)

Many factors contribute to changing distributions, some internal to the animals (e.g., variations in their genetic compositions or in the epigenetics of gene expression, either or both of which may lead to variations in either or both their performance capacities and in their tolerances, preferences, or both for particular combinations of environmental conditions), and others the result of any of many important external factors (e.g., climate change, availability of food, season of the year). Changes can be characterized by wide ranges of both magnitudes and time constants, varying from small and short term to large and long term. Both short-term and long-term historical factors can be influential, if not determinative. A consequence of this complexity is that there are many mathematical and theoretical approaches that have been used to study and understand animal distributions. Useful references include May and McLean (2007), Upchurch et al. (2011), and Nosil (2012).

One widely used approach is the concept of the ecological niche. Niches may be viewed mathematically as n-dimensional spaces. The dimensions are the large number of different properties that characterize a particular species. Those properties fall into three major categories: performance capacities (e.g., how fast and how far it can swim, fly, or move on the land), tolerances for variations in abiotic variables (e.g., ranges of environmental temperatures, salinities, and partial pressures of oxygen), and tolerances with respect to biotic variables (e.g., food supply, predation, parasites, and disease). Performance capacities are the focus of a substantial part of this volume.

Two evolutionary concepts are useful here. The first is the concept of resistance adaptations. Animals can evolve changes in their tolerance

TABLE 2.1
Phyla of living metazoan animals.

Phylum	Common name(s)	Diversity
	Phyla with poorly defined tissues	
Porifera	Sponges	15,000?
Placozoa	Placozoans	2
	Diploblastic (two tissue layers)	
Ctenophora	Comb jellies	150
Cnidaria	Corals, anemones, jellies	9,000?
	Triploblastic (three tissue layers): Protostomes	
Myxozoa	Myxosporidia	2,200
Brachiopoda	Lamp shells	300
Phoronida	Horseshoe worms	12
Entoprocta	Goblet worms	180
Nemertea	Ribbon worms	900
Mollusca	Snails, clams, chitons, octopus, squid	50,000?
Sipuncula	Peanut worms	250
Bryozoa	Moss animals	5,000
Annelida	Segmented worms, leeches	13–22,000?
Platyhelminthes	Flatworms, tapeworms	20,000
Mesozoa*	Mesozoans	50
Cycliophora	Cycliophorans	3
Micrognathozoa	Micrognathozoans	1
Gnathostomulida	Jaw worms	100
Rotifera	Rotifers	2,000
Gastrotricha	Gastrotrichs	700
Chaetognatha	Arrow worms	125
Priapulida	Penis worms	20
Loricifera	Brush heads	100
Kinorhyncha	Mud dragons	150
Nematomorpha	Horsehair worms	300
Nematoda	Roundworms	80,000?
Tardigrada	Water bears	750
Onychophora	Velvet worms	90
Arthropoda	Scorpions, spiders, centipedes, insects, crustaceans	1,000,000?
	Triploblastic: Deuterostomes	
Echinodermata	Sea stars, sea urchins, sea cucumbers, crinoids	7,000
Hemichordata	Acorn worms, pterobranchs	110
Chordata	Tunicates, fishes, amphibians, reptiles, birds, mammals	57,000?

NOTE: This list represents the current consensus of biological professionals with respect to the number of phyla (32) and their scientific names. Common names are those most widely used in the scientific literature; in major phyla a few of the best-known subgroups included are listed. Diversity is best current estimates of numbers of species in each phylum. Numbers followed by "?" indicate groups including large numbers of still undescribed or unknown species. For further scientifically reliable information online consult www.tol.org. See text for additional explanation.

*Phylogenetic relationships are not clear for this group with two tissue layers.

limits for environmental variables. Some animals can tolerate (survive and possibly reproduce) only within narrow ranges of a variable, whereas others can tolerate wider ranges. The widths of these tolerances can change as results of mutations and selective processes.

The second is the concept of capacity adaptations. Capacity here refers to rate processes (e.g., how frequently, how fast, and how far can an animal move its limbs). Capacities can also change as results of mutation and selection.

For any given species, its niche can be defined as the set of physical locations and environmental conditions within which the species lives during specific stages of its life history. This definition implies that the species in question may live in different niches at different stages of its life (e.g., for a fish, while it is a developing embryo in an egg, a larva, a juvenile, a young adult, and an older adult).

The definition also implies that there are at least two different niches to consider for each animal. The first is the potential niche: the n-dimensional space defined by the resistance and capacity properties of the species (e.g., the highest and lowest environmental temperatures it can survive over an annual cycle; how far a forest dwelling animal can jump; the highest and lowest parasite loads it can tolerate). The second is the realized niche: the n-dimensional space actually occupied by the species in the real world. Many investigations of many different species indicate that, in the vast majority of situations, the realized niches of species are significantly smaller than their potential niches.

The variety of interactions that occur between animals and their abiotic and biotic environments are major contributors to the differences between the potential and realized niches of species. As environmental conditions vary over space and time, both niches also change. As lineages evolve, the niches of the species evolve as well. These changes in the distribution, properties, and abundance of species occupying different niches are major determinants of the properties of the ecosystems in which they live. Thus as animals evolve, their ecosystems also evolve (Flatt and Heyland 2011).

2.3.2 Habitats, Environments, and Ecosystems

The answers to questions relating to where animals live and how they got to where they are vary depending upon both the physical scale and the time scale being discussed. On the largest physical scales and over the longest time periods, the answers fall into biogeography. Some clades are where they are as a result of continental drift (plate tectonics) over multiple millions of years (e.g., Australian marsupials). Other distributions (e.g., reef-building corals in the tropical eastern Pacific Ocean) are at least partly dependent upon relatively short-period fluctuations in patterns and intensities of ocean currents due to geophysical and meteorological cycles having variable periods of tens to hundreds of years. Thus present-day geographic distributions are resultants of many factors and influences, each having its own time frame (Harte 2011; Upchurch et al. 2011).

Depending upon which animal taxon is being considered, what its life history is like, whether it is aquatic or terrestrial, etc., chances are high that local distributions of the taxon (distances involved ranging from meters to kilometers) are more or less patchy. Few animals live everywhere in a local area. This means that most animals in most places live in specific habitats (e.g., tide pools along a rocky coast; a forest). The chances are also high that they actually live in even smaller microhabitats within the larger area (e.g., rodent burrows; quiet pools in freshwater streams).

An evolutionarily significant consequence of this fine-grained variability within habitats is that there is associated variability with respect to the selective pressures exerted on the realized niches of the different subpopulations of the species. Variations like these, over time, can initiate speciation processes (Lynch 2007; May and MacLean 2007; Vinicius 2010; Flatt and Heyland 2011; Nosil 2012).

On larger spatial scales, habitats are subtypes of environments. Environments include larger, usually more temporally persistent, places such as mountain ranges, rivers, plains, forests, deserts, beaches, rocky coastlines, the open sea, and the deep sea. Each environment exerts its own unique array of selective pressures on the animals living within it. These effects are major generators of homoplasies of many kinds in many groups (Sanderson and Hufford 1996; McGhee 2011; Wake et al. 2011; Dial et al. 2015; Gordon and Notar 2015).

The arrays of species living within particular environments interact in complex ways with both the abiotic and biotic aspects of the environment and with each other (Boucot and Poinar 2010; Upchurch et al. 2011; Dial et al. 2015). The results of all of these interactions are ecosystems. Ecosystems also vary in their properties over space and time. As environmental and ecological conditions change organisms evolve and so do ecosystems. The five major extinction events that have played huge roles in the evolutionary history of life on Earth are all examples of macroscale results of ecosystem changes (Levinton 2001; Wagner 2011; Hazen 2012; Shapiro 2012; Erwin and Valentine 2013; Dial et al. 2015). The ecosystem during the Jurassic Period in a tropical forest in the region that later became South America was very different from the ecosystem now operating in the Amazon forest (Ebach and Tangney 2006; Upchurch et al. 2011).

2.4 THE RECORD OF ANIMAL EVOLUTION

2.4.1 The Geological Timescale

Even a summary discussion of the nature and content of the fossil record requires a general familiarity with the major divisions of geologic time and with some of the most biologically important events that occurred. Here we discuss only the Phanerozoic Eon, the most recent 542 million years (Erwin and Valentine 2013). The Phanerozoic Eon includes the three major eras in the history of metazoan animal life: the Paleozoic (began 542 million years ago [mega-annums; My]), the Mesozoic (251 My), and the Cenozoic (66 My). The terminology and the dates used here are those approved by the Geological Names Committee of the U.S. Geological Survey in 2010 (Figure 2.3).

The major periods within the eras, with their start dates (My, rounded to the nearest million) are as follows:

Paleozoic Era	
Cambrian	542
Ordovician	488
Silurian	444
Devonian	416
Carboniferous	359
Permian	299
Mesozoic Era	
Triassic	251
Jurassic	200
Cretaceous	146
Cenozoic Era	
Paleogene	66
Neogene	23
Quaternary	3

Continued

There are variable numbers of named epochs within these periods and of ages within many of the epochs. A variety of geologically and biologically important events were associated with the transitions between some of these periods. That was particularly the case for the boundaries between eras.

2.4.2 Plate Tectonics and Biogeography

Earth is a dynamic planet. Its constant activity with respect to changing the locations of the continental land masses, the process called plate tectonics, apparently has gone on for more than 3 Gy. The movements are slow by human standards, but during the almost 0.6 Gy duration of the Phanerozoic Eon, they have resulted in massive redistributions of the continental plates. These redistributions have had major biogeographic and other evolutionary consequences. Here we mention only a few salient points.

Both in the early Phanerozoic and for much of the Mesozoic most of the continental plates were closely packed together and formed huge supercontinents. The Mesozoic supercontinent (Pangaea) was centered at high latitudes near the South Pole. This location means that evolutionary processes at work on land animals during that period of time occurred under climate and photoperiod regimes characteristic of high latitudes. Climates even far inland may have been relatively cool. Animals were exposed annually to long periods of both continuous daylight and continuous darkness. Those were the general conditions under which much of dinosaur evolution occurred.

An important result of the long existence of Pangaea was that many groups of animals, both invertebrates and vertebrates, had wide and continuous distributions on the continent during much of the Mesozoic. Fossil animals dating from that period are similar to identical on both sides of the divides that opened up. As the continent broke up, the populations of those animals were divided. Environmental conditions

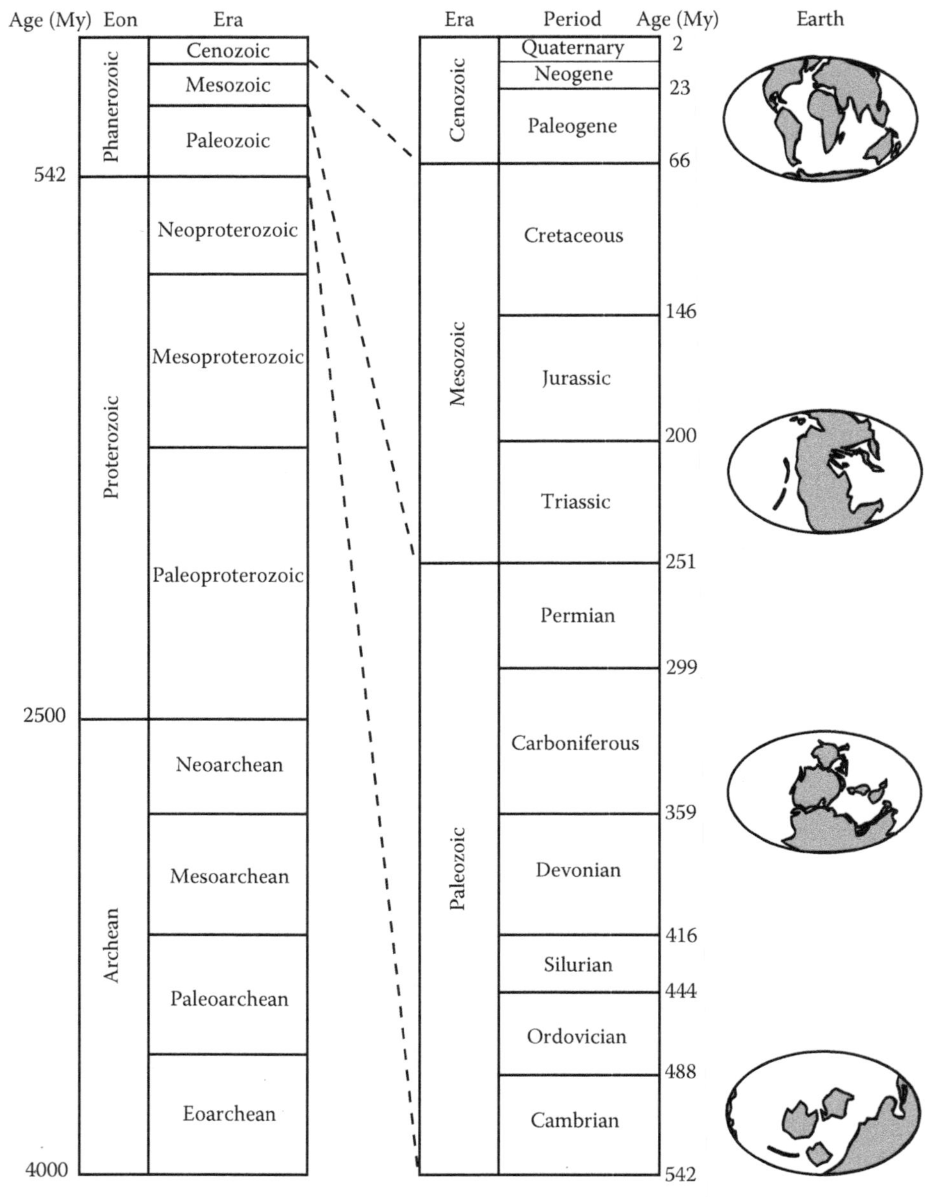

Figure 2.3. Schematic simplified geological timescale for the past 4 Gy. Left column: Complete time line showing absolute ages (in millions of years) and ratified standard names for the major eons and eras. Center column: Expanded time line for the Phanerozoic Eon (most recent 542 My) showing ratified standard names for included eras and periods. Absolute ages indicated for the beginnings of each of the periods. Each period includes several to many shorter named epochs not included in this figure. Right column: Simplified maps of general distributions of continental plates on Earth around times indicated in center column. The North Pole is at the top of each map. (Redrawn and modified from U.S. Geological Survey Geologic Names Committee, Fact Sheet 2010-3059 [July 2010], and the CHRONOS Program.)

and selective pressures became different for the subdivisions of the populations. As examples we now find many animal groups in Africa and in South America that have differentiated, but that still have close affinities to the related groups on the other continent.

The ocean during that period was also very different from the way it is now. The North Pole was far out in the open sea and the distribution of deep sea environments bore no resemblance to the present. Environmental conditions in the deep sea were probably very different from present conditions. The deep sea then was probably much warmer than it is now.

Given the lack of land barriers in the Northern Hemisphere (the reverse of the present situation), the system of oceanic currents was completely different from what we have now. An important result of that was that Earth's climate regime was also very different.

2.4.3 Paleontology, Ichnology, and Taphonomy

Paleontology classically centered on finding and studying fossils. That is still a large part of the work being done (Benton 2014; Dial et al. 2015). Newly discovered fossils are used to verify, revise, and occasionally dramatically change understanding of cladistic relationships in different lineages. Some of the most dramatic recent revisions of ideas of reptile evolution, for example, derived from studies of amazingly complete Mesozoic fossils from very fine-grained deposits in northeastern China. Those fossils showed, among other things, that many dinosaurs had well-developed feathers. Specialized methods have made it possible to determine the colors of some of those feathers. Combining results like these with multiple other lines of evidence has established the fact that living birds, once thought to be a separate subclass of vertebrates, are actually a derived group of theropod dinosaurs. They are the only group of dinosaurs that survived the asteroid collision that ended the Cretaceous Period and began the Cenozoic Era. Useful references include Boucot and Poinar (2010), Sanchez-Villagra (2012), and Benton (2014).

More recent fossils, especially those dating from the later Pleistocene and early Holocene epochs (the last 700,000 years or less), sometimes still contain identifiable ancient DNA (Section 2.2.2). With great care that DNA can be isolated from more recent contaminants and can be partially sequenced. Studies of ancient DNA obtained from mammoths frozen in Arctic tundra have clarified the phylogenetic relationships of mammoths and living elephants. The oldest DNA so far successfully and reliably sequenced came from 700,000-year-old fossil horse bones continuously frozen in Arctic permafrost. DNA isolated from a single human finger bone found in a cave in Siberia has established two important new facts concerning human evolution. First, the bone came from an unknown paleohuman population (now named the Denisovans, from the name of the cave). Second, the Denisovan nuclear genome shows some of the clearest evidence yet found that there was active interbreeding between paleohumans and modern humans (Orlando and Willerslev 2014).

Fossils are not always the actual physical remains of organisms. There is an active subfield of paleontology, ichnology, that studies trace fossils—remains of the activities of animals rather than of the animals themselves (Seilacher 2007). Animal trackways are among the best-known trace fossils. Remains of trackways and sometimes associated burrows of soft-bodied benthic marine invertebrates are important evidence for the existence of metazoan animals long before the beginning of the Phanerozoic. The abundances of burrows, also the physical shapes and dimensions of those burrows, have been combined with estimates of the physical properties of the sediments in which they occur (from grain sizes and shapes), also with studies of burrowing mechanisms in living species of benthic invertebrates, to make inferences about multiple aspects of the lives, behavior, and ecology of animals that have never been seen.

Large-scale trackways made by groups of dinosaurs moving about during the Mesozoic have provided insights into the sizes and body masses of the dinosaurs that made the tracks. A surprising variety of biomechanical inferences can be made from the dimensions and depths of the impressions made in what was mud at the time. Physical properties of the mud were estimated from the properties of the rocks containing the tracks. Other features of the tracks, such as stride lengths, made it possible to estimate speeds of movement. The relative numbers of tracks made by different species of dinosaurs provided insights into possible social behaviors of the animals.

Taphonomy, a subfield of paleoecology, studies other aspects of fossil environments and biological associations. The physical and chemical properties of the rocks in which the fossils are found, also in some cases aspects of the stable isotopic compositions of fossilized hard parts (shells, bones), can all be combined with statistical analyses of the populations of fossils of other organisms that were found in the same deposits as the subject animals. The overall result often is a fairly detailed understanding of both the general features of the paleoenvironments involved and of the paleoecosystems in which the animals lived (Raff et al. 2006).

2.4.4 Major Features of the Record

Both the fossil record and the inferred evolutionary histories of living animals derived from cladistic analyses of molecular evolution include evidence for the occurrence of many other phenomena and processes than we have discussed so far. Here we mention a few of the more significant of these features (Levinton 2001; Dial et al. 2015; Eldredge 2015).

Perhaps the most general concern with the fossil record is statistical. It relates to the extent to which the fossils found and described by paleontologists accurately represent the assemblages of organisms that lived in particular times and places. The concern also extends to the accuracy of inferences of paleoecological conditions. The best response to this concern is to say it depends. Multiple factors are involved in each situation (Raff et al. 2006; Seilacher 2007). The following list is not comprehensive:

1. The ages of the fossils and the deposits in which they were found are a prime concern. Older rocks are generally harder to find than younger rocks. If one is studying the early Cambrian, they are rare indeed. This is due partly to greater losses of older rocks to erosion and partly to greater losses due to a wide range of other destructive processes. Whatever the causes may be the question must be asked: How representative are the smaller numbers of older deposits of the diversity of animals and of the range of conditions that existed at the times they were laid down?
2. A second universal concern about sampling accuracy derives from the fact that environmental conditions conducive to the formation and preservation of fossils are also rare at any point in geological time. Most fossils occur in sedimentary deposits, usually deposits laid down under water that was either not moving at all or moved very slowly. Wherever an animal died that might be fossilized intact or in pieces, the remains also had to be removed from other conditions that could further degrade or destroy it (e.g., scavenger animals, bacteria of decay, and mechanical upheavals of many possible kinds). Animals that die in other places, times, and conditions do not get fossilized.
3. A third factor is the absolute abundance of particular kinds of animals at specific times and places. Common animals are more likely to be preserved, and rare animals much less likely to be preserved. Fossil assemblages are also likely to be skewed with respect to representation of different life history stages.
4. A fourth factor is that there are not now and never have been sufficient numbers of paleontologists with sufficient exploration and research budgets to be able to investigate all promising rock exposures at all times. Thus, there is a substantial element of stochasticity in what has been found and further variability in what has been properly studied and described.

The bottom line is that the fossil record as we know it contains large amounts of uncertainty and statistical noise. It is structurally incomplete and imprecise. It remains, however, the only direct evidence we have as to what lived, where, when, and under what conditions. Paleontologists are acutely aware of these issues and work hard to reduce their effects.

Molecular phylogenies currently have comparable statistical sampling issues. Recognition of the importance of epigenetic and epigenomic phenomena further complicates the picture. For many groups of animals, too few species have been studied. The species that have been studied have only been partially studied (too few complete genomes, complete gene sequences, representative samples of different categories of genes). In principle, unlike the structural problems of the fossil record, these issues may eventually be resolved. As nucleotide sequencing speeds increase and costs decrease, complete sequences for larger numbers of genes and genomes from larger numbers and species diversities of organisms will become available. Bioinformatic methods for analyzing, verifying, and testing the huge data sets that are generated by these studies will be further developed and perfected (Lynch 2007; Dunn et al. 2008; Cannarozzi and Schneider 2012; Stevens 2013; Gerstein et al. 2014).

There are many other significant problem areas relating to our knowledge and understanding of the evolutionary histories and relationships of animals. There also are many interesting phenomena that are only poorly or partly understood (Avise 2008 2011; Asher and Sumner 2015).

2.5 ORGANIC EVOLUTION

Here we describe central concepts fundamental to what may be called the classical (Darwin–Wallace) selection-based model of biological evolution. This model is also sometimes called the modern synthesis of evolutionary theory. It was largely developed between the later 1930s and the 1950s.

These concepts are directly relevant to a substantial majority of sexually reproducing metazoans. There are, however, multiple alternate patterns that characterize specific groups, some of which are quite large and ecologically significant. Avise (2008, 2011) summarizes information about many of these variations on the theme and other special situations (e.g., clonal, nonsexually reproducing animals).

Ongoing developments in systems biology, epigenetics, and epigenomics make it probable that a modified model of evolutionary mechanisms and processes will be recognized relatively soon. This new model will take into account the effects of changes in gene expression, not only changes in gene frequencies. This model is unlikely to completely supersede the classical model; rather, it will expand and further refine it. The principles deriving from the classical model will suffice for this discussion (Noble 2011, 2012, 2013; Avise and Ayala 2013; Eldredge 2015).

2.5.1 Evolution Defined

For most kinds of animals, individuals are the actors in evolution, but the evolutionarily significant results of their actions manifest themselves at the organizational levels of populations and above. Thinking about populations, and genetically based variations between individuals within populations, is central to understanding biological evolution.

A population is a group of animals of a specific kind that live close enough to each other geographically that they have reasonable chances of meeting others in the group and, under appropriate circumstances, reproducing with them. The biological boundaries of a population are reached at those points beyond which the chances of two animals meeting within their reproductive lifetimes are zero.

Genetically based variations between individuals within a population may occur at any or all of the levels of structural and organizational complexity in their bodies. They may involve appearance, structure, or function. The important question with respect to evolutionary significance is the extent to which the variation results from genetically determined differences compared to nongenetically based (environmentally determined) variations. Genetically based variations are the only variations likely to persist from one generation to the next (note that this statement has been shown to be incomplete and inaccurate by recent work in evolutionary epigenetics and epigenomics).

Biological evolution may be defined, as a first approximation, as changes in the genetic compositions of populations through time. Genetic composition in most contexts refers to the relative abundances (frequencies) of genes in the population. Organic evolution has classically been defined as change in gene frequencies within populations over time.

2.5.2 Evolutionary Process

Classic Darwin–Wallace evolutionary process is based on natural selection. The results of selection are manifested as differential reproductive success (varying reproductive fitness) within a population of animals having different genotypes. Animals with higher fitness contribute more strongly to the genetic composition of the next generation than do those with lower fitness. Biologists say that animals with higher fitness are selected for, and those with lower fitness are selected against. The most important consequence of selection operating differentially on animals with different genotypes is change in gene frequencies within the population.

There are three ultimate sources for genotypic variation between individuals in metazoan populations: mutation, migration, and genetic drift. Mutation refers to changes in genetic composition resulting from any of a wide range of either physical or chemical events or processes (e.g., exposure to ionizing radiation; exposure to mutagenic chemicals). Migration refers to the physical movement of individual animals with varying genotypes between different parts of a single population or between more or less isolated parts of a population. Genetic drift is a stochastic change in the genetic composition of a population that may result from an external event (e.g., an earthquake) that kills a significant fraction of the population without regard to the genetic fitnesses of the individuals. Most clear examples of genetic drift involve small populations.

The directions in which natural selective pressures push the genotypic compositions, and therefore the gene frequencies, of real populations vary continuously in both space and time. For most animals in most environments most of the time, these changes in direction may occur on time scales ranging from daily to seasonally to interannually to geologically significant periods. As a result, evolutionary change usually occurs relatively slowly and gradually. There have been major exceptions to this statement (Section 2.5.4; Eldredge 2015).

2.5.3 Speciation

A major result of the accumulation of evolutionary changes in different parts of animal populations (parts here referring to both temporally and spatially separable parts) is usually genetic subdivision and eventual differentiation of the parts. The general term for the process is speciation. The description of speciation that follows is based upon what evolutionary biologists call the biological species concept (BSC). In its simplest form, the BSC describes speciation on the basis of processes of allopatric differentiation of geographically separated subpopulations (Nosil 2012).

Subdivision of populations distributed over significant geographic areas also usually occurs more or less gradually. Environmental conditions often vary significantly in different parts of the range of a particular population. These variations result in varying selective pressures. When differing selective pressures in different regions are accompanied by reduced frequencies of reproduction between subpopulations, the stage is set for the initiation of speciation. Speciation occurs when gene flows between subpopulations become very low and the selective pressures on the now relatively genetically isolated groups produce sufficient changes in gene frequencies such that reproductive success of subsequent matings between members of the differentiated groups declines substantially or goes to zero (the fitness of hybrid matings declines to low levels or goes to zero). A functionally comparable isolation of two differentiating populations also occurs if hybrid offspring result from matings, but those offspring are selected against and are removed from the next generation.

Biologists describing and classifying the diversity of living animals usually use this definition of the BSC to decide at what point two clearly related groups of animals have become two different species. That is the point at which they may be given different scientific names. This definition has several important implications. The most relevant here is that it highlights a significant problem with how to define species over evolutionary time. The BSC defines living species only— it is a definition relating to our present time horizon. The BSC is a time horizontal definition.

At least in principle the correctness of a designation of an animal as belonging to one species or another can be tested experimentally: Can matings between the two forms produce viable offspring that will or will not be selected against under natural circumstances? No such experiments are possible with fossil animals. Thus the BSC literally cannot apply to the fossil record. There is continuing debate and discussion about exactly what constitutes a species in the fossil record. We do not have a widely applicable definition of time vertical species.

We also must mention that the BSC is not the only definition of a species that is used by evolutionary biologists. It is one of a group of four major types of what are termed pattern-oriented definitions of species (the other three are cohesion, phenetic, and recognition). There are also four major types of process-oriented species definitions (cohesion, ecological, evolutionary, and phylogenetic). Each of these other types of definitions has been applied to ranges of special circumstances (Ruse and Travis 2009).

2.5.4 Phylogeny

Long-term, large-scale continuation of speciation processes combine with accumulation of new mutations in differentiating groups to produce new organisms. These ancestral organisms are then the basal forms for new evolutionary lineages (clades) that lead to the hierarchy of higher level systematic categories: genera, families, orders, and the rest.

A second implication of the BSC is that there is no experimental way to test the characteristics used to define any of the higher categories. Because species cannot hybridize, or hybrid offspring have zero fitness, there is no way to directly determine the limits of a genus, much less a family or order. This is one of the main areas in evolutionary biology where molecular phylogenetic approaches have been of great value.

Relatively recent studies of the fossil record, using a range of new technologies, have shown that on occasion catastrophic events (e.g., widespread major volcanic eruptions, the breakup of supercontinents, collisions with asteroids) occur that produce widespread, large-scale, and rapid biological evolutionary changes. Paleontologists now recognize at least five major planet-wide, environmentally induced extinction events in the history of life. Each of these events led to large changes in the species composition of the surviving biodiversity. Each event also subsequently provided a range of opportunities for new diversifications of many different phylogenetic lineages. All of these events resulted in the extinction of high percentages (some more than 70%) of the existing life forms on Earth (Wagner 2011; Dial et al. 2015).

2.5.5 Adaptation

A final point relates to the use in evolutionary contexts of the word "adaptation." Adaptation is a word that should be carefully used but often is not. Readers should remember that not everything that one observes about an animal, whatever it may be, is demonstrably an adaptation, or significantly adaptive, whether positively or negatively. Animal features or properties may be correctly called adaptations only if they demonstrably have detectable influence upon the lifetime fitness of that animal. Substantial amounts of research are needed to reach that standard.

The recognition that most small-scale variations seen within groups of animals have little or no direct selective significance or adaptive value led to the development of a theory of neutral selection. That theory has wide application in genomics, bioinformatics, systems biology, and epigenetics.

REFERENCES

Asher, C., and S. Sumner. 2015. The genetics of society. *The Scientist* 29: 39–45.

Avise, J.C. 2008. *Clonality: The Genetics, Ecology and Evolution of Sexual Abstinence in Vertebrate Animals*. Oxford University Press, New York, 237 pp.

Avise, J.C. 2011. *Hermaphroditism: A Primer on the Biology, Ecology and Evolution of Dual Sexuality*. Columbia University Press, New York, 232 pp.

Avise, J.C., and F.J. Ayala (eds.). 2013. *Essential Readings in Evolutionary Biology*. Johns Hopkins University Press, Baltimore, MD.

Bell, M.A., D.J. Futuyma, W.F. Eanes, and J.S. Levinton (eds.). 2010. *Evolution since Darwin: The First 150 Years*. Sinauer Associates, Sunderland, MA, 688 pp.

Benton, M.J. 2014. *Vertebrate Paleontology*, 4th edn. Wiley-Blackwell, Chichester, UK, 480 pp.

Boucot, A.J., and G.O. Poinar. 2010. *Fossil Behavior Compendium*. CRC Press, Boca Raton, FL, 416 pp.

Boyle, A.P., et al. 2014. Comparative analysis of regulatory information and circuits across distant species. *Nature* 512: 453–456. DOI: 10.1038/nature13668.

Broom, M., and J. Rychtar. 2013. *Game-Theoretical Models in Biology*. Chapman & Hall/CRC Press, Boca Raton, FL, 520 pp.

Bustin, M., and T. Misteli. 2016. Nongenetic functions of the genome. *Science* 352, aad6933. DOI: 10.1126/science.aad6933.

Cannarozzi, G.M., and A. Schneider. 2012. *Codon Evolution: Mechanisms and Models*. Oxford University Press, Oxford, UK, 320 pp.

Ci, W., and J. Liu. 2015. Programming and inheritance of parental DNA methylomes in vertebrates. *Physiology* 30: 63–68. DOI: 10.1152/physiol.00037.2014.

Costello, M.J., R.M. May, and N.E. Stork. 2013. Can we name Earth's species before they go extinct? *Science* 339: 413–416.

Dial, K.P., N. Shubin, and E.L. Brainerd (eds.). 2015. *Great Transformations in Vertebrate Evolution*. University of Chicago Press, Chicago, IL, 424 pp.

Dubitzky, W, O. Wolkenhauer, H. Yokota, and K.-H. Cho (eds.). 2013. *Encyclopedia of Systems Biology*. Springer-Verlag, New York, 2367 pp.

Dunn, C.W., et al. 2008. Broad phylogenomic sampling improves resolution of the animal tree of life. *Nature* 452: 745–749.

Ebach, M.C., and R.S. Tangney (eds.). 2006. *Biogeography in a Changing World*. CRC Press, Boca Raton, FL, 232 pp.

Eldredge, N. 2015. *Eternal Ephemera: Adaptation and the Origin of Species from the Nineteenth Century through Punctuated Equilibria and Beyond*. Columbia University Press, New York, 398 pp.

Erwin, D.H. and J.W. Valentine. 2013. *The Cambrian Explosion: The Construction of Animal Biodiversity*. Roberts and Company, Greenwood Village, CO, 416 pp.

Flatt, T., and A. Heyland. 2011. *Mechanisms of Life History Evolution: The Genetics and Physiology of Life History Traits and Trade-Offs*. Oxford University Press, Oxford, UK, 504 pp.

Gerstein, M.B. et al. 2014. Comparative analysis of the transcriptome across distant species. *Nature* 512: 445–448. DOI: 10.1038/nature13424.

Gilbert, S.F., and D. Epel. 2009. *Ecological Developmental Biology: Integrating Epigenetics, Medicine and Evolution*. Sinauer, Sunderland, MA, 480 pp.

Gordon, M.S. 1999. The concept of monophyly: A speculative essay. *Biol Philos* 14: 331–348.

Gordon, M.S., and J.C. Notar. 2015. Can systems biology help to separate evolutionary analogies (convergent homoplasies) from homologies? *Progr Biophys Mol Biol*. 117:19–29. DOI: 10.1016/j.pbiomolbio.2015.01.005.

Hall, B.K., and W.M. Olson (eds.). 2006. *Key Words and Concepts in Evolutionary Developmental Biology*. Harvard University Press, Cambridge, MA, 496 pp.

Harte, J. 2011. *Maximum Entropy and Ecology: A Theory of Abundance, Distribution and Energetics*. Oxford University Press, Oxford, UK, 280 pp.

Hazen, R.M. 2012. *The Story of Earth: The First 4.5 Billion Years, from Stardust to Living Planet*. Penguin Academic, New York, 320 pp.

Hedges, S.B. and S. Kumar. 2009. *The Timetree of Life*. Oxford University Press, Oxford, UK, 576 pp.

Huson, D.H., R. Rupp, and C. Scornavacca. 2011. *Phylogenetic Networks: Concepts, Algorithms and Applications*. Cambridge University Press, Cambridge, UK, 374 pp.

International Code of Zoological Nomenclature. 2016. http://www.iczn.org/iczn/index.jsp

Kitching, I.J., P.L. Forey, C.J. Humphries, and D.M. Williams. 1998. *Cladistics: The Theory and Practice of Parsimony Analysis*, 2nd edn. Oxford University Press, Oxford, UK, 230 pp.

Koonin, E.V. 2011. *The Logic of Chance: The Nature and Origins of Biological Evolution*. FT Press Science, Upper Saddle River, NJ, 528 pp.

Koonin, E.V. 2012. A half-century after the molecular clock: New dimensions of molecular evolution. *EMBO Reports* 13: 664–666.

Lemey, P., M. Salemi, and A.-M. Vandamme. 2009. *The Phylogenetic Handbook: A Practical Approach to Phylogenetic Analysis and Hypothesis Testing*, 2nd edn. Cambridge University Press, Cambridge, UK, 750 pp.

Levinton, J.S. 2001. *Genetics, Paleontology and Macroevolution*, 2nd edn. Cambridge University Press, Cambridge, UK, 617 pp.

Lynch, M. 2007. *The Origins of Genome Architecture*. Sinauer Associates, Sunderland, MA, 389 pp.

MacLeod, N. (ed.). 2008. *Automated Taxon Identification in Systematics: Theory, Approaches and Applications*. CRC Press, Boca Raton, FL, 368 pp.

Map of Life. Map of Life project is an on-line platform for species distributions. 2016. https://mol.org/

Mattick, J.S. 2012. Rocking the foundations of molecular genetics. *Proc Natl Acad Sci USA* 109: 16400–16401.

May, R., and A. McLean (eds.). 2007. *Theoretical Ecology: Principles and Applications*, 3rd edn. Oxford University Press, Oxford, UK, 272 pp.

McGhee, G. 2011. *Convergent Evolution: Limited Forms Most Beautiful*. MIT Press, Cambridge, MA, 325 pp.

Nelson, J.S. 2006. *Fishes of the World*, 4th edn. Wiley, Hoboken, NJ, 601 pp.

Nelson, V.R., and J.H. Nadeau. 2010. Transgenerational genetic effects. *Epigenomics* 2: 797–806.

Nielsen, C. 2012. *Animal Evolution: Interrelationships of the Living Phyla*, 3rd edn. Oxford University Press, Oxford, UK, 464 pp.

Noble, D. 2011. Neo-Darwinism, the modern synthesis and selfish genes: Are they of use in physiology? *J Physiol* 589 (Pt 5): 1007–1015.

Noble, D. 2012. A theory of biological relativity: No privileged level of causation. *Interface Focus* 2: 55–64.

Noble, D. 2013. Physiology is rocking the foundations of evolutionary biology. *Exp Physiol* 98 (8): 1235–1243. DOI: 10.1113/expphysiol.2012.071134.

Nosil, P. 2012. *Ecological Speciation*. Oxford University Press, Oxford, UK, 280 pp.

Open Tree of Life. A comprehensive, dynamic and digitally-available tree of life that synthesizes published phylogenetic trees along with taxonomic data. tree.opentreeoflife.org/

Orlando, L., and E. Willerslev. 2014. An epigenetic window into the past? *Science* 345: 511–512. DOI: 10.1126/science.1256515.

Pechenik, J.A. 2014. *Biology of the Invertebrates*, 7th edn. McGraw-Hill Higher Education, New York, 624 pp.

Pough, F.H., C.M. Janis, and J.B. Heiser. 2013. *Vertebrate Life*, 9th edn. Pearson, Boston, MA, 720 pp.

Pross, A. 2012. *What Is Life? How Chemistry Becomes Biology*. Oxford University Press, Oxford, UK, 256 pp.

Raff, E.C., J.T. Villinski, F.R. Turner, P.C.J. Donoghue, and R.A. Raff. 2006. Experimental taphonomy shows the feasibility of fossil embryos. *Proc Natl Acad Sci USA* 103: 5846–5851. DOI: 10.1073/pnas.0601536103.

Roff, D.A. 2010. *Modeling Evolution: An Introduction to Numerical Methods*. Oxford University Press, Oxford, UK, 352 pp.

Rogers, S.O. 2016. *Integrated Molecular Evolution*, 2nd edn. CRC Press, Boca Raton, FL, 574 pp.

Ruse, M., and J. Travis (eds.). 2009. *Evolution: The First Four Billion Years*. Harvard University Press, Cambridge, MA, 979 pp.

Sanchez-Villagra, M. 2012. *Embryos in Deep Time: The Rock Record of Biological Development*. University of California Press, Berkeley, CA, 265 pp.

Sanderson, M.J., and L. Hufford (eds.). 1996. *Homoplasy: The Recurrence of Similarity in Evolution*. Academic Press, San Diego, CA, 339 pp.

Schuh, R.T., and A.V.Z. Brower. 2009. *Biological Systematics: Principles and Applications*, 2nd edn. Cornell University Press, Ithaca, NY, 328 pp.

Seilacher, A. 2007. *Trace Fossil Analysis*. Springer-Verlag, Berlin, 220 pp.

Shapiro, J.A. 2012. *Evolution: A View from the 21st Century*. FT Press Science, Upper Saddle River, NJ, 272 pp.

Sisu, C., et al. 2014. Comparative analysis of pseudogenes across three phyla. *Proc Natl Acad Sci USA* 111(37): 13361–13366. DOI: 10.1073/pnas.1407293111.

Stevens, H. 2013. *Life Out of Sequence: A Data-Driven History of Bioinformatics*. University of Chicago Press, Chicago, IL, 302 pp.

Tollefsbol, T.O. 2014. *Transgenerational Epigenetics: Evidence and Debate*. Academic Press, Waltham, MA, 413 pp.

Tree of Life. 2016. D.R. Maddison and K-S. Schultz (eds.). A collaborative effort of biologists and nature enthusiasts from around the world. More than 10,000 web pages provide information about biodiversity, characteristics of organisms and their evolutionary history. tolweb.org/tree/

Upchurch, P., A.J. McGowan, and C.S.C. Slater (eds.). 2011. *Paleogeography and Paleobiogeography: Biodiversity in Space and Time*. CRC Press, Boca Raton, FL, 239 pp.

Vinicius, L. 2010. *Modular Evolution: How Natural Selection Produces Biological Complexity*. Cambridge University Press, Cambridge, UK, 248 pp.

Wagner, A. 2011. *The Origins of Evolutionary Innovations: A Theory of Transformative Change in Living Systems*. Oxford University Press, Oxford, UK, 264 pp.

Wake, D.B., M.H. Wake, and C.D. Specht. 2011. Homoplasy: From detecting pattern to determining process and mechanism of evolution. *Science* 331: 1032–1035.

Wiley, E.O., and B.S. Lieberman. 2011. *Phylogenetics: The Theory of Phylogenetic Systematics*, 2nd edn. Wiley-Blackwell, Hoboken, NJ, 424 pp.

CHAPTER THREE

Swimming

JOHN O. DABIRI and MALCOLM S. GORDON

3.1 INTRODUCTION

This chapter is about how metazoan animals swim in the diversity of aquatic environments—the oceans, brackish waters, and fresh waters. The wide variety of animals discussed spend at least significant parts of their respective life histories moving through waters that can vary widely in temperature, solute content, and hydrostatic pressure and therefore in density, viscosity, buoyancy, and surface tension. They also vary widely in flow characteristics such as velocities of bulk flows and levels and scales of vorticity (turbulence) (Denny 1988, 1993; Kaiser et al. 2011; additional references in Chapter 1).

The largest part of this chapter is structured as a logical mathematical development of the basic principles of hydrodynamics as applied to the shared general properties of swimming animals that underlie their biodiversity (Lighthill 1975; Hildebrand 1976; Saffman 1992). The chapter discusses hydrodynamically important aspects of anatomy, functional morphology, biomechanics, kinematics, energetics, and macroscopic bioengineering (Vogel 1994; Wainwright and Reilly 1994; Liem et al. 2001; Shadwick and Lauder 2006; Lauder 2011; Vogel 2013). More detailed and specific discussions of multiple related topics (e.g., stability, maneuverability, control, and aspects of relevant methods and technologies) may be found in the papers and books listed in the references section. We also do not consider here the large, active, and diverse literature that has developed in recent years around the development and use of bioinspired and biomimetic robots based on swimming animals.

We begin with a summary of biologically and evolutionarily important points and issues that are background for the detailed physical, mathematical, and analytical discussions that follow. More general points and issues relating to basic concepts and to biological evolution are discussed in Chapters 1 and 2.

3.1.1 Evolutionary Origins of Swimming

Leaving aside speculative theories of possible extraterrestrial origins for life on Earth, all direct and indirect evidence in the fossil record supports aquatic origins for living organisms (Bada and Lazcano 2009; Lane 2009; Nielsen 2012; additional references in Chapter 1). This is also the case for the origins of metazoan animals, a major set of evolutionary events that occurred during the Neoproterozoic Era about 800 million years ago. With respect to metazoans, the fossil evidence is strongly supported by extensive inferential evidence from many other fields in biology, especially developmental biology, life history studies, and phylogenetic systematics (Hedges and Kumar 2009; Calcott and Sterelny 2011; Dial et al. 2015; additional references in Chapter 2).

There is no definitive evidence that establishes where the earliest metazoans lived, in fresh water, brackish water, the ocean, or some combination of these environments. Based on general considerations such as the relative sizes and scales of these broad category environments, the odds are favorable to marine origins. Presently, the most fully developed and best documented general theory of metazoan origins (the trochaea theory) indicates that the earliest forms were probably planktonic marine organisms. These basal forms are thought to have diversified into three major lineages: the Cnidaria, the Protostomia, and the Deuterostomia. These lineages further diversified as summarized in the classification of living phyla in Table 2.1 and the cladogram in Figure 2.2. References include Nielsen (2012), Pough et al. (2013), and Pechenik (2010); additional references are in Chapter 2.

There is also no definitive evidence establishing how the earliest metazoans moved around. They could have been burrowing benthic sediment dwellers, negatively buoyant benthic surface dwellers, or neutrally buoyant swimmers. The trochaea theory supports the third option. Chances are good that the earliest metazoans were small (Seilacher 2007).

Wherever the earliest metazoans lived, how large they were, and whatever their initial modes of locomotion may have been, there can be no doubt that the first forms that evolved that had increased mobility had to swim. Based on the varied body shapes, sizes, and locomotor modes of the larvae of almost all presently aquatic phyla of metazoans, it seems likely that the evolutionary opportunities (unoccupied aquatic niches and habitats) that opened up as soon as the first swimming forms appeared triggered rapid diversification of later swimming forms. Swimming is therefore evolutionarily the most basal form of animal locomotion. Nature has had long periods and a wide variety of raw materials available for design and process experimentation, refinement, and sophistication.

This primacy of swimming in overall animal phylogeny has not, however, meant primacy forever. The evolutionary history of swimming includes multiple examples of homoplasy, especially of convergences in both morphologies and in swimming methods (Dial et al. 2015; Gordon and Notar 2015; additional references in Chapter 2: Sanderson and Hufford 1996, McGhee 2011).

Multiple lineages in only 7 of the 32 phyla left the water and successfully invaded the land (see Chapters 2 and 5). Then, after varying periods, new clades arose within some of those lineages that reversed course and secondarily reinvaded aquatic environments.

The clearest examples of secondary reinvasion among the invertebrates are found among the arthropod insects. Among the multiple insect orders involved are caddis flies, mayflies, dragonflies, and mosquitoes, all of which live in fresh waters and have aquatic larvae with varied morphologies, swimming capacities, and methods. There also are groups having swimming and diving adult life history stages. Orders included are water bugs (Hemiptera) and water beetles (Coleoptera).

The vertebrates include multiple major groups that have secondarily reinvaded aquatic environments. The five Mesozoic reptilian lineages that led to the metriorhynchid archosaurs, spinosaurid dinosaurs, mosasaurs, plesiosaurs, and ichthyosaurs were examples. The living sea snakes and Galapagos Islands marine iguanas (lepidosaur lizards) are others. Because all birds are now considered to be highly derived theropod dinosaurs, which were terrestrial, the many groups of aquatic (swimming and diving) opposite birds that lived during the later Mesozoic as well as the modern groups of aquatic birds (e.g., loons, grebes, penguins, cormorants, ducks, and geese) are also all examples.

Multiple similar examples also occurred among the Mesozoic groups of basal mammals. Among living mammals four well-known groups are examples: sea otters; the sirenian dugongs and manatees; the pinniped seals, sea lions, and walruses; and the cetacean dolphins, porpoises, and whales. The polar bear is a special case among the bears.

3.1.2 Phylogeny, Biodiversity, and Environments

Only one phylum of living animals is completely terrestrial (Onychophora, the velvet worms, considered to be related to the Arthropoda). All 31 of the other phyla include aquatic representatives, most often marine animals but with a substantial subset also including freshwater forms. Significant subsets of the aquatic groups are completely sessile and benthic, or are burrow dwellers, or are endoparasites, as adults. All of these latter groups, however, achieve wide geographic distributions by means of planktonic swimming larvae. Most of those larvae are ciliated to varying extents.

Ciliary propulsion is also used extensively, sometimes exclusively, by adult forms in multiple phyla. These phyla include the Ctenophora (comb jellies) and the nonvertebrate chordate salps (Thaliacea, a class of urochordates, phylum Chordata; Neilsen 2012; Pechenik 2010).

Further summarization of the diversity of swimming animals is best done by considering major aquatic environments (and included ecosystems) and the body sizes of the animals considered (Kaiser et al. 2011). The major environments are the pelagic (open waters), near benthic (near the bottom or shore), and fluviatile (flowing waters). Body sizes range from microscopic (submillimeter to a few millimeters) to small (1–10 cm), medium (10 cm–1 m), and large (1–40 m).

Pelagic environments are prevalent both in oceans and large freshwater lakes. Actively swimming animals in oceanic pelagic environments range in size from microscopic to large. The phylogenetically and morphologically diverse ciliate larvae just described are mostly microscopic and live and swim in multiple ecosystems within this environment. Pelagic ecosystems range from the open seas far from land to inshore bays and from the tropics to the polar regions. The hydrodynamic environments range from the almost stationary unmixed waters of the deep seas far from the surface to the often highly turbulent surface waters.

The small pelagic swimmers are mostly members of the zooplankton. The zooplankton includes a large variety of animals belonging to many of the phyla. The major groups, on the basis of biomass, are often arthropod crustaceans (especially copepods and assorted shrimp), cnidarians, and juvenile stages of both mollusks and fishes. The phylogenetic composition of the zooplanktonic ecosystem varies widely both seasonally and regionally. Substantial subsets of the zooplankton are active vertical migrators.

Medium-sized actively swimming pelagic forms include many cnidarian jellyfish, ctenophore comb jellies, molluscan (cephalopod) squid, molluscan (gastropod) pteropods, urochordate salps, and many kinds of chordate fishes. Substantial subsets of these groups are both horizontal and vertical migrators. Most pelagic oceanic birds belong in this category (e.g., penguins, diving petrels, albatross, shearwaters, petrels, auks, and murres).

Large actively swimming pelagic forms include some cnidarian jellyfish, many kinds of squid, and many kinds of chordates. The largest pelagic chordates include such animals as oarfish, marlin, and tunafish (bony fishes); large sharks (elasmobranchs); sea turtles (chelonian reptiles); and many kinds of cetacean dolphins, porpoises, and whales. The blue whale is the largest animal known to have ever lived on Earth.

Freshwater pelagic swimmers are much less phylogenetically diverse. Only a few major lakes around the world are large enough and geologically old enough to have evolved true pelagic faunas. Lake Baikal in Siberia and the North American Great Lakes are the two best examples. In both cases the pelagic swimmers are primarily small crustaceans and fishes.

Near benthic environments occur worldwide in both marine and fresh waters. They include a huge range of different habitats and ecosystems. In the marine realm, they vary from the bottoms of deep-sea trenches to the abyssal plains; to localized features such as the deep-sea vents and seamounts; to the continental slopes and shelves; to inshore reefs of many kinds; to shallow bays and open coastlines with muddy, sandy, or rocky bottoms and shores.

The phylogenetic diversity of swimming animals in this diversity of near benthic environments includes essentially all phyla and all size ranges except for the very largest. Many of these animals are familiar to many people.

The hydrodynamic conditions occurring in near benthic environments also vary dramatically. The deep-sea waters close to active vents are highly turbulent, whereas the vast majority of the deep sea is almost stationary, with water velocities below a few centimeters per second. Inshore waters near exposed shores are also often highly turbulent, with frequent large waves breaking and producing high water velocities. Tidal channels in inshore areas having large tidal ranges often experience highly turbulent flows, with bulk velocities up to 2–3 m/s.

Fluviatile fresh waters also vary widely. Some large streams and rivers flow slowly and smoothly, whereas others are rapid. Even largely slow rivers may have significant stretches of increased slopes and rapid turbulent flows. The actively swimming animals living in major fluviatile environments often show a variety of specialized adaptations to these differing conditions. Examples include the torrent faunas, both invertebrate and fish, that live in the major streams draining high mountain watersheds in both the Asian Himalayas and the south American Andes.

3.1.3 Anatomy, Functional Morphology, Biomechanics, Kinematics, and Physiology

How animals swim must be described, analyzed, and modeled in the contexts of the five subject areas listed. Each of the areas has attracted the attention of many researchers beginning at least with Aristotle and other ancient Greek scientists. The references section for this chapter includes major summaries of each area. Here we highlight some of the more general issues and also some striking phenomena and processes.

3.1.3.1 Anatomy

The features of aquatic metazoan anatomies that relate to swimming generally seem to represent evolutionary efforts to optimize performance (i.e., maximize thrust, minimize drag, maximize energy efficiency, balance maneuverability and stability, automate and simplify control). The specific morphological structures that have evolved to do these things vary radically between lineages having different bauplans (body plans), having different body sizes, living in hydrodynamically different environments, and varying widely in both levels and kinds of activity.

In the context of this chapter, variation in body shapes, sizes, and levels of performance are among the most important factors (Wainwright and Reilly 1994; Liem et al. 2001; Pechenik 2010; Pough et al. 2013). Very small animals with complex body shapes and multiple limbs (e.g., many planktonic crustaceans) often move slowly in short bursts and are otherwise stationary. Thus, they live most of the time in low Reynolds number flow regimes. The structures of the limbs they use for both propulsion and feeding reflect this—they usually bear numbers of hair-like structures that increase the thickness of their boundary layers (Vogel 1994, 2008, 2013).

Intermediate ranges of Reynolds numbers characterize the flows around most actively swimming mid-sized animals (Rivera 2008; Tytell et al. 2010). Streamlining of body shapes is important in many groups (e.g., squid, most bony fishes, and most sea turtles). Fins of varying shapes, sizes, and aspect ratios are placed strategically to function in thrust generation, maneuverability, and maintenance of dynamic stability.

High Reynolds number regimes implying substantial turbulence in flows are often associated with large, high-speed swimmers such as marlins, tuna fishes, mackerel sharks, manta rays, and many dolphins and porpoises. The great baleen whales such as finback, humpback, and blue operate in flow regimes comparable to those applicable to large boats and small ships (Fish and Lauder 2006; Fish et al. 2008).

3.1.3.2 Functional Morphology, Biomechanics, and Kinematics

Functional morphology, biomechanics, and kinematics of locomotion are inseparably intertwined in most groups (Wu et al. 1975; Weihs 1977; Vogel 2008; Long et al. 2010; Tytell et al. 2010; Lauder 2011). For the invertebrate groups, there is substantial literature on swimming biomechanics, kinematics, and energetics in the cnidarian jellyfishes (Dabiri et al. 2007; Costello et al. 2008; Feitl et al. 2009), the arthropod crustaceans, and the pteropod and cephalopod mollusks (Bartol et al. 2008; Chang and Yen 2012). The many species of actively swimming, open water squid are particularly interesting. Many of those squid (cephalopod mollusks) are close to optimally designed as highly maneuverable, high-performance autonomous underwater vehicles. They have streamlined, cylindrical bodies fitted with actively controllable fins and powered for locomotion by steerable pulse jet engines (Weihs 1977; Bartol et al. 2008).

Here we provide more detailed information about the functional morphology and biomechanics of swimming in the chordate fishes (Bartol et al. 2005; Fish and Lauder 2006; Shadwick and Lauder 2006; Syme 2006; Lauder 2011; Miller et al. 2012). We discuss only older juvenile and adult animals. The larvae and young juvenile stages of the groups mentioned are comparably varied and diverse in body shapes and swimming modes. Swimming kinematics and hydrodynamics are discussed below.

Fish locomotion is one of the largest and most active areas of research in animal hydromechanics and hydrodynamics. It also has substantial practical consequences for the nutrition of many humans (Palstra and Planas 2013). Recall that there are more species of fishes (more than 30,000) than there are of all other vertebrate taxa combined (Nelson 2006). The highest order fish taxa include the hagfishes, lampreys, elasmobranchs, and bony fishes. The hagfishes and lampreys are both small groups including relatively few species. They are all shaped like eels and swim actively relatively infrequently. There is a huge diversity of body shapes and sizes, fin sizes and distributions, and patterns and mechanisms of locomotor mechanisms among the elasmobranchs and bony fishes.

A complex classification system is used to characterize fish swimming modes and propulsion mechanisms (Lauder 2006; Irschick and Higham 2016). In broad terms, the many thousands of species of actively swimming fishes that spend most of their lives moving at moderate speeds (cruising speeds) in the water column can be divided into two functional groups: body and caudal fin (BCF) swimmers and median and paired fin (MPF) swimmers. BCF swimmers derive most of their propulsive thrust from varying amounts of lateral (side-to-side) sinusoidal oscillation of their bodies and tail fins. The higher performance groups using BCF modes of swimming often have well-streamlined body shapes. Most MPF swimmers keep their bodies relatively stiff and straight under most swimming conditions. They generate their thrust primarily from well-coordinated fin movements. MPF swimmers have very varied body shapes that are often not streamlined at all.

There also are thousands of fish species that rarely swim, if they swim at all. These other forms are usually benthic, with many living in enclosed spaces in geometrically complex habitats such as coral reefs, or they are burrowing forms that spend most of their lives below the surface in sandy or muddy bottom habitats.

The BCF swimmers include most of the fishes familiar to most people. There are four broad categories. The first category includes the eels and other eel-shaped fishes. Many eels have near circular body cross sections, but there are important groups (e.g., moray eels) that are variably laterally compressed and deep bodied. Cruising speed swimming (anguilliform swimming) in these animals involves lateral sinusoidal movements of the entire body, beginning at the nose and progressing with little change in amplitude to the end of the tail. Several groups of fishes having very different body shapes (e.g., flatfishes and many blennies) also swim in the anguilliform mode. Lateral sinusoidal movements of progressively more posterior parts of the bodies characterize cruising speed swimming in the other important BCF groups.

The second BCF group is the subcarangiform swimmers (e.g., herring, trout, carp, and goldfish) that often have oval to near-cylindrical body cross sections. These fishes hold the forward 20–30% of their bodies stiff; lateral sinusoidal movements begin at that position and progress posteriorly, increasing in amplitude, until they reach the end of the tail. Subcarangiform swimmers have relatively broad and deep bases of their tails (caudal peduncles).

The third BCF group is the carangiform swimmers (e.g., jacks and pompanos) that usually are relatively laterally compressed and deep bodied. They hold the forward 50% of their bodies stiff; lateral sinusoidal movements begin there and progress, increasing in amplitude, to the end of the tail. Carangiform swimmers have narrow and slender caudal peduncles.

The fourth BCF group is the thunniform swimmers (e.g., mackerel sharks and tunafishes) that are again fairly heavy bodied and near cylindrical in cross section. Lateral sinusoidal movements in these forms are confined to the most posterior 10–15% of their bodies. Their caudal peduncles are narrow, and their tail fins are strongly forked with the lobes shaped like high aspect ratio wings. The tails oscillate rapidly under cruising speed conditions, acting like ship propellers.

Fishes swimming in MPF modes are highly varied in almost all important morphological respects. Many different groups are recognized. Here we mention three:

Labriform swimmers (e.g., wrasses, surfperches, surgeonfishes, and butterflyfishes) are probably the most abundant and also the best studied of the MPF groups. Labriform swimmers fly through the water by using their pectoral fins as primary sources of thrust. Most of these fishes have relatively slender and deep, laterally compressed body cross sections.

Tetrodontiform swimmers (e.g., puffers and swellfishes) are often heavy bodied, often almost circular or box shaped in cross section. They propel themselves using synchronized, well-coordinated movements of both their pectoral fins and the main vertical fins, their dorsal and anal fins.

One of the most highly evolved groups of MPF swimming bony fishes deserves special mention. They are the ostraciiform swimmers. The tropical marine boxfishes (family Ostraciidae) have the forward 65–75% of their bodies encased in rigid bony carapaces. The shapes of these carapaces have been subjected to extensive natural selection, one result of which has been the development in different species of varying numbers of bony keels placed in strategic locations. Hydrodynamic studies of models of the carapaces show that their shapes have unusually low drag and that the keels act as vorticity generators (leading edge vortices).

The vortices play significant roles in maintaining dynamic stability of the fishes during rectilinear swimming, even in turbulent environments. Thrust for forward motion is generated in these fishes by synchronized and coordinated movements of all major fins.

Publication of the results of the research work on boxfishes attracted the attention of the international automobile industry. One consequence of that attention was the development by several manufacturers of automobile designs having body shapes similar to those of particular boxfishes. Significant numbers of those vehicles are presently in use.

The cetacean dolphins, porpoises, and whales have convergently evolved swimming modes like those of BCF swimming fishes. Note that the planes of oscillation of their tails are vertical rather than horizontal.

3.1.3.3 Physiology

Physiological properties of swimming animals play important roles in many contexts. One of the most significant is the connection between major sources of metabolic energy and swimming performance (Lauder 2006; Syme 2006; Palstra and Planas 2013).

Locomotor performance of essentially all animals, whether aquatic or not, can be described by two sets of mathematical relationships: the relationships between speed and endurance and between speed and metabolic rates. To ensure comparability of the data for particular organisms, care must be taken in performing these measurements to control for substantial numbers of other variables that can also affect the results. These other variables include abiotic factors such as ambient temperatures and oxygen partial pressures. Biotic factors include, but are not restricted to, body size, life history stage of the animals, nutritional status, sex, and age.

Assuming suitably controlled conditions, the shapes and other properties of both relationships are ultimately determined by two other interrelated factors. These factors are whether or not fatigue limits endurance in different ranges of speeds and the metabolic causes of fatigue, when it develops. In the context of this discussion, fatigue may be due to the animal running out of energy while using aerobic metabolic processes, or it may be due to the accumulation of waste products of anaerobic metabolism.

The interactions between these factors lead to a three-category general classification of performance states:

1. Low levels of speed can be maintained on aerobic metabolism for indefinitely long times. This range of speeds is called sustained activity.
2. Intermediate ranges of speeds that are supported entirely or primarily by aerobic processes can end after progressively shorter durations, with fatigue resulting from shortage of fuel. These activity levels are called prolonged activity.
3. High speeds usually can endure for relatively short times that rapidly become shorter as speeds increase further. Fuel stores may still be adequate in these situations, but fatigue results from some combination of tissue level oxygen shortage (hypoxia) and the accumulation of waste products of anaerobic metabolism. These activity levels are called burst activity.

Most swimming animals spend most of their lives using sustained levels of activity. They use prolonged levels of activity as needed while doing such things as swimming against moderate-speed water currents and foraging for food. They use burst activity for emergencies, such as evading a predator.

3.1.4 Swimming Modes

There are so many different modes of swimming used by aquatic metazoans that there is little value to be gained by trying to list them. Here again we highlight some partial generalizations and some special cases.

Several major classes in multiple phyla use ciliary propulsion as adults, not only as larvae. The ctenophore comb jellies and the urochordate salps have been mentioned previously. Ciliary propulsion is characterized by low, usually fairly steady speeds in usually rectilinear trajectories.

Pulse jet propulsion is near universal among both the medusan cnidarian jellyfish and the cephalopod octopuses and squid. The mechanisms that produce the pulses of water jets are completely different between these two phyla. This propulsion mode usually results in rectilinear trajectories covered at oscillating variable speeds.

Many groups swim using oscillatory movements of their bodies. These groups include invertebrates such as platyhelminth flatworms and nudibranch gastropod mollusks. Among protochordate vertebrates, the larvacean urochordates look similar to and swim in the same mode as vertebrate amphibian tadpoles. The BCF swimming fishes described in the previous section belong here as do archosaurian reptiles (alligators, caimans, and crocodiles), all aquatic snakes, and the lepidosaur Galapagos marine iguana. All the completely aquatic groups of marine mammals (sea otters, sirenians, pinnipeds, and cetaceans) use this mode.

Many other groups swim with their bodies held relatively straight and rigid. Different lineages use varied methods of propulsion. The arthropod insect diving beetles and water bugs use their legs for thrust generation. Decapod crustacean arthropods (shrimp, krill, and lobsters) use their swimming legs (swimmerets) for prolonged and sustained levels of activity. Speeds in these modes are usually slow and trajectories are rectilinear. They use their entire caudal body in large-scale repeated contractions for high-speed escape, burst activities that often involve rapid changes in direction.

The MPF swimming fishes described above fall into this category. The multiple subgroups of MPF swimming fishes use a wide range of different combinations of fins and patterns and coordination of fin movements for propulsion.

Swimming and diving birds such as penguins, diving petrels, auks and murres, cormorants, and diving ducks all generally keep their bodies straight and rigid. They fall into two broad groups with respect to propulsion: underwater fliers that use their wings for thrust and steer with their bodies and feet and underwater paddlers that use their legs and feet for thrust generation.

Special types of aquatic mammals (polar bears, muskrats, capybaras, and beavers) most often use their legs and feet for propulsion.

Natural selective pressures for improving energy efficiency in locomotion have led to the development of a variety of different patterns of limb movements (called gaits) that are used for swimming at different speeds in many of the groups that keep their bodies rigid. Different gaits are often characteristic of activity levels ranging from sustained through prolonged to burst. Gaits also have been described for some groups of undulatory swimmers.

Animal swimming poses challenges not found in either aerial or terrestrial locomotion. Unlike aerial locomotion, the mass density of fluid surrounding an aquatic organism is comparable to that of the animal itself. And, unlike most terrestrial locomotion, the motion of the medium in contact with the organism is as significant as the animal's own movements. In this chapter, these properties are exploited to model and analyze the physical processes that give rise to aquatic locomotion. The models anticipate the rich diversity of design solutions that animals use to address the challenge of aquatic locomotion. When freed from the constraints of biological evolution, the models also suggest new propulsion strategies for engineered systems. Useful references to methodological and technological approaches to this area include Lighthill (1975), Lu et al. (2008), Flammang et al. (2011), Lauder (2011), and Miller et al. (2012).

3.2 QUANTIFYING FLUID MOTION: THE LAGRANGIAN PERSPECTIVE

Sir Isaac Newton's third law—"For every action, there is an equal and opposite reaction"—dictates that the fluid surrounding an aquatic animal moves in response to the applied locomotive forces. Because the animal and the fluid have similar mass densities, the resulting motions of the animal and fluid are also of a similar order of magnitude. Thus, the motion of the surrounding fluid provides an observable measure of aquatic locomotion that can be quantified and modeled.

Fluid motion is commonly described from one of two complementary viewpoints, Lagrangian and Eulerian. From a **Lagrangian** perspective, the entire fluid medium is treated as a collection of infinitesimally small parcels, typically called **fluid particles**. Each fluid particle possesses a momentum per unit mass—its velocity—that can be altered by local spatial imbalances in the fluid pressure or by external forces applied to the fluid. The interactions among fluid particle velocity, fluid pressure, and external force are governed by Newton's second law:

$$\sum_{n=1}^{n=N} \mathbf{f}_n = \frac{D\mathbf{u}}{Dt} \tag{3.1}$$

In words, Equation 3.1 states that the momentum of a fluid particle changes over time due to the action of externally applied forces on the fluid particle. The capital sigma, Σ, denotes a summation of the applied forces $\mathbf{f}_n$, beginning with the first force (with index $n = 1$) and ending with the Nth and final force acting simultaneously on the fluid particle. The lowercase **f** is used to indicate that the force is given per unit mass, to be consistent with the momentum per unit mass (i.e., the velocity **u**) on the right-hand side of the equation. Bold symbols indicate vector-valued variables that possess a magnitude and a direction of action. Finally, the capitalized derivative D/Dt is the convention that will be used to indicate that the property in the numerator of the derivative is being tracked over time for a specific fluid particle. This distinction becomes important when the Eulerian perspective is introduced later.

In practice, it is usually more straightforward to track the position of a fluid particle instead of its momentum. Because the fluid particle velocity is given by the time derivative of its position, Equation 3.1 can be rewritten as

$$\sum_{n=1}^{n=N} \mathbf{f}_n = \frac{\partial^2 \mathbf{x}}{\partial t^2} \tag{3.2}$$

where the notation $\partial/\partial t$ is used to indicate that although we are only computing temporal changes of the property in the numerator, that property is also dependent on parameters other than time. Specifically, the particle position **x** is dependent not only on time, but also on its initial position when tracking began (say, $\alpha = a\mathbf{i} + b\mathbf{j} + c\mathbf{k}$, in a three-dimensional Cartesian coordinate system, where **i**, **j**, and **k** are vectors in perpendicular directions with magnitude equal to 1; see Figure 3.1), and on the time t_0 at which tracking began. Note that the fluid particle position is also treated as a vector. The magnitude of the position vector is equal to the distance from the fluid particle to the origin of the coordinate system, and the position vector points from the origin to the fluid particle (Figure 3.1).

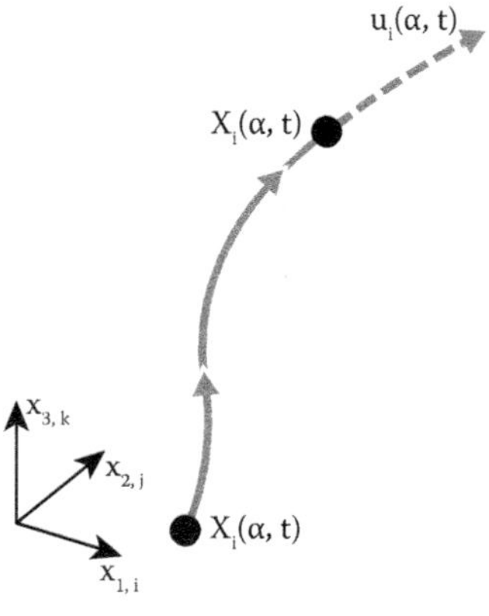

Figure 3.1. Fluid particle moving along an arbitrary trajectory in a Cartesian coordinate system.

The external forces of primary interest in the present context are due to local spatial imbalances in the fluid pressure and due to the motion of the swimming animal. The effect of fluid viscosity is neglected in this initial discussion; we see how this approximation can become problematic in Section 3.4. To model the effect of fluid pressure, let us consider a simple geometric model of a fluid particle: a cube. The shape we assume is not important, but given our previous choice of a Cartesian coordinate system, a cubical fluid particle simplifies the analysis. Figure 3.2 illustrates the fluid particle. Its dimensions are dx, dy, and dz along the **i**, **j**, and **k** axes, respectively; and the cube dimensions are much smaller than the characteristic length scale of the fluid flow

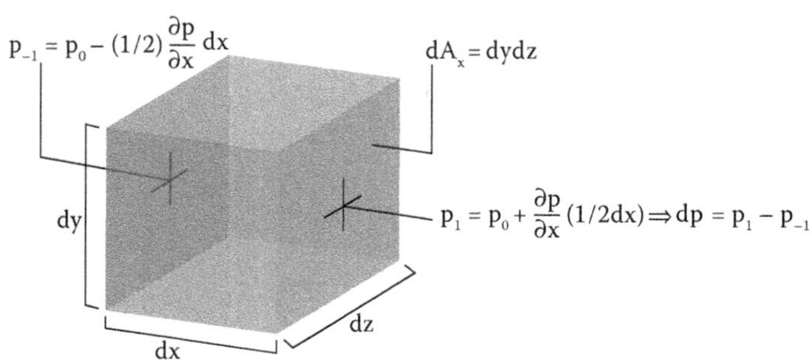

Figure 3.2. An infinitesimal cubic fluid particle.

(e.g., the animal appendages) because the fluid particle is assumed to be infinitesimal.

The local fluid pressure, p, acts perpendicular to the outer surface of each face of the cube. Because the outer surfaces of the two faces perpendicular to each axis are facing in opposite directions, the resulting forces due to the local fluid pressure also act in opposite directions. When the fluid pressure is uniform surrounding the fluid particle, these forces cancel along all three coordinate axes. However, any spatial imbalance in the local fluid pressure will lead to net forces acting on the fluid particle. Along the **i**-axis, this pressure imbalance leads to a net force per unit mass $\mathbf{f}_{px}$ along the axis that depends on the area ($dA_x = dy \cdot dz$) of the cube surfaces that are perpendicular to the **i**-axis and on the volume dxdydz of the fluid particle:

$$\mathbf{f}_{px} = \frac{-\left(p_+ dA_x - p_- dA_x\right)}{\rho dxdydz}\mathbf{i} \tag{3.3}$$

where p_+ is the pressure on the side of the cube facing the positive **i**-axis, p_- is the pressure on the side of the cube facing the negative **i**-axis, and ρ is the fluid mass density. The negative sign in front of Equation 3.3 arises because the pressure acts on the outer surfaces of the fluid particle; therefore, the pressure p_+ acts in the negative **i**-direction and p_- acts in the positive **i**-direction.

The local pressure on each side of the cubical fluid particle is related to the pressure p_0 in the center of the fluid particle by the spatial pressure gradient. For example,

$$p_+ = p_0 + \frac{\partial p}{\partial x}\left(\frac{dx}{2}\right) \tag{3.4}$$

$$p_- = p_0 - \frac{\partial p}{\partial x}\left(\frac{dx}{2}\right) \tag{3.5}$$

Equations 3.4 and 3.5 implicitly assume that the pressure varies linearly with position across the fluid particle from p_- to p_+, and therefore the pressure at each face of the fluid particle can be inferred from the slope of the pressure variation at the center of the fluid particle (i.e., the pressure gradient) and the distance from the center of the fluid particle (dx/2). This linear assumption is justified given the small size of the fluid particle relative to the length scales of the fluid flow. A Lagrangian abstraction that uses larger fluid particles could follow a similar approach, but it would need to incorporate higher order terms in the Taylor series expansion of p(x), to account for nonlinear variations in pressure across the fluid particle.

Substituting the pressure definitions in Equations 3.4 and 3.5 and the definition $dA_x = dy \cdot dz$ into Equation 3.3 gives the net force per unit mass on the fluid particle along the **i**-axis:

$$\mathbf{f}_{px} = -\frac{1}{\rho}\frac{\partial p}{\partial x}\mathbf{i} \tag{3.6}$$

Pressure imbalances along the **j**- and **k**-axis lead similarly to

$$\mathbf{f}_{py} = -\frac{1}{\rho}\frac{\partial p}{\partial y}\mathbf{j} \tag{3.7}$$

$$\mathbf{f}_{pz} = -\frac{1}{\rho}\frac{\partial p}{\partial z}\mathbf{k} \tag{3.8}$$

we sometimes refer to the vector sum of these pressure imbalances by using the vector gradient symbol ∇:

$$\mathbf{f}_p = -\frac{1}{\rho}\frac{\partial p}{\partial x}\mathbf{i} + -\frac{1}{\rho}\frac{\partial p}{\partial x}\mathbf{j} + -\frac{1}{\rho}\frac{\partial p}{\partial x}\mathbf{k} = -\frac{1}{\rho}\nabla p \tag{3.9}$$

From where do the pressure imbalances ∇p arise? They are often a reaction to externally applied forces, such as the forces applied to the fluid by the moving appendages of an organism. So, each fluid particle can be influenced both by the direct action of the aquatic animal and by the resulting imbalances in fluid pressure. In terms of the force equation (Equation 3.2), these dynamics can be expressed as

$$\mathbf{f}_p + \mathbf{f}_l = \frac{\partial^2 \mathbf{x}}{\partial t^2} \tag{3.10}$$

where $\mathbf{f}_l$ is the locomotive force applied by the animal on the fluid, per unit mass of the fluid. Decomposing the position vector **x** into components x**i** +y**j** + z**k** and using Equations 3.6 through 3.8 lead to the set of equations below:

$$\frac{\partial^2 x}{\partial t^2} = f_{lx} - \frac{1}{\rho}\frac{\partial p}{\partial x} \tag{3.11}$$

$$\frac{\partial^2 y}{\partial t^2} = f_{ly} - \frac{1}{\rho}\frac{\partial p}{\partial y} \tag{3.12}$$

$$\frac{\partial^2 z}{\partial t^2} = f_{lz} - \frac{1}{\rho}\frac{\partial p}{\partial z} \tag{3.13}$$

These equations completely describe the fluid dynamics in principle, but they hide the important dependence of each Lagrangian trajectory on the initial position ($\alpha = a\mathbf{i} + b\mathbf{j} + c\mathbf{k}$) and time at which each fluid particle is tracked. This dependence can be made explicit by combining Equations 3.11 through 3.13 to form an equivalent set of governing equations, namely,

$$\left(\frac{\partial^2 x}{\partial t^2} - f_{lx}\right)\frac{\partial x}{\partial \alpha} + \left(\frac{\partial^2 y}{\partial t^2} - f_{ly}\right)\frac{\partial y}{\partial \alpha} + \left(\frac{\partial^2 z}{\partial t^2} - f_{lz}\right)\frac{\partial z}{\partial \alpha} - \frac{1}{\rho}\frac{\partial p}{\partial \alpha} = 0 \tag{3.14}$$

where Equation 3.14 represents three independent equations with $\alpha = a$, b, or c. The spatial derivatives $\partial/\partial\alpha$ quantify the proportional change in the component of the subsequent fluid particle position given in the numerator, due to small differences in the initial position component α.

3.2.1 Lagrangian Coherent Structures

By observing the motion of *all* of the fluid particles surrounding a swimming animal, Equation 3.14 can be solved for the force $\mathbf{f}_l$ applied by the animal on the fluid. The locomotive force that propels the animal is identically equal to $-\mathbf{f}_l$ (i.e., Newton's third law). In theory, the analysis is complete at this point. In practice, of course, it is not feasible to track the infinite assemblage of fluid particles that comprise the fluid medium, except in very special cases such as spatially uniform flow. However, by following a small subset of the fluid particles, it is possible to quantitatively categorize the fluid motion into a small number of finite regions within which the particles all possess similar behavior. These regions of the flow are known as *Lagrangian coherent structures* (LCSs). The structures are Lagrangian, as their motion is described on the basis of individual fluid particle motions. Because the fluid particles within each structure exhibit similar behavior, the structures formed by the fluid particles are described as coherent. The concept of LCSs provides a more economical way to quantify fluid motion from a Lagrangian perspective.

Qualitatively, the objective of the LCS analysis is to distinguish between pairs of closely spaced fluid particles whose spacing remains relatively constant over time, and pairs whose spacing increases significantly over time. Those fluid particle pairs in the former group are expected to belong to the same LCSs, whereas the latter group characterizes pairs of particles that likely straddle the boundaries between adjacent LCSs. Because the pairs of fluid particles in the analysis are closely spaced, only a small number of fluid particle pairs in the flow will straddle the LCS boundaries. Therefore, the trajectories of this small subset of fluid particle pairs can be used to reveal the LCS boundaries in the flow.

As an example of the LCS analysis, consider the motion of fluid particles in the flow of *Hill's spherical vortex*, a simple model for the vortices shed by a swimming animal. Figure 3.3a illustrates representative Lagrangian fluid particle trajectories in a reference frame moving with the vortex ring. An *a priori* evaluation suggests four distinct regions of fluid particle motion: (I) flow outside of the vortex on the upper side; (II) flow inside the vortex on the upper side; (III) flow inside the vortex on the lower side; and (IV) flow outside the vortex on the lower side. The LCS analysis reveals the boundaries between these regions based on the behavior of fluid particle pairs, as opposed to a priori knowledge of the flow structure.

To see this, let us track the motion of three fluid particles located near the rear stagnation point of the vortex ring (Figure 3.3b). Fluid particles *A* and B are both located in region I and separated by a distance δ_{AB}; fluid particle C is located in region II and is separated by a distance δ_{BC} from fluid particle B. At time $t = t_0$, the initial distances between the fluid particles are assumed to be infinitesimal, that is, much smaller than the characteristic length scale of the vortex ring. A time $t = t_0 + T$ later, the fluid particles will have moved to the new positions *A'*, B', and C' and will be separated by the new distances δ'_{AB} and δ'_{BC}. Because the fluid particles in region I all move downstream past the vortex ring, fluid particles *A* and B move together, and $\delta'_{AB} \approx \delta_{AB}$. In contrast, particle C circulates in region II while fluid particle B moves downstream; hence $\delta'_{BC} >> \delta_{BC}$. This behavior suggests that fluid particles B and C were initially located on opposite

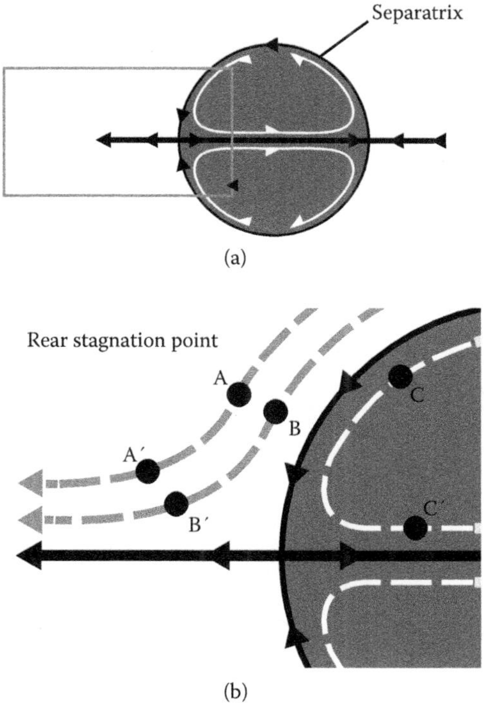

Figure 3.3. (a) Streamlines in a vortex ring. (b) Zoom in on fluid particle trajectories inside the highlighted rectangle in (a).

sides of an LCS boundary. Furthermore, the initial position of fluid particles B and C indicates the precise location of the LCS boundary at time $t = t_0$.

This process of LCS boundary identification can be formalized by considering an arbitrary fluid particle $\mathbf{x}$. The fluid flow can be thought of as a *flow map* f_{t0}^{t0+T} that moves particles from their positions at time t_0 to new positions at time $t_0 + T$. This is indicated symbolically as

$$f_{t0}^{t0+T}(\mathbf{x}) : \mathbf{x}(t_0) \rightarrow \mathbf{x}(t_0 + T) \qquad (3.15)$$

Now, let us consider an infinitesimal fluid particle separation $\delta\mathbf{x}_0$. This fluid particle separation is vector valued because it has both a magnitude (initially small) and a direction that points from the fluid particle at one endpoint of the separation to the fluid particle at the other endpoint. Similar to our calculation of the change in pressure across a fluid particle by using the pressure gradient in Equations 3.4 and 3.5, we can compute the change in position of the endpoints of δx_0 by using the gradient of the flow map. Hence, after a time T has passed, the fluid particle separation is given by

$$\delta\mathbf{x} = \frac{df_{t0}^{t0+T}(\mathbf{x})}{d\mathbf{x}}\delta\mathbf{x_0} \qquad (3.16)$$

or, using vector gradient notation,

$$\delta\mathbf{x} = \nabla f_{t0}^{t0+T}(\mathbf{x})\delta\mathbf{x_0} \qquad (3.17)$$

In Equations 3.16 and 3.17, we have used the same linear assumption that was applied in the pressure derivation above so that the higher order terms in the Taylor series expansion of $f_{t0}^{t0+T}(\mathbf{x})$ can be neglected.

The magnitude of the fluid particle separation $\delta\mathbf{x}$ depends on the initial direction of $\delta\mathbf{x}_0$, that is, its orientation. To see this more clearly, consider the behavior of the separation $\delta\mathbf{x}_0$ in the linear shear flow given by $u = y$ (Figure 3.4). If the direction of $\delta\mathbf{x}_0$ is aligned with the flow direction, the magnitude of $\delta\mathbf{x}$ after any time T will be unchanged, because both endpoints of $\delta\mathbf{x}_0$ are located at the same y-position and therefore move at the same velocity. Conversely, if $\delta\mathbf{x}_0$ is oriented perpendicular to the flow direction, the magnitude of $\delta\mathbf{x}$ will increase indefinitely as T increases, because the endpoints are located at different y-positions and therefore move at different velocities. We must therefore determine the maximum fluid particle separation for all possible initial orientations of $\delta\mathbf{x}_0$. This can be accomplished by considering an *eigenvalue* problem. First, let us express the magnitude of $\delta\mathbf{x}$ in terms of the flow map $f_{t0}^{t0+T}(\mathbf{x})$ by computing the magnitude of Equation 3.17:

$$\|\delta\mathbf{x}\| = \sqrt{\left\langle \delta\mathbf{x}_0, [\nabla f(\mathbf{x})]^* \nabla f(\mathbf{x})\delta\mathbf{x}_0 \right\rangle} \qquad (3.18)$$

where the asterisk denotes a transpose operator, and the angled brackets denote an inner product, for example, $\langle \mathbf{x}_1, \mathbf{x}_2 \rangle = x_1x_2 + y_1y_2 + z_1z_2$. The product $[\nabla f(\mathbf{x})]^* \nabla f(\mathbf{x})$ is the *Cauchy–Green deformation tensor*, and it subsequently denoted Δ. The maximum fluid particle separation is associated with the maximum eigenvalue of Δ, that is,

$$\|\delta x\|_{max} = \sqrt{\lambda_{max}(\Delta)}\,\|\delta x_0\| \qquad (3.19)$$

Equation 3.19 has an implicit dependence on the duration over which fluid particle motion is tracked; for fluid particles that are diverging,

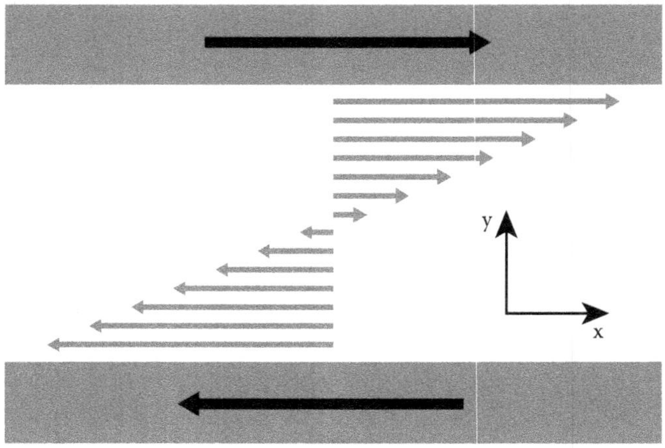

Figure 3.4. Shear flow between horizontally moving walls.

$\|\delta\mathbf{x}\|_{max}$ will increase as T increases. In addition, the fluid particle separation is often exponential, which can lead to large differences in the value of $\|\delta\mathbf{x}\|_{max}$ at different locations in the flow and can make analyzing the values over an entire flow field challenging. For these reasons, we define a *finite-time Lyapunov exponent* (FTLE), $\sigma_t^{t+T}(\mathbf{x})$, that is normalized for the effects of increasing observation time and exponential fluid particle separation that are implicit in Equation 3.19:

$$\sigma_t^{t+T}(\mathbf{x}) = \frac{1}{|T|}\ln\sqrt{\lambda_{max}(\Delta)} \qquad (3.20)$$

or

$$\sigma_t^{t+T}(\mathbf{x}) = \frac{1}{|T|}\ln\frac{\|\delta\mathbf{x}(T)\|}{\|\delta\mathbf{x}(0)\|} \qquad (3.21)$$

where, in Equation 3.21, it is understood that $\delta\mathbf{x}$(0 and $\delta\mathbf{x}$(T)) correspond to the directions of maximum particle separation. When the FTLE is calculated throughout a fluid flow, ridges of local maxima in a contour plot of the FTLE indicate the boundaries of LCSs in the flow.

The FTLE is a function of the position **x** where the fluid particle pair is initialized, and although its value is based on the fluid motion over the time interval from t to t+T, the value of the FTLE is associated with the instantaneous time t. The absolute value of the duration of particle observation, $|T|$, appears in Equations 3.20 and 3.21 because, in general, the fluid particles can be tracked either forward or backward in time. When tracked forward in time (i.e., $T > 0$), the LCS boundaries that are revealed are called *repelling* LCSs, as particles on either side of the boundary diverge on average. Conversely, when tracked backward in time (i.e., $T < 0$), the particles on either side of the associated LCS boundary converge on the boundary in forward time (i.e., they diverge in backward time). These boundaries are called *attracting* LCSs for this reason.

The set of repelling and attracting LCSs together identify the major fluid structures in the flow. In the vortex ring example in Figure 3.3, the attracting LCS boundaries reveal the upstream side of the vortex, whereas the repelling LCS boundaries reveal the downstream side. In real fluid flows, the LCSs can be significantly more complex, as shown in the LCS analysis of a laboratory-generated vortex ring in Figure 3.5a and the LCSs around a freely swimming jellyfish in Figure 3.5b. Nonetheless, these structures represent a more compact description of the fluid motion compared to the infinite set of fluid particle trajectories that comprise the flow.

3.3 QUANTIFYING FLUID MOTION: THE EULERIAN PERSPECTIVE

As an alternative to tracking the motion of individual fluid particles, one can instead observe changes in the properties of the fluid at fixed locations in space. In this *Eulerian* (pronounced with

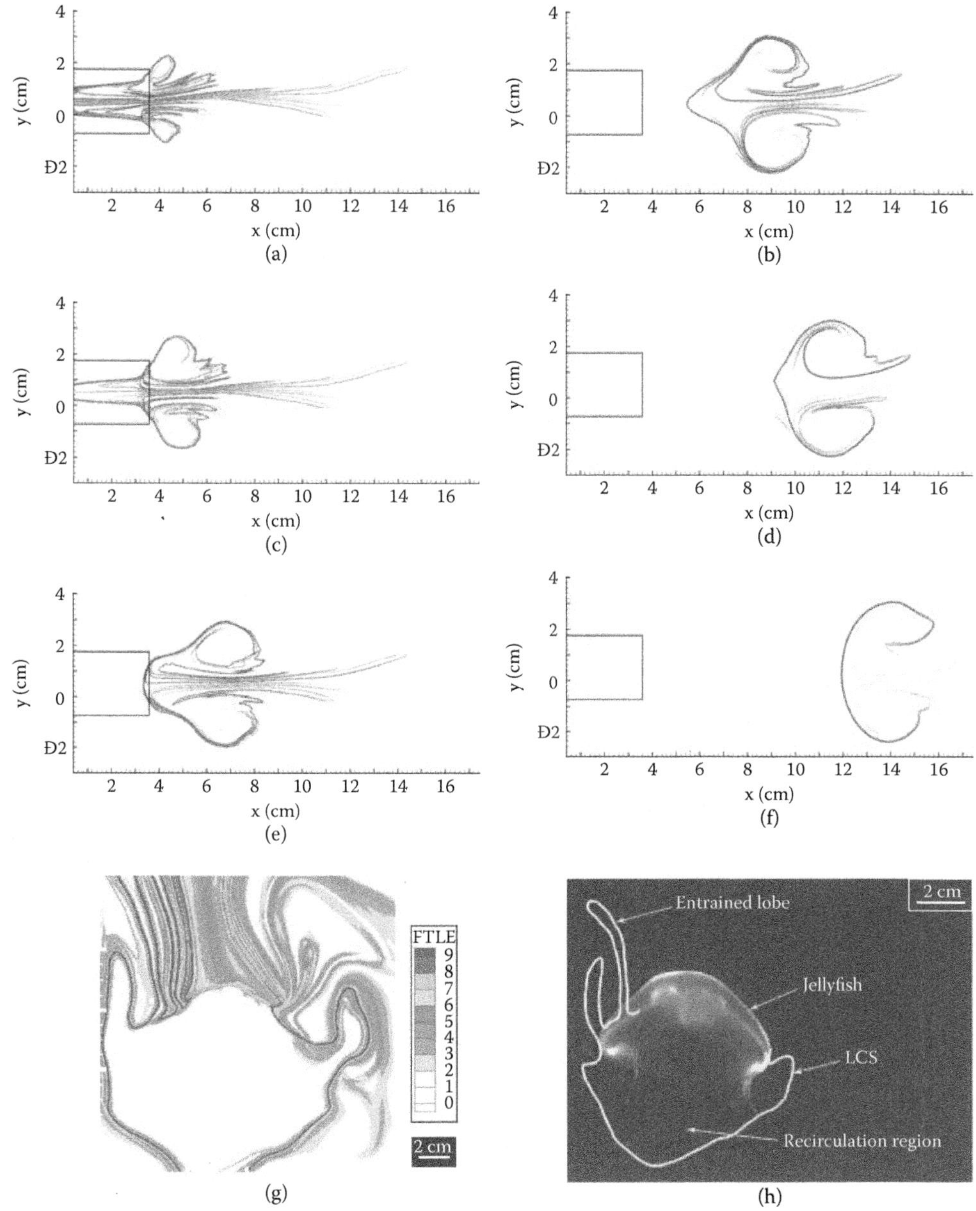

Figure 3.5. Lagrangian coherent structure (LCS) analysis of (a–f) a forming vortex ring; (g) freely swimming jellyfish; (h) overlay of instantaneous LCS with jellyfish. (Based on Shadden, S.C., et al., *Phys. Fluids*, 18(4), 047105, 2006; Shadden, S.C., et al., *J. Fluid Mech.*, 593, 315–331, 2007.)

an "Oy") approach, the forces acting on the fluid due to local pressure imbalances and due to the animal motion are the same as in the Lagrangian analysis above. As such, the right-hand sides of the force equations (Equations 3.11 through 3.13) is unchanged. And, as before, these applied forces result in the acceleration of the fluid particles that comprise the fluid medium. However, our present goal is to express the fluid particle motion in terms of the fluid momentum at fixed locations in space, so that the flow can be analyzed from the latter perspective. Because the fluid momentum per unit mass is given by its velocity, our focus is on changes to the velocity of the fluid in particular.

Let us consider how the velocity of a fluid particle changes as it moves through a sequence of fixed locations in the flow. If the flow is *steady*, meaning that the velocity at a given fixed location in space does not change over time, then the fluid particle velocity can only change by moving to a new location in the flow. For example, for a small change dx in position along the **i**-axis, the change in the u-component of velocity (i.e., in the x-direction along the **i**-axis) will be

$$du = \frac{\partial u}{\partial x} dx \tag{3.22}$$

Similarly, a small change dy or dz in position along the **j**- or **k**-axis, respectively, will change the u-component of fluid particle velocity by amounts

$$du = \frac{\partial u}{\partial y} dy \tag{3.23}$$

$$du = \frac{\partial u}{\partial z} dz \tag{3.24}$$

The total change in the u-component velocity due to arbitrary motion in three dimensions is given by the sum of Equations 3.22 through 3.24:

$$du = \frac{\partial u}{\partial x} dx + \frac{\partial u}{\partial y} dy + \frac{\partial u}{\partial z} dz \tag{3.25}$$

Fluid flow, especially that created by swimming animals, is typically *unsteady*; the fluid momentum at each fixed location in space varies over time. In this case, the fluid particle momentum can change independent of any changes in position, due to temporal changes in the fluid velocity at the fixed locations in the flow. Over a short time dt, this temporal change in fluid velocity is given by

$$du = \frac{\partial u}{\partial t} dt \tag{3.26}$$

The changes in fluid particle velocity due to changes in position (Equation 3.25) and due to changes in time (Equation 3.26) can occur simultaneously. Combining both equations and using the capitalized derivative notation introduced in Equation 3.1, the total rate of change of fluid particle momentum can be expressed in terms of the Eulerian velocities:

$$\frac{Du}{Dt} = \frac{\partial u}{\partial t} + u\frac{\partial u}{\partial x} + v\frac{\partial u}{\partial y} + w\frac{\partial u}{\partial z} \tag{3.27}$$

where $u = dx/dt$, $v = dy/dt$, and $w = dz/dt$ effectively measure how rapidly the fluid particle changes position along the **i**-, **j**-, and **k**-axis, respectively. The derivative D/Dt is appropriately called the "total derivative" or "substantial derivative," as it captures both spatial and temporal variations of the fluid velocity.

Equation 3.27 and analogous expressions for the v- and w-components of velocity can be used to write an Eulerian form of Newton's second law for the balance of forces acting on the fluid. Replacing the left-hand sides of Equations 3.11 through 3.13 gives

$$\frac{\partial u}{\partial t} + u\frac{\partial u}{\partial x} + v\frac{\partial u}{\partial y} + w\frac{\partial u}{\partial z} = f_{lx} - \frac{1}{\rho}\frac{\partial p}{\partial x} \tag{3.28}$$

$$\frac{\partial v}{\partial t} + u\frac{\partial v}{\partial x} + v\frac{\partial v}{\partial y} + w\frac{\partial v}{\partial z} = f_{ly} - \frac{1}{\rho}\frac{\partial p}{\partial y} \tag{3.29}$$

$$\frac{\partial w}{\partial t} + u\frac{\partial w}{\partial x} + v\frac{\partial w}{\partial y} + w\frac{\partial w}{\partial z} = f_{lz} - \frac{1}{\rho}\frac{\partial p}{\partial z} \tag{3.30}$$

or, in more compact vector form,

$$\frac{\partial \mathbf{u}}{\partial t} + (\mathbf{u} \cdot \nabla)\mathbf{u} = \mathbf{f}_l - \frac{1}{\rho}\nabla p \tag{3.31}$$

Equation 3.31 is commonly referred to as *Euler's equation*.

Although the Lagrangian expression of Newton's second law for the fluid dynamics (Equation 3.14) is different in appearance from Euler's equation for the same force balance, the present derivations show that the two perspectives are intimately related. The methods are further compared and contrasted at the end of this section.

3.3.1 Bernoulli's Equation

In the preceding discussion of the Lagrangian perspective on fluid flows, a method was described that uses a subset of the fluid particle trajectories to infer global properties of the flow structure (i.e., Lagrangian coherent structures). A similar reduction is possible for the Eulerian equations of motion. To begin, recall that local changes in any fluid property q are reflected in its spatial gradient:

$$\nabla q = \frac{\partial q}{\partial x}\mathbf{i} + \frac{\partial q}{\partial y}\mathbf{j} + \frac{\partial q}{\partial z}\mathbf{k} \tag{3.32}$$

By deriving a gradient form of Equation 3.31 and identifying where the gradient is zero, we can find locations in the flow where key properties of the flow are constant. The full Eulerian flow field can then be characterized on the basis of those zero-gradient locations.

To proceed, we must make approximations to the real animal–fluid interactions that occur during aquatic locomotion. First, we assume that no energy is lost due to the applied locomotive force $\mathbf{f}_l$. In this case, the force is called *conservative* (i.e., it conserves energy), and the force vector can then be expressed as the gradient of a *potential function*:

$$\mathbf{f}_l = \nabla \phi_l \tag{3.33}$$

where the potential function ϕ_l is a solution of Laplace's equation:

$$\frac{\partial^2 \phi_l}{\partial x^2} + \frac{\partial^2 \phi_l}{\partial y^2} + \frac{\partial^2 \phi_l}{\partial z^2} = \nabla^2 \phi_l = 0 \tag{3.34}$$

Determining ϕ_l in practice usually requires knowledge of its value (or of $\mathbf{f}_l$, via Equation 3.32) at relevant boundaries of the flow, such as at the animal–fluid interface.

Second, we assume that the fluid is *irrotational* everywhere; in other words, the fluid particles are nonrotating. In this case, the Eulerian velocity field $\mathbf{u}$ can also be expressed as the gradient of a potential function:

$$\mathbf{u} = \nabla \phi_u \tag{3.35}$$

where ϕ_u can be determined in a similar manner as ϕ_l. Given these two assumptions, whose veracity will be examined subsequently, Euler's equation becomes

$$\nabla\left(\frac{\partial \phi_u}{\partial t}\right) + (\nabla\phi_u \cdot \nabla)\nabla\phi_u = \nabla\left(\phi_l - \frac{p}{\rho}\right) \tag{3.36}$$

The second term on the left-hand side of Equation 3.36 can be replaced by using a *vector identity* to incorporate a new term that is more convenient for achieving a gradient form of Euler's equation. This vector identity is one of two that will be used frequently in this chapter; as such, the reader is encouraged to become familiar with it. Given a vector $\mathbf{a}$,

$$(\mathbf{a} \cdot \nabla)\mathbf{a} = \nabla\left(\frac{\mathbf{a} \cdot \mathbf{a}}{2}\right) - [\mathbf{a} \times (\nabla \times \mathbf{a})] \tag{3.37}$$

where the symbol $\times$ denotes a *vector product*, for example,

$$\mathbf{x}_1 \times \mathbf{x}_2 = (y_1 z_2 - y_2 z_1)\mathbf{i} + (x_2 z_1 - x_1 z_2)\mathbf{j} + (x_1 y_2 - x_2 y_1)\mathbf{k} \tag{3.38}$$

If we let $\mathbf{a} = \nabla\phi_u$ in Equation 3.37 and note that $\nabla \times \nabla\phi_u = 0$ by definition (the curious reader can use Equation 3.37 to verify this), the second term on the left-hand side of Equation 3.36 becomes

$$(\nabla\phi_u \cdot \nabla)\nabla\phi_u = \nabla\left[\frac{(\nabla\phi_u)^2}{2}\right] \tag{3.39}$$

Using this result and replacing $\nabla\phi_u$ with $\mathbf{u}$ for convenience, we arrive at a gradient form of Euler's equation:

$$\nabla\left(\frac{\partial \phi_u}{\partial t} + \frac{u^2}{2} + \frac{p}{\rho} - \phi_f\right) = 0 \tag{3.40}$$

This equation is satisfied if, along a given path in the flow,

$$\frac{\partial \phi_u}{\partial t} + \frac{u^2}{2} + \frac{p}{\rho} - \phi_f = C \tag{3.41}$$

where C is a constant. Equation 3.41 has the form of *Bernoulli's equation*, and it determines the relationship between velocity and pressure along any path in the flow. It will be a powerful tool in our analysis of animal swimming because examination of a few key flow paths by using Equation 3.41 can be used to deduce global flow properties.

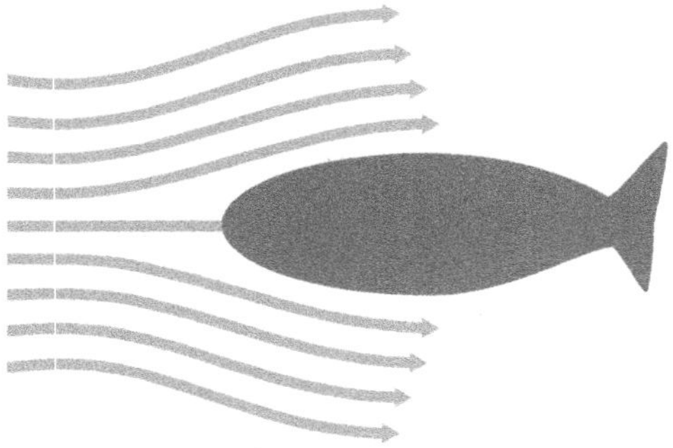

Figure 3.6. Schematic of streamlines around a swimming fish.

For example, Figure 3.6 illustrates a model of the flow created by a suction feeding fish. If we approximate this flow as steady and neglect the forces due to locomotion, Equation 3.41 reduces to

$$\frac{u^2}{2} + \frac{p}{\rho} = C \tag{3.42}$$

Because Equation 3.42 holds along any path in the flow, let us consider a path from the mouth of the fish to a region far away. In the distant ambient fluid at rest, the flow speed $u_{ambient} = 0$. The fluid pressure in the region far from the fish is balanced with the local gravitational forces; let us denote this *hydrostatic* (literally, "water at rest") pressure as p_0. Given this information regarding the flow at the end the path far from the fish, Equation 3.42 tells us that the value of the Bernoulli constant at every point along the flow path leading to the fish is

$$C = \frac{(0)^2}{2} + \frac{p_0}{\rho} = \frac{p_0}{\rho} \tag{3.43}$$

Hence, we can immediately relate the flow velocity u_{fish} at the mouth of the fish to the suction pressure p_{fish} that the fish creates:

$$\frac{u_{fish}^2}{2} + \frac{p_{fish}}{\rho} = \frac{p_0}{\rho} \tag{3.44}$$

or

$$u_{fish} = \sqrt{\frac{2}{\rho}\left(p_0 - p_{fish}\right)} \tag{3.45}$$

In steady Eulerian flows, each streamline (i.e., each line tangent to the local Eulerian velocity vectors) corresponds identically to a trajectory of fluid particles in the flow and is therefore a flow path on which Bernoulli's equation can potentially be evaluated. However, in unsteady flows that occur more commonly in animal swimming, fluid particle trajectories and instantaneous streamlines can differ significantly. We must use greater caution in applying Bernoulli's equation in unsteady flows because not all flow paths will satisfy Equation 3.41. This concern warrants that we examine how to quantify flow unsteadiness in the first place. By answering this question, we also see that many of the flows of interest in aquatic locomotion are better suited for either a Lagrangian or an Eulerian analysis.

3.4 QUANTIFYING FLOW UNSTEADINESS

A fish coasting at constant velocity seems to be an intuitive example of a steady flow. Indeed, the fluid particle trajectories and streamlines are consistent with one another when viewed in a frame of reference fixed on the fish (Figure 3.7a). If we move into a frame of reference fixed in space so that the fish is now moving, a different picture emerges. The streamlines closest to the body form closed loops that terminate on the surface of the fish, and the fluid particle trajectories exhibit looping paths (Figure 3.7c,d). The discrepancy between streamlines and fluid particle trajectories suggests that the flow is unsteady.

The preceding example illustrates an important principle that we use in analyzing animal swimming: the steadiness of the flow depends on the frame of reference in which it is observed. Because steady flows will often allow for more liberal use of analytical tools such as Bernoulli's equation, it is to our advantage to identify a frame of reference in which the flow is steady and to conduct the analysis from there.

Moving from one frame of reference to another can be accomplished by adding or subtracting a constant velocity U from the entire Eulerian velocity field to create a new velocity field $\mathbf{u}_{new}(\mathbf{x})$. Accordingly, we can identify two possible scenarios regarding the effect of the frame transformation $\pm U$:

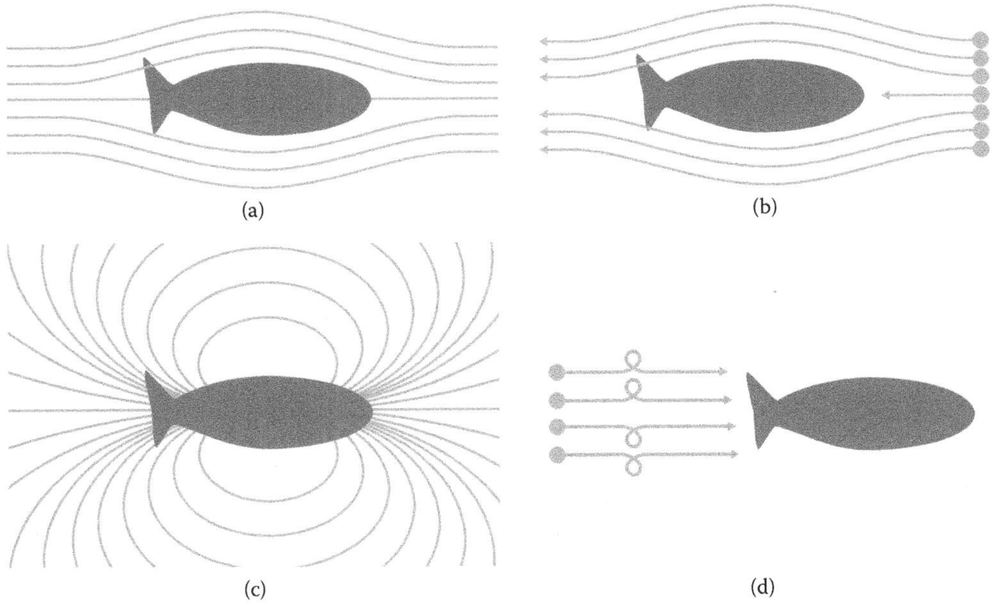

(a) (b) (c) (d)

Figure 3.7. (a) Schematic of streamlines in the frame of reference of the swimming fish. (b) Schematic of fluid particle trajectories in the frame of reference of the swimming fish. (c) Schematic of streamlines in a stationary reference frame. (d) Schematic of fluid particle trajectories in a stationary reference frame.

Scenario #1: There exists some value of U such that for all locations **x** in the flow, the velocity $\mathbf{u}_{new}(\mathbf{x})$ is constant over time. In this case, the entire flow is steady in at least one frame of reference, and the flow should be analyzed in that frame. Because the Lagrangian and Eulerian equations of fluid motion are identical in this scenario, either set can be used for the analysis.

Scenario #2: For all possible values of U, there exists a location **x** in the flow such that the local Eulerian velocity $\mathbf{u}_{new}(\mathbf{x})$ varies over time. In this case, the flow is inherently unsteady. The Lagrangian and Eulerian perspectives on this flow will be different, and more study will be required to determine the most effective method of analysis. As we will see later in this chapter, the two most common approaches are (I) to approximate the unsteady flow by an equivalent *quasi-steady* flow, thereby enabling use of tools like Bernoulli's equation; or (II) to transform the Eulerian equations into a *vorticity* form and track the regions of major vortical motion (i.e., fluid particle rotation) to infer the locomotive forces. This latter approach can make use of the LCS concept described in Section 3.1 to identify the relevant vortical regions of the flow.

3.5 SCALING IN ANIMAL SWIMMING

Our discussion of animal swimming has thus far omitted any mention of the effect of animal size on locomotion. In fact, by requiring that the fluid particles are always much smaller than the swimming organism, the governing equations of fluid motion derived in the preceding section imply that animal–fluid interactions are independent of animal size. The swimming of a bacterium and a whale would equally be described by Equations 3.14 and 3.31.

We can see this more clearly by attempting to extract characteristic length and time scales of the organism from the equations of motion. For convenience, let us consider only the x-component of the Eulerian form of the equations of motion, given in Equation 3.28. A new *dimensionless* form of each term in the equation, that is, having no physical units, can be created by expressing each term relative to the animal length and time scales.

This process is called **normalization**. For example, the ratio of the velocity component u to the animal swimming speed U_0 defines a normalized, dimensionless speed u^*:

$$u^* = \frac{u}{U_0} \tag{3.46}$$

Similarly, the perpendicular velocity components v and w can be normalized as

$$v^* = \frac{v}{U_0} \tag{3.47}$$

$$w^* = \frac{w}{U_0} \tag{3.48}$$

The spatial derivatives $\partial/\partial x$, $\partial/\partial y$, and $\partial/\partial z$ can each be normalized using the characteristic length scale L_0 of the animal:

$$\frac{\partial}{\partial x^*} = L_0 \frac{\partial}{\partial x} \tag{3.49}$$

$$\frac{\partial}{\partial y^*} = L_0 \frac{\partial}{\partial y} \tag{3.50}$$

$$\frac{\partial}{\partial z^*} = L_0 \frac{\partial}{\partial z} \tag{3.51}$$

To normalize the time derivative $\partial/\partial t$, we will need a characteristic timescale. Because the characteristic speed U_0 has physical units of length per unit time, it can be combined with the characteristic length L_0 to create a timescale for normalization:

$$\frac{\partial}{\partial t^*} = \frac{L_0}{U_0} \frac{\partial}{\partial t} \tag{3.52}$$

To normalize the pressure in Equation 3.28, note from Bernoulli's equation (e.g., Equation 3.41) that the physical units of pressure are the same as ρu^2, given that the terms are added together there. Therefore, if we also define a characteristic fluid density ρ_0, the pressure can be made dimensionless as

$$p^* = \frac{p}{\rho_0 U_0^2} \tag{3.53}$$

The locomotive force per unit mass f_{lx} has the physical dimensions of acceleration. Hence, it can be normalized by using the characteristic speed and length of the animal:

$$f_{lx}^* = \frac{L_0}{U_0^2} f_{lx} \tag{3.54}$$

Last, the density can be normalized as $\rho^* = \rho/\rho_0$.

The characteristic length, speed, and time scales used here are not unique. Proper selection of scales for normalization will depend on the particular aspects of locomotion that are of interest. In any case, the process of normalization will follow that done presently.

Each term in the equation of fluid motion (Equation 3.28) can now be replaced by the respective dimensionless parameters and characteristic scales in Equations 3.46 through 3.54. The resulting form of Euler's equation is

$$\frac{U_0}{L_0} U_0 \frac{\partial u^*}{\partial t^*} + U_0 u^* \frac{U_0}{L_0} \frac{\partial u^*}{\partial x^*} + U_0 v^* \frac{U_0}{L_0} \frac{\partial u^*}{\partial y^*} + U_0 w^* \frac{U_0}{L_0} \frac{\partial u^*}{\partial z^*} = \frac{U_0^2}{L_0} f_{lx}^* - \left(\frac{1}{\rho_0}\right) \frac{1}{\rho^*} \left(\frac{\rho_0 U_0^2}{L_0}\right) \frac{\partial p^*}{\partial x^*} \tag{3.55}$$

Equation 3.55 has a curious, but revealing, property: when each term is divided by U_0^2/L_0, we recover a dimensionless analog of Equation 3.28, that is,

$$\frac{\partial u^*}{\partial t^*} + u^* \frac{\partial u^*}{\partial x^*} + v^* \frac{\partial u^*}{\partial y^*} + w^* \frac{\partial u^*}{\partial z^*} = f_{lx}^* - \frac{1}{\rho^*} \frac{\partial p^*}{\partial x^*} \tag{3.56}$$

Thus, despite our deliberate attempt to extract physical scales from the equation of fluid motion, we have instead further demonstrated the scale independence of Euler's equation.

This result is inconsistent with intuition. Indeed, a causal examination of the flow field of a bacterium and a whale confirms that there are profound implications of the length scale of the animal in the flow. It is clear that we have neglected important, scale-dependent forces in our analysis. The following sections examine some of these forces.

3.5.1 *Viscosity*

A common, scale-dependent force that emerges in animal–fluid interactions is that due to the viscosity of the fluid. For the majority of flows created by aquatic organisms, this force is linearly proportional to the spatial gradients of flow speed. Fluids that exhibit this behavior are called *Newtonian*, and the relationship between the *viscous shear* force per unit area τ_v and the velocity gradients is given by, for example,

$$\tau_{v,xy} = \mu \frac{\partial u}{\partial y} \tag{3.57}$$

Referencing our fluid particle model in Figure 3.2, this force is applied in the x-direction on the faces of the cube that are normal to the j-axis, that is, perpendicular to the direction along with the velocity is changing. To balance rotational torques on the fluid particle, it is essential that $\tau_{v,xy} = \tau_{v,yx}$ locally.

To determine the viscous force per unit mass on a fluid particle in the flow, we can use the same approach applied to the pressure imbalance in Section 3.2. The resulting force term is

$$\mathbf{f}_v = \frac{\mu}{\rho}\nabla^2\mathbf{u} \tag{3.58}$$

where μ is the *dynamic viscosity*, which has physical dimensions of *mass/(length × time)*. The operator ∇^2 is the *Laplacian*, an example of which was described in Equation 3.34.

The ratio of dynamic viscosity to fluid density is called the *kinematic viscosity*, as it normalizes for the effect of fluid density, which gives the fluid its inertia (and hence its dynamics). Using the kinematic viscosity and adding the viscous force in Equation 3.58 to Euler's equation (Equation 3.31) leads to the *Navier–Stokes equation*:

$$\frac{\partial \mathbf{u}}{\partial t} + (\mathbf{u}\cdot\nabla)\mathbf{u} = \mathbf{f}_l - \frac{1}{\rho}\nabla p + \nu\nabla^2\mathbf{u} \tag{3.59}$$

Let us attempt to again extract characteristic scales from this equation. The Laplacian term in the x-component of viscous force can be normalized as

$$\nabla^{*2}u^* = \left(\frac{L_0^2}{U_0}\right)\nabla^2 u \tag{3.60}$$

Substituting this dimensionless parameter and the characteristic scales U_0 and L_0 into Equation 3.55, and dividing each term by U_0^2/L_0 again, we get

$$\begin{aligned}\frac{\partial u^*}{\partial t^*} + u^*\frac{\partial u^*}{\partial x^*} + v^*\frac{\partial u^*}{\partial y^*} + w^*\frac{\partial u^*}{\partial z^*} \\ = f_{lx}^* - \frac{1}{\rho^*}\frac{\partial p^*}{\partial x^*} + \frac{\nu}{U_0L_0}\nabla^{*2}u^*\end{aligned} \tag{3.61}$$

In this case, our final result does indeed depend on the characteristic scales U_0 and L_0. Specifically, if the ratio U_0L_0/ν is large, the viscous force term vanishes from the equation of fluid motion. This ratio is called the *Reynolds number* (denoted Re), and its value depends on our choice of characteristic speed and length scales, as well as on the kinematic viscosity of the fluid. Because the real fluid environment of aquatic animals always has nonzero viscosity, strictly speaking we can only neglect viscous forces for very large, very fast-moving, or both types of organisms. Nonetheless, we will find that in many unsteady flows, the effect of viscous forces remains secondary to the dynamics of the pressure imbalances in the flow.

3.5.2 Buoyancy

Aquatic organisms whose average mass density is different from the surrounding fluid must contend with gravitational forces. In addition, the hydrostatic pressure of the fluid (see Equation 3.42 and accompanying text for an example) increases with depth in the fluid due to gravity. Per unit of fluid mass, the gravitational force is given by

$$f_g = -g \tag{3.62}$$

where the negative sign accounts for the downward direction of gravity. Because gravity has units of acceleration, it can be normalized in the same manner as the locomotive force in Equation 3.54:

$$f_g^* = \frac{L_0 g}{U_0^2} \tag{3.63}$$

The ratio in Equation 3.63 is the inverse of the *Froude number* (denoted Fr) and plays a role similar to that of the Reynolds number, in the sense that large values of Froude number indicate that gravitational effects can be neglected in the equations of fluid motion. The effect of Froude number is even more direct than the Reynolds number, as it is occurs as a standalone term.

3.5.3 Surface Tension

Small organisms at the surface of a body of water can experience the effects of surface tension. At the interface between air and water, intermolecular forces are unbalanced due to the differential pull of air and water on each surface water molecule. Thus, increasing the interfacial surface area, for example, by pressing down on the water surface, requires an input of energy. This energy per unit of interface area has the physical dimensions of a force per unit of interface length; hence, it is referred to as surface tension. The effect can be a nuisance, such as when air bubbles become trapped within the breathing apparatus of a swimming organism; or it can be exploited, as in the propulsion of water striders (Figure 3.8).

Quantifying the effect of surface tension per unit mass of fluid is awkward given that the fluid interface is an area rather than a volume. Nonetheless, we can define a force per unit mass:

$$f_{st} = \frac{\gamma}{\rho_0 L_0^2} \tag{3.64}$$

Figure 3.8. Water strider on the free surface of a pond. (From Lou Murray, Earth Day and a birding trip to the Eastern Sierras, http://greenlifeinsocal.wordpress.com/2010/04/21/earth-day-and-a-birding-trip-to-the-eastern-sierras/. With permission.)

where γ is the surface tension, which has a value of approximately 0.07 N/m for a clean air–water interface. Following the procedure for viscosity and buoyancy, the normalized surface tension per unit mass is

$$f_{st}^* = \frac{L_0}{U_0^2}\frac{\gamma}{\rho_0 L_0^2} = \frac{\gamma}{\rho_0 U_0^2 L_0} \tag{3.65}$$

This ratio is inverse of the *Weber number* (denoted We). As in the case of the Froude number, its effect in the equations of motion is direct, and only for large values of the Weber number can the surface tension be neglected. One should use caution in selecting the appropriate speed and length scales to define the Weber number. Because we are interested in changes to the air–water interface, the parts of the organism directly in contact with the interface are most relevant for determining the characteristic scales. For example, the water strider in Figure 3.8 is best characterized on the basis of the size and motion of the swimming legs.

3.5.4 Thermal Convection

There are a few instances in which temperature differences between different parts of the ocean can be utilized by swimming organisms to enhance locomotion. In the vicinity of hydrothermal vents on the sea floor, for example, large differences between the ambient water temperature and the vented water lead to the formation of buoyant plumes that can transport organisms away from the benthos.

The change in fluid density due to changes in its temperature is quantified by the coefficient of thermal expansion β:

$$\beta = -\frac{1}{\rho}\left(\frac{\partial \rho}{\partial \Theta}\right)_p \tag{3.66}$$

where Θ is temperature and the subscript p denotes that the derivative is evaluated at constant pressure. Values of β for various temperatures and pressures can be found in engineering reference textbooks. Because the coefficient of thermal expansion has units of inverse temperature, it is often presented as the product $\beta\Theta$.

The buoyancy force per unit of fluid mass that arises due to the temperature-induced change in fluid density is

$$f_{tc} = -\beta(\Theta - \Theta_0)g \quad (3.67)$$

where Θ_0 is the temperature of the ambient fluid. Normalizing as in the previous cases gives

$$f_{tc}^* = -\frac{L_0}{U_0^2}\beta(\Theta - \Theta_0)g \quad (3.68)$$

The dimensionless force in Equation 3.68 can be related to the Froude number by

$$f_{tc}^* = \frac{\beta(\Theta - \Theta_0)}{Fr} \quad (3.69)$$

The effect of thermal convection can therefore be neglected if the Froude number is large or the change in fluid temperature is small. Furthermore, by defining an additional dimensionless quantity, the *Grashof number* (denoted Gr),

$$Gr = -\frac{L_0^3}{v_0^2}\beta(\Theta - \Theta_0)g \quad (3.70)$$

the Reynolds number and the Froude number can be related:

$$\frac{Gr}{Re^2} = \frac{\beta(\Theta - \Theta_0)}{Fr} \quad (3.71)$$

Equation 3.71 allows for the comparison of two dimensionless quantities (i.e., Gr and Re) to determine the importance of thermal convection. This is useful when comparing aquatic organisms that differ in scale and are located in environments with different thermal conditions.

To be sure, this is not an exhaustive list of relevant hydrodynamic forces or of the dimensionless ratios that can be used to quantify animal–fluid interactions. They do, however, represent the most common physical forces occurring in aquatic locomotion.

3.6 SCALING UNSTEADY FLOWS

The starting point for the scaling analyses in the previous section was the identification of characteristic length and velocity scales. Although the length is typically a static property and therefore straightforward to define, the velocity can change substantially in unsteady flows. Consider, for example, the jet efflux from a swimming squid. To determine the Reynolds number of the jet, should we use the maximum jet speed, which occurs only once per jet pulse, to describe the entire swimming cycle; or the instantaneous jet speed, even though this allows for the possibility of zero Reynolds number in between jet pulses? Furthermore, it is not clear that a time-averaged Reynolds number physically is meaningful, given that the Navier–Stokes Equation 3.61 in which the Reynolds number appears describes the *instantaneous* balance of forces.

Because we lack a definitive solution to this issue, let us consider a rule of thumb to bear in mind when conducting scaling analyses for unsteady flows. If we proceed with a time-averaged Reynolds number, for example,

$$\overline{Re} = \frac{1}{T}\int_{t_0}^{t_0+T} \frac{U(t)L_0}{\nu}dt \quad (3.72)$$

where the velocity scale is averaged over a time T after some initial time t_0, then we cannot use large values of $\overline{Re}$ alone as sufficient justification to neglect viscous forces. The preceding example of the squid jet shows that within the swimming cycle it is feasible that the Reynolds number becomes small, at which time viscous forces can become nonnegligible. Therefore we must show that Re(t) is small for *all* values of t to neglect viscous forces.

More fundamentally, we must concern ourselves with whether the spatial extent of the region in which viscous forces act is limited. If so, then we can often neglect the effects of viscous forces despite their existence in the flow. In a steady flow, large Reynolds numbers are indicative of such *vorticity confinement*. The basic limitation of the time-averaged Reynolds number is that it is by itself insufficient to deduce the existence of vorticity confinement.

For example, compare the flow around the fluke of gliding dolphin with that around an actively flapping fluke (Figure 3.9). Viscous forces act at the surface of the fluke, but in the gliding dolphin the resulting velocity gradients at the water–skin interface are transported in a thin

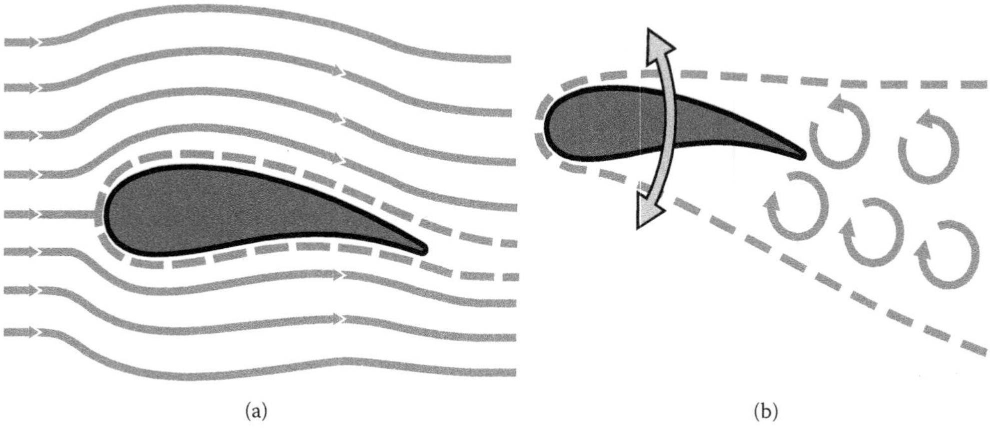

Figure 3.9. (a) Schematic of flow past a stationary airfoil. Boundary layer indicated in dashed lines. (b) Schematic of flow past oscillating airfoil. Wake vortices indicated by circular arrows.

boundary layer. The majority of the flow can therefore be analyzed without taking into account the viscous forces. In contrast, the flapping fluke motion distributes the fluid affected by viscous forces over a larger region of the flow. Hence, we cannot make an a priori claim of vorticity confinement even if the time-averaged Reynolds number of the flow is large in this case.

Our rule of thumb is then to examine both the time-averaged Reynolds number and the relative timescales of body motion and fluid shear. The timescale of the body motion T_b is that of the swimming appendage of interest, and it is given by the inverse of the temporal frequency F_b of the motion:

$$T_b \sim \frac{1}{F_b} \tag{3.73}$$

A timescale of the fluid shear can be derived from the length scale of the region affected by viscous forces, such as wake width W, and the average velocity $\overline{U_w}$ at which the viscosity-affected fluid is distributed into the wake. For the flapping fluke, the velocity of the fluke would be an appropriate surrogate for $\overline{U_w}$. The corresponding fluid timescale is

$$T_f \sim \frac{W}{\overline{U_w}} \tag{3.74}$$

The ratio of these timescales is given by

$$\frac{T_f}{T_b} \sim \frac{F_b W}{\overline{U_w}} \tag{3.75}$$

and is a type of **Strouhal number** (denoted St). To neglect viscous forces in unsteady flows, we require that the time-averaged Reynolds number is large and that the Strouhal number in Equation 3.75 is small. Small values of Strouhal numbers occur for low frequencies of body motion (i.e., quasi-steady motion); for small wake widths (i.e., vorticity confinement); or large values of the wake velocity, which lead to rapid transport of viscosity-affected flow away from the animal of interest.

In the following section, we present examples of these modeling concepts applied to practical problems of biological importance.

3.7 PRACTICAL EXAMPLES OF BIOLOGICAL SCALING

Here we examine two practical applications of the scaling concepts that were introduced in the previous section. In the first example, we see that predictions regarding the relationship between animal structure and function can be made without explicit use of the governing equations of fluid motion. However, the second example presents a cautionary tale regarding the limits of such ad hoc scaling analyses. This motivates a more in-depth study of animal–fluid interactions that makes use of the full repertoire of the governing equations.

3.7.1 Ontogenetic Scaling

Changes in animal properties that occur concurrently with growth are described by *ontogenetic* scaling laws. An example is the change in body shape that occurs during the transition of a jellyfish from its juvenile stage as a *ephyra* to its adult form. During this process of morphogenesis, the animal increases in size by up to 100-fold, from as small as a millimeter to well more than 10 cm in many cases. As the animal increases in overall diameter, gaps between the arm-like *lappets* of the ephyra are filled by the growth of new tissue (Figure 3.10).

We have already seen that for appendages moving at low Reynolds numbers, the effect of viscous forces—and hence, the presence of the animal body—is felt throughout a *boundary layer* region surrounding the moving appendages. In fact, for closely spaced appendages at low Reynolds numbers, the boundary layer fluid between the appendages moves as is if it were in direct contact with the appendages themselves. To understand why, recall that spatial gradients in fluid velocity are limited by the presence of fluid viscosity, which resists shearing of the flow. Fluid in contact with a solid appendage moves at the speed of the appendage: the *no-slip* condition. Because spatial gradients of the fluid velocity are limited by viscosity, the nearby fluid in the gaps between appendages moves at a speed similar to the fluid in contact with the appendage. Hence, the appendages and the fluid between them move as if they were a single, solid body (with the exception that the fluid in the gap does not impose a no-slip condition on the surrounding flow).

This concept of *equivalent surfaces* has been demonstrated in a variety of biological contexts, including aquatic chemoreception by antennae and swimming of microorganisms. In jellyfish ephyra, this suggests a mechanism that would enable the animal to affect a larger region of the surrounding flow, by using the boundary layers between the lappets to move water instead of actual tissue. The potential advantage of this approach is that the animal can sustain a smaller tissue mass, thereby decreasing metabolic requirements, while still affecting a region of the flow commensurate with its maximum body diameter.

The jellyfish ephyrae does grow additional tissue between the lappets as it grows in size (Figure 3.10). Is this growth correlated with the dynamics of the equivalent fluid surfaces? To answer this question, let us consider a simplified model of the ephyra (Figure 3.11a). From an aboral (i.e., dorsal) view, the fraction of the circumscribed circle of diameter D that is filled with tissue is

$$S = \frac{A_{D0} + A_{\text{lappets}}}{A_{\text{total}}} \tag{3.76}$$

where $A_{D0} = \pi D_0^2/4$; $A_{\text{lappets}} = Nbl$; and $A_{\text{total}} = \pi D^2/4$. The diameter of the central disc of tissue is D_0. The number of lappets is given by N, which is 8 in the present model but more generally is a multiple of 4. The lappet width is b and its length is l. These latter two parameters are approximated as

$$b \approx \pi D_0/N \tag{3.77}$$

Figure 3.10. Changes in jellyfish ephyra morphology with increasing body size.

$$l \approx (D - D_0)/2 \quad (3.78)$$

Substituting for the area variables in Equation 3.76 gives

$$S \approx \frac{\pi D_0^2/4 + \pi D_0 (D - D_0)/2}{\pi D^2/4} \approx \frac{2D_0}{D} - \left(\frac{D_0}{D}\right)^2 \quad (3.79)$$

For the boundary layers around the lappets to completely fill the adjacent gaps, let us require that the gap area A_{gap} be less than or equal to the area of the boundary layer on the corresponding side of the lappet (Figure 3.11b). The gap area is given by the remainder of A_{total} after subtracting A_{D0} and $A_{lappets}$:

$$A_{gap} = \frac{\pi}{4}(D - D_0)^2 \quad (3.80)$$

The boundary layer thickness will vary with radial distance r along the lappet, because the lappet motion is faster at the tip than it is at the junction with the central disc. Hence, we must integrate along the lappet length to determine the total boundary layer area:

$$A_{BL} = 2N\int_0^l \delta(r)\,dr \quad (3.81)$$

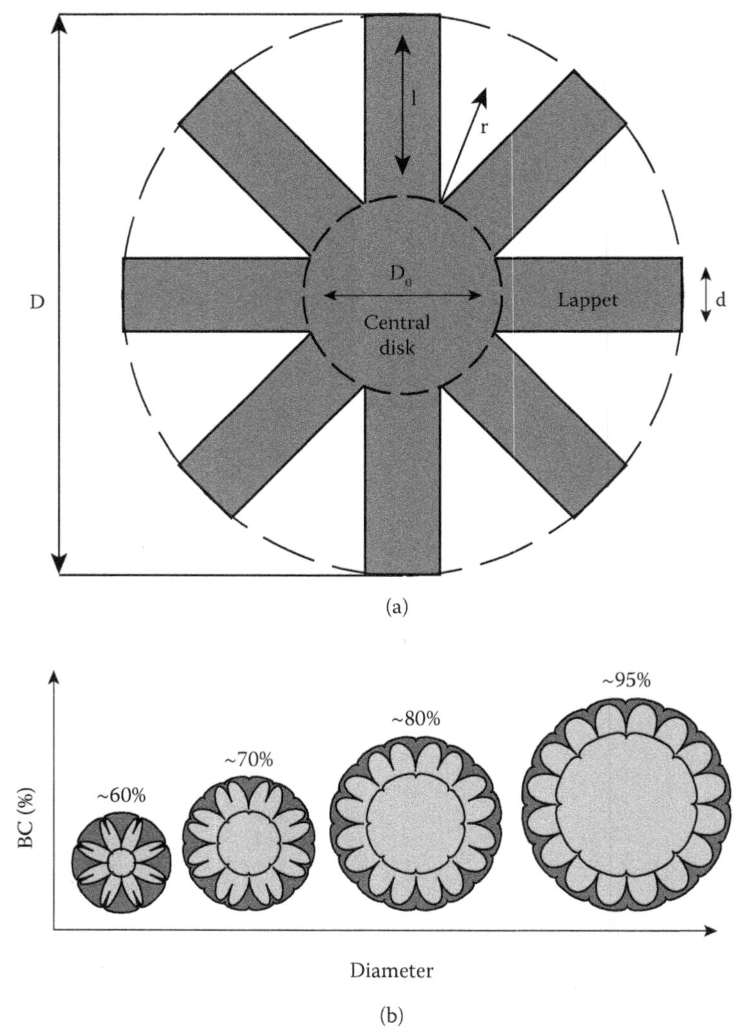

Figure 3.11. (a) Geometric model of a jellyfish ephyra. (b) Illustration of bell continuity (BC) versus bell diameter.

At each radial station, the boundary layer thickness increases with distance from the initial point of contact between the fluid and the appendage. In the standard fluid dynamics example of parallel flow past a flat plate, contact is initiated at the upstream edge of the plate, and the region influenced by viscous forces (i.e., the boundary layer) increases with downstream distance. The rate at which the boundary layer grows is determined by the balance of viscous forces acting at the fluid–solid interface and inertial forces that transport the fluid downstream. An approximate form of the Navier–Stokes equation can be solved to show that this balance results in a boundary layer profile given by the equation

$$\delta = \frac{Cx}{\sqrt{Re}} \tag{3.82}$$

where the constant C depends on the particular flow geometry. For a parallel flow past a flat plate, $C \approx 5.5$. Figure 3.12 illustrates the analogy that we draw between the flat plate flow and the flow past the lappet. The lappet width b becomes the characteristic length scale of the flow, replacing x in Equation 3.82. In addition, we can approximate the lappet motion as rigid-body rotation at angular speed Ω about a hinge located at the junction with the central disc. The characteristic velocity is therefore Ωr. The velocity increases from the central disc at $r = 0$ to the lappet tip at $r = l$. Hence the model captures the qualitative motion of the real ephyra lappets. Using this velocity, we can define a local Reynolds number

$$Re(r) = \frac{\omega r b}{\nu} \tag{3.83}$$

Substituting this Reynolds number definition into Equation 3.82 provides a model for the radial profile of the boundary layer thickness:

$$\delta(r) = C\sqrt{\frac{b\nu}{\omega r}} \tag{3.84}$$

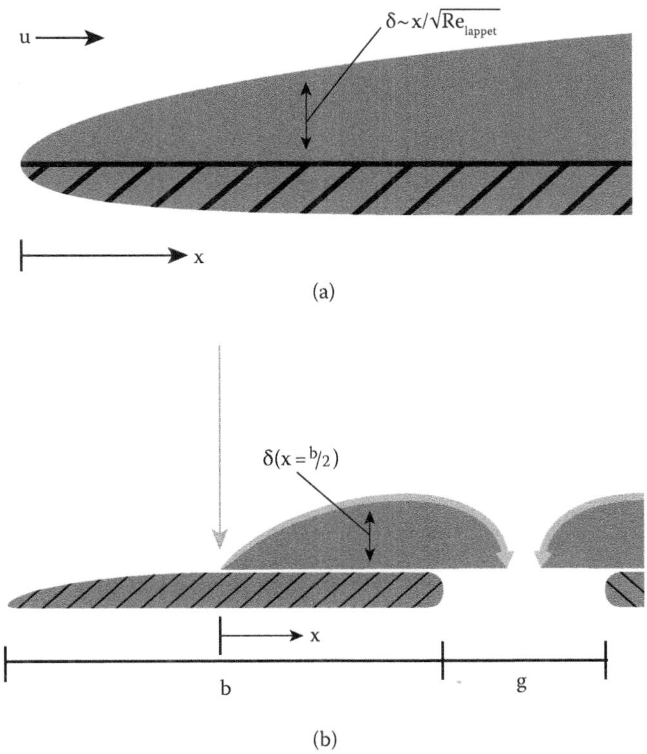

Figure 3.12. (a) Schematic of boundary layer growth over a solid surface (hatched region). (b) Model of boundary layer growth over adjacent jellyfish lappets.

Inspection of this boundary layer expression confirms our intuition: the boundary layer increases with increasing fluid viscosity or lappet thickness, the latter allowing the layer additional downstream distance to grow (Figure 3.12). Conversely, the boundary layer becomes thinner when velocities (and associated inertial forces) are higher, namely at the tip and when the lappet flaps rapidly.

Solving Equation 3.81 by using Equation 3.84, we arrive at an expression for the area of the boundary layer:

$$A_{BL} = C\sqrt{\frac{8\pi N\nu D_0 (D - D_0)}{\omega}} \qquad (3.85)$$

Using Equations 3.80 and 3.85, the condition $A_{gap} < A_{BL}$ for the formation of an equivalent surface can therefore be expressed as the inequality

$$128C^2 N\nu D_0 - \pi\omega (D - D_0)^3 \geq 0 \qquad (3.86)$$

By making assumptions on C and Ω, this equation can be solved for D as a function of D_0. Substituting the result into Equation 3.76 gives a *prediction* for the minimum shape fraction S that can use boundary layer dynamics to form an equivalent circular surface. Figure 3.13 plots an example of the model versus morphogenesis measurements from two lineages of jellyfish, *Aurelia* and *Chrysaora*. The trends for both groups agree well with the prediction of the model, indicating that the process of morphogenesis may be tuned to exploit the animal–fluid interactions that lead to the formation of an effective circular surface.

It is noteworthy that the preceding model has made only informal use of the governing equations of fluid motion. We have benefited from a simplified geometrical model of the ephyrae that possessed sufficient fidelity to capture the essential aspects of locomotion. Models of this sort will be invaluable for improving our understanding of aquatic locomotion. However, as we discuss next, the ad hoc modeling approach is not without its limitations.

3.7.2 Phylogenetic Scaling

Comparisons of animal properties across multiple species can be described by **phylogenetic** scaling laws. Returning to jellyfish as an example, more than 1000 extant species of jellyfish range in size from less than 1 mm to more than 1 m, a 1000-fold range of body size. The **fineness ratio** of the animals, a measure of the body shape based on the height-to-diameter ratio, varies from 0.1 to 3. A **morphospace** plot of fineness ratio versus body diameter shows a hyperbolic trend, with small animals at all fineness ratios, but only low-fineness shapes at large body diameters (Figure 3.14). In addition, across the various jellyfish lineages the layer of swimming muscle remains a single cell layer thick. Might this constraint on swimming muscle capacity explain the unique morphospace of jellyfish?

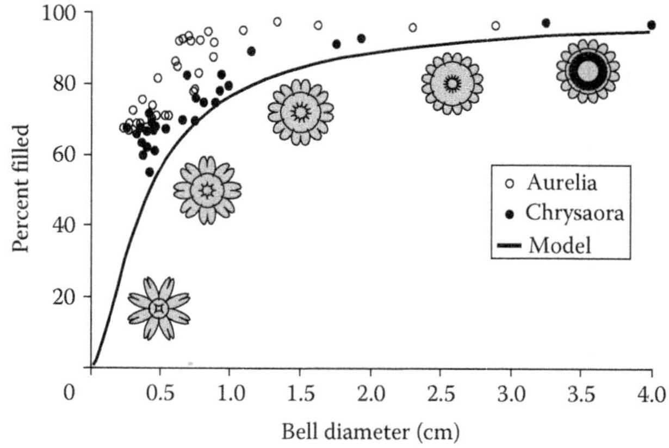

Figure 3.13. Comparison of theoretical morphogenesis model with measurement data. (Based on Feitl, K.E., et al., *Biol. Bull.*, 217, 283–291, 2009.)

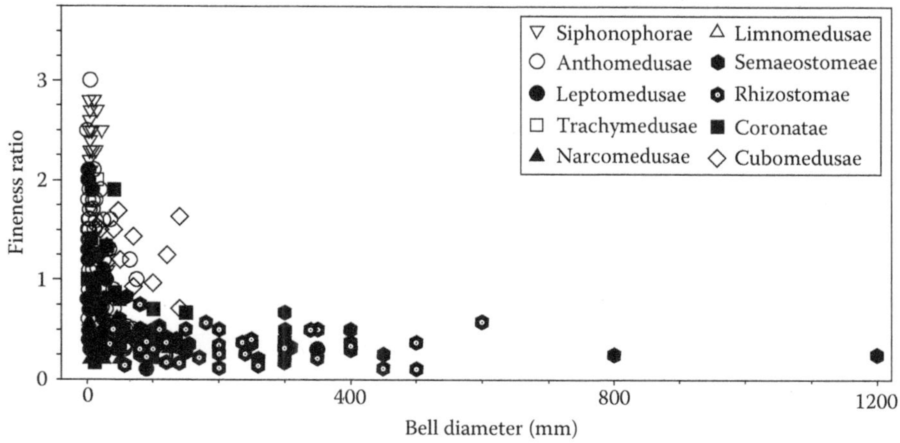

Figure 3.14. Measurements of jellyfish fineness ratio versus bell diameter. (Based on Costello, J.H., et al., *Invertebr. Biol.*, 127, 265–290, 2008.)

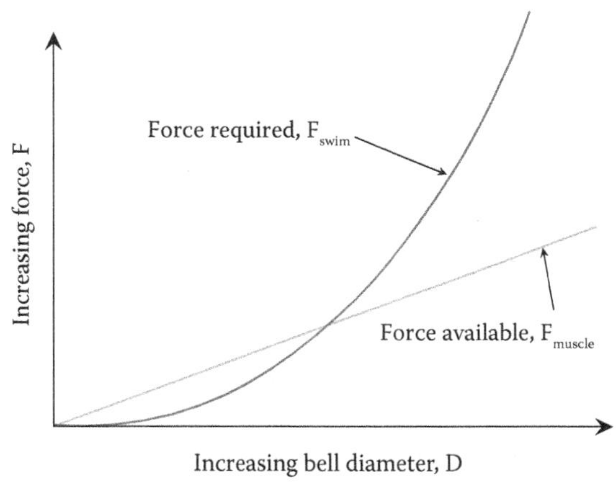

Figure 3.15. Comparison of scaling laws for force available and force required for jellyfish locomotion.

Let us consider that to achieve locomotion, the force F_m supplied by the muscle should exceed the force F_l required to swim. This latter force can be expected to scale with the mass m_f of fluid moved by the animal during locomotion and the forward acceleration a that results. If we assume that the animal acceleration is nominally constant and that the mass of water moved by the animal is proportional to its own volume, the scaling of swimming force with body diameter is

$$F_l = m_f a \sim D^3 \tag{3.87}$$

The force supplied by the muscles is proportional to the contractile stress that it can generate and to the cross-sectional area of the muscle layer, which is given by the product of the layer thickness τ and the layer width w (Figure 3.15). Across the various jellyfish lineages, the width of the muscle layer is approximately equal to one half of the body diameter. Hence,

$$F_m = \sigma\tau D/2 \sim D \tag{3.88}$$

Figure 3.15 plots the trends given by Equations 3.87 and 3.88. For small body diameters, the condition for achieving locomotion is satisfied, because the muscle force exceeds that required for locomotion. However, the force required for locomotion increases more rapidly with body diameter than the muscle force, leading to a critical body diameter above which the preceding model would predict that jellyfish cannot swim due to lack of sufficient muscle force. Substituting in measured muscle properties for σ and τ, this phylogenetic model predicts that no species of jellyfish with body diameters greater than approximately 10 cm should exist. The measurements shown in Figure 3.14 demonstrate the fallacy of that prediction.

To resolve this discrepancy, we must delve deeper into the physical mechanisms of aquatic locomotion. We begin by exploring the role of fluid rotation in animal–fluid interactions.

3.8 VORTICITY AND AQUATIC LOCOMOTION

The process of aquatic locomotion involves the transfer of both linear and angular momentum from the animal to the fluid. The linear momentum takes the form of the fluid velocity, and the angular momentum of the fluid is manifested in its *vorticity*. In essence, vorticity is measure of a fluid particle's tendency to rotate about an axis through its center. In this section, we see that the velocity and vorticity of the fluid share deep physical and mathematical connections. Furthermore, we find that our analysis of animal locomotion proceeds more effectively in many cases by tracking changes in the fluid vorticity instead of the velocity. This approach allows us, for example, to resolve the shortcomings of the phylogenetic scaling analysis in the previous section.

To begin, let us first extract vorticity from the Navier–Stokes equation (Equation 3.59), which ultimately governs the creation and transport of vorticity in the fluid. The viscous force term $\nu\nabla^2\mathbf{u}$ can be written in a different, equivalent form by using another vector identity. Given a vector **a**,

$$\nabla^2\mathbf{a} = \nabla(\nabla\cdot\mathbf{a}) - \nabla\times(\nabla\times\mathbf{a}) \quad (3.89)$$

where an example the vector product operation $\times$ was given in Equation 3.38. Because we can approximate the fluid medium as *incompressible*, the Eulerian velocity field **u** is by definition divergence free, that is, $\nabla\cdot\mathbf{u} = 0$. The viscous force term can therefore be expressed equivalently using Equation 3.89 as

$$\nabla^2\mathbf{u} = -\nabla\times(\nabla\times\mathbf{u}) \quad (3.90)$$

The vector product of the gradient operator ∇ and the velocity vector **u** is called the *curl* of the velocity field. This is the mathematical definition of vorticity ω:

$$\omega = \nabla\times\mathbf{u} = \left(\frac{\partial w}{\partial y} - \frac{\partial v}{\partial z}\right)\mathbf{i} + \left(\frac{\partial u}{\partial z} - \frac{\partial w}{\partial x}\right)\mathbf{j} + \left(\frac{\partial v}{\partial x} - \frac{\partial u}{\partial y}\right)\mathbf{k} \quad (3.91)$$

in Cartesian coordinates. Physically, the vorticity is defined as twice the angular velocity of the local fluid. The direction of the vorticity vector is along the axis of rotation, with counterclockwise rotation denoted with positive values and clockwise rotation being negative. The connection between vorticity and viscous forces is clear: vorticity depends on spatial gradients of velocity, which are resisted by viscous forces. Conversely, the application of shear forces on a viscous fluid, as occurs during aquatic locomotion, results in the creation of vorticity.

Substituting the vector identity in Equation 3.90 and the vorticity definition in Equation 3.91 into the Navier–Stokes equation (Equation 3.59) leads to a new form of the governing equation:

$$\frac{\partial\mathbf{u}}{\partial t} + (\mathbf{u}\cdot\nabla)\mathbf{u} = \mathbf{f}_l - \frac{1}{\rho}\nabla p - \nu(\nabla\times\omega) \quad (3.92)$$

Now, let us extract a vorticity from the term $(\mathbf{u}\cdot\nabla)\mathbf{u}$ in Equation 3.92. Using the same vector identity that led us to Bernoulli's equation (see Equation 3.36), we now have

$$\frac{\partial\mathbf{u}}{\partial t} + \nabla\left(\frac{\mathbf{u}\cdot\mathbf{u}}{2}\right) - \mathbf{u}\times\omega = \mathbf{f}_l - \frac{1}{\rho}\nabla p - \nu(\nabla\times\omega) \quad (3.93)$$

In light of the definition of vorticity in Equation 3.91, we can replace the temporal variation in velocity **u** with the temporal variation in vorticity ω by taking the curl of each term in Equation 3.93. Because the curl of each term is taken, the following equality holds:

$$\frac{\partial\omega}{\partial t} + \nabla\times\nabla\left(\frac{\mathbf{u}\cdot\mathbf{u}}{2}\right) - \nabla\times(\mathbf{u}\times\omega) = \nabla\times\mathbf{f}_l - \frac{1}{\rho}(\nabla\times\nabla p) - \nu\nabla\times(\nabla\times\omega) \quad (3.94)$$

By definition, the curl of the gradient of a vector is zero (as in Section 3.2, the curious reader can use Equation 3.38 to verify this). Hence, the second term on both the left- and right-hand sides of Equation 3.94 can be removed, leaving

$$\frac{\partial \omega}{\partial t} = \nabla \times (\mathbf{u} \times \omega) + \nabla \times \mathbf{f}_l - \nu \nabla \times (\nabla \times \omega) \quad (3.95)$$

This equation provides a budget that identifies mechanisms capable of changing the local vorticity over time: advective transport of the vorticity by the local flow velocity field (first term on right-hand side), creation of new vorticity due to the locomotive forces applied by the animal (second term), and diffusion of vorticity at a rate determined by the fluid viscosity (third term). To see this more clearly, consider a *control volume* surrounding a swimming organism and including a surface at the animal–fluid interface (Figure 3.16); let us keep track of the vorticity within that volume of fluid.

Equation 3.95 can be applied to the control volume by evaluating it on the surface of the control volume, because any vorticity entering or exiting the control volume must cross its surface. Integrating over the surface S gives

$$\frac{\partial}{\partial t}\int_S \omega \cdot \mathbf{dS}$$

$$= \int_S \left[\nabla \times (\mathbf{u} \times \omega) + \nabla \times \mathbf{f}_l - \nu \nabla \times (\nabla \times \omega)\right] \cdot \mathbf{dS} \quad (3.96)$$

where the subscript S on each integral denotes the region of interest, and the dot (·) denotes an inner or *scalar product* (see below Equation 3.18 for an example). The variable **dS** denotes a small area of the control volume surface; it is a vector that is aligned perpendicular to the local surface.

The *Stokes theorem* relates the integral of the curl of a vector on a surface to the integral of that vector around any closed contour C that wraps the surface, for example,

$$\int_S (\nabla \times \mathbf{a}) \cdot \mathbf{dS} = \oint_C \mathbf{a} \cdot \mathbf{dc} \quad (3.97)$$

where each small segment of the closed contour C is denoted **dc**. The direction of the vector **dc** is locally tangent (parallel) to C. The Stokes theorem allows us to simplify Equation 3.96 as

$$\frac{\partial}{\partial t}\int_S \omega \cdot \mathbf{dS} = \oint_C (\mathbf{u} \times \omega + \mathbf{f}_l - \nu \nabla \times \omega) \cdot \mathbf{dc} \quad (3.98)$$

The total quantity of vorticity within the control volume, $\int_S \omega \cdot \mathbf{dS}$, is called the *circulation* of the flow. The term $\mathbf{u} \times \omega$ is called the *vorticity flux*, a measure of vorticity transported by the flow. Hence, Equation 3.98 states that the rate of change of fluid circulation is given by the net effect of vorticity fluxes through the surface of the control surface, the creation of new vorticity by locomotive forces, and the diffusion of vorticity by the fluid viscosity.

Note that in the absence of external forces applied on the fluid by the swimming organism, the circulation in the control volume can only change due to advection or diffusion of *preexisting* vorticity; vorticity cannot be created in the fluid, only at the animal–fluid interfaces. This property of vorticity makes it a powerful

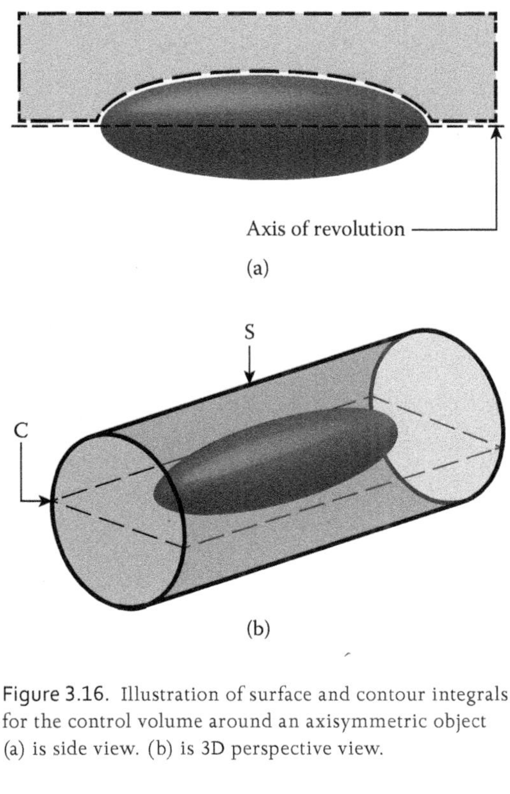

Figure 3.16. Illustration of surface and contour integrals for the control volume around an axisymmetric object (a) is side view. (b) is 3D perspective view.

tool for studying aquatic locomotion, because we can immediately identify and characterize the presence of locomotive forces on the basis of changes to the circulation of the surrounding fluid. We take full advantage of this property later in this chapter.

Before proceeding, it is instructive to examine the concept of conservation of circulation more carefully. This exercise highlights the role of viscosity in affecting the fluid circulation, and it also provides greater intuition regarding when neglect of viscous forces might be acceptable. Let us begin by using the Stokes theorem to define fluid circulation in terms of the Eulerian velocity field **u**:

$$\Gamma = \oint_{C_m} \mathbf{u} \cdot \mathbf{dc} \tag{3.99}$$

where C_m is any closed *material curve* surrounding the fluid of interest. The term *material* indicates that the curve is transported by the local flow velocity. Taking the time derivative of this circulation definition gives

$$\frac{\partial \Gamma}{\partial t} = \oint_{C_m} \frac{D\mathbf{u}}{Dt} \cdot \mathbf{dc} + \oint_{C_m} \mathbf{u}(\mathbf{dc} \cdot \nabla \mathbf{u}) \tag{3.100}$$

Two terms arise in the time derivative, following application of the *chain rule* for differentiation of multiplied quantities, namely **u** and **dc**. The total derivative $D\mathbf{u}/Dt$ is required for the first integral on the right-hand side of Equation 3.100 because we are interested in the velocity on the contour C_m, which is itself moving with the local flow. Hence, we must track the fluid in a Lagrangian sense. The second term on the right-hand side arises because the small contour element **dc** is also changing as a function of time due to the motion of C_m as a material curve. The time derivative $\partial \mathbf{dc}/\partial t$ can be expressed equivalently as $\mathbf{dc} \cdot \nabla \mathbf{u}$, using logic analogous to that which leads to the $(\mathbf{u} \cdot \nabla)\,\mathbf{u}$ term in the Euler equation (Equation 3.31). Substituting for $D\mathbf{u}/Dt$ from the Navier–Stokes equation (Equation 3.59), the time derivative of circulation in Equation 3.100 becomes

$$\frac{\partial \Gamma}{\partial t} = \oint_{C_m} \left(\mathbf{f}_l - \frac{\nabla p}{\rho} + \nu \nabla^2 \mathbf{u} \right) \cdot \mathbf{dc} + \oint_{C_m} \mathbf{u}(\mathbf{dc} \cdot \nabla \mathbf{u}) \tag{3.101}$$

The last term in Equation 3.101 can be replaced using the same vector identity (Equation 3.37) that we previously applied to the term $(\mathbf{u} \cdot \nabla)\mathbf{u}$:

$$\frac{\partial \Gamma}{\partial t} = \oint_{C_m} \left(\mathbf{f}_l - \frac{\nabla p}{\rho} + \nu \nabla^2 \mathbf{u} \right) \cdot \mathbf{dc} + \oint_{C_m} \left[\nabla \left(\frac{\mathbf{u} \cdot \mathbf{u}}{2} \right) - \mathbf{u} \times \omega \right] \cdot \mathbf{dc} \tag{3.102}$$

At this stage, we can significantly simplify our expression for the rate of change of fluid circulation, by noting two mathematical properties. First, the integral of the gradient of a parameter along a closed path is identically zero. To see why, consider that the gradient is a measure of the change in a parameter as one moves in space. Those changes accumulate as we integrate along a path (Figure 3.17a). However, along a closed path, we must return to the initial value of the parameter, lest we have the contradiction that the parameter possesses two different values at the same location in space. Hence, the total gradient integrated along the closed path must be zero. This leads us to eliminate the pressure gradient from the first integral in Equation 3.102, and to eliminate the gradient of **u·u** from the second integral.

Second, we note a geometric identity:

$$(\mathbf{a} \times \mathbf{b}) \cdot \mathbf{c} = 0 \quad \text{if} \quad (\mathbf{a} \times \mathbf{c}) = 0 \tag{3.103}$$

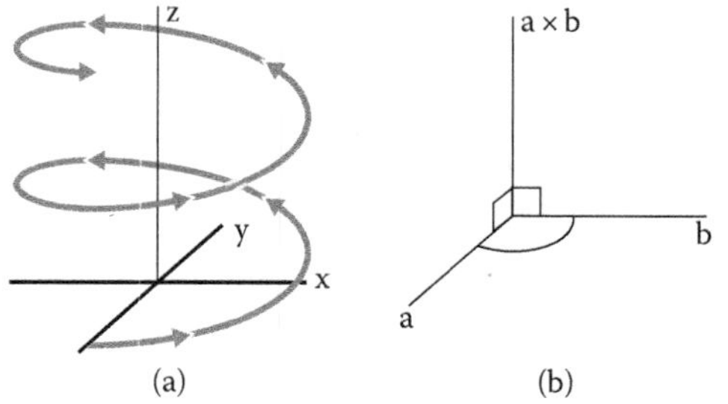

Figure 3.17. Illustrations of vector identities. (a) Integration along a continuous path. (b) Orthogonality of the cross product.

As illustrated in Figure 3.17b, the vector product of **a** and **b** is perpendicular to both **a** and **b** by definition. If $(\mathbf{a} \times \mathbf{c}) = 0$, then vectors **a** and **c** are parallel. It is therefore impossible for the vector product of **a** and **b**, which is perpendicular to **a**, to also be parallel to **c**. This further simplifies Equation 3.102 because $(\mathbf{u} \times \mathbf{dc}) = 0$; hence, $(\mathbf{u} \times \omega) \cdot \mathbf{dc}$ must be equal to zero.

Using the vector identity (Equation 3.89) to replace $\nabla^2\mathbf{u}$, our equation for the rate of change of circulation becomes

$$\frac{\partial \Gamma}{\partial t} = \oint_{C_m} (\mathbf{f}_l - \nu \nabla \times \omega) \cdot \mathbf{dc} \qquad (3.104)$$

The fluid circulation in the region bounded by the contour C_m can change over time by only two possible mechanisms: the creation of new vorticity via the animal–fluid interaction, or the action of viscosity. Viscous effects can reduce the circulation by diffusion of vorticity out of the region bounded by C_m. In addition, *vorticity cancellation* can occur between regions of the flow with opposite directions of rotation.

Fluxes of vorticity can rearrange the total circulation, but cannot result in a change of the circulation. The distinction between this result and what we previously found in Equation 3.98 is that the latter control volume was Eulerian and fixed in space, whereas the contour C_m is advected along with the vorticity in the flow. Hence, it is not possible for vorticity to cross C_m except via diffusion.

Equation 3.104 indicates that the timescale over which the circulation changes is determined by the time scales of the locomotive forces and of viscous diffusion of vorticity. We previously identified the timescale of locomotion in the context of the Strouhal number (see Section 3.5). If this timescale of the animal body motion, T_b:$1/F_b$ (see Equation 3.73), is comparable to the diffusion timescale for vorticity to travel a characteristic diffusion length L_0:

$$T_d \sim \frac{L_0^2}{\nu} \qquad (3.105)$$

then both terms of the integral in Equation 3.104 must be included in the analysis. However, if the diffusive processes are slow relative to the locomotion, such that $T_d \gg T_b$, then for practical purposes the viscous term can be neglected as a slow modification on the approximated dynamics. The ratio of T_d to T_b defines a *frequency Reynolds number*:

$$\mathrm{Re}_F = \frac{F_0 L_0^2}{\nu} \qquad (3.106)$$

As with our previous Reynolds number definitions, large values enable us to neglect the effect of viscous forces. At high-frequency Reynolds number and in the absence of external forces applied by the swimming organism

$$\frac{\partial \Gamma}{\partial t} = 0 \qquad (3.107)$$

Equation 3.107 is a statement of *Kelvin's circulation theorem* for *inviscid* fluid.

3.9 A MODEL OF VORTICITY CREATION DURING AQUATIC LOCOMOTION

In the previous section, we saw that vorticity is only created at the animal–fluid interface during locomotion. Here, we use a simplified geometric model to quantify the vorticity production in terms of the motion of the animal appendages. Figure 3.18 illustrates a fin-like appendage that we use to model the process of locomotion. Let us assume that the fin is moving at a constant velocity so that, in the reference frame of the fin, the flow moves steadily from left to right at speed U_0.

As a fluid particle approaches the fin from upstream, the side of the particle adjacent to the animal may come into contact with the fin surface. The no-slip condition (see Section 3.6) requires that the side of the fluid particle in contact with the solid fin surface must now move at the same speed as the fin (on average). The opposite side of the fluid particle away from the fin surface initially continues moving downstream due to its linear momentum. Thus, the particle proceeds to pivot around the side in contact with the fin, leading to clockwise rotation of the fluid particle around an axis perpendicular to the flow (i.e., out of the page). Vorticity has been created.

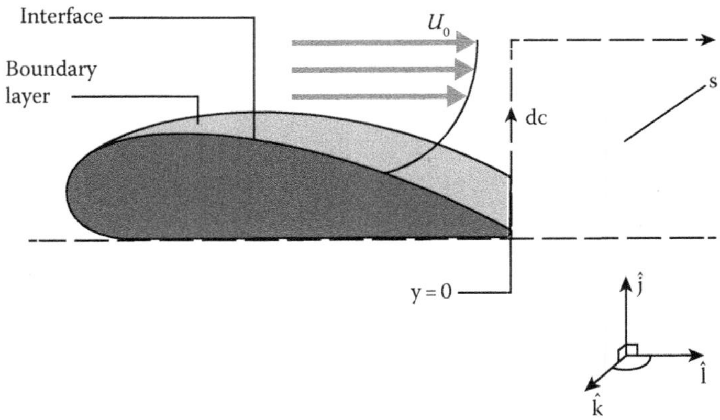

Figure 3.18. Boundary layer and control volume downstream from solid body.

In general, fluid particles near the fin surface move more slowly than the fluid particles in the free stream. Because the fluid viscosity resists spatial gradients in the velocity, fluid particles that barely avoid contact with the fin cannot move at the full free-stream speed due to the large velocity gradient that would result. As a consequence, these fluid particles are also slowed, but to a lesser degree than the fluid particles at the fin surface. The presence of the wall is felt at increasing distances from the animal–fluid interface as the flow progresses downstream along the fin. We have previously discussed this boundary layer growth in the context of ontogenetic scaling (see Section 3.6). Now, we can use the concepts developed in the previous section to quantify the change in fluid circulation due to the shedding of this boundary layer at the trailing edge of the fin.

For mathematical convenience only, let us assume (I) the flow is globally two-dimensional, (II) vorticity advection is large relative to vorticity diffusion, (III) the flow is locally one-dimensional near the fin surface, and (IV) the velocity gradients in the direction perpendicular to the fin surface are larger than those parallel to the fin.

Assumption I implies that the vorticity vector has only one component, $\boldsymbol{\omega} = -\omega\mathbf{k}$, where the minus sign is consistent with our convention that counterclockwise rotation is positive. Assumption II implies that, for Equation 3.98 evaluated in the region S indicated in Figure 3.18,

$$\nu\nabla \times \omega << \mathbf{u} \times \omega \tag{3.108}$$

Assumption III determines the relative magnitudes of the flow velocity in the direction parallel and perpendicular to the fin: $\mathbf{u}\cdot\mathbf{j} << \mathbf{u}\cdot\mathbf{i}$. Finally, assumption IV implies that, for example, $\partial v/\partial x << \partial u/\partial y$. With these assumptions, the vorticity of Equation 3.98 evaluated downstream of the fin is

$$\frac{\partial}{\partial t}\int_S \omega \cdot \mathbf{dS} = \frac{d\Gamma_w}{dt} = \int_0^\delta u\frac{\partial u}{\partial y}dy \tag{3.109}$$

where Γ_w is the circulation in the wake of the fin. To arrive at Equation 3.109, we have noted that $\mathbf{f}_I = 0$ in the wake (because there is no animal–fluid interface there), and we have used the z-component of the vorticity definition (Equation 3.91). The integral over the closed contour surrounding the wake is reduced significantly because the integrand of Equation 3.98 need only be evaluated where the vorticity is nonzero. As long as the surface S encloses the entire wake with room to spare, the only portion of the contour C with nonzero vorticity is the intersection with the boundary layer. Hence, we only integrate across the boundary layer, which is located between $y = 0$ and $y = \delta$, the boundary layer thickness.

Outside of the boundary layer, $u(y > \delta) = U_0$. Therefore, the integral in Equation 3.109 becomes

$$\begin{aligned}\frac{d\Gamma_w}{dt} &= \int_0^\delta u\,du = \frac{u^2}{2}\Big|_0^\delta \\ &= \frac{\left[u(y=0)\right]^2 - \left[u(y=0)\right]^2}{2} = \frac{1}{2}U_0^2\end{aligned} \tag{3.110}$$

or, allowing for time dependence,

$$\frac{d\Gamma_w}{dt}(t) = \frac{1}{2}U_0^2(t) \tag{3.111}$$

Equation 3.111 is often called the *slug model*, from its origin in the study of jets (i.e., ejected "slugs" of fluid). It is a simple, but powerful, tool that connects the motion of the aquatic animal to the vorticity that it creates. In the following sections, we explore the behavior of the vorticity once it is shed by the animal and the consequences for locomotion.

3.10 BOUNDARY LAYER SEPARATION FROM AQUATIC ANIMALS

Before we can take advantage of the connection between animal swimming kinematics and wake circulation in Equation 3.111, we need to develop a conceptual picture of how the wake is formed. We have seen that the source of circulation in the wake is vorticity created at the animal–fluid interfaces and subsequently transported in the boundary layers surrounding the animal appendages. Formation of the wake begins with separation of those boundary layers from the animal surface. Hence, our first task is to predict under what conditions *boundary layer separation* occurs.

As the name implies, boundary layer separation is the process in which the vorticity created at the animal–fluid interface detaches from the animal surface and enters the bulk fluid. A more precise, physical definition can be achieved by considering the Lagrangian trajectories of fluid particles on the surface of the animal appendage (Figure 3.19). Given the surface path $\mathbf{P}_s$ along which fluid particles move on the animal surface, a flow *separation point* $\mathbf{x}_s$ exists if the nearby fluid particle velocities $\mathbf{u}(\mathbf{x}_s - \mathbf{dP}_s)$ and $\mathbf{u}(\mathbf{x}_s + \mathbf{dP}_s)$ and the point $\mathbf{x}_s$ satisfy each of the following conditions:

1. $\mathbf{u}(\mathbf{x}_s - \mathbf{dP}_s)$ and $\mathbf{u}(\mathbf{x}_s + \mathbf{dP}_s)$ are antiparallel (i.e., parallel but with opposite directions)
2. $\mathbf{u}(\mathbf{x}_s - \mathbf{dP}_s)$ and $\mathbf{u}(\mathbf{x}_s + \mathbf{dP}_s)$ are directed toward $\mathbf{x}_s$
3. $\mathbf{x}_s$ is dependent on the spatial distribution of vorticity

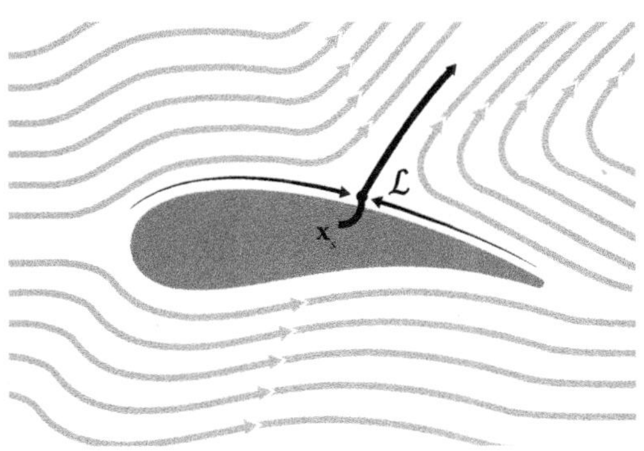

Figure 3.19. Schematic of flow separation from a solid body.

The first two criteria are illustrated in Figure 3.19; they require that the fluid particles located a small distance $\mathbf{dP}_s$ from the separation point are converging toward the separation point before detaching from the animal–fluid interface. The third criterion eliminates from our consideration the *kinematic separation* that occurs in inviscid flows to satisfy mass conservation.

Because the flows associated with aquatic locomotion are typically unsteady, it is often difficult to precisely identify the separation point $\mathbf{x}_s$ in practice. However, in many cases it is possible to deduce the occurrence of flow separation without pinpointing $\mathbf{x}_s$, by using only knowledge of the surface shape (via $\mathbf{P}_s$) or the local spatial distribution of vorticity.

3.10.1 Surface Shape and Motion Indicators of Flow Separation

The most reliable indicator of flow separation is the existence of a spatial discontinuity in the slope of the surface path $\mathbf{P}_s$. The sharp trailing edge of the appendage in Figure 3.20a is a clear example of a scenario in which the boundary layer will separate from the animal surface rather than make an abrupt change of direction around the trailing edge. The requirement that the velocity at the trailing edge remains finite is called the *Kutta condition*. An important subtlety of this indicator of flow separation is that the separation point $\mathbf{x}_s$ need not coincide with the location of the discontinuity in surface shape. For example, the low Reynolds number flow past a crescent-shaped object will separate along the concave surface downstream from the sharp leading edge (Figure 3.20b). Nonetheless, the flow will often separate in the vicinity of the discontinuity. In these cases, the local flow speed tangent to the surface near the discontinuity is useful for estimating the resulting wake circulation via Equation 3.111.

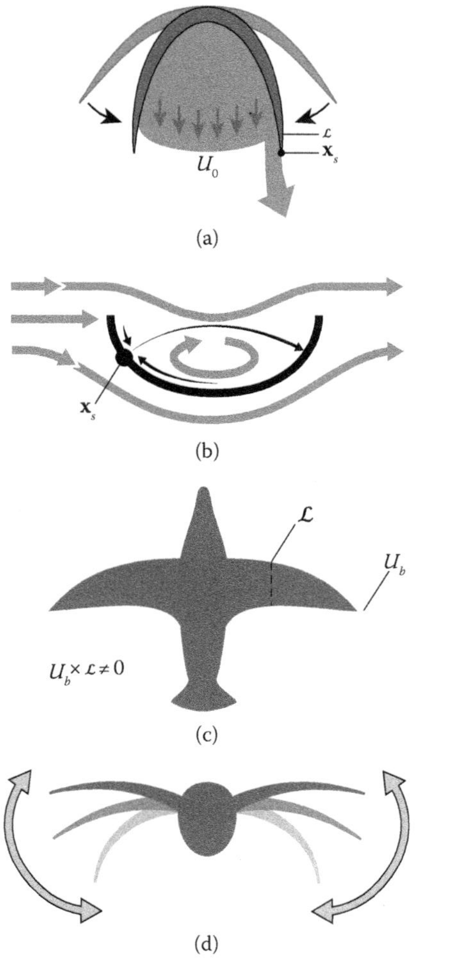

Figure 3.20. Examples of flow separation. (a) Flow separation from a sharp edge. (b) Flow separation at low Reynolds number. (c-d) Flow separation due to body oscillation.

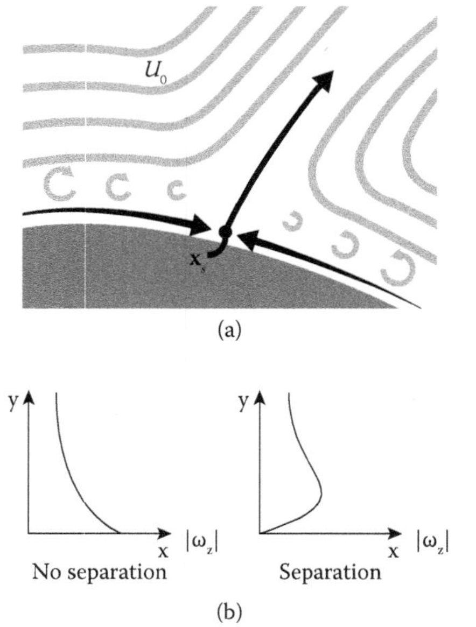

Figure 3.21. Schematic (a) and plot (b) of near-wall vorticity for flow separation.

A second indicator of flow separation is motion of the relevant appendage perpendicular to the surface path $\mathbf{P}_s$. This type of motion characterizes many flapping modes of aquatic locomotion (Figure 3.20c) and can be described quantitatively as $\mathbf{U}_b \times \mathbf{P}_s \neq 0$, where the magnitude of $\mathbf{U}_b$ is the speed of the flapping motion. For flow separation due to flapping, the flow velocity $-\mathbf{U}_b$ in the reference frame of the appendage is most appropriate for estimating the wake circulation in Equation 3.111.

3.10.2 Surface Vorticity Indicators of Flow Separation

Most aquatic animals exhibit one or both of the surface indicators described above. Deducing the characteristic velocity to be used in Equation 3.111 is relatively straightforward in these cases. In the absence of evidence of flow separation based on the appendage shape and motion, one can resort to an examination of the spatial distribution of boundary layer vorticity. This can be challenging in practice and often requires complementary empirical measurements or numerical simulations. Nonetheless, we can identify vorticity-based criteria for flow separation.

First, the vorticity at the separation point goes to zero from both $\mathbf{x}_s - \mathbf{dP}_s$ and $\mathbf{x}_s + \mathbf{dP}_s$. As shown in Figure 3.21a, the opposite directions of fluid particle rotation on either side of the separation point $\mathbf{x}_s$ are only compatible if the fluid particle rotation goes to zero at the separation point. More quantitatively, note that dimensional analysis suggests that the magnitude of boundary layer vorticity goes as $|\omega|:U_0/\delta$, where U_0 is the local flow velocity and is the boundary layer thickness (see, e.g., the model assumptions that led to Equation 3.116). Because the velocities $\mathbf{u}(\mathbf{x}_s - \mathbf{dP}_s)$ and $\mathbf{u}(\mathbf{x}_s + \mathbf{dP}_s)$ are antiparallel, U_0 must go to zero at $\mathbf{x}_s$; therefore, the vorticity also goes to zero.

A consequence of this vorticity criterion is that flow is spatially decelerating near the separation point. Indicators of flow deceleration, such as diverging streamlines (Figure 3.21a), can therefore be used as an additional indicator of flow separation.

The second indicator of flow separation is related to the first indicator: because the vorticity at the surface goes to zero at the separation point, the location of maximum vorticity is no longer at the animal–fluid interface where vorticity is generated. The presence of a peak in vorticity magnitude away from the surface of the appendage can indicate nearby flow separation (Figure 3.21b).

It is worth reiterating that evaluation of vorticity-based criteria for flow separation can require considerably more effort than examination of criteria based on the shape and kinematics of the swimming organism. Hence, our modeling efforts in this chapter rely primarily on the latter properties of aquatic animals.

If it can be confirmed that flow separation does not occur, much of the locomotion can be analyzed using the wealth of analytical tools from classical fluid mechanics. In the more likely event that flow separation does occur, we must consider the kinematics and dynamics of the resulting *wake vortices*.

3.11 KINEMATICS AND DYNAMICS OF AQUATIC ANIMAL VORTEX WAKES

What happens to the boundary layer vorticity once it is shed into the wake? Answering this question requires an understanding of the relationship between vorticity and velocity. Let us first consider an abstraction of the separated boundary layer, namely a single row of fluid particles leaving the surface of the animal appendage and entering the wake. For a boundary layer emanating from the upper surface of the appendage, as in Figure 3.22, each fluid particle that comprises the boundary layer rotates in the clockwise direction around its z-axis perpendicular to the plane of the page (see Section 3.8). In a reference frame that translates downstream with one of the fluid particles, say X_0, the local flow velocity on the upstream side of the fluid particle is directed upward, whereas the velocity on the downstream side of the fluid particle is directed downward. The adjacent fluid particles upstream (X_{-1}) and downstream (X_{+1}) of the fluid particle X_0 must therefore be moving upward and downward, respectively, relative to the position X_0. This motion of the adjacent fluid particles is a consequence of the fact that vorticity (i.e., each rotating fluid particle) is advected at the local fluid velocity. Furthermore, recall that the fluid viscosity resists large spatial gradients in flow velocity; hence, the velocity of adjacent fluid particles cannot be significantly different at their interface. In this sense, the presence of vorticity at X_0 *induces* a local flow velocity in its vicinity.

If the upstream fluid particle X_{-1} is still attached to the surface of the appendage, its motion is constrained by the no-slip condition: in the reference frame of the appendage, it cannot move. Therefore, the fluid particle X_0 immediately downstream must rotate downward relative to the appendage for the relative motion of X_{-1} and X_0 to be consistent with their clockwise rotation and induced velocity fields. The next downstream fluid particle X_{+1} must rotate downward further still relative to the appendage, so that the relative motion of X_0 and X_{+1} are consistent with their clockwise rotation and induced velocities; and so on. The net effect is that the boundary layer that has separated from the animal–fluid interface rolls up like a scroll. The direction of rolling up matches the rotation of the individual fluid particles (Figure 3.22).

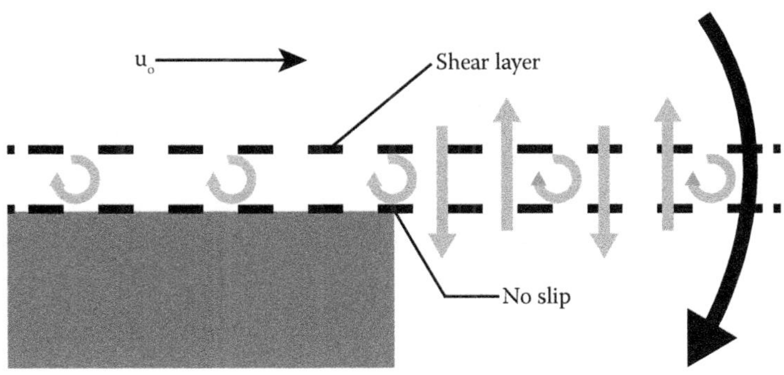

Figure 3.22. Kinematics boundary layer roll-up.

The relationship between vorticity and induced velocity can be generalized to an arbitrary spatial distribution of vorticity. First, recall that because the flow is incompressible, the velocity field $\mathbf{u}_v$ induced by vorticity is divergence free, that is, $\nabla \cdot \mathbf{u} = 0$. We can therefore define an arbitrary function ψ such that

$$\mathbf{u}_v = \nabla \times \psi \tag{3.112}$$

Because the divergence of the curl of a function is identically equal to zero, Equation 3.112 automatically satisfies the incompressibility condition. The vorticity can be recovered from the velocity field $\mathbf{u}_v$ by taking its curl:

$$\omega = \nabla \times (\nabla \times \psi) \tag{3.113}$$

This relationship can be simplified by using the vector identity (Equation 3.89) to replace the right-hand side of Equation 3.113, giving

$$\omega = \nabla(\nabla \cdot \psi) - \nabla^2 \psi \tag{3.114}$$

Because the function is arbitrary thus far, let us choose it such that its divergence is zero. Hence, the first term in Equation 3.114 vanishes, resulting in a *Poisson equation* for the function ψ:

$$\nabla^2 \psi = -\omega \tag{3.115}$$

The solution of Equation 3.115 has the form of a *Green's function* (the reader is encouraged to consult an advanced calculus textbook such as Hildebrand, 1976, for details):

$$\psi(\mathbf{x}) = \frac{1}{4\pi} \int_V \frac{\omega(\chi)}{r} dV \tag{3.116}$$

According to this solution, the value of the function at any point $\mathbf{x}$ in space depends on an integral of the vorticity distribution within the volume of interest V. The symbol χ is a surrogate or *dummy variable* for the positions in V that are being integrated. The parameter r is the distance between the location $\mathbf{x}$ at which the function is being evaluated and the location χ of the vorticity. The 1/r dependence indicates that the effect of vorticity on the function ψ—and hence on the induced velocity—decreases the farther the vorticity is located away from the location of interest. Note that the function has a *singularity* at r = 0, where the integrand becomes infinite; therefore, the velocity induced by the vorticity of a fluid particle at its own location is undefined.

Equation 3.112 allows us to recover the induced velocity field from Equation 3.116 as the *Biot–Savart equation*,

$$\mathbf{u}_v(x) = \nabla \times \psi(x) = \frac{1}{4\pi} \int_V \frac{\omega(\chi) \times (\mathbf{x} - \chi)}{|\mathbf{x} - \chi|^3} dV \tag{3.117}$$

where the distance r in Equation 3.116 has been written more explicitly as the vector magnitude $|\mathbf{x} - \chi|$. For two-dimensional flow in a plane, the integral over the volume in Equation 3.117 can be replaced by an integral over the area S of interest:

$$\mathbf{u}_v(x) = \frac{1}{2\pi} \int_S \frac{\omega(\chi) \times (\mathbf{x} - \chi)}{|\mathbf{x} - \chi|^2} dS \tag{3.118}$$

As an example of the application of Equation 3.118, let us calculate the velocity field induced by a single point of vorticity with circulation Γ_{pv} (see Figure 3.22 for example). The induced velocity is confined to the plane perpendicular to the axis of fluid particle rotation. The spatial distribution of vorticity of the *point vortex* $\boldsymbol{\omega}_{pv}(\chi)$ can therefore be expressed as

$$\omega_{pv}(\chi) = \Gamma_{pv} \delta(\chi - \chi_0) \tag{3.119}$$

where $\mathbf{d}\chi$ is directed along the (infinitesimal) length of the point vortex normal to the plane of fluid rotation, and Γ_{pv} is the circulation of the point vortex (see Equation 3.98). Substituting Equation 3.119 into Equation 3.118, the induced velocity is

$$\mathbf{u}_{pv}(\mathbf{x}) = \frac{1}{2\pi} \int_S \frac{\Gamma_{pv} \delta(\chi - \chi_0) \times (\mathbf{X} - \chi)}{|\mathbf{x} - \chi|^2} dS = \frac{\Gamma_{pv}}{2\pi |\mathbf{x} - \chi|} \tag{3.120}$$

If the vorticity in the fluid is approximated by a curve instead of a single point, the resulting *vortex filament* can be evaluated using an approach similar to that used to arrive at Equation 3.120, by integrating along $\mathbf{d}\chi$. More complex vorticity distributions often require numerical analysis. Hence, in much of our analyses of animal swimming in the following sections, we approximate the real vorticity distribution by simple geometric models. These primarily involve combinations of point vortices and vortex filaments.

3.11.1 Rules Governing Wake Vortex Motion

The mathematical models presented above lead to general principles regarding the behavior of vorticity in fluids. These rules constrain the possible configurations that can be achieved by wake vorticity; as a consequence, they provide a priori insight into the construction of effective models of aquatic locomotion.

First, as we saw in Section 3.7, in the absence of viscous diffusion the circulation of the wake vortices is conserved:

$$\left.\frac{d\Gamma}{dt}\right|_{wake} = 0 \tag{3.121}$$

This constraint provides some flexibility during measurements of aquatic locomotion, because the wake can be observed downstream from its point of creation without losing dynamical information that is implicit in the wake structure.

To appreciate the second rule, we need to first define the concept of a vortex tube. In a three-dimensional flow, curves that are tangent to the local vorticity vectors are called *vortex lines*. In close analogy to the concepts of streamlines and streamtubes, groups of parallel vortex lines form *vortex tubes* in the flow (Figure 3.23). Recall that the divergence of the curl of a vector is identically zero (see Equation 3.122). Within the volume of a segment of the vortex tube, we may therefore write

$$\int_{vt} \nabla\cdot(\nabla\times\mathbf{u})dV = \int_{vt} \nabla\cdot\boldsymbol{\omega}dV = 0 \tag{3.122}$$

The divergence of vorticity in the vortex tube volume can be related to integrals on the surfaces of the vortex tube by using *Gauss' law*:

$$\int_{vt} \nabla\cdot\boldsymbol{\omega}dV = \int_{A_c} \boldsymbol{\omega}\cdot\mathbf{n}_c dS + \int_{A_1} \boldsymbol{\omega}\cdot\mathbf{n}_1 dS + \int_{A_2} \boldsymbol{\omega}\cdot\mathbf{n}_2 dS = 0 \tag{3.123}$$

Everywhere on the curved surfaces A_c of the vortex tube section,

$$\boldsymbol{\omega}\cdot\mathbf{n}_c = 0 \tag{3.124}$$

where $\mathbf{n}_c$ is a unit vector perpendicular to the local curved surface. Equation (3.124) holds by definition, because the vortex lines that comprise the vortex tube are parallel to the local vorticity vectors. Combining Equations 3.123 and 3.124, leaves

$$\int_{A_1} \boldsymbol{\omega}\cdot\mathbf{n}_1 dS + \int_{A_2} \boldsymbol{\omega}\cdot\mathbf{n}_2 dS = 0 \tag{3.125}$$

or, equivalently

$$\Gamma_1 = \Gamma_2 \tag{3.126}$$

In other words, the strength of the vortex tube does not vary along its length. This property has major consequences for the possible configurations of wake vorticity, because the vortex lines cannot terminate in the fluid. They can only terminate at the animal–fluid interface (during vortex formation) or else form closed loops in the wake. It is for this reason that ring and chain arrangements of vorticity are consistently observed in aquatic animal wakes (Figure 3.23).

As an aside, it is worth noting that the preceding developments caution against strictly two-dimensional models of aquatic locomotion. The wake dynamics demand a three-dimensional wake configuration, which cannot be properly accounted using only two-dimensional constructs such as point vortices.

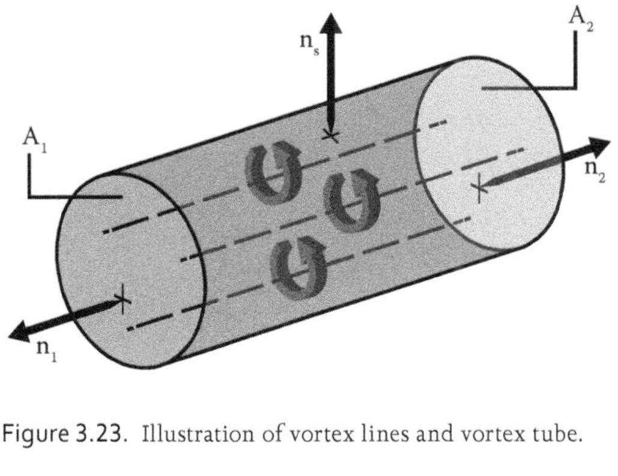

Figure 3.23. Illustration of vortex lines and vortex tube.

3.12 ESTIMATING SWIMMING FORCES BASED ON THE ANIMAL WAKE

Now that we have developed a physical picture of how a swimming animal creates motion in the surrounding fluid, we are prepared to examine how that fluid motion leads to locomotion. Rather than beginning with the most general case including unsteady wake vortex formation, let us consider the case of steady flow first.

3.12.1 Steady Flow

Figure 3.24 illustrates two possibilities that exist for the steady (or time-averaged) velocity profile across the wake of body in the flow. In the first case, the flow speed decreases immediately behind the body. This *velocity defect* in the wake profile is an indication that the body experiences a net drag that would tend to move the body downstream. In contrast, a *velocity surplus* in the downstream vicinity of the body is indicative of net thrust production. Both of these scenarios can be quantified by analyzing a control volume surrounding the body.

Within the control volume *V*, conservation of fluid mass can be expressed as

$$\int_S \rho(\mathbf{u}\cdot\mathbf{n})dS = 0 \qquad (3.127)$$

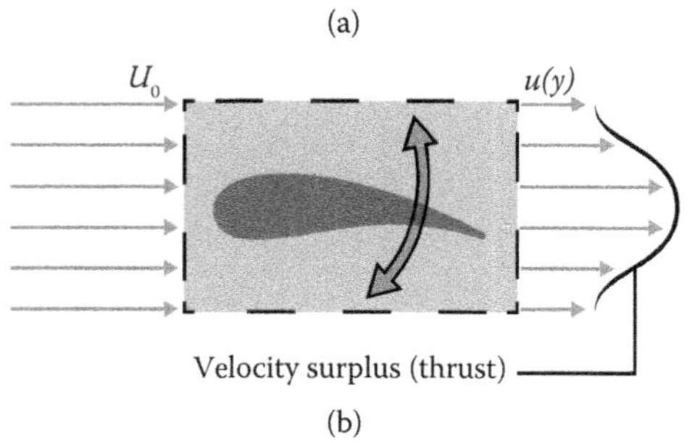

Figure 3.24. Schematic of wake velocity deficit (a) and surplus (b).

In words, Equation 3.127 states that an input flux of fluid mass through one surface of the control volume must be compensated by a flow of fluid mass out of the control volume elsewhere, so that the total mass of fluid in the control volume remains constant.

Momentum is also conserved in the control volume, as described by the steady form of the Navier–Stokes equation:

$$\mathbf{f}_l = \int_S \rho\mathbf{u}(\mathbf{u}\cdot\mathbf{n})dS - \int_S p\mathbf{n}dS \qquad (3.128)$$

As before, $\mathbf{f}_l$ is the force exerted on the fluid by the animal.

Equations 3.127 and 3.128 are sufficient in principle to determine the locomotive forces. However, their practical implementation typically requires additional assumptions such as two-dimensional flow and negligible spatial gradients of pressure on the surface of the control volume. The former assumption of two-dimensional flow is implicit in most force estimation models used to study aquatic locomotion. These include blade-element airfoil theories for flapping locomotion, slender-body and asymptotic (i.e., small-amplitude) analyses of undulatory locomotion (see Section 3.15), and linear stability methods. Nevertheless, as we saw in the previous section, the animal–fluid interactions underlying aquatic locomotion are fully three-dimensional.

The latter assumption regarding the spatial distribution of fluid pressure is a consequence of the difficulty in determining fluid pressure in the bulk flow. In cases of quasi-steady flow (see Section 3.3), the time-averaged pressure field can often be approximated in this way; the instantaneous pressure gradients tend to vanish when averaged over a cycle of appendage oscillation.

However, the time-averaging used to create a quasi-steady flow can result in the loss of key information regarding locomotion. Consider, for example, the net vertical force created by a negatively buoyant swimming organism. The time-averaged vertical force is defined as

$$\overline{F}_V = \frac{1}{T}\int_0^T F_V(t)dt \qquad (3.129)$$

It is apparent from Equation 3.129 that the time-averaged force is not unique; an infinite set of temporal force profiles F_V (t) can possess the same time-averaged force. Figure 3.25 shows three profiles, each with a time-averaged force equal to zero. Yet, their common time-averaged force hides significant differences in the resultant vertical trajectory of the organism: two of the trajectories move in opposite directions; the third does not change vertical position at all! Time-averaged analyses should therefore be used with caution, especially when the trajectory of the swimming animal is of interest.

3.12.2 Unsteady Flow in an Infinite Volume

Strictly speaking, the equations of fluid motion 3.127 and 3.128 can be converted to unsteady equations by simply adding time derivatives of the control volume mass and momentum, respectively. However, the issues of two-dimensionality and the pressure field persist. A solution is to express the conservation of momentum in terms of an infinitely large control volume that contains all of the vorticity created at the animal–fluid interface:

$$\mathbf{f}_l = \frac{d}{dt}\left(\frac{\rho}{2}\int_{V\to\infty} \mathbf{x}\times\omega dV\right) \tag{3.130}$$

The integral in Equation 3.130 is the *vortex impulse*, effectively the momentum of the vortex wake. Its derivation is presented in the next section. For now, let it suffice to note that the primary benefit of this approach is that it eliminates explicit dependence on the pressure field. To be sure, the new formulation typically requires assumptions regarding the shape of the wake vortices. In addition, it assumes that the integration volume is infinitely large, and that the fluid velocity is zero far from the animal. These requirements may not be satisfied in practice, for example, when studying aquatic organisms in a finite-size water tank or in a reference frame in which the free-stream fluid velocity far from the animal is nonzero. With some additional analysis, the following section describes an approach that avoids these limitations.

3.12.3 Unsteady Flow in a Finite Volume

The starting point for this approach to wake-based force estimation is the concept of vortex impulse introduced above:

$$\mathbf{I}_v = \frac{\rho}{2}\int_{V_v} \mathbf{x}\times\omega dV \tag{3.131}$$

where the integral is evaluated over the finite volume V_v of the wake vortex of interest (Figure 3.26). The locomotive forces ultimately relate to the time derivative of the vortex impulse.

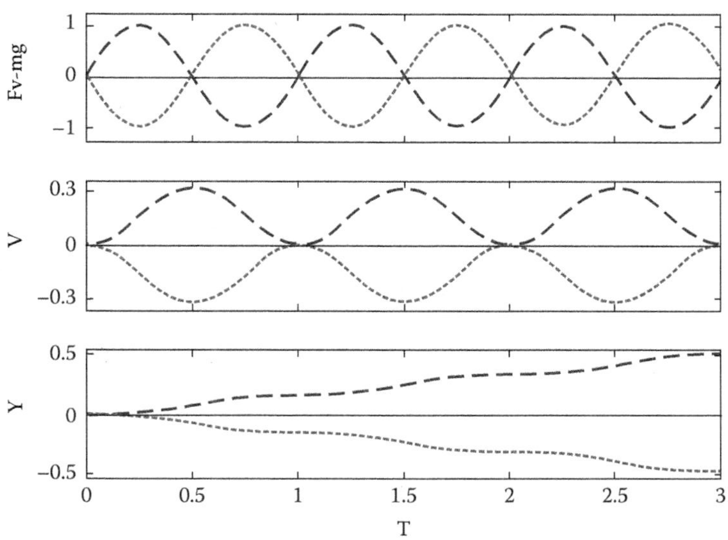

Figure 3.25. Differences between time-averaged and instantaneous quantities. Top panel, three instantaneous net force profiles with equivalent time averages (equal to zero). Corresponding vertical velocity (middle panel) and position (bottom panel). (Based on Dabiri, J.O., *J. Exp. Biol.*, 208, 3519–3532, 2005.)

Hence, taking the time derivative of Equation 3.131 yields

$$\frac{d\mathbf{I}_v}{dt} = \frac{\rho}{2}\int_{V_v} \mathbf{x} \times \frac{D\omega}{Dt} dV + \frac{\rho}{2}\int_{V_v} \mathbf{u} \times \omega dV \quad (3.132)$$

Now, let us replace the total derivative of vorticity $\mathbf{D\omega}/\mathrm{Dt}$ by using the vorticity Equation 3.95. In the present analysis, we neglect the action of viscosity. Hence,

$$\frac{\partial \omega}{\partial t} - \nabla \times (\mathbf{u} \times \omega) = \nabla \times \mathbf{f}_l \quad (3.133)$$

The vector products in the second term on the left-hand side of Equation 3.133 can be replaced with scalar products by using the following vector identity:

$$\nabla \times (\mathbf{u} \times \omega) = \mathbf{u}(\nabla \cdot \omega) - \omega(\nabla \cdot \mathbf{u}) - (\mathbf{u} \cdot \nabla)\omega + (\omega \cdot \nabla)\mathbf{u} \quad (3.134)$$

As described previously, the velocity and vorticity fields are divergence free; therefore, the first two terms on the right-hand side of Equation 3.134

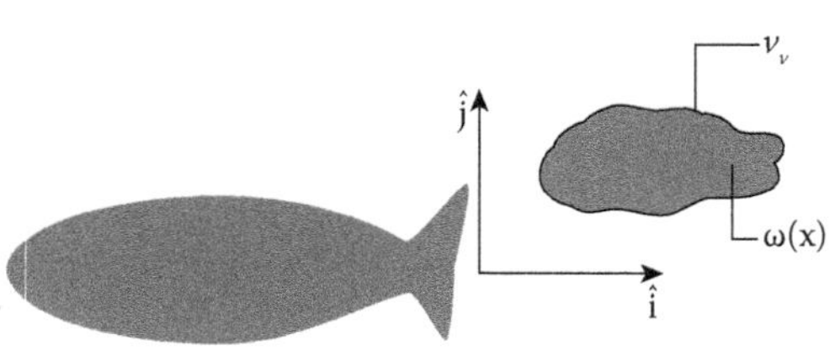

Figure 3.26. Schematic of vortex patch shed behind a swimming fish.

are equal to zero. Substituting the remaining terms into Equation 3.133 gives

$$\frac{\partial \omega}{\partial t} + (\mathbf{u} \cdot \nabla)\omega = (\omega \cdot \nabla)\mathbf{u} + \nabla \times \mathbf{f}_l \quad (3.135)$$

or, equivalently

$$\frac{D\omega}{Dt} = (\omega \cdot \nabla)\mathbf{u} + \nabla \times \mathbf{f}_l \quad (3.136)$$

Equation 3.136 is the *Helmholtz equation*. It states that the vorticity of a fluid particle can change either due to the forces applied by the animal on the fluid (via $\nabla \times \mathbf{f}_l$) or due to velocity gradients aligned with the local vorticity vector [giving a nonzero scalar product in $(\omega \cdot \nabla)\mathbf{u}$]. This latter effect is known as *vortex stretching*. Conceptually, the velocity field stretches the local vortex tubes; being divergence free, the vortex tubes narrow and elongate in response (in close analogy to the behavior of an incompressible mass of fluid that is stretched). To conserve angular momentum, the fluid particles in the narrower, stretched vortex tube must rotate faster, leading to stronger vorticity in the tube (Figure 3.27).

The expression for $D\omega/Dt$ in Helmholtz's equation can be used in Equation 3.132, leading to

$$\frac{d\mathbf{I}_v}{dt} = \frac{\rho}{2}\int_{V_v} \left[\mathbf{x} \times (\omega \cdot \nabla)\mathbf{u} + \mathbf{x} \times (\nabla \times \mathbf{f}_l) + \mathbf{u} \times \omega\right] dV \quad (3.137)$$

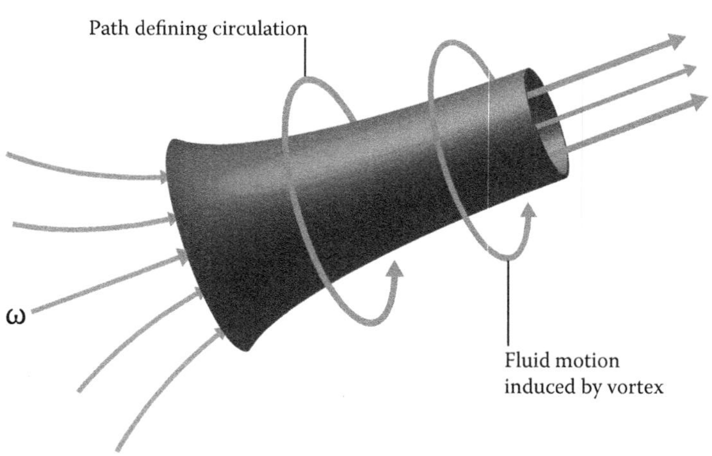

Figure 3.27. Illustration of relationship among velocity, vorticity, and circulation.

The first two terms on the right-hand side of Equation 3.137 can be simplified by using two additional vector identities. First,

$$\int_{V_V} \mathbf{x}\times(\boldsymbol{\omega}\cdot\nabla)\mathbf{u}dV = \int_{V_V} \mathbf{u}\times\boldsymbol{\omega}dV - \int_{S_V} (\mathbf{u}\times\mathbf{x})\boldsymbol{\omega}\cdot\mathbf{n}dS \tag{3.138}$$

where the second integral is evaluated on the surface of the wake vortex. This surface integral is identically zero in our case, because $\boldsymbol{\omega}\cdot\mathbf{n} = 0$ on the surface of the vortex (see Equation 3.123 for example).

The second vector identity extracts the locomotive force:

$$\int_{V_V} \mathbf{x}\times(\nabla\times\mathbf{f}_l)dV = 2\int_{V_V} \mathbf{f}_l dV + \int_{S_V} \mathbf{x}\times(\mathbf{n}\times\mathbf{f}_l)dS \tag{3.139}$$

The surface integral in Equation 3.139 also vanishes on the surface of the wake vortex. This occurs by our construction; we model the locomotive force as being applied normal to the vortex volume, so that $\mathbf{n}\times\mathbf{f}_l = 0$. In reality, of course, the locomotive forces occur at the animal–fluid interface.

Nonetheless, the results from Equations 3.138 and 3.139 can be substituted into Equation 3.137 and solved for the total locomotive force $\mathbf{F}_l = \rho\int_{V_V}\mathbf{f}_l dV$. The result is

$$\mathbf{F}_l = \frac{d\mathbf{I}_v}{dt} - \rho\int_{V_V} \mathbf{u}\times\boldsymbol{\omega}dV \tag{3.140}$$

The last term in Equation 3.140 is called the *vortex force*. The associated velocity field $\mathbf{u}$ includes contributions from external factors such as wake interactions with the animal appendages after vortex formation (i.e., *wake capture*), interactions with other animals or objects in the vicinity, and ambient flow currents. If and only if these phenomena are absent and thus the velocity field in the wake is wholly determined by the vorticity distribution, then the last term in Equation 3.140 becomes

$$\rho\int_{V_V} \mathbf{u}_v\times\boldsymbol{\omega}dV = 0 \tag{3.141}$$

where $\mathbf{u}_v$ is the vorticity-induced velocity field (see Equation 3.117). In this case, the locomotive force is given rather simply as

$$\mathbf{F} = \frac{d\mathbf{I}_v}{dt} \tag{3.142}$$

which is consistent with Equation 3.130.

In reality, neglecting the second term in Equation 3.140 can cause us to lose much of the interesting swimming dynamics, just as taking the time-average did in at the beginning of this section. More importantly, within the vortex force lies many of the design strategies used by biological systems to adapt their locomotion to the local fluid environment. Hence, in the next section we examine that contribution more carefully and achieve a straightforward model to better illuminate its significance.

3.13 THE VORTEX FORCE AND WAKE VORTEX ADDED-MASS

The vortex force introduced in the previous section arises when swimming animals interact with vortices in the flow. The vortices may be created by the animal itself or by neighboring animals, or even by stationary objects (e.g., a logjam in stream). To model the vortex force, we can interpret it in terms of the shape and motion of the vortices. First, let us use the vector identity (3.37):

$$\mathbf{u}_e\times\boldsymbol{\omega} = \nabla\left(\frac{\mathbf{u}_e\cdot\mathbf{u}_e}{2}\right) - (\mathbf{u}_e\cdot\nabla)\mathbf{u}_e \tag{3.143}$$

where in Equation 3.143 and what follows, we use $\mathbf{u}_e$ to explicitly denote that the flow velocity of interest is that which is external and in addition to the vorticity-induced velocity field. Integrating this equation over the vortex volume gives

$$\int_{V_V} \mathbf{u}_e\times\boldsymbol{\omega}dV = \int_{V_V} \nabla\left(\frac{\mathbf{u}_e\cdot\mathbf{u}_e}{2}\right) - (\mathbf{u}_e\cdot\nabla)\mathbf{u}_e dV \tag{3.144}$$

Gauss' law (3.123) allows us to evaluate the integrals in Equation 3.144 only on the surface of the vortex instead of through its entire volume. Hence,

$$\int_{V_V} \mathbf{u}_e\times\boldsymbol{\omega}dV = \int_{S_V} \frac{1}{2}u_e^2\mathbf{n} - (\mathbf{u}_e\cdot\mathbf{n})\mathbf{u}_e dS \tag{3.145}$$

Recall that, for our conceptual model of the vortex wake, the vorticity is confined to the finite volume V_v and the locomotive force is applied within that volume. Therefore, we can approximate the flow outside of the wake vortex as being *irrotational*, that is, having no vorticity. With these assumptions, we can use Bernoulli's equation (see Section 3.2) to relate the external flow speed u_e on the surface of the vortex to the local pressure p and external flow velocity potential ϕ_e:

$$\frac{u^2}{2} = C - \frac{\partial \phi_e}{\partial t} - \frac{p}{\rho} \tag{3.146}$$

Conceptually, the velocity potential is the impulsive pressure p_i applied for an infinitesimal duration dt that would be required to achieve the observed flow starting from fluid at rest. In mathematical terms,

$$\phi_e = \frac{p_i}{\rho} dt \tag{3.147}$$

and

$$\nabla \phi_e = \mathbf{u}_e \tag{3.148}$$

The physical significance of the velocity potential as an analog to pressure becomes clear later in this section.

Replacing $\mathbf{u}_e$ in Equation 3.145 with the Bernoulli relationship leads to

$$\int_{V_V} \mathbf{u}_e \times \boldsymbol{\omega} dV = \int_{S_V} \left[\left(C - \frac{\partial \phi_e}{\partial t} - p \right) \mathbf{n} - (\mathbf{u}_e \cdot \mathbf{n}) \nabla \phi_e \right] dS \tag{3.149}$$

The first term in the surface integral reduces to $C\int_{S_V} \mathbf{n} dS$, which is identically equal to zero because the normal vectors $\mathbf{n}$ cancel each other on any closed surface. The pressure term in the surface integral is also equal to zero, because the pressure is balanced over any closed surface in inviscid flow. This result leads to *D'Alembert's paradox*, which is that a body in inviscid flow experiences no drag.

With these simplifications, we are left with terms that depend on the velocity potential:

$$\int_{V_V} \mathbf{u}_e \times \boldsymbol{\omega} dV = -\int_{S_V} \left[\frac{\partial \phi_e}{\partial t} \mathbf{n} + (\mathbf{u}_e \cdot \mathbf{n}) \nabla \phi_e \right] dS \tag{3.150}$$

or, equivalently

$$\int_{V_V} \mathbf{u}_e \times \boldsymbol{\omega} dV = -\frac{d}{dt} \int_{S_V} \phi_e \mathbf{n} dS \tag{3.151}$$

The simplification in Equation 3.151 is achieved noting that the two terms in Equation 3.150 are related by the chain rule for differentiation (see Equation 3.100 for a similar example).

The vortex force term in Equation (3.140) can therefore be replaced as

$$\mathbf{F} = \frac{d\mathbf{I}_v}{dt} + \frac{d}{dt} \int_{S_V} \phi_e \mathbf{n} dS \tag{3.152}$$

Comparison with the Navier–Stokes formulation in Equation 3.128 shows that force estimation based on the vortex wake replaces the reliance on velocity and pressure with a dependence on vorticity (via the vortex impulse) and a pressure-like velocity potential term. The advantage is that velocity potential becomes evident as we introduce the concept of *added-mass*.

3.13.1 *Added-Mass*

Physically, the added-mass of a body is the mass of fluid surrounding the body that, due to the pressure field on the surface of the body, is set into motion along with the body. The concept is subtle in that whereas the added-mass is a constant fraction of the fluid mass displaced by the body, the added-mass is not comprised of the same set of fluid particles at all times. This is illustrated in Figure 3.28, which shows the trajectories of fluid particles surrounding a sphere in inviscid fluid (see also Section 3.3). Individual fluid particles

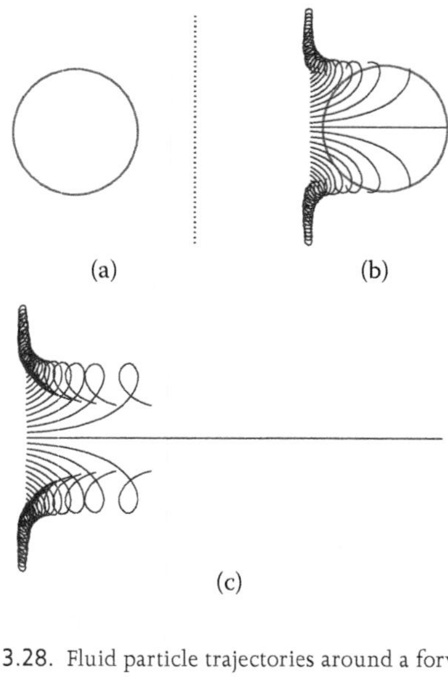

Figure 3.28. Fluid particle trajectories around a forward translating sphere. (Based on Dabiri, J.O., J. Exp. Biol., 208, 3519–3532, 2005.)

are constantly entrained by the sphere and subsequently released. However, the total mass of fluid in motion at any time remains constant.

For unidirectional motion in the x-direction, the added-mass is given by

$$M_{Ax} = \frac{\rho}{U_0}\int_{S_b} \phi \mathbf{n} \cdot \mathbf{i} dS \tag{3.153}$$

where U_0 is the speed of the body and the integral is evaluated over the surface S_b of the body. The force required to accelerate the body must overcome the inertia of both the body M_b and its added-mass:

$$F_x = \frac{dU_0}{dt}(M_b + M_{Ax}) \tag{3.154}$$

In general, the magnitude of the added-mass depends on both the direction of the acceleration relative to the body axes and on whether the acceleration is linear or angular. Each of the possible combinations of body motion and resultant body force can be assembled into a 6 by 6 matrix called the *added-mass tensor*. It is often normalized by mass of fluid displaced by the body, that is,

$$\alpha_b = \frac{\mathbf{M}_A}{\rho V_b} \tag{3.155}$$

where $\boldsymbol{\alpha}_b$ is a 6 by 6 matrix of dimensionless numerical coefficients. For bodies with geometric symmetry, many of the added-mass coefficients are zero. We focus on such cases, wherein the remaining coefficients relate the linear body motion in a given direction to the forces in the same direction.

For example, let us consider the added-mass of the sphere illustrated in Figure 3.28. Its velocity potential can be determined by solving Equation 3.34 by using knowledge of the flow velocity at the sphere boundary and far from the sphere. The result is

$$\phi_e(r,\theta) = \frac{U_0 a^3 \cos\theta}{2r^2} \tag{3.156}$$

where U_0 is the speed of the sphere, a is its radius, and (r, θ) is the coordinate system in Figure 3.28. On the surface of the sphere, the radius r is equal to a. Hence,

$$\phi_e(a,\theta) = \frac{U_0 a \cos\theta}{2} \tag{3.157}$$

Substituting into the added-mass definition (Equation 3.153) gives

$$M_{Ax} = \frac{\rho}{U_0}\int_{\beta=0}^{2\pi}\int_{\theta=0}^{\pi} \frac{U_0 a \cos\theta}{2} \cos\theta a^2 \sin\theta d\theta d\beta \tag{3.158}$$

or, simplifying,

$$M_{Ax} = \frac{2}{3}\rho\pi a^3 = \frac{1}{2}\rho V_b \tag{3.159}$$

In other words, the added-mass of the sphere is equal to one-half of the mass of fluid displaced by the fluid. Per the normalization in Equation 3.155, the added-mass coefficient is 1/2.

As a second example, let us consider the flow past a spherical vortex ring moving at speed U_0 (see Figure 3.3a for example). Incidentally, the velocity potential of the flow external to the spherical vortex is the same as that given in Equation 3.156. Therefore, the vortex ring also has an added-mass that can be computed in the same manner as the solid body, and the added-mass coefficient is identical!

An important caveat in applying the vortex added-mass concept to evaluate the force equation (Equation 3.152) is that only the flow not induced by the vortex itself should be included in the calculation. The vortex impulse term can be decomposed using *Burgatti's theorem* as

$$\int_{V_v} \mathbf{x} \times \boldsymbol{\omega} dV = \int_{V_v} \mathbf{u}_v dV + \int_{S_v} \phi_v \mathbf{n} dS \tag{3.160}$$

where the subscript v denotes properties of the vortex. Thus, the vortex impulse term implicitly accounts for the self-induced motion of the vortex and its added-mass. This self-induced motion of the vortex should therefore be subtracted from U_0 before the added-mass analysis.

3.14 A MODEL FOR LOCOMOTIVE FORCES BASED ON WAKE VORTEX IMPULSE AND ADDED-MASS

Evaluating the force equation (Equation 3.152) requires a kinematic model for the structure of the wake vortices. As discussed in Section 3.10, the physical constraints on vorticity lead to the

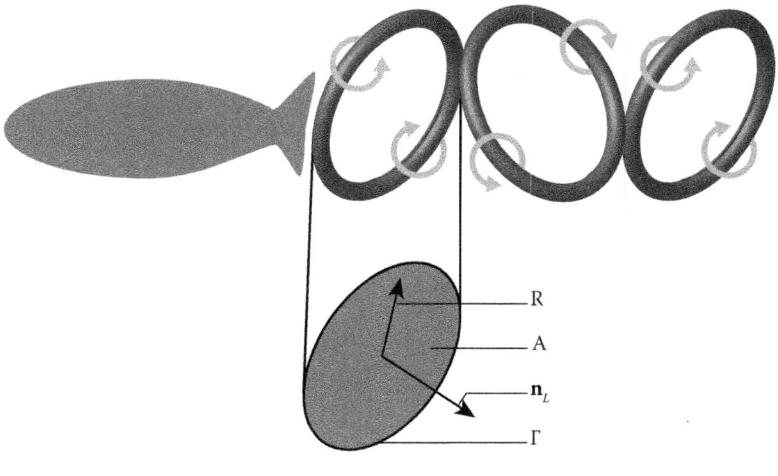

Figure 3.29. Schematic of vortex ring wake and corresponding geometric model.

formation of vortex loops in the wake, as illustrated in Figure 3.29. Let us approximate these loops as being infinitesimally thin, such that the diameter of the vortex tubes is negligible. The geometry of each vortex loop can then be fully described by its circulation Γ, the area A enclosed by the vortex loop, and the vector $\mathbf{n}_L$ oriented normal to the plane containing the vortex loop. Using a vortex filament model for the vorticity as in Equation 3.119, the vortex impulse can be calculated as

$$\frac{\rho}{2}\int_{\text{loop}} \mathbf{x}\times\omega dV = \rho\Gamma A\mathbf{n}_L \qquad (3.161)$$

The result in Equation 3.161 immediately leads to well-known expression for the steady hydrodynamic lift created by a steadily translating fin. If we approximate the fin as having finite span W, the generated vortex loop area increases over time as $dA/dt = WU_0$ (see Figure 3.29). Hence, the lift is

$$\mathbf{L} = \frac{d\mathbf{I}}{dt} = \rho\Gamma W U_0 \mathbf{n}_L \qquad (3.162)$$

This force is applied to the fluid and is directed downward (the reader is encouraged to apply the *right-hand rule* for vector orientation to Equation 3.161, where $\mathbf{x}$ is most conveniently evaluated relative to an origin inside the vortex loop); therefore, the animal experiences an upward force.

From the discussion in the preceding section, the force due to vortex added-mass is given by

$$\mathbf{F}_A = \rho\frac{d}{dt}\left[V_v(1+\alpha_v)\mathbf{u}_e\right] \qquad (3.163)$$

where V_v is the volume of the vortex loop, $\boldsymbol{\alpha}_v$ is the added-mass tensor of the vortex loop, and $\mathbf{u}_e$ is the component of vortex loop motion that is not self-induced. This last parameter can be evaluated as the difference between the observed vortex loop motion and the vortex velocity predicted using the Biot–Savart Equation 3.117.

Using Equations 3.161 and 3.163 in the force equation (Equation 3.152) provides a comprehensive model for locomotive forces during animal swimming:

$$\mathbf{F}_l = -\rho\frac{d}{dt}\left[\Gamma A\mathbf{n}_L + V_v(1+\boldsymbol{\alpha}_v)\mathbf{u}_e\right] \qquad (3.164)$$

Our challenge in studying the dynamics of a swimming organism is reduced to the task of creating a model for each of the parameters in Equation 3.164. The vortex circulation can be approximated by using the slug model, Equation 3.111, and assumptions on the kinematics of the swimming appendages. The area A of the vortex loop can also be modeled based on the shape and kinematics of the appendages, as was demonstrated in Figure 3.29. The stroke plane of the appendages suggests the orientation $\mathbf{n}_L$ of the vortex loops.

Estimation of the vortex volume and the vortex added-mass coefficient can be more difficult and may require some qualitative knowledge of the wake flow structure. Similarly, estimates of $\mathbf{u}_e$ are most effective if they are based on a priori information regarding the external flow environment. It is these aspects that add a subjective, even artistic, component to the modeling of aquatic animal locomotion. Experience and validation by empirical observations are the best arbiters of the quality of a model of aquatic locomotion. A good model can be a powerful aid in understanding how animals can adapt their animal–fluid interactions to achieve desired functions (see next section).

3.15 JELLYFISH PHYLOGENETIC SCALING REVISITED

Now that we have developed a more sophisticated understanding of the animal–fluid interactions underlying locomotion, let us reconsider the phylogenetic scaling illustrated in Figure 3.14. Because we are concerned with the scaling of the force required for locomotion and not specific details of the animal trajectory, we can make use of the time-averaged force over the swimming cycle duration T (see Equation 3.129):

$$\overline{F_l} = \frac{T_J}{T} F_J - \frac{T_R}{T} F_R \qquad (3.165)$$

where T_J and T_R are the durations of the locomotive force during the jetting phase of the swimming cycle (F_J) and during the relaxation or recovery phase (F_R), respectively. The recovery phase forces are oriented opposite to those during the jetting phase and lead to the observed deceleration of the animals at the end of each swimming cycle. Their effect is to reduce the average force required for locomotion.

The jetting phase force can be approximated by the flux of momentum exiting the control volume surrounding the animal (Figure 3.30). Hence,

$$F_J \approx \rho A_J U_J^2 \qquad (3.166)$$

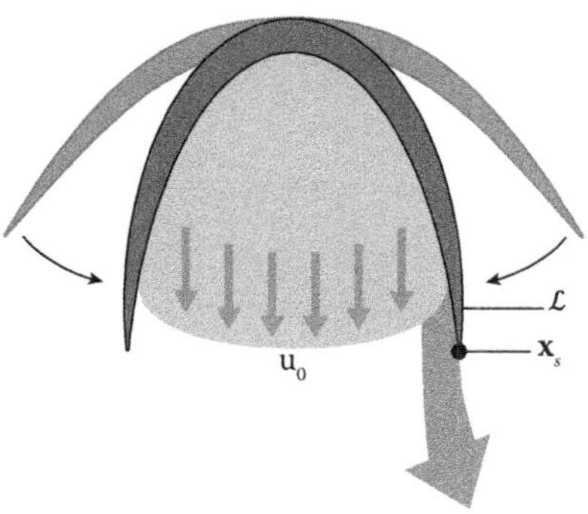

Figure 3.30. Illustration of jellyfish during bell contraction.

where the jet speed U_J is related to the jellyfish bell volume V and oral aperture area A_J as

$$U_J \approx \frac{1}{A_J}\frac{dV}{dt} \qquad (3.167)$$

by the requirement that mass is conserved in the control volume (Equation 3.127).

Using Equation 3.167 in 3.166 provides an estimate of the jetting force in terms of only the animal shape and kinematics:

$$F_J \approx \frac{\rho}{A_J}\left(\frac{dV}{dt}\right)^2 \qquad (3.168)$$

where the oral aperture area is related to the aperture diameter as $A_J = \pi D^2/4$.

If we approximate the animal bell as a hemiellipsoid, its volume is given by

$$V = \frac{\pi}{6} HD^2 \qquad (3.169)$$

where H is the height of the bell along the axis of symmetry. Taking the time derivative leads to

$$\frac{dV}{dt} = \frac{\pi}{6}\left(2HD\frac{dD}{dt} + \frac{dH}{dt}D^2\right) \qquad (3.170)$$

It is empirically observed that the height changes to a lesser degree than the bell diameter over the course of a swimming cycle. Therefore, if $dH/dt \ll dD/dt$

$$\frac{dV}{dt} \approx \frac{\pi}{3} HD\frac{dD}{dt} \qquad (3.171)$$

During bell relaxation, a prominent *stopping vortex ring* is formed by the animal. Here, we can make use of the modeling tools developed in the preceding sections. Figure 3.30 illustrates our model of the stopping vortex ring. We assume that the vortex tube that comprises the ring is infinitesimally thin (see Figure 3.29 for comparison) and is oriented with its normal vector $\mathbf{n}_L$ pointed in the direction of forward swimming. This vortex orientation is consistent with the force of the fluid on the animal in the backward direction, as in Equation 3.165.

Following Equation 3.161, the locomotive force during the recovery phase due to stopping vortex formation can be estimated as

$$F_R \approx \rho\left(\frac{\pi D^2}{4}\right)\frac{\Gamma}{T_R} \qquad (3.172)$$

The slug model (Equation 3.111) can be used to estimate the circulation of the stopping vortex ring:

$$\Gamma \approx \frac{1}{2} U_c^2 T_R \qquad (3.173)$$

where U_c is the characteristic velocity at the edges of the bell where the vortex ring is formed. By modeling the bell edge as moving radially with speed $U_c = (1/2)dD/dt$, the circulation in Equation 3.172 is

$$\Gamma \approx \frac{1}{8}\left(\frac{dD}{dt}\right)^2 T_R \qquad (3.174)$$

The model estimates in Equations 3.168, 3.171, 3.172, and 3.174 can be substituted into the force equation (Equation 3.165), giving

$$\overline{F_l} \approx \rho\left(\frac{4\pi}{9} f^2 - \frac{\pi}{32}\right)\left(D\frac{dD}{dt}\right)^2 \qquad (3.175)$$

where f is the fineness ratio H/D.

The time dependence of the bell diameter D(t) can be modeled as a simple sinusoid with maximum and minimum values given by the nominal bell diameter D and D/2, respectively:

$$D(t) = \frac{D}{4}(3 + \cos\omega t) \qquad (3.176)$$

where ω is the swimming frequency in radians per second.

Returning to the requirement that the locomotive force $\overline{F_l}$ must not exceed the available muscle force F_m given in Equation 3.88, we arrive at a prediction for the maximum allowable fineness ratio of a jellyfish for a given body diameter D:

$$f_{max} \approx \sqrt{\frac{9}{16\pi\rho}\left(\frac{\pi\rho}{32} + \frac{\sigma\tau}{g(\omega)D^3}\right)} \qquad (3.177)$$

The frequency-dependent function $g(\Omega)$ is given by the time average of $\left(D\frac{dD}{dt}\right)^2$ over the duration of one swimming cycle. For the function used presently (Equation 3.185), this parameter is given by

$$g(\omega) = \frac{\omega}{2\pi}\int_0^{s\pi/\omega} \frac{1}{256}\omega^2 \sin^2\omega t(3 + \cos\omega t)^2\, dt \qquad (3.178)$$

Figure 3.31 compares the model prediction with the morphospace of jellyfish measurements. The agreement is good and, unlike our first scaling attempt in Section 3.6, the new model correctly anticipates the existence of low-fineness ratio body shapes at large body diameters. In fact, the allowable morphospace given by Equation 3.177 predicts that arbitrarily large jellyfish can exist as long as the fineness ratio is no greater than $\sqrt{9/128} \approx 0.265$. This particular value of limiting fineness ratio is the result of the various geometric and kinematic assumptions in the model.

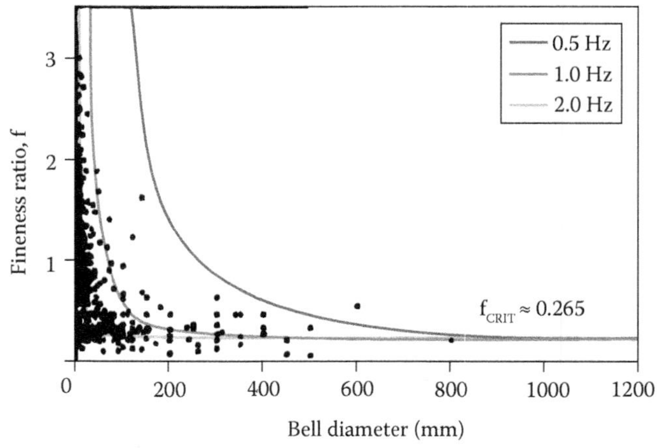

Figure 3.31. Comparison of theoretical morphospace model with measurements. (Based on Dabiri, J.O., et al., J. Exp. Biol., 210, 1868–1873, 2007.)

Nonetheless, it is supported by known jellyfish species.

Perhaps most striking about this model is that it has not made use of any free parameters, because we did in the ontogenetic scaling example in Section 3.6. This is but one example of the utility of vortex wake modeling.

3.16 WAKE-BASED MODELS OF ANIMAL SWIMMING WITHOUT VORTICITY

The modeling approaches described thus far have focused primarily on the properties of vorticity created by the swimming animal. In the next few sections, we examine methods that acknowledge the presence of vorticity in the flow but do not require explicit knowledge of the vorticity field.

Our starting point is Burgatti's theorem, introduced in Section 3.12:

$$\int_{V_V} \mathbf{x} \times \boldsymbol{\omega} dV = \int_{V_V} \mathbf{u}_v dV + \int_{S_V} \phi_v \mathbf{n} dS \quad (3.179)$$

We previously used this relationship to explain the proper application of the added-mass concept to vortex loops. However, a closer examination shows that the right-hand side of the equation can be used in its entirety to replace the vortex impulse in the force equation (Equation 3.152). If we can define a boundary surrounding the wake vortex (e.g., using the LCS concept introduced in Section 3.1), then the first term on the right-hand side is just the linear momentum of the vortex "body" per unit density; it is given by the product of the vortex volume and the velocity of its center of mass. The second term on the right-hand side remains the added-mass of the "body." The advantage of this approach is that we no longer need to distinguish between the self-induced motion of the wake vortices and the motion caused by external flow sources, as is required for the velocity potential ϕ_e in Equation 3.152.

Furthermore, assuming diffusion of vorticity is limited, the hydrodynamic forces and moments on the vortex body—and the resulting locomotive forces on the animal—can be computed in the same manner as on a deformable solid body in inviscid fluid. Consider, for example, the wake vortex illustrated in Figure 3.32. Let ϕ_1 be the velocity potential of flow surrounding the vortex body, that is, $\mathbf{U}_1 = \nabla\phi_1(\mathbf{X},t)$. The capitalized position vector $\mathbf{X}$ denotes locations relative to a stationary frame of reference, whereas the variable $\mathbf{x}$

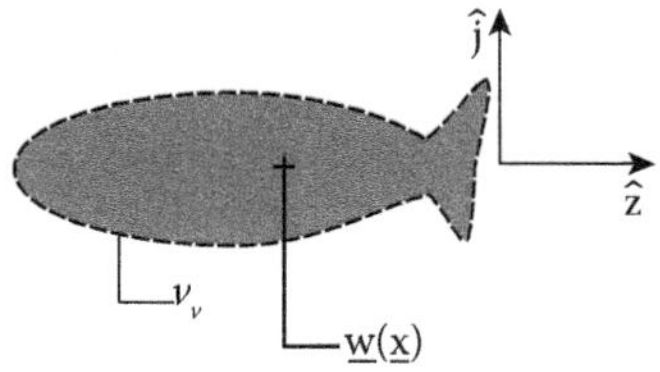

Figure 3.32. Schematic of vortex patch shed behind a swimming fish.

is measured relative to the centroid of the vortex body. The velocity potential due to motion of the vortex body is denoted ϕ_2 and has components

$$\phi_2 = \mathbf{U}_2 \cdot \boldsymbol{\Phi} + \boldsymbol{\Omega} \cdot \boldsymbol{\Psi} + \phi_d + \phi_0 \quad (3.180)$$

where the terms in Equation 3.180 account for, in order, the linear velocity of the vortex body centroid; the angular velocity of the principal axes of the vortex body; deformation of the vortex body shape; and the requirement that the external flow $\mathbf{U}_1$ cannot pass through the vortex body. The potential functions $\boldsymbol{\Phi}$, $\boldsymbol{\Psi}$, ϕ_d, and ϕ_0 each satisfy Laplace's Equation 3.34, with the following boundary conditions on the surface of the vortex body $S_v(\mathbf{x}, t) = 0$:

$$\left.\frac{\partial \boldsymbol{\Phi}}{\partial \mathbf{n}}\right|_{S_V} = \mathbf{n}; \left.\frac{\partial \boldsymbol{\Psi}}{\partial \mathbf{n}}\right|_{S_V} = \mathbf{x} \times \mathbf{n};$$
$$\left.\frac{\partial \phi_d}{\partial \mathbf{n}}\right|_{S_V} = \frac{-\partial S/\partial t}{|\nabla S|}; \left.\frac{\partial \phi_0}{\partial \mathbf{n}}\right|_{S_V} = -\mathbf{U}_1 \cdot \mathbf{n} \quad (3.181)$$

Given the total velocity potential $\phi = \phi_1 + \phi_2$, the hydrodynamic force and moment on the vortex body are given by

$$\mathbf{F} = \rho \frac{d}{dt} \int_{S_V} \phi \mathbf{n} dS - \rho \int_{S_V} \nabla\phi (\nabla\phi \cdot \mathbf{n}) dS + \frac{\rho}{2} \int_{S_V} (\nabla\phi)^2 \mathbf{n} dS \quad (3.182)$$

$$\mathbf{M} = \rho \frac{d}{dt} \int_{S_V} \phi \mathbf{x} \times \mathbf{n} dS + \rho \mathbf{U}_2 \times \int_{S_V} \phi \mathbf{n} dS - \rho \int_{S_V} \mathbf{x} \times \mathbf{U}_1 (\nabla\phi \cdot \mathbf{n}) dS + \frac{\rho}{2} \int_{S_V} (\nabla\phi)^2 (\mathbf{x} \times \mathbf{n}) dS \quad (3.183)$$

These are the hydrodynamic force and moment that the animal must overcome to create the observed vortex wake shape and kinematics. The reader is encouraged to consult the primary literature (Galper and Miloh 1995) for a details regarding the derivation of Equations 3.182 and 3.183.

In many cases, these equations of motion require numerical solution. However, let us consider an example where the equations becomes simplified in form. The linear motion of a single animal appendage in the absence of ambient flow can produce a vortex body that does not rotate. Hence, we can approximate that $\mathbf{U}_1 = \phi_0 = 0$. If the deformation of the vortex body shape is negligible, for example $\phi_d << \mathbf{U}_2 \cdot \mathbf{\Phi}$, then Equation 3.182 simplifies to

$$\mathbf{F} = \frac{d}{dt}\int_{S_V} (\mathbf{U}_2 \cdot \mathbf{\Phi})\mathbf{n}dS \tag{3.184}$$

Per Newton's second law, this sum of this hydrodynamic force and the locomotive force exerted by the animal on the fluid ($\mathbf{F}_l$) leads to the observed acceleration of the vortex body:

$$\mathbf{F} + \mathbf{F}_l = \rho\frac{d}{dt}(V_v\mathbf{U}_2) \tag{3.185}$$

Therefore, the locomotive force can be deduced from the calculated hydrodynamic force and the observed (or modeled) motion of the vortex body. All of this is accomplished without explicit reference to the vorticity in the flow.

3.17 ENERGETIC EFFICIENCY OF SWIMMING

Aquatic locomotion requires an expenditure of energy by the organism. Because this energy is limited by the animal's capacity to collect and store it, animal survival can often depend on efficient use of the available energy. Generally, the energy available to the animal is either put to a useful purpose or it is lost back to the environment. During locomotion, we consider "useful" energy as energy that goes toward propelling the animal through the water. The remaining energy is lost to the environment as kinetic energy in the wake, and it is subsequently converted to heat by viscous forces acting on the spatial gradients of velocity.

Note that we are not concerned at present with a formal thermodynamic definition of efficiency, for which the concept of useful work has a narrower meaning. Here, forward motion of the animal is deemed useful regardless of the effect of locomotion on the disorder or *entropy* of the surrounding fluid environment.

Given the diversity of modes of aquatic locomotion, let us begin by considering the "black box" organism in Figure 3.33 that moves at speed U_0 and creates a velocity surplus in the wake of $U_j - U_0$ (see Section 3.11). In a quasi-steady approximation (used here because we are not concerned with details of the swimming trajectory, as in Section 3.11), the thrust F_l produced by the organism to propel itself forward can be estimated by the product of the rate $\dot{m}$ at which surrounding

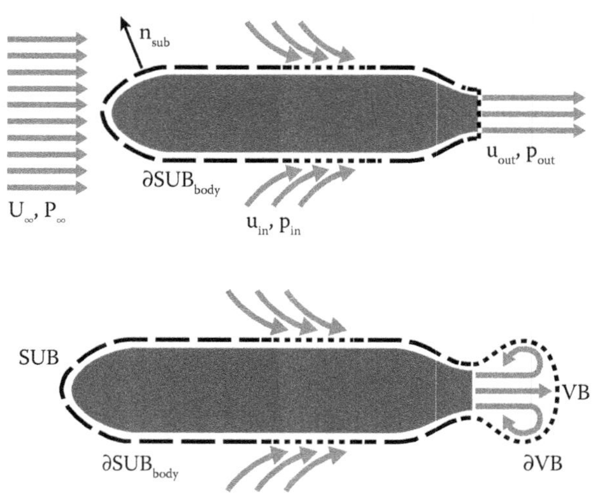

Figure 3.33. Top panel: Control volume around self-propelled body. Bottom panel: Control volume around self-propelled body and forming wake vortex.

fluid mass is affected by the animal and the velocity surplus imparted to that fluid:

$$F_l \approx \dot{m}\left(U_j - U_0\right) \qquad (3.186)$$

This result is consistent with the steady flow momentum Equation 3.128, where the effect of spatial imbalances in the fluid pressure is assumed to be negligible. This assumption is relaxed later in this section.

The useful work done by the organism on the fluid per unit time is then given by the product of the locomotive force and the speed at which the organism encounters new fluid, that is,

$$P_l = F_l U_0 \qquad (3.187)$$

The rate at which energy is lost to the wake is proportional the kinetic energy of each fluid particle in the wake, which is itself proportional to the square of the velocity surplus, and the rate $\dot{m}$ at which fluid enters the wake:

$$P_{loss} = \dot{m}\left[\frac{1}{2}\left(U_j - U_0\right)^2\right] \qquad (3.188)$$

The energetic efficiency of swimming can be defined as the ratio between the useful work done by the organism on the fluid per unit time, divided by the total amount of energy going toward locomotion and lost in the wake. This *Froude efficiency* is defined as

$$\eta_f = \frac{P_l}{P_l + P_{loss}} = \frac{1}{1 + P_{loss}/P_l} \qquad (3.189)$$

or, equivalently

$$\eta_f = \frac{2}{1 + U_j/U_0} \qquad (3.190)$$

The Froude efficiency highlights a critical trade-off that is faced by any animal as it swims. Whereas the locomotive force depends on a large velocity surplus in the wake (Equation 3.186), swimming efficiency is maximized as the velocity surplus goes to zero, so that $U_j/U_0 = 1$ in Equation 3.190.

If animals were indeed constrained to quasi-steady flow, this trade-off would severely limit the opportunities to adapt locomotion for both speed and efficiency. However, as we have emphasized in this chapter, aquatic locomotion is inherent unsteady. Let us therefore consider a more complete picture of swimming efficiency that includes effects such as pressure imbalances in the flow, which are lost when viewing aquatic locomotion a quasi-steady perspective.

Figure 3.33 illustrates a control volume around the fictitious organism that serves as a model for our study of swimming efficiency. The control volume is displaced by a small distance δ from the animal–fluid interface so that we can avoid including viscous stresses in the equations of motion. The subsequent equations implicitly assume that the fluxes of mass, momentum, and energy through the small region of thickness δ are small relative to the total volume of surrounding fluid that interacts with the animal.

For mathematical convenience we presume that flow approaches the locomotory appendage from the sides and is propelled backward at the rear of the animal. Equations expressing the conservation of mass, momentum in the swimming direction, and energy can be written, respectively, as

$$\rho\int_{S_{a,out}} \mathbf{u}_{out} \cdot \mathbf{n}_a dS + \rho\int_{S_{a,in}} \mathbf{u}_{in} \cdot \mathbf{n}_a dS = 0 \qquad (3.191)$$

$$F_l = \rho\frac{d}{dt}\int_{V_a} \mathbf{u} dV + \rho\int_{S_{a,out}} u_{out}^2 dS + \int_{S_a} p\mathbf{n}_a \cdot \mathbf{i} dS \qquad (3.192)$$

$$P_{in} = \frac{\rho}{2}\frac{d}{dt}\int_{V_a} \mathbf{u}\cdot\mathbf{u} dV + \frac{\rho}{2}\int_{S_{a,in}} (\mathbf{u}_{in}\cdot\mathbf{u}_{in})\mathbf{u}_{in}\cdot\mathbf{n}_a dS$$
$$+ \frac{\rho}{2}\int_{S_{a,out}} (\mathbf{u}_{out}\cdot\mathbf{u}_{out})\mathbf{u}_{out}\cdot\mathbf{n}_a dS$$
$$+ \int_{S_{a,out}} p(\mathbf{u}_{out}\cdot\mathbf{n}_a) dS \qquad (3.193)$$

where the subscript *a* denotes properties on the surface of the animal and the subscripts in and out denote properties at locations where flow approaches and exits the animal appendage, respectively. The input power P_{in} is the sum of the useful work and lost wake energy per unit time.

As described above, neglect of the pressure terms in these equations leads to the Froude efficiency defined in Equation 3.190. However, by creating a model of vortex formation in the wake, we can incorporate the contribution of the pressure field and also draw connections to the modeling accomplished in the preceding sections. First, let us amend the control volume around the animal to include the outer surface of a vortex ring forming at its rear (Figure 3.33). This rear portion of the control volume is time varying as the vortex ring grows due to the fluid entering it from upstream and from local entrainment of the surrounding fluid. The new control volume corresponds to a new set of conservation equations:

$$\rho\frac{\partial V_v}{\partial t}+\rho\int_{S_{a,in}}\mathbf{u}_{in}\cdot\mathbf{n}_a dS-\dot{m}_{entrain}=0 \quad (3.194)$$

$$F_l=\rho\frac{d}{dt}\int_{V_a+V_v}\mathbf{u}dV+\int_{S_a}p\mathbf{n}_a\cdot\mathbf{i}dS+\int_{S_v}p\mathbf{n}_v\cdot\mathbf{i}dS \quad (3.195)$$

$$P_{in}=\frac{\rho}{2}\frac{d}{dt}\int_{V_a+V_v}\mathbf{u}\cdot\mathbf{u}dV+\frac{\rho}{2}\int_{S_{a,in}}(\mathbf{u}_{in}\cdot\mathbf{u}_{in})\mathbf{u}_{in}\cdot\mathbf{n}dS \quad (3.196)$$

The new conservation equations replace the dependence on the pressure of the exiting flow with a dependence on properties of the vortex wake. If the diffusion of vorticity past the vortex boundary S_v is limited, as in the case of high-frequency Reynolds numbers (see Equation 3.106), then we can apply Bernoulli's Equation 3.41 on the surface of the new control volume. The pressure at any point along the control volume is related to the local flow speed u and the velocity potential ϕ by

$$p=\rho\left[C(t)-\frac{\partial\phi}{\partial t}-\frac{u^2}{2}\right] \quad (3.197)$$

where the Bernoulli constant C is allowed to vary with time during the unsteady locomotion. Substituting for the pressure in the momentum conservation Equation 3.195 gives

$$\begin{aligned}F_l=&\rho\frac{d}{dt}\int_{V_a+V_v}\mathbf{u}dV-\rho\int_{S_a}\frac{\partial\phi}{\partial t}\mathbf{n}_a\cdot\mathbf{i}dS\\&-\frac{\rho}{2}\int_{S_a}u^2\mathbf{n}_a\cdot\mathbf{i}dS\\&-\rho\int_{S_v}\frac{\partial\phi}{\partial t}\mathbf{n}_v\cdot\mathbf{i}dS-\frac{\rho}{2}\int_{S_v}u^2\mathbf{n}_v\cdot\mathbf{i}dS\end{aligned} \quad (3.198)$$

where the integral of the Bernoulli constant over the closed control volume surface is zero (see Equation 3.149 below for example).

If we assume that while the wake vortex is attached to the swimming animal, the velocity along the entire control volume is U_0, then the squared velocity u^2 in the surface integrals in Equation 3.198 are constants and therefore sum to zero in the same manner as the Bernoulli constant. The momentum conservation equation simplifies to

$$\begin{aligned}F_l=&\rho\frac{d}{dt}\int_{V_a+V_v}\mathbf{u}dV-\rho\int_{S_a}\frac{\partial\phi}{\partial t}\mathbf{n}_a\cdot\mathbf{i}dS\\&-\rho\int_{S_v}\frac{\partial\phi}{\partial t}\mathbf{n}_v\cdot\mathbf{i}dS\end{aligned} \quad (3.199)$$

Recall from Section 3.12 that the surface integral of the velocity potential is the added-mass of the body enclosed by the surface. Hence, the second term on the right-hand side of Equation 3.199 is the added-mass of the animal, and the third term on the right-hand side is the added-mass of the wake vortex. Denoting the added-mass coefficient for unidirectional motion along the **i**-axis as α_x, and replacing the volume integrals of velocity with the centroid velocities, the momentum and energy conservation equations become, respectively,

$$F_l=\rho(1+\alpha_{x,a})V_a\dot{U}_0+\rho\frac{\partial}{\partial t}\left[(1+\alpha_{x,v})V_v\mathbf{U}_v\cdot\mathbf{i}\right] \quad (3.200)$$

$$P_{in}=\rho\left(U_0\dot{U}_0V_a+U_v\dot{U}_vV_v+\frac{1}{2}U_v^2\dot{V}_v\right) \quad (3.201)$$

where the overdots denote a time derivative, and it is assumed that the rate of change of the vortex volume is primarily due to the entrainment of surrounding fluid; hence, the second term on the right-hand side of Equation 3.196 vanishes according to the mass conservation Equation 3.194. This assumption is often true for vortex wakes formed at high-frequency Reynolds number. However, in circumstances where that approximation is inappropriate, Equation 3.201 is an overestimate of the power input by the animal during swimming.

In analogy with the Froude efficiency, we can define an unsteady swimming efficiency as

$$\eta=\frac{\overline{F_lU_0}}{\overline{P_{in}}} \quad (3.202)$$

or, using the equations we have developed,

$$\eta = \frac{\overline{U_0\left\{(1+\alpha_{x,a})V_a\dot{U}_0 + \rho\frac{\partial}{\partial t}\left[(1+\alpha_{x,v})V_v\mathbf{U}_v\cdot\mathbf{i}\right]\right\}}}{\overline{U_0\dot{U}_0V_a + U_v\dot{U}_vV_v + \frac{1}{2}U_v^2\dot{V}_v}} \tag{3.203}$$

The efficiency of animals that swim without large accelerations or decelerations can be approximated more simply, by using the fact that $\dot{U}_0$ is approximately zero:

$$\eta = \frac{\overline{U_0\frac{\partial}{\partial t}\left[(1+\alpha_{x,v})V_v\mathbf{U}_v\cdot\mathbf{i}\right]}}{\overline{U_v\dot{U}_vV_v + \frac{1}{2}U_v^2\dot{V}_v}} \tag{3.204}$$

We can further simplify this unsteady efficiency by approximating the time derivative of a given property ξ as

$$\frac{\partial\xi}{\partial t} \approx \Omega_l\xi(T) \tag{3.205}$$

where $\xi(T)$ is the value of the property at the end of each vortex formation event and Ω_l the number of vortex formation events per unit time. This quasi-steady approximation of Equation 3.204 is simply

$$\eta = \frac{2}{3}(1+\alpha_{x,v})\frac{U_0}{U_v} \tag{3.206}$$

Figure 3.34 compares this quasi-steady swimming efficiency with the Froude efficiency for $\alpha_{x,v} = 0.5$, corresponding to a spherical wake vortex. The two curves have the same values at the limits of zero and perfect swimming efficiency, but the Froude efficiency is larger at intermediate values of efficiency. In general, the upper bounds do not coincide due to the different assumptions used in their derivation.

The most general form of the unsteady efficiency in Equation 3.203 allows for a wide range of strategies to improve swimming efficiency. By direct count, 10 different parameters can be manipulated to affect the swimming efficiency: the animal's swimming speed U_0 and acceleration $\dot{U}_0$; the animal's displaced fluid volume V_a, and shape, via $\alpha_{x,a}$; the wake vortex shape $\alpha_{x,v}$, volume V_v, and speed U_v; and the rates of change of those wake vortex properties. This in sharp contrast with the Froude efficiency, which suggests only the swimming speed and wake speed as means of affecting swimming efficiency. With the tools developed in this section, we can now evaluate the various swimming modes found in nature to identify which subsets of the 10 parameters are most commonly adapted by animals, and in what manner.

The final three sections of this chapter review some of the classic models of animal locomotion found in the primary literature. These reviews are introductory and are meant to expose the reader to the salient physical features of the models. References at the end of the chapter provide more comprehensive details on the variants of the models that have been developed in recent years.

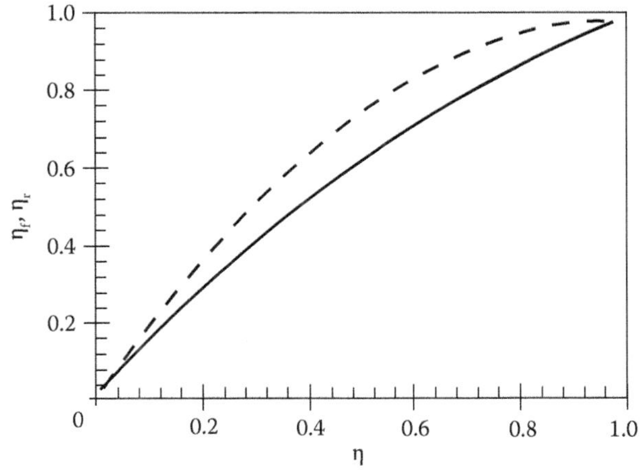

Figure 3.34. Plot of Froude and rocket efficiencies versus wake-based efficiency. (Based on Ruiz, L.A., et al., J. Fluid Mech., 668, 5–32, 2011.)

3.18 CLASSIC SWIMMING MODELS: UNDULATORY LOCOMOTION

A broad range of aquatic animal species have independently converged on the use of undulatory locomotion, including eels, water snakes, and water worms. Figure 3.35 illustrates the body shape and kinematics that we use to construct a model of the swimming dynamics.

Let us assume that the transverse motion of the body $h(x, t)$ is small relative to the length l of the animal. Furthermore, we assume that the flow created by the body motion is constrained to two-dimensional y-z planes perpendicular to the swimming direction x and that the fluid has no viscosity. Finally, the body shapes in cross sections near the animal head are approximated as circular, whereas those near the tail are thin ellipses.

With these assumptions, the physical mechanism that leads to locomotion is the transfer of momentum from the animal body to the fluid in each transverse cross section. The fluid momentum $I(x, t)$ is wholly characterized by the velocity $V(x, t)$ of the added-mass of fluid (per unit depth) $A(x)$ surrounding each two-dimensional cross section of the animal:

$$I(x,t) = \rho V(x,t) A(x) \qquad (3.207)$$

where ρ is the fluid density. The velocity of each cross section relative to flow past the animal at velocity U_0 is

$$V(x,t) = \frac{Dh}{Dt} = \frac{\partial h}{\partial t} + U_0 \frac{\partial h}{\partial x} \qquad (3.208)$$

As in our discussion of Eulerian acceleration in Section 3.2, the total derivative is need to account for variations of the transverse motion in time and along the length of the body. Similarly, Newton's second law equates that Lagrangian rate of change of fluid momentum in each transverse cross section to the lateral force exerted by the fluid on the animal, i.e.,

$$L(x,t) = -\rho\left(\frac{\partial}{\partial t} + U_0 \frac{\partial}{\partial x}\right)\left[V(x,t)A(x)\right] \qquad (3.209)$$

To obtain the perpendicular thrust force, let us consider the energetics of the swimming motion. The power input required to create the lateral forces is given by the product of the lateral force and the lateral body velocity:

$$P_{in} = -\int_0^l \frac{\partial h}{\partial t} L(x,t)\,dx \qquad (3.210)$$

where the negative value arises because we are now considering the force of the animal on the fluid, the converse of Equation 3.209. Using the lateral force equation (Equation 3.209) gives

$$P_{in} = \rho \int_0^l \frac{\partial h}{\partial t}\left(\frac{\partial}{\partial t} + U_0 \frac{\partial}{\partial x}\right)\left[V(x,t)A(x)\right]dx \qquad (3.211)$$

or, using the time derivative of Equation 3.208 and with some algebra,

$$P_{in} = \rho \int_0^l \left(\frac{\partial}{\partial t} + U_0 \frac{\partial}{\partial x}\right)\left[\frac{\partial h}{\partial t} V(x,t)A(x)\right]dx \\ -\rho \int_0^l \frac{\partial V}{\partial t} V(x,t)A(x)\,dx \qquad (3.212)$$

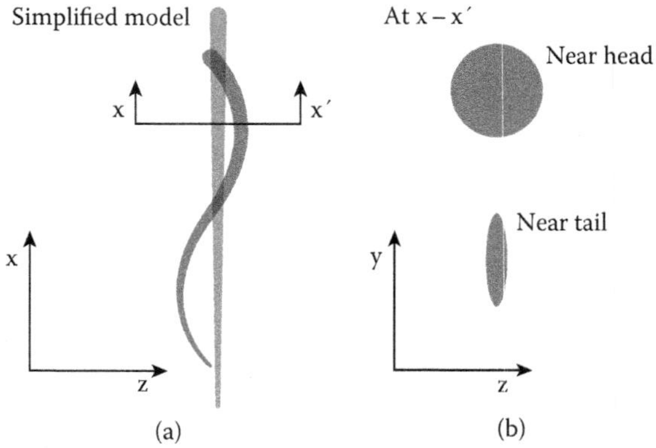

Figure 3.35. Geometric model of swimming eel. (a) Dorsal view. (b) Anterior and posterior views.

The second term in Equation 3.212 arises because we have included the lateral body velocity $\partial h/\partial t$ inside the total derivative, implicitly creating an additional term on the right-hand side of the equation. This additional term is spurious and is therefore cancelled by the new second term.

The following mathematical relationships hold by definition:

$$\frac{d}{dt}\left(\frac{V^2}{2}\right) = V\frac{\partial V}{\partial t} \tag{3.213}$$

$$\int_A^B \frac{\partial}{\partial x}(\cdots)\,dx = [\cdots]_A^B = [\cdots](B) - [\cdots](A) \tag{3.214}$$

where the ellipses represent an arbitrary function. Hence, we can combine the time-derivative terms in Equation 3.212, leading to

$$P_{in} = \frac{\partial}{\partial t}\left(\rho\int_0^l \frac{\partial h}{\partial t} VA dx - \frac{\rho}{2}\int_0^l V^2 A dx\right) + \rho U_0\left[\frac{\partial h}{\partial t}VA\right]_0^l \tag{3.215}$$

where the dependence of V and A on x and t is implied but has been omitted for clarity. As in the preceding section, let us consider the time-averaged power:

$$\overline{P_{in}} = \frac{1}{T}\int_0^T \frac{\partial}{\partial t}\left(\rho\int_0^l \frac{\partial h}{\partial t} VA dx - \frac{\rho}{2}\int_0^l V^2 A dx\right)dt + \frac{1}{T}\int_0^T \rho U_0\left[\frac{\partial h}{\partial t}VA\right]_0^l dt \tag{3.216}$$

If the duration of the time integration is taken to be infinite, that is, $T \to \infty$, the first term on the right-hand side of Equation 3.216 goes to zero because the term in parentheses is bounded while the coefficient $1/T$ goes to zero. Replacing the cross-section velocity by using Equation 3.208, and assuming that the anterior animal body shape terminates at a point such that $A(x = 0) = 0$,

$$\overline{P_{in}} = \rho U_0 A(l)\overline{\left[\frac{\partial h}{\partial t}\left(\frac{\partial h}{\partial t} + U_0\frac{\partial h}{\partial x}\right)\right]_{x=l}} \tag{3.217}$$

Physically, Equation 3.217 represents the product of the tail trailing edge lateral velocity $\partial h/\partial t$ and the lateral momentum $U_0 AV$ shed at the tail trailing edge; that is, it is the total rate at which work is done by the animal on the fluid during locomotion. As we saw in Section 3.16, the total work goes to useful work in swimming and to kinetic energy that is lost in the wake:

$$\overline{P_{in}} = \overline{F_l}U_0 + \frac{\rho}{2}\overline{V^2}AU_0 \tag{3.218}$$

or, using Equation 3.217 and rearranging the terms

$$\overline{F_l} = \frac{\rho}{2}A(l)\overline{\left[\left(\frac{\partial h}{\partial t}\right)^2 - U_0^2\left(\frac{\partial h}{\partial x}\right)^2\right]_{x=l}} \tag{3.219}$$

Interestingly, the undulatory swimming model in Equation 3.219 determines the thrust generated by the animal solely on the basis of the animal–fluid interactions at the tail. This implies that the dynamics at the tail represent the cumulative effect of the hydrodynamics of the upstream flow past the rest of the body.

We can now examine the efficiency of undulatory swimming by using the Froude efficiency concept in Equation 3.202. The mathematical details can be found elsewhere (Lighthill 1975). The final result is

$$\eta_f = 1 - \frac{\frac{1}{2}\overline{\left[\left(\frac{\partial h}{\partial t} + U_0\frac{\partial h}{\partial x}\right)^2\right]_{x=l}}}{\overline{\left[\frac{\partial h}{\partial t}\left(\frac{\partial h}{\partial t} + U_0\frac{\partial h}{\partial x}\right)\right]_{x=l}}} \tag{3.220}$$

Remarkably, the swimming efficiency depends only on the kinematic waveform $h(x, t)$. The cross-sectional shape occurs in both the thrust and the work and is therefore eliminated from the equation when their ratio is taken.

The need for only $h(x, t)$ to determine the swimming efficiency is perhaps the greatest triumph of this model of undulatory locomotion. That said, the waveform is not arbitrary. First, the waveform must satisfy a lateral force balance given by Newton's second law:

$$\rho\int_0^l S(x)\frac{\partial^2 h}{\partial t^2}dx = -\rho\int_0^l\left(\frac{\partial}{\partial t} + U_0\frac{\partial}{\partial x}\right)\left[A(x)\left(\frac{\partial h}{\partial t} + U_0\frac{\partial h}{\partial x}\right)\right]dx \tag{3.221}$$

where $S(x)$ is the cross-sectional area of the body at a given location x along its length. The left-hand side of Equation 3.221 is the product of the mass and acceleration of the animal body (assuming neutral buoyancy); the right-hand side is the fluid force on the body.

Second, the waveform must balance the angular acceleration of the animal body in yaw (i.e., rotation around a dorsoventral axis) with the hydrodynamic torque on the body:

$$\rho\int_0^l xS(x)\frac{\partial^2 h}{\partial t^2}dx = -\rho\int_0^l x\left(\frac{\partial}{\partial t}+U_0\frac{\partial}{\partial x}\right)\left[A(x)\left(\frac{\partial h}{\partial t}+U_0\frac{\partial h}{\partial x}\right)\right]dx \quad (3.222)$$

To achieve high swimming efficiency, Equation 3.220 suggests that the cross-sectional velocity V of the fluid should be less than the lateral body velocity at the tail, that is,

$$|V|_{x=l} = \left|\frac{\partial h}{\partial t}+U_0\frac{\partial h}{\partial x}\right|_{x=l} < \left|\frac{\partial h}{\partial t}\right|_{x=l} \quad (3.223)$$

However, Equation 3.219 indicates that the time-averaged swimming thrust scales as

$$|\bar{T}| \sim \left|\frac{\partial h}{\partial t}+U_0\frac{\partial h}{\partial x}\right| \quad (3.224)$$

As in Section 3.16, we are once again presented with a trade-off between swimming efficiency and swimming thrust.

Let us consider two example kinematic waveforms to improve our intuition regarding the undulatory swimming efficiency. Perhaps the simplest motion one can conceive is a *standing wave* for which the lateral displacement of the body occurs in phase along the length of the body instead of in sequence:

$$h(x,t) = H(x)\cos\omega t \quad (3.225)$$

where $H(x)$ is the amplitude of the lateral motion and is the angular frequency of the motion. Substituting this waveform into the Froude equation (Equation 3.220) gives

$$\eta_f = 1-\frac{\omega^2H^2(l)+U_0^2H'^2(l)}{2\omega^2H^2(l)} \quad (3.226)$$

where $H' = \partial H/\partial x$. The maximum efficiency of the standing wave kinematics is only 50%, but the fact that this motion is predicted to produce forward thrust at all is interesting in itself.

A more realistic swimming motion is a *traveling wave* with an amplitude that varies along the length of the body:

$$h(x,t) = f(x)g\left(t-\frac{x}{c}\right) \quad (3.227)$$

The function f contains the spatial variation in the wave amplitude, whereas the function g accounts for the motion of the wave from head to tail at speed c. The swimming efficiency of this waveform is

$$\eta_f = 1-\frac{\left(1-\frac{U_0}{c}\right)^2 f^2(l)\overline{g'^2}+U_0^2f'^2(l)\overline{g^2}}{2\left(1-\frac{U_0}{c}\right)f^2(l)\overline{g'^2}} \quad (3.228)$$

where the primed quantities denote differentiation with respect to along the body. The reader is encouraged to consult Lighthill (1975) for details of the mathematics.

Implicit in Equation 3.228 are several clues for enhancing the swimming efficiency of an undulatory waveform. First, positive thrust is only generated for $c > U_0$; the maximum speed at which the wave can be passed down the body determines the maximum possible swimming speed. Second, efficiency is increased if the tail is aligned with the swimming direction such that $f'(l) = 0$; this eliminates the last negative term in the numerator of Equation 3.228. Generally, it is desirable to confine large amplitude motions to the tail region to reduce lateral recoil and associated yawing of the head caused by the hydrodynamic torque (see Equation 3.222). The recoil can be further limited by simultaneously causing portions of the tail region to move in opposite directions, so that their torques oppose one another. The net result of these considerations is a undulatory body motion similar to that illustrated in Figure 3.36. Observations of undulatory swimmers in nature confirm that this is the prevalent motif for the body kinematics.

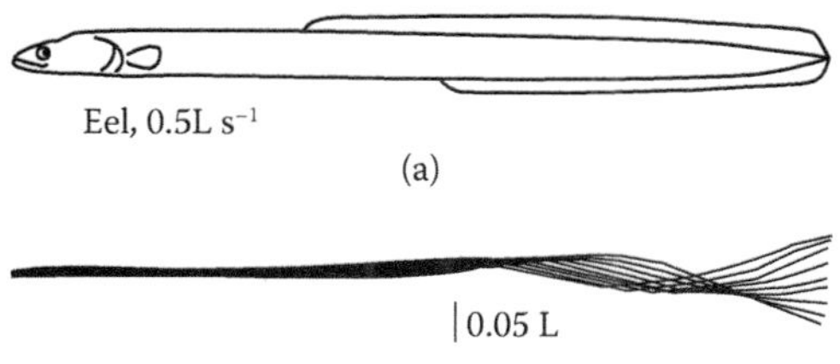

Figure 3.36. Measurements of eel kinematics. (a) Side view of eel. (b) Outlines of body centerline during swimming cycle. (Based on Tytell, E.D., et al., *J. Exp. Biol.*, 211, 187–195, 2008.)

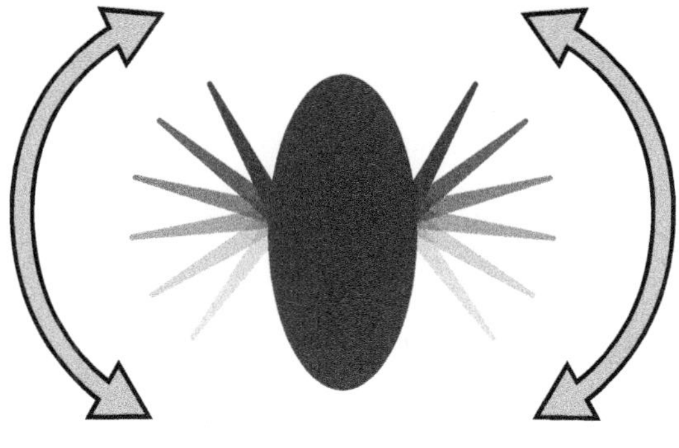

Figure 3.37. Illustration of reciprocal flapping motion.

3.19 CLASSIC SWIMMING MODELS: LOCOMOTION AT LOW REYNOLDS NUMBERS

Microorganisms such as bacteria, sperm, and ciliates function in hydrodynamic environments where the fluid inertia is negligible. An immediate consequence of locomotion at low Reynolds numbers is that to be effective, the propulsive and recovery strokes cannot be reciprocal motions along the same trajectory. This constraint arises in part because the hydrodynamic force acting on a moving appendage at low Reynolds number is proportional to the velocity of the appendage instead of the square of the velocity as in high Reynolds number flows (see Equation 3.128 for example). Let us consider for example the reciprocal flapping motion of the fictitious organism illustrated in Figure 3.37.

The change in forward momentum during a cycle of propulsion and recovery can be estimated as

$$\Delta I \approx F_p T_p - F_r T_r \tag{3.229}$$

where F and T are the force and duration of the propulsive (subscript p) and recovery (subscript r) strokes, respectively. The duration of each stroke is given approximately by the stroke length L divided by the speed of the stroke, V_p or V_r. Combined with the linear relationship between force and velocity at low Reynolds number, Equation 3.267 can be approximated as

$$\Delta I \approx V_p \frac{L}{V_p} - V_r \frac{L}{V_r} = 0 \tag{3.230}$$

The result in Equation 3.230 is particularly striking because it holds regardless of the difference in the duration of the propulsive and recovery strokes. Time has no dynamical significance for swimming at low Reynolds number; the spatial configuration of the swimming kinematics plays a dominant role.

The equations of fluid motion reflect the unique dynamics that occur at a low Reynolds number. To incorporate the effect of scale, recall the dimensionless Navier–Stokes Equation 3.61:

$$\frac{\partial u^*}{\partial t^*} + u^* \frac{\partial u^*}{\partial x^*} + v^* \frac{\partial u^*}{\partial y^*} + w^* \frac{\partial u^*}{\partial z^*} = f_{lx}^* - \frac{1}{\rho^*}\frac{\partial p^*}{\partial x^*} + \frac{\nu}{U_0 L_0}\nabla^{*2} u^* \tag{3.231}$$

To arrive at this form of the Navier–Stokes equation, the fluid pressure was normalized by using the dynamic fluid pressure, that is, $p^* = p/(\rho_0 U^2)$. However, because we anticipate that viscous forces play a dominant role at low Reynolds numbers, a more appropriate normalization in this case is $p^* = pL_0/(\mu U_0)$. Similarly, the locomotive force f_{lx} is more appropriately normalized as $f_{lx}^* = f_{lx}\rho L_0^2/\mu U_0$. With these new normalizations, the dimensionless Navier–Stokes equation becomes

$$\frac{\partial u^*}{\partial t^*} + u^* \frac{\partial u^*}{\partial x^*} + v^* \frac{\partial u^*}{\partial y^*} + w^* \frac{\partial u^*}{\partial z^*} = \frac{1}{Re}\left(f_{lx}^* - \frac{1}{\rho^*}\frac{\partial p^*}{\partial x^*} + \nabla^{*2} u^*\right) \tag{3.232}$$

where $Re = U_0L_0/\nu$. For small Reynolds numbers, the right-hand side of Equation 3.232 is much larger than the left-hand side. This scenario is described by the *Stokes equation*:

$$f_{lx}^* - \frac{1}{\rho^*}\frac{\partial p^*}{\partial x^*} + \nabla^{*2} u^* = 0 \tag{3.233}$$

where we have incorporated the locomotive force. The Stokes equation describes the static equilibrium of the flow at any point in time. In words, the sum of the forces acting on the fluid is always equal to zero. Because the fluid has no inertia, the fluid motion ceases as soon as the locomotive force ceases.

3.20 FLAGELLAR PROPULSION

Let us apply these concepts to model the propulsion of a swimming flagellum. Figure 3.38 shows a simplified flagellum consisting of an infinitesimally thin tail and a head of arbitrary shape. Our strategy for modeling the flagellum is to represent it as a collection of point forces $\mathbf{F}(\mathbf{x})$.

In dimensional form, the Stokes equation for a single point force is

$$-\nabla p + \mu\nabla^2\mathbf{u} + \mathbf{F}\delta(x) = 0 \tag{3.234}$$

where the *delta function* $\delta(x)$ is infinite at the location of interest $x = 0$ and has a value of zero everywhere else:

$$\begin{aligned} \delta(x) &= \infty \quad \text{if} \quad x = 0 \\ \delta(x) &= 0 \quad \text{if} \quad x \neq 0 \end{aligned} \tag{3.235}$$

The integral of the delta function is particularly useful, as it gives

$$\int_{-\infty}^{\infty} \delta(x)\,dx = 1 \tag{3.236}$$

By using the vector identity (Equation 3.89) and recalling that the fluid is incompressible and

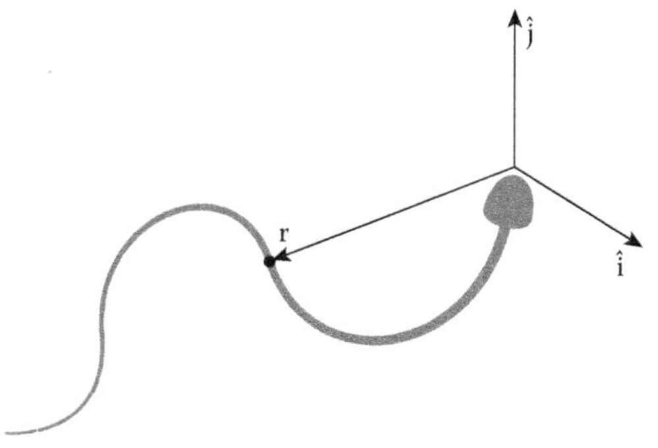

Figure 3.38. Coordinate system for analysis of low Reynolds number undulatory swimming.

hence divergence free, the viscous force term can be replaced as

$$-\nabla p + \mu\nabla \times (\nabla \times \mathbf{u}) + \mathbf{F}\delta(x) = 0 \tag{3.237}$$

Taking the divergence of Equation 3.237 and replacing the pressure term by using the vector identity (Equation 3.89) again, the Stokes equation becomes

$$\nabla^2 p = \nabla \cdot [\mathbf{F}\delta(x)] \tag{3.238}$$

Equation 3.238 can be solved analytically (the reader is encouraged to consult an advanced calculus textbook), giving

$$p(r) = -\nabla \cdot \left(\frac{\mathbf{F}}{4\pi r}\right) \tag{3.239}$$

where r is the distance from the point force. The location of the point force is a singularity in the flow because the pressure is undefined there. However, at any nonzero distance from the point force, the fluid velocity can be determined by solving the equation

$$\mu\nabla^2\mathbf{u} = \nabla p \tag{3.240}$$

For example, if $\mathbf{F} = F\mathbf{i}$, so that the force is only applied in the x-direction, Equation 3.239 can be solved to show that the resulting pressure field is

$$p(r, x) = \frac{Fx}{4\pi r^3} \tag{3.241}$$

where x and r are measured relative to the location of the point force. The corresponding velocity field is given by solution of Equation 3.240:

$$\nabla^2\mathbf{u} = \frac{\nabla p}{\mu} = \frac{F}{4\pi\mu}\left[\left(\frac{1}{r^3} - \frac{3x^2}{r^5}\right)\mathbf{i} - \frac{3xy}{r^5}\mathbf{j} - \frac{3xz}{r^5}\mathbf{k}\right] \tag{3.242}$$

This equation has the solution [see Lighthill (1975) for details]

$$\mathbf{u} = \frac{F}{8\pi\mu}\left[\left(\frac{x^2 + r^2}{r^3}\right)\mathbf{i} + \frac{xy}{r^3}\mathbf{j} + \frac{xz}{r^3}\mathbf{k}\right] \tag{3.243}$$

The velocity field in Equation 3.243 is called the *Stokeslet* velocity field and is plotted in Figure 3.39.

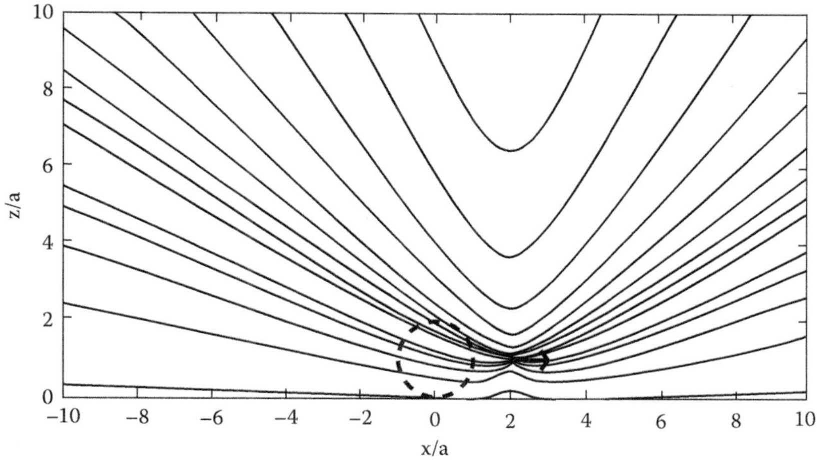

Figure 3.39. Streamlines corresponding to a Stokeslet.

The Stokeslet is a powerful tool for modeling low Reynolds number locomotion because the effect of the appendages on the fluid can be represented by collections of the Stokeslets. To start, let us consider low Reynolds number flow past a slender cylinder (Figure 3.40). Our goal is to replace the real cylinder with a distribution of Stokeslets. For brevity, the reader is again referred to the references section for details of the derivation. The resulting Stokeslet force per unit length normal to the cylinder axis is

$$\mathbf{f}_N = \frac{8\pi\mu\mathbf{U}_N}{1+\ln\left(\frac{4cb}{a^2}\right)} \tag{3.244}$$

where $\mathbf{U}_N$ is the free-stream flow velocity normal to the cylinder, a is the cylinder radius, and the geometric parameters b and c are indicated in Figure 3.40. The force $\mathbf{f}_N$ is balanced by the hydrodynamic drag on the cylinder due to the flow at velocity $\mathbf{U}_N$. Equivalently, $\mathbf{f}_N$ is the force that must be applied to the cylinder to move it through the fluid at velocity $\mathbf{U}_N$. It is notable that the force in Equation (3.244) decreases without limit for large values of c and b. This inconsistency is related to the *Stokes paradox*, which presents difficulties when computing two-dimensional low Reynolds number flows. Because the organisms of interest to us are three-dimensional, our analysis need not be affected by this nuance in the equations of fluid motion.

Using similar methods, the Stokeslet force for flow tangential to a slender cylinder is found to be

$$\mathbf{f}_T = \frac{8\pi\mu\mathbf{U}_T}{-2+2\ln\left(\frac{4cb}{a^2}\right)} \tag{3.245}$$

As was suggested at the beginning of this section, the forces on the cylinder are proportional to the velocity at which the cylinder moves. We can normalize the forces by the velocities to create *resistance coefficients*:

$$k_N = \frac{\mathbf{f}_N}{\mathbf{U}_N};\ k_T = \frac{\mathbf{f}_T}{\mathbf{U}_T} \tag{3.246}$$

Given the resistance coefficients, we can model flagellar swimming as a superposition of normal and tangential Stokeslet forces acting on the approximately cylindrical appendage. To do this, we must first prescribe the waveform of the flagellum. Let us parameterize the body shape as

$$(x,y,z)=(X(s),Y(s),Z(s))$$

where s is the distance from head along the flagellum (Figure 3.38). The waveform has a spatial period Λ such that, in the transverse directions $Y(s+\Lambda) = Y(s)$ and $Z(s+\Lambda) = Z(s)$. The spatial periodicity along the length of the body is complicated by the transverse displacements decreasing the length of the body along the x-direction. If the length of the transversely displaced body along the x-direction is α times the straight body length L_0, then the spatial periodicity of the body is given by $X(s+\Lambda) = X(s) + \alpha\Lambda$.

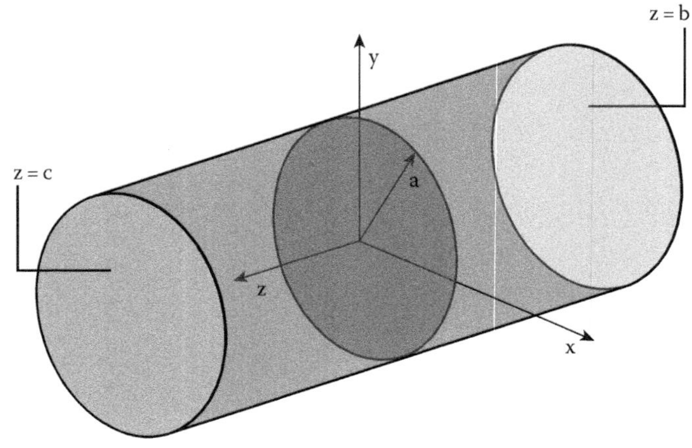

Figure 3.40. Geometric model of a cylinder in Stokes flow.

In the reference frame of the flagellum, the waveform is typically a traveling wave of the form

$$(x, y, z, t) = \left(X(s - ct), Y(s - ct), Z(s - ct)\right) \quad (3.247)$$

where c is the wave speed along the curved body.

Relative to the fluid, the propulsive wave travels backward along the body at a speed $V - U_0$, where $V = c$. Therefore, we can decompose the velocity into normal and tangential components by using the geometry of the relevant velocity vectors. The result is

$$U_N = (V - U_0)\left[1 - X'^2(s - ct)\right]^{\frac{1}{2}} \quad (3.248)$$

$$U_T = (V - U_0)X'(s - ct) - c \quad (3.249)$$

where the primed quantities indicate differentiation with respect to position s along the flagellum. The corresponding forces can be determined by using the resistance coefficients in Equation 3.246. When integrated along the length of the body, the components of those forces in the swimming direction determine the thrust for locomotion:

$$F_l = \int_0^{L_0} (\mathbf{f}_N \cdot \mathbf{i} + \mathbf{f}_T \cdot \mathbf{i})\,ds \quad (3.250)$$

Similarly, the power input required for locomotion can be computed from the produce of the force and the local velocity:

$$P_{in} = \int_0^{L_0} \left\{k_N (V - U_0)^2\left[1 - X'^2(s - ct)\right] + k_T\left[(V - U_0)X'(s - ct) - c\right]^2\right\} ds \quad (3.251)$$

Equations 3.250 and 3.251 can be used to define a swimming efficiency as in the previous two sections. Swimming at low Reynolds numbers is generally inefficient due to the greater contribution of viscous dissipation as an energy sink in the flow. In addition, animals at low Reynolds numbers cannot benefit from inertial mechanisms such as unpowered coasting through the fluid. Nonetheless, the equations derived here are amenable to optimization, and we expect that animals that function at low Reynolds numbers have converged on swimming modes that address the unique challenges of Stokes flow.

3.21 CLASSIC SWIMMING MODELS: GROUP DYNAMICS IN AQUATIC LOCOMOTION

Our focus in this chapter has been primarily on the dynamics of individual animals—even individual appendages—creating single wake vortices. To close the chapter, we consider the interactions among vortices in a wake and among animals is a group. Both examples are inspired by studies published in the primary literature in the 1970s (see references section).

3.21.1 Vortex Interactions

The nature of vortex interactions in an animal wake can be illustrated in the context of a comparison between steady jetting and pulsed jetting of an animal. This mode of locomotion is used to varying degrees by squid, salps, jellyfish, and other marine invertebrates. Let us define a control volume surrounding one of these animals, as

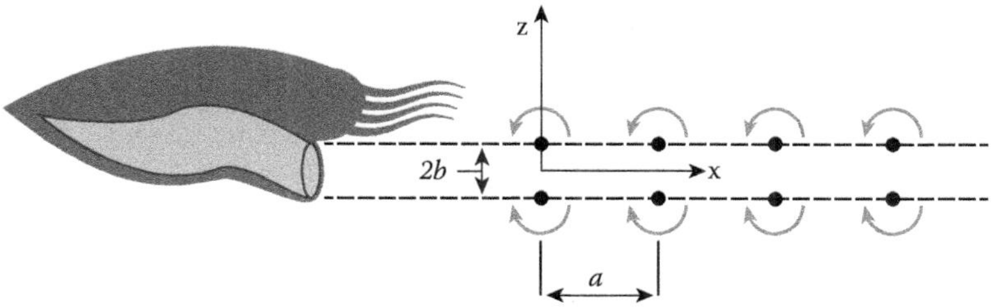

Figure 3.41. Illustration of aquatic animal and vortex ring wake.

shown in Figure 3.41. The animal creates a train of vortex rings in its wake, which propagate downstream and out of the control volume. In a quasi-steady-state approximation, the thrust generated by the animal is equal to the vortex impulse I_v that leaves the control volume per jet cycle:

$$\overline{F_{l,p}} \approx \frac{I_v}{T} \tag{3.252}$$

where the subscript p indicates the pulsed nature of the jet. The period T of each jet cycle can be approximated as the vortex ring spacing a divided by the vortex velocity U_v. The vortex velocity itself consists of a self-induced component U_{si} and component induced by the presence of neighboring vortices U_{ni}.

The self-induced component can be determined from application of the Biot–Savart Equation 3.117:

$$U_{si} = \frac{\Gamma}{4\pi b}\left[\ln\frac{8b}{\varepsilon} - 0.25\right] \tag{3.253}$$

where Γ is the vortex ring circulation, b is the vortex ring diameter, and ε is the vortex core thickness (i.e., diameter of the vortex tube containing vorticity), where it is assumed that $\varepsilon \ll b$. Note that the subsequent analysis is valid for thick rings as well, although the self-induced velocity is modified from Equation 3.253.

To determine the contribution to vortex velocity from neighboring vortices in the wake, let us assume that the wake consists of an infinite row of equally spaced vortex rings. In this case, the neighbor-induced vortex velocity is

$$U_{ni} = \frac{\Gamma}{2\pi b}\sum_{n=1}^{\infty} D_n\left[F(D_n) - E(D_n)\right] \tag{3.254}$$

where $D_n = (1 + na/(2b))^{-1/2}$. The functions F and E are elliptic integrals of the first and second kind, respectively. The reader is encouraged to consult the primary literature (Levy and Forsdyke 1927) for details of the mathematical derivation. Given the vortex velocity as the sum of Equations 3.253 and 3.254, the locomotive force generated by pulsed jetting can be estimated as

$$\overline{F_{l,p}} \approx nU_v m(1+\alpha_x) \tag{3.255}$$

where n is the number of vortex rings that leave the control volume per unit time, m is the mass of each vortex ring, and α_x is the added-mass coefficient of the vortex ring (see Section 3.12). In comparison, the thrust produced by steady jetting can be approximated as

$$\overline{F_{l,s}} \approx nmU_s \tag{3.256}$$

where U_s is the velocity of the steady jet flow. To accommodate an objective comparison between pulsed and steady jet flow, the mass flux of the steady is set equal to nm, the mass flux of the pulsed jet. The ratio of pulsed-to-steady jet thrust is therefore

$$\frac{\overline{F_{l,p}}}{\overline{F_{l,s}}} \approx \frac{U_v}{U_s}(1+\alpha_x) \tag{3.257}$$

or, replacing the vortex velocity with its components

$$\frac{\overline{F_{l,p}}}{\overline{F_{l,s}}} \approx \frac{U_{si}}{U_s}\left(1+\frac{U_{ni}}{U_{si}}\right)(1+\alpha_x) \tag{3.258}$$

Equation 3.258 shows that even without the presence of neighboring vortices (i.e., $U_m = 0$), a benefit for locomotive thrust exists in the pulsed jet if $U_{si} > U_s$ or if the added-mass coefficient α_x is large enough. It can be shown numerically that the ratio U_{si}/U_s varies from 0.439 for a thin vortex ring to 1.301 for Hill's spherical vortex.

Furthermore, the added-mass coefficient of the vortex ring ranges from 0.5 for Hill's spherical vortex to 1.0 in the thin ring limit. Therefore, in the absence of neighboring vortices, the force ratio in Equation 3.258 has the following limits:

$$0.878 < \frac{\overline{F_{l,p}}}{F_{l,s}} < 1.952 \qquad (3.259)$$

The model predicts that a train of isolated, thin vortex rings produces less thrust than an equivalent steady jet; however, a train of thick vortex rings produces nearly twice as much thrust.

Given this result, it seems that an optimal scenario for pulsed jetting would be the formation of one large vortex ring in the wake. Unfortunately, this is not possible because maintaining a stable vortex ring requires a minimum threshold of energy input. When this threshold is not met, the vortex ring stops forming and smaller vortices form at the lee side of the vortex ring. The time t after jet initiation at which vortex ring growth is limited typically occurs when the dimensionless ratio $U_j t/D_j$ approaches 4, where U_j is the jet speed and D_j is the jet diameter. The implications of this limit for aquatic locomotion are not yet fully understood.

Returning to the neighbor-induced vortex motion, Figure 3.42a plots the ratio of neighbor-induced vortex speed to self-induced vortex speed for a range of intervortex spacing. As expected, the effect of neighbor-induced motion becomes pronounced when the vortex rings are in proximity. Consequently, the thrust of pulsed jetting can be further enhanced by creating closely spaced vortex rings (Figure 3.42b). In practice, we must also concern ourselves with the stability of the train of vortex rings. Perturbation of any of the vortex rings off the axis of symmetry can destabilize the entire wake.

3.21.2 Animal Interactions

Aquatic organisms often find themselves in groups. The vertical migrations of zooplankton and the long-distance migrations of bony fishes are just two common examples. The arrangement of animals in the group is determined by a variety of behavioral factors. Hydrodynamic interactions have been hypothesized as an additional consideration, because they might provide propulsive benefits for the group of swimming organisms beyond what can be achieved by swimming in isolation. For example, the vortices shed by each animal induce a local velocity field that could potentially benefit adjacent animals.

Let us consider an inviscid model of this interaction to explore the possibility that animal interactions within a group can benefit individual animal locomotion. Figure 3.43 illustrates a two-dimensional model of a school of fish. Our goal is to determine an optimal angle θ between the adjacent rows of fish to maximize any enhancements in swimming thrust from the nearby wake vortices.

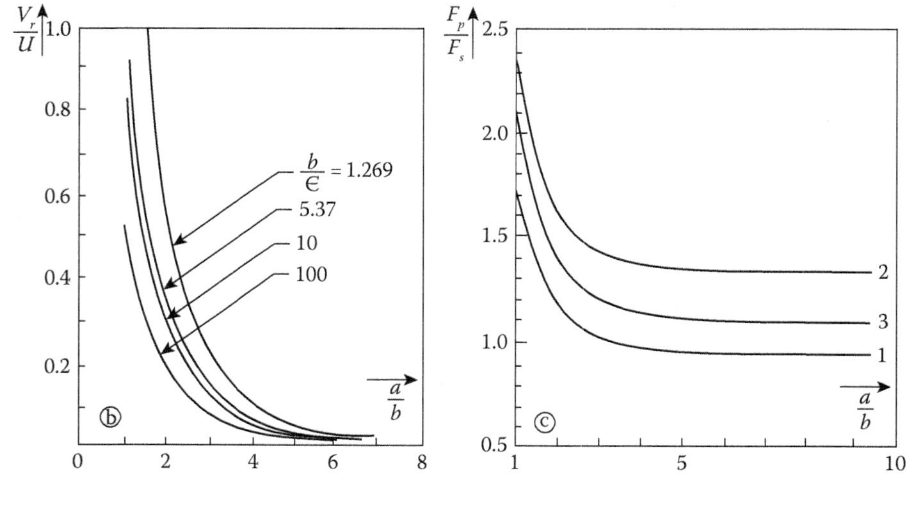

Figure 3.42. (a) Enhancement of wake vortex velocity due to motion induced by neighboring vortex rings. (b) Ratio of pulsed jet force to steady jet force. (Based on Weihs, D., *Fortschritte der Zoologie*, 24, 171–175, 1977.)

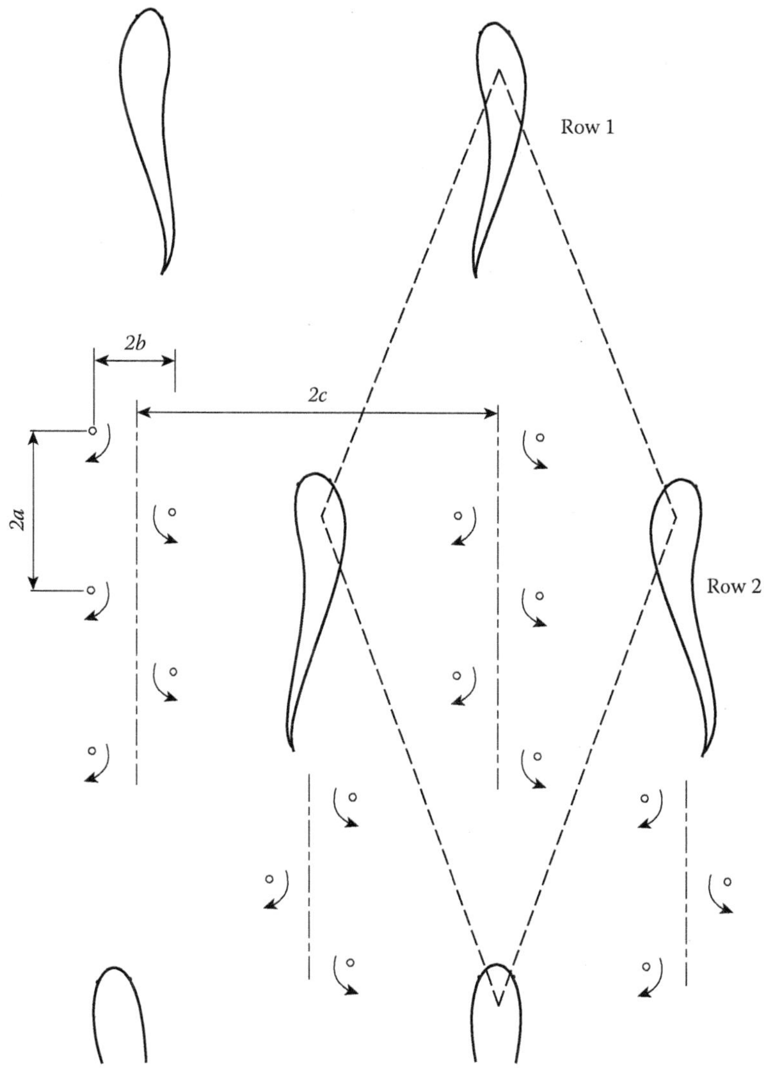

Figure 3.43. Geometric model of fish school and wake vortices. (Based on Weihs, D., 1975, Some hydrodynamical aspects of fish schooling, *Swimming and Flying in Nature*, T. Wu, C. Brokaw, and C. Brennen, eds., vol. 2, pp. 703–718, Plenum, New York.)

Placing the vortex wakes in a complex plane where $z = x + iy$ (see Figure 3.43), counterclockwise-rotating vortices are located at positions

$$A^{1+}_{m,n} = -a/2 + 2na + i(4mc + b) \quad (3.260)$$

$$A^{2+}_{m,n} = a/2 + 2na + i((4m + 2)c + b) \quad (3.261)$$

and clockwise-rotating vortices are located at positions

$$A^{1-}_{m,n} = a/2 + 2na + i(4mc - b) \quad (3.262)$$

$$A^{2-}_{m,n} = -a/2 + 2na + i((4m + 2)c - b) \quad (3.263)$$

where m is the set of integers from $-\infty$ to ∞ and n is the set of integers from 0 to ∞. The complex potential of each vortex is

$$W = -i\frac{\Gamma}{2\pi}\ln(z - z_v) \quad (3.264)$$

where Γ is the vortex circulation, z_v is the location of each vortex, and the velocity field induced by each vortex is

$$dW/dz = u - iv \quad (3.265)$$

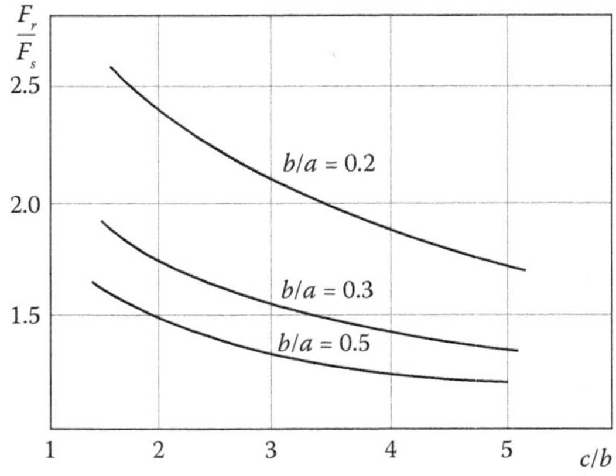

Figure 3.44. Ratio of schooling fish thrust to isolated fish thrust for varying geometric parameters. (Based on Weihs, D., 1975, Some hydrodynamical aspects of fish schooling, *Swimming and Flying in Nature*, T. Wu, C. Brokaw, and C. Brennen, eds., vol. 2, pp. 703–718, Plenum, New York.)

The complex potential of the entire array of vortices can be calculated by adding each of the individual vortex potentials. Thus,

$$W_{\text{total}} = -i\frac{\Gamma}{2\pi}\sum_{m=-\infty}^{\infty}\sum_{n=0}^{\infty}\Big[\ln\left(z - A_{m,n}^{1+}\right) + \ln\left(z - A_{m,n}^{2+}\right) - \ln\left(z - A_{m,n}^{1-}\right) - \ln\left(z - A_{m,n}^{2-}\right)\Big] \quad (3.266)$$

or, equivalently

$$W_{\text{total}} = i\frac{\Gamma}{2\pi}\sum_{n=0}^{\infty}\ln\frac{\sinh\frac{\pi}{4c}(z+\alpha-2na)\sinh\frac{\pi}{4c}(z+2ic-\alpha-2na)}{\sinh\frac{\pi}{4c}(z-\alpha-2na)\sinh\frac{\pi}{4c}(z+2ic+\alpha-2na)} \quad (3.267)$$

where $\alpha = (a/2) - ib$, and sinh is the hyperbolic sine function.

Optimal arrangements of the animals are those that maximize the component of induced fluid velocity in the swimming direction. The complexity of Equation 3.267 makes it necessary to proceed using numerical methods of calculation. Figure 3.44 illustrates the results. The thrust generated by a fish in the school can be more than double that of an isolated animal. As expected, this effect becomes more pronounced as the animals become more closely spaced. For typical observed values of fish spacing and length L in a school, for example, $a/b = 10/3$ and $b/L = 0.3$, the optimal configuration angle is approximately 30°. This prediction is consistent with the sparse measurements available on fish schools. Of course, this agreement does not prove a hydrodynamic benefit of fish schooling; however, this simplified model does indicate the feasibility of animal interactions improving aquatic locomotion.

REFERENCES

Bada, J.L., and A. Lazcano. 2009. The origin of life. In *Evolution: The First Four Billion Years* (M. Ruse and J. Travis, eds.), pp. 49–79. Harvard University Press, Cambridge, MA.

Bartol, I.K., M. Gharib, P.W. Webb, D. Weihs, and M.S. Gordon. 2005. Body-induced vortical flows: A common mechanism for self-corrective trimming control in boxfishes. *J Exp Biol* 208: 327–344.

Bartol, I.K., P.S. Krueger, J.T. Thompson, and W.J. Stewart. 2008. Swimming dynamics and propulsive efficiency of squids throughout ontogeny. *Integr Comp Biol* 48: 720–733.

Calcott, B., and K. Sterelny (eds.). 2011. *The Major Transitions in Evolution Revisited*. MIT Press, Cambridge, MA, 329 pp.

Chang, Y., and J. Yen. 2012. Swimming in the intermediate Reynolds range: Kinematics of the pteropod *Limacina helicina*. *Integr Comp Biol* 52: 597–615.

Costello, J.H., S.P. Colin, and J.O. Dabiri. 2008. The medusan morphospace: Phylogenetic constraints, biomechanical solutions and ecological consequences. *Invertebr Biol* 52: 265–290.

Dabiri, J.O. 2005. On the estimation of swimming and flying forces from wake measurements. *J Exp Biol* 208: 3519–3532.

Dabiri, J.O., S.P. Colin, and J.H. Costello. 2007. Morphological diversity of medusan lineages constrained by animal-fluid interactions. *J Exp Biol* 210: 1868–1873.

Denny, M. 1988. *Biology and the Mechanics of the Wave-Swept Environment*. Princeton University Press, Princeton, NJ, 329 pp.

Denny, M. 1993. *Air and Water: The Biology and Physics of Life's Media*. Princeton University Press, Princeton, NJ, 341 pp.

Dial, K.P., N. Shubin, and E.L. Brainerd (eds.). 2015. *Great Transformations in Vertebrate Evolution*. University of Chicago Press, Chicago, IL, 424 pp.

Feitl, K.E., A.H. Millett, S.P. Colin, J.O. Dabiri, and J.H. Costello. 2009. Functional morphology and fluid interactions during early development of scyphomedusae. *Biol Bull* 217: 283–291.

Fish, F., and G.V. Lauder. 2006. Passive and active flow control by swimming fishes and mammals. *Ann Rev Fluid Mech* 38: 193–224.

Fish, F.E., L.E. Howle, and M.M. Murray. 2008. Hydrodynamic flow control in marine mammals. *Integr Comp Biol* 48: 788–800.

Flammang, B.E., G.V. Lauder, D.R. Troolin, and T.E. Strand. 2011. Volumetric imaging of fish locomotion. *Biol Lett* 7: 695–698.

Galper, A., and T. Miloh. 1995. Dynamic equations of motion for a rigid or deformable body in an arbitrary non-uniform potential flow field. *J Fluid Mech* 295: 91–120.

Gordon, M.S., and J.C. Notar. 2015. Can systems biology help to separate evolutionary analogies (convergent homoplasies) from homologies? *Progr Biophys Mol Biol* 117: 19–29. doi: 10.1016/j.pbiomolbio.2015.01.005.

Hedges, S.B., and S. Kumar (eds.). 2009. *The Timetree of Life*. Oxford University Press, Oxford, UK, 572 pp.

Hildebrand, F.B. 1976. *Advanced Calculus for Applications*. Prentice-Hall, Englewood Cliffs, NJ, 733 pp.

Irschick, D., and T. Higham. 2016. *Animal Athletes: An Ecological and Evolutionary Approach*. Oxford University Press, Oxford, UK. 255 pp.

Kaiser, M.J., M.J. Attrill, S. Jennings, D.N. Thomas, D.K.A. Barnes, A.S. Brierley, J.G. Hiddink, H. Kaartokallio, N.V.C. Polunin, and D.G. Raffaelli. 2011. *Marine Ecology: Processes, Systems and Impacts*, 2nd edn. Oxford University Press, Oxford, UK, 501 pp.

Lane, N. 2009. *Life Ascending: The Ten Great Inventions of Evolution*. WW Norton, New York, 344 pp.

Lauder, G.V. 2006. Locomotion. In *The Physiology of Fishes*, 3rd edn. (D.H. Evans and J.B. Claiborne eds.), pp. 3–46. CRC Press, Boca Raton, FL.

Lauder, G.V. 2011. Swimming hydrodynamics: Ten questions and the technical approaches needed to resolve them. *Exp Fluids* 51: 23–35.

Levy, H., and A.G. Forsdyke. 1927. The stability of an infinite system of circular vortices. *Proc Roy Soc A*: 594–604.

Liem, K.F., W.E. Bemis, W.F. Walker, and L. Grande. 2001. *Functional Anatomy of the Vertebrates: An Evolutionary Perspective*, 3rd edn. Harcourt College, Fort Worth, TX.

Lighthill, J. 1975. *Mathematical Biofluiddynamics*. Society for Industrial and Applied Mathematics, 287 pp.

Long, J.H., Jr., M.E. Porter, R.G. Root, and C.W. Liew. 2010. Go reconfigure: How fish change shape as they swim and evolve. *Integr Comp Biol* 50: 1120–1139.

Lu, X.-Y., X.-Z. Yin, and B.-G. Tong. 2008. Studies of hydrodynamics in fishlike swimming propulsion. In *Bio-mechanisms of Swimming and Flying* (N. Kato and S. Kamimura, eds.), pp. 143–154. Springer, Tokyo.

Miller, L.A., D.I. Goldman, T.L. Hedrick, E.D. Tytell, Z.J. Wang, J. Yen, and S. Alben. 2012. Using computational and mechanical models to study animal locomotion. *Integr Comp Biol* 52: 553–575.

Nelson, J.S. 2006. *Fishes of the World*, 4th edn. Wiley, Hoboken, NJ, 601 pp.

Nielsen, C. 2012. *Animal Evolution: Interrelationships of the Living Phyla*, 3rd edn. Oxford University Press, Oxford, 464 pp.

Palstra, A.P., and J.V. Planas (eds.). 2013. *Swimming Physiology of Fish: Towards Using Exercise to Farm a Fit Fish in Sustainable Aquaculture*. Springer-Verlag, Berlin.

Pechenik, J.A. 2010. *Biology of the Invertebrates*, 6th edn. McGraw-Hill Higher Education, Boston, MA, 606 pp.

Pough, F.H., C.M. Janis, and J.B. Heiser. 2013. *Vertebrate Life*, 9th edn. Benjamin Cummings, New York, 720 pp.

Rivera, G. 2008. Ecomorphological variation in shell shape of the freshwater turtle *Pseudemys concinna* inhabiting different aquatic flow regimes. *Integr Comp Biol* 48: 769–787.

Ruiz, L.A., R.W. Whittlesey, and J.O. Dabiri. 2011. Vortex-enhanced propulsion. *J Fluid Mech* 668: 5–32.

Saffman, P.G. 1992. *Vortex Dynamics*. Cambridge University Press, Cambridge, UK, 328 pp.

Seilacher, A. 2007. *Trace Fossil Analysis*. Springer-Verlag, Berlin, 220 pp.

Shadden, S.C., J.O. Dabiri, and J.E. Marsden. 2006. Lagrangian analysis of fluid transport in empirical vortex ring flows. *Phys Fluids* 18(4): 047105.

Shadden, S.C., K. Katija, M. Rosenfeld, J.E. Marsden, and J.O. Dabiri. 2007. Transport and stirring induced by vortex formation. *J Fluid Mech* 593: 315–331.

Shadwick, R.E., and G.V. Lauder (eds.). 2006. *Fish Biomechanics*. Elsevier Academic Press, San Diego, CA, 540 pp.

Syme, D.A. 2006. Functional properties of skeletal muscles. In *Fish Biomechanics* (R.E. Shadwick and G.V. Lauder, eds.), pp. 179–240. Elsevier Academic Press, Amsterdam.

Tytell, E.D., E.M. Standen, and G.V. Lauder. 2008. Escaping flatland: Three-dimensional kinematics and hydrodynamics of median fins in fishes. *J Exp Biol* 211: 187–195.

Tytell, E.D., I. Borazjani, F. Sotiropoulos, T.V. Baker, E.J. Anderson, and G.V. Lauder. 2010. Disentangling the functional roles of morphology and motion in the swimming of fish. *Integr Comp Biol* 50: 1140–1154.

Vogel, S. 1994. *Life in Moving Fluids: The Physical Biology of Flow*, 2nd edn. Princeton University Press, Princeton, NJ.

Vogel, S. 2008. Modes and scaling in aquatic locomotion. *Integr Comp Biol*. 48: 702–712.

Vogel, S. 2013. *Comparative Biomechanics*, 2nd edn. Princeton University Press, Princeton, NJ.

Wainwright, P.C. and S.M. Reilly. 1994. *Ecological Morphology: Integrative Organismal Biology*. University of Chicago Press, Chicago, IL.

Weihs, D. 1975. Some hydrodynamical aspects of fish schooling. In *Swimming and Flying in Nature* (T. Wu, C. Brokaw, and C. Brennen, eds.), vol. 2, pp. 703–718. Plenum, New York.

Weihs, D. 1977. Periodic jet propulsion of aquatic creatures. *Fortschritte der Zoologie* 24: 171–175.

Wu, T., C. Brokaw, and C. Brennen (eds.). 1975. *Swimming and Flying in Nature*. Plenum Press, New York, 1005 pp.

CHAPTER FOUR

Natural Flight

JOHN J. VIDELER

4.1 INTRODUCTION

This chapter deals with flight in nature, summarizing the current knowledge of aerodynamics and biomechanics, biodiversity, functional morphology, flight energetics, and evolution.

Chapters 1 and 2 provide a relevant background with respect to basic concepts and evolutionary biology.

Unlike the ancient basal evolutionary position of animal swimming as described in Chapter 3, aerial flight is highly derived, mainly because gravity is a major factor in the aerodynamics of flight in thin air in contrast with locomotion while submerged in dense water. In contrast, the laws of fluid dynamics governing swimming and flying are principally the same. The discussion of the aerodynamics of natural flight in this chapter largely assumes as background information the discussion of the principles and mathematics of fluid dynamics summarized in Chapter 3.

Lift, gravity, thrust, and drag are the main forces governing flight. A history of the development of knowledge leading to the understanding of flight shows how the basic principles used in nature were discovered.

Nature evolved many solutions to generate lift and thrust forces to keep organisms airborne during gliding or active flapping flight. An overview provides insight into the diversity of the lift-generating mechanisms discovered so far. Many secrets of, in particular, insect flight still await discovery.

Flight biodiversity encompasses both passive and powered flight. Gliding probably preceded flapping flight. Natural selection for effective gliding strengthened morphological and behavioral changes that made powered flight possible.

Powered flight is restricted to animals from only two of 32 phyla: the arthropods (specifically the insects) and the chordate vertebrates.

Many orders of insects include flying forms; indications are that flight arose independently in each of these orders. There have been three independent origins of flight in vertebrate groups: in the Mesozoic reptilian (not dinosaurian) pterosaurs; in the birds, which are derived theropod dinosaurs; and in the mammalian bats. True powered flight is clearly evolutionarily extremely difficult to achieve. The physical, morphological, physiological, and behavioral requirements are severe. Powered flight abilities are among the clearest examples of convergent evolution among animals.

The functional morphology of the flight apparatus of birds is best understood, although only a handful of species have extensively been studied. In birds lift, thrust, and flight control are provided by feathered front legs and an array of tail feathers. Wings consist of three functionally different parts: the arm section close to the body, the hand wing, and the alula or bastard wing.

Bats and pterosaurs fly with skin flaps mainly attached to the skeleton of the front legs. Despite this resemblance, details of the wing structures show that skinned wings were reinvented separately in each group.

Wings of insects come in one or two pairs, and these are complex skin folds and not modified legs. Structure and function of muscles and wings as well as the kinematics and lift and thrust-generating techniques are extremely diverse among the flying orders.

The energetic cost of flight is an important evolutionary selection criterion. Powered flight is energetically extremely demanding. Actual measurements of the metabolic costs of animals in flight, accumulated over the past decennia, show not only how much energy is required but also general trends with respect to the weight carried.

The last section of this chapter summarizes and discusses how evolution by natural selection achieved powered flight in the main flying groups of animals. The summary of the current knowledge given in this chapter does not mean that the evolution of natural flight has reached a final stage. Natural selection is still ongoing, and we may expect even better designs in the future; however, we are not able to make good use of these because of the time it will take. Instead, we can use the current knowledge to find engineering solutions.

Alexander (2015) provides a popular treatment of many of these topics.

4.2 HISTORY OF ANIMAL FLIGHT RESEARCH

4.2.1 Early Observations and Ideas

When did people start to have an interest in animal flight? Knowledge of the behavior and especially of the locomotion performance of prey animals would have increased the chance of survival of our hunter-gatherer ancestors.

Prehistoric art occasionally shows depictions of birds, whereas representations of insects are rare and those of bats nonexistent (Dale Guthrie 2005). Half-open caves in the Sierra de San Francisco, a deep canyon in the central desert of Baja California, Mexico, contain prime examples in wall paintings that are more than 100 centuries old. Figure 4.1 shows impressions of gliding birds of prey (possibly condors) with wings and tails spread, and Figure 4.2 shows a picture of a cormorant preparing for landing. The details of the latter drawing suggest that the artist observed the behavior closely and may have interpreted it using some knowledge of the functional anatomy of the bird. It is unlikely to be a coincidence that there is an emphasis on the large swept-back hand wings. These mainly consist of the large primary feathers attached to the hand skeleton. Hand wings are important in generating lift and drag forces at low speed (Videler 2006). The arm wings, where the secondary flight feathers are connected to the skeleton of the upper and lower arm, are very short and kept close to the body. It is the oldest known drawing where the bastard wing or alula is depicted. An alula consists of a few flight feathers in connection with the first digit of the hand skeleton. The alula and the sharp leading edge hand wings generate leading edge vortices (LEVs) to generate lift and drag at decreasing speeds during landing, as described in Section 4.3. The artist also seems to have appreciated that cormorants use their spread-out feet as airbrakes.

About 50 centuries ago, written information in the Western world started to emerge during the ancient Assyrian–Babylonian, Egyptian, and the Minoan civilizations. The emphasis was on religious, medical, and astronomical topics, but also included the emergence of sophisticated mathematical principles as well as information about food. Birds were often depicted showing feather arrangements of the

Figure 4.1. Prehistoric paintings of gliding birds in the central desert of Baja California, Mexico.

Figure 4.2. Prehistoric picture of a landing cormorant.

wings in great detail, clearly distinguishing hand wings with large primary feathers from the arm part. Alulae were often faithfully depicted. Ancient Egyptian culture considered bats as a type of birds. Fruit-eating bats were distinguished from insectivores (in ancient Egyptian language Zedjakhem and Daqi, respectively). Amulets with bat remains were used, presumably to ward off evil spirits. Cases containing both a mummified owl and a bat have been found in tombs, presumably to give the deceased the ability to fly in the dark (Houlihan 1997).

The scarab beetle often with extra bird-like wings is the most prominently depicted insect during the time of the pharaohs; but butterflies, dragonflies, house flies, locusts, and the praying mantis also occur. However, despite detailed interest in flying animals, nothing points at knowledge of principles of flight.

The Minoan culture, contemporary with Ancient Egypt, was advanced in navigational skills, but there is no written account of the state of scientific knowledge on flight. From the story of Daedalus and Icarus told later by the Greeks, we can extrapolate that there was interest in the art of flight of birds. The enigmatic Phaistos disk from Crete, second millennium BC, contains pictures of bees, pigeons, and a small falcon (possibly a kestrel) in flight.

In Europe, written history starts with Greek civilization. About 24 centuries ago, Aristotle summarized the then existing knowledge and added to it using observations and a philosophical approach (Barnes 1991).

4.2.2 Ancient Natural History

Aristotle's ideas were accepted as truth for more than 20 centuries. He proposed that natural life is based on a hierarchy of three souls. Human beings possess a rational soul (psyché) that makes it possible to grow, to reproduce, to move, and to think. The animal soul lacks the capacity to think, whereas the soul of a plant only allows for growth and reproduction. The lowest level in the hierarchy is matter without a soul; above that in increasing order are plants, animals, and people. He considered the human male to be at the highest possible level and by nature superior to the female as, contrary to her, he is warm and active. Women were cold and born in northerly winds. Mammals other than man were considered below man, then egg-producing animals, and crustaceans below that. In Parts of Animals, book IV, Aristotle says,

> In birds the arms or forelegs are replaced by a pair of wings, and this is their distinctive character. For it is part of the substance of a bird that it shall be able to fly; and it is by the extension of wings that this is made possible. Moreover, birds cannot as a fact fly if their legs be removed, nor walk without their wings.

He described how he thinks that wings enable the bird to fly in the second chapter of his book on the Movement of Animals. He tries to match observations with theory:

> For just as there must be something immovable within the animal, if it is to be moved, so even more must there be without it something immovable, by supporting itself upon which that which is moved moves. For were that something always to give way (as it does for tortoises walking on mud or persons walking in sand) advance would be impossible, and neither would there be any walking unless the ground were to remain still, nor any flying or swimming were not the air and the sea to resist.

This notion of resistance or drag approaches Newtonian mechanics very closely. However, in the next statement Aristotle ruins it all by returning to his philosophical concept on the mover and the moved:

> And this which resists must be different from what is moved, the whole of it from the whole of that, and what is thus immovable must be no part of what is moved: otherwise there will be no movement.

Evidence for this statement comes from the fact that

> one only can move a boat by pushing it while standing on the shore, not by pushing the mast while standing on the ship.

Aristotle's opinion about the function of the tail,

> In winged creatures the tail serves like a ships rudder, to keep the flying thing in its course,

was accepted as the truth for many ages. Although not the right explanation, it indicates that Aristotle recognized the problem of dynamic stability inherent in moving in a fluid medium.

His description of the function of the crest on the breastbone in birds and of the large pectoral muscles is highly confusing:

> The breast in all birds is sharp-edged, and fleshy. The sharp edge is to minister to flight, for broad surfaces move with considerable difficulty, owing to the large quantity of air which they have to displace; while the fleshy character acts as a protection, for the breast, owing to its form, would be weak, were it not amply covered.

Does the sharp edge refer to the carina on the sternum?

Aristotle closely approached the concepts of inertia and momentum, but he looked for the cause of continuation of movement outside the moving object. Two centuries later the Greek Hipparchus describes that the thrown object itself possesses the throwing force. However, Johannes Philoponus from Alexandria reasoned in sixth century AD, that the power of movement is transferred to the object on its release and momentum keeps it going in the same direction (Sambursky 1987). The knowledge of acceleration required for a full explanation only started to emerge another 1000 years later.

4.2.3 Emerging Concepts

It took until around 1500 AD for new thoughts about flight to emerge. Leonardo da Vinci (1452–1519) kept a notebook about flight of birds (and some bats): *Sul volo degli Uccelli*. It disappeared soon after his death and was lost for centuries. It was rediscovered at the turn of the nineteenth century, and the Old Italian text written in mirror image was transcribed and translated. [Marinoni's (1976) transcription is used here; quotations are translated from modern Italian into English]. Other codices he left behind contain loose notes about the flight of birds. Da Vinci really was ahead of his time, but his brilliant scientific ideas could not be used by succeeding generations because he did not publish his findings accessibly. Leonardo wrote in Italian as he lacked the ability to write in Latin, the lingua franca of science in those days.

Leonardo da Vinci was interested in the interaction between birds and air because he dreamt about man-powered flight. In *Sul volo degli Uccelli* he wrote:

> From the mountain which bears the name of the big bird the famous bird will fly and fill the world with its great fame" and "For the first time the big bird will fly from the back of the mighty swan, filling the whole world with stupor and all writings with his fame, bringing eternal glory to the nest where it was borne.

(The mountain indicated here is thought to be the Monte Ceceri near Fiesolo close to Florence.) Several sketches in his notebook show designs for essential parts of artificial wings (Figure 4.3). The structures are not based on the anatomy of

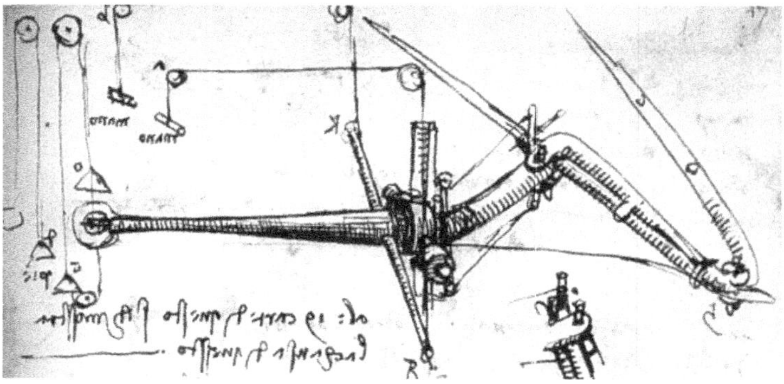

Figure 4.3. Artificial wing design by Leonardo da Vinci.

Figure 4.4. Airflow and forces on a flying bird according to Leonardo da Vinci.

bird wings; they are inspired by his interpretation of the forces on the wings of birds in flight. Most drawings of birds in the notebook show tracings of the presumed air movements and indications of the forces involved (Figure 4.4). Leonardo must have had a very quick eye; his sketches resemble tracings from high-speed film pictures of birds in flight. He translated his observations into laws governing the flight of birds, these translations sound like instructions for birds. For example, on page 6 verso of *Sul volo degli Uccelli*, he wrote:

> If the wing tip is beaten by the wind from below, the bird could be overturned if it does not use one of two remedies. It must either immediately lower the beaten wing tip under the wind or it should beat the distal half of the opposite wing down.

He was intrigued by flow in water and air and made sketches of flow patterns with photographic precision. He observed that rapids occur in shallow or narrow parts of a river and that flow in wider, deeper parts is slower. His conclusion was that the product of the cross-sectional area and the flow velocity had to be constant to keep the passing mass of water per unit time constant. He formulated the continuity equation for an incompressible fluid 230 years before Daniel Bernoulli put it on paper. Leonardo da Vinci rediscovered after Aristotle that air induces resistance, and he initially reached the false conclusion that the wings in flight compress the air to generate lift forces. (We now know that to compress air in open space, velocities approaching the speed of sound are required.) Six years before his death he seems to have changed his mind when he wrote in Codex E:

> What quality of air surrounds birds in flying? The air surrounding birds is above thinner than the usual thinness of the other air, as below it is thicker than the same, and it is thinner behind than above in proportion to the velocity of the bird in its motion forwards, in comparison with the motion of its wings towards the ground; and in the same way the thickness of the air is thicker in front of the bird than below, in proportion to the said thinness of the two said airs.

Anderson (1997) indicates that if the words "thinness" and "thickness" are replaced by "pressure" and "thinner" and "thicker" by "lower and higher pressure," we obtain a clear explanation of the pressure distribution around a conventional wing. That implies that Leonardo understood the causes of lift and pressure drag forces on a wing. He also formulated the idea, still applied in water and wind tunnels today that it makes no difference aerodynamically whether flow passes a stationary body or a body moves through a stationary flow. In the Codex Atlanticus, he wrote:

> As it is to move the object against the motionless air so it is to move the air against the motionless object" and "the same force as is made by the thing against air, is made by air against the thing.

He believed this force was proportional to the surface area and the velocity of the body. As becomes clear later on, he was right about the surface area, but not about velocity. In flying and swimming animals and aircraft, drag forces are proportional to the square of the velocity. Da Vinci also realized that not only the surface area but also the shape of the body determines resistance to fluid flow, as demonstrated by various drawings of streamlined bodies and projectiles based on the shapes of fish. Da Vinci realized that in flying birds the center of gravity does not coincide with the center of lift. He was also convinced that bird wings consist of various parts, each with its own function.

Falling bodies and the laws of gravity were certainly within the broad range of Leonardo da Vinci's interests. However, it was Galileo Galilei (1564–1642) who first used experimental results to calculate the acceleration of falling bodies. He falsified Aristotle's idea that a heavy spherical body uses less time to fall from a given height than a lighter spherical body of the same size. He did find that the resistance of objects varying in shape causes differences in falling periods. Galileo discovered that aerodynamic resistance is proportional to the density of the air, but he considered it small enough to ignore.

Giovanni Alfonso Borelli (1608–1679) was a South Italian mathematician who is often regarded as the founder of biomechanics. Toward the end of his life, he wrote his major opus: *De motu animalium*. The Vatican withheld approval for the printing of the book that was released 1 year after his death (Borelli 1680; translated by Maquet 1989). It is structured around propositions: Propositions 182 to 204 of chapter 22 discuss flight. Borelli describes the structure and function of the flight apparatus of birds as follows:

> The wings have a stiff skeleton on the front side and are covered by flexible wind-tight feathers. The body has heavy pectoral muscles, in strength comparable with the heart muscle, but it is otherwise lightly built containing hollow bones and air sacs and it is covered with light feathers. The ribs, shoulders and wings contain little flesh. The muscles of the hind legs are weakly developed. The pectoral muscles are four times as strong. Birds fly by beating the air with their wings. They jump as it were through the air just as a person can jump on the ground. The wings compress the air making it to react to the wing beat as solid ground reacts to the push off of feet. The air offers resistance because it does not want to be displaced and mixed with stationary air. The air particles rub against each other and that causes resistance. Apart from that is air elastic. Wing beats compress the air and the air bounces back. The jumps occur during the down strokes. During the upstrokes, the wings move with the stiff leading edge forward, followed by the flexible feathers. They then do not meet resistance similar to a sword that moves with the sharp edge forward.

Borelli developed the "plum stone theory" to answer the question how the wing movement can generate propulsive forces. The explanation is illustrated in frames 2 and 3 of his Table 13 (Figure 4.5).

> The stiff leading edges of the wings with the flexible planes of feathers behind them form a wedge that is driven through the air during the down stroke. The oblique trailing edges of the wedge formed by the flexible feathers push the air backward and the bird forward. The action can be compared with shooting a slippery plum stone from between the thumb and the index finger. The stone shoots off in a direction perpendicular to the direction of compression due to its wedge shape.

Aristotle's idea that the tail of a bird functioned like a ship's rudder still had many followers, but Borelli thought the function of the tail was to steer the bird upward and downward rather than laterally. An experiment with a model in water proves his hypothesis (table 13, pictures 4 and 5) (Figure 4.5):

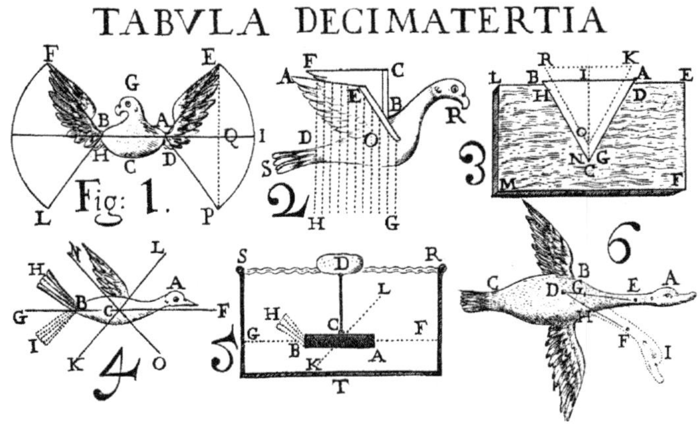

Figure 4.5. Bird flight principles depicted by Borelli (1680).

To function as the rudder of a boat, the tail would have to be implanted vertically. Birds change their horizontal direction by beating the left and right wing at different speeds. The action can be compared with the way a rower alters coarse by pulling harder on one oar than on the other. Birds with long necks can use these to steer up- and downwards but not to the left or the right. Moving the head to the left or the right would change the position of the centre of gravity to a position next to the flight direction and that would cause serious unbalance.

Borelli comments that such behavior would be useless and stupid, unworthy of the cleverness of nature (*indigna naturae solertia*).

Borelli also paid attention to landing.

The impetus must disappear during landing otherwise accidents will happen. Birds can avoid these accidents in different ways. Wings and tail can be spread and kept perpendicular to the flight direction. Just prior to landing, the wings can beat actively against the flight direction and the bending legs may absorb the remaining bit of impetus.

Borelli's fascinating ideas regrettably lacked some fundamental insights in fluid mechanical principles. It is a tragedy that these very principles were discovered during his lifetime, but he was unaware of them or did not realize they applied.

4.2.4 The Discovery of Aerodynamic Principles

Air is a gas mixture, at sea level roughly consisting of 78% nitrogen, 21% oxygen, and 1% traces of other gases including carbon dioxide (Fahy 2009). It is matter consisting of freely moving, nonpacked molecules that can fill any space evenly and completely and deform under the slightest pressure. The three laws of motion put forth by Newton (1642–1727) describe the behavior of flying objects in air reasonably well. The first law—the law of inertia—states that an object stays at rest or in uniform motion along a straight path as long as it is not subjected to external forces. Newton's second law quantifies external force as equal to the mass of the object times the acceleration induced. The third law states that for every action, there is an equal and opposite reaction.

Christiaan Huygens (1629–1695) discovered (while inventing the pendulum clock) that the drag of a moving object in air is not proportional to its speed but to its speed squared. Huygens also started to consider conservation of energy during collisions. But it was his pupil Gottfried Leibniz (1646–1716) who concluded that energy is conserved by transfer from one body to the other and from one form of energy to another. The pendulum of a clock continuously exchanges between gravitational potential energy (equal to the product of its height above the ground [h], its mass [m] and the gravitational acceleration [g]) and kinetic energy. Leibnitz called the latter form of energy vis viva and believed that it equals the mass of the pendulum times the velocity squared. Only near the end of the nineteenth century did Lord Kelvin introduce the terms potential energy ($E_p = mhg$) and kinetic energy. Kinetic energy is the amount of work a moving body does when it is stopped, which equals half of Leibnitz's vis viva ($E_k = \frac{1}{2}mv^2$). Why half and why velocity squared? The force (F) required to decelerate mass m traveling at speed v to a halt is according to Newton's second law mass times the deceleration. The average speed between v and 0 is ½v. The distance d covered during deceleration equals ½vt. Where t, the time it takes to decelerate from v to 0, is the starting velocity divided by the average deceleration (va^{-1}). The kinetic energy (E_k) involved is force times distance, so

$$\begin{aligned} E_k &= Fd = m\,a\,d = m\,a\,\tfrac{1}{2}v\,t \\ &= m\,a\,\tfrac{1}{2}v\,v\,a^{-1} = \tfrac{1}{2}mv^2 \end{aligned} \tag{4.1}$$

Note that the outcome is independent of the deceleration, the distance covered, and the time taken.

The molecules in gases and liquids can move about freely; both substances are fluids and obey similar physical rules. Early on, rules were developed for imaginary ideal nonviscous fluids. This inviscid assumption is more realistic for gases than for liquids, but without the viscosity of air flight would be impossible.

Daniel Bernoulli (1738) found that the pressures in an ideal fluid are inversely related to the speed of flow. His underlying assumptions were that the density of the fluid must be constant, the viscosity negligibly small, and the flow laminar. In flowing fluid, energy is conserved through exchange of dynamic and static pressure. The dynamic pressure is the kinetic energy per unit volume, $\frac{1}{2}\rho v^2$, with the density of the fluid ρ instead of the mass m. (v is the speed of the flow; ρ is mass per unit volume.)

The static pressure is the total of the ambient pressure plus the pressure due to the elevation of the fluid h. This potential pressure is the potential energy per unit volume and analogously the equation is ρhg instead of mhg. Bernoulli discovered that the sum of static and dynamic pressures, P, is constant under laminar flow conditions:

$$\tfrac{1}{2}\rho v^2 + \rho hg = P = \text{constant} \qquad (4.2)$$

This law turned out to be robust and works well as long as the density of the fluid is constant even if laminar flow conditions are not met and viscosity is not negligible. The explanation of animal flight is partly based on Bernoulli's law.

Leonhard Euler (1707–1783) developed differential equations relating pressures and velocities in a fluid in three dimensions based on Newton's second law, the law of conservation of mass, and on the principle of continuity. Euler's model considers an arbitrary small cubic volume within a fluid. The ribs of the cube run orthogonally in x-, y-, and z-directions. The fluid moves and passes through the cube in any direction. Components of movement can be resolved in each of the three directions x, y, and z. The velocity of the mass of fluid is allowed to change while passing through the imaginary cube. This implies that there are velocity gradients in the three orthogonal directions. By assuming that compression effects are negligibly small, Euler was able to show that the sum of the velocity changes in each direction was zero, satisfying the requirement that the volume and mass of the fluid in the cube had to be constant. Under the conditions indicated, Euler's equations can describe fluid flow quantitatively. The equations can also predict the forces on an object in flow. However, these predictions are not very realistic because viscous forces close to the object are neglected. Jean Claude Saint-Venant (1797–1886), in a 1843 paper based on earlier work of Claude-Louis Navier (1785–1836), added the effect of viscosity to the Euler equations, grossly increasing their complexity. In 1845, George Gabriel Stokes (1819–1903) independently derived and published the same equations. The Navier–Stokes equations are still widely used to model flow phenomena.

Halfway through the nineteenth century, the basic physical principles determining the interaction between moving objects in air and water were understood, but the application to the understanding of flight required more empirical approaches.

4.2.5 Understanding Flight

George Cayley (1773–1857) carefully observed flight in nature because he wanted to copy it (Gibbs-Smith 1962). His are the first quantitative kinematic data regarding the flapping flight of a bird (although it is not clear how the measurements were done). He recorded that a rook flew at 34.5 ft s^{-1} and covered 12.9 ft during one wing beat cycle. He estimated the vertical wing excursion at 0.75 ft and the vertical speed of the wing at 4 ft s^{-1}. He discovered that an oblique airflow on feathers and wings generates lift force that varies with the square of the relative airspeed multiplied by the density. Sketches in his notebook show the longitudinal and chord-wise profiles of a heron wing. The cross section shows the shape of a conventional cambered wing with a rounded leading edge (Figure 4.6). In *On Aerial Navigation*, part III (1810) Cayley describes "solids of least resistance" using the body shapes of trout, dolphin, and the Eurasian woodcock as examples. These resembled ideal streamlined bodies with a round front part, a pointed end, and the largest width at about one-third of the length from the front. Such bodies, with a width-over-length ratio close to one quarter, provide the lowest resistance for the largest volume, but it is not clear that Cayley realized this. Cayley designed and built the first manned airplanes (Figure 4.7). At least two were tested. A young groom flew the first flapping machine. It crashed and he got hurt after flying over a short distance. Cayley is said to have blamed the boy for the failure because he did not keep the wings flapping fast enough, was too fat, and got frightened. Cayley's coachman crashed in

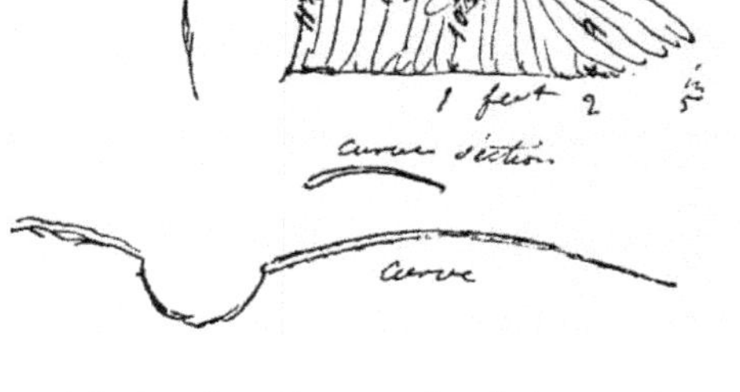

Figure 4.6. Aerodynamic properties of a heron wing drawn by Sir George Cayley.

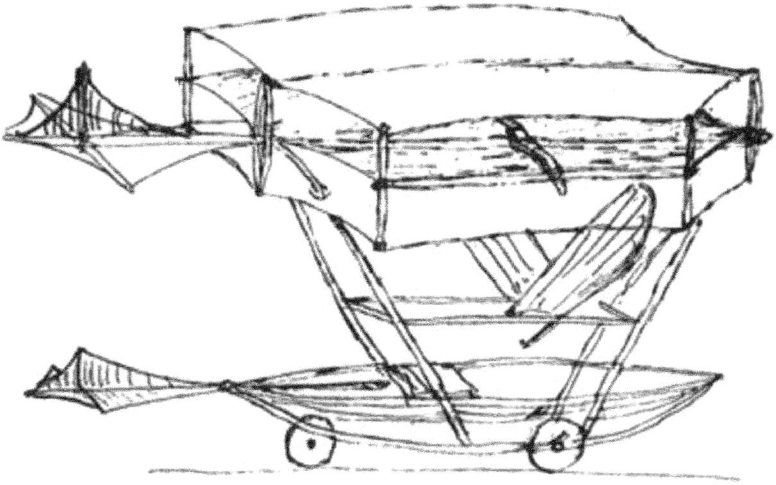

Figure 4.7. Cayley's flying machine with flapping wings and streamlined fuselage.

the second attempt in an improved version in 1853. He is said to have complained that "he was hired to drive and not to fly." In 1846, Cayley predicted,

> A hundred necks have to be broken before all the sources of accident can be ascertained and guarded against.

Unlike the next student of flight, he made sure that it was not his own neck.

Otto Lilienthal (1848–1896) published his book *Der Vogelflug als Grundlage der Fliegekunst (Birdflight as the Basis for the Art of Flight)* (Berlin) in 1889. Accurate analysis of the external anatomy of bird wings inspired his experimental work with model wings. He discovered that cambered wings with a rounded leading edge produce upward-directed force when moved through the air under a small angle of incidence. He designed a carousel force balance to measure lift and drag on model wings. Lilienthal focused his attention on the arm part of large birds (the white stork (Figure 4.8) was his main object of study) because lift forces are mainly generated there. The hand wing produces thrust during flapping flight and is lightly built to avoid large moments of inertia about the axis of rotation in the shoulder joint. Lilienthal succeeded in that he built and flew the first airplanes, but he was fatally injured in a crash with one of his gliders at the age of 48. The brothers Orville and Wilbur Wright in the United States (who helped Lilienthal's widow financially) continued along the same path and managed to master the problems with flight control. The start of the era of aviation is marked by their first flights in an engine-powered plane on December 14, 1903.

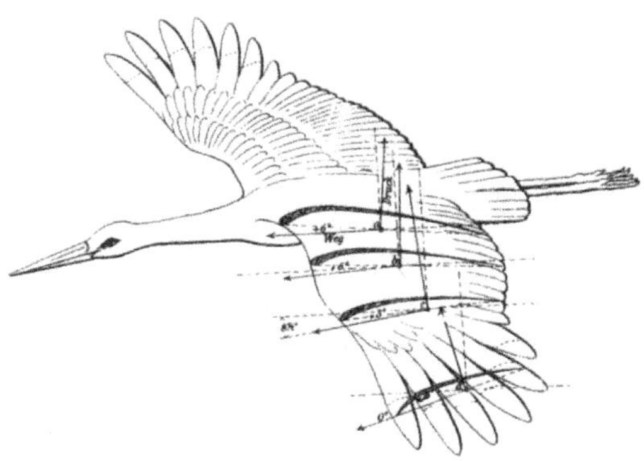

Figure 4.8. A white stork with cross-sectional profiles. (From Lilienthal, O., *Der Vogelflug als Grundlage der Fliegekunst*, 3rd edn., Oldenbourg, München, 1889.)

Etienne-Jules Marey (1830–1904) was a French physiologist who designed experiments to study the activity pattern of flight muscles and wing beat parameters of flying birds. His major opus is *Le Vol des Oiseaux*. The first three-dimensional, high-speed films of flying birds and detailed analysis of the kinematics of the wing beats and measurements of lift forces during flapping movements are among his important contributions. He measured pressure changes and drag with ingenious experimental equipment, with numerous manometers (Figure 4.9). The results supported the direct relationship between the drag and the surface area of an object, but they also revealed that an additional

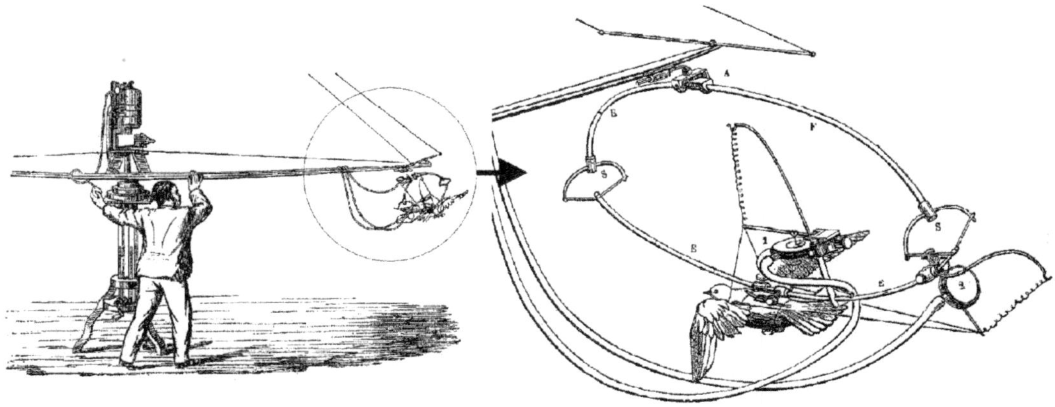

Figure 4.9. Marey's (1890) experimental setup to measure flight characteristics of birds.

coefficient is required. This drag coefficient, defined as the ratio of the drag of an object and that of a flat plate of 1 m^2 moving at 1 m s^{-1}, has to be empirically determined.

Understanding flight requires insight in problems of scaling. The laws of similitude discovered by Reynolds (1842–1912) provide the basic understanding to treat scaling properly. In fluid flow, four parameters determine the nature of the flow: the length dimension in the direction of the flow, the flow velocity, the density of the fluid, and the viscosity of the fluid. The behavior of the fluid impacts lift and drag forces in a complex way. In a steady and smooth flow, particles are following one direction, the flow regime is direct or laminar. With increasing length, velocity, or both, the flow (sometimes rather abruptly) becomes sinuous or turbulent. The pattern of turbulent flow is chaotic and full of rotation. A dimensionless number can be calculated for each flow situation. The magnitude of that Reynolds number (Re) number gives an indication of the flow regime. It is the ratio of density over viscosity, multiplied by the velocity of the flow and the length. As both the numerator and the denominator of this fraction have the SI units N s m^{-2} (momentum per unit area), the Re number is a dimensionless number. Viscous forces in laminar flow dominate at Re numbers smaller than 1. At higher values, inertia becomes more important and the transition to a turbulent regime occurs.

The Navier–Stokes equations were originally only applicable to calculate flow phenomena assuming laminar flow conditions, and they often provided results that deviated from measured values. In 1904, Ludwig Prandtl (1875–1953) suggested a practical solution for these problems by dividing the flow near a solid object into two regions: a layer of fluid close to the object where viscosity is a dominant factor and the flow outside that boundary layer where Euler's equations could be applied. A rough estimate of the depth of the boundary layer can be obtained from the ratio of the length over the square root of the Re number (Lighthill 1990).

4.2.6 An Explosion of Scientific Research

Many scientists, mainly from Germany, France, the UK, the USSR, and the United States, contributed to the knowledge of animal flight during the first half of the twentieth century. Comparative and functional anatomy, kinematics, and ornithopters were the main targets. High-speed cameras had become available and there was an interest in flight, due to air forces becoming increasingly important in wars during that period. Several hundred articles on the subject were published in scientific journals and some 25 books (Slijper 1950). During the second half of last century and the first decade of the twenty-first century, novel physiological methods to measure muscle performance, quantitative flow visualization techniques, and computational fluid dynamic approaches emerged and flourished, resulting in thousands of publications. Much of that knowledge is now easily accessible through the World Wide Web.

4.3 HOW NATURE CREATES LIFT, THRUST (AND DRAG)

Understanding natural flight requires a basic knowledge of the mechanisms involved in keeping flyers aloft. Gliding is the simplest form of steady flight. The necessary lift is produced by wing-like structures, and thrust comes from gravitational forces by exchanging altitude for speed: potential energy for kinetic energy. Both animals and plants use gliding as a means of transport. Powered flight is restricted to the animal kingdom. In addition to the principles used to generate lift in gliding, animals also need thrust forces for take-off and powered horizontal flight. Flapping wings can generate both lift and thrust at once. Here I explore the various solutions to flapping flight that evolution has produced.

4.3.1 A General Aerodynamic Model for an Animal in Steady Powered Flight

A simple model approach (Videler 2006, using Sunada and Ellington 2000) considers a bird flying at a constant speed in still air at constant altitude (Figure 4.10). Assuming left-right symmetry, the resultant forces on, for example, a bird such as the kestrel can be regarded in a vertical median plane. The lift force (L) counteracts the weight (*W*), or the mass of the bird (m_b) times the acceleration due to gravity (*g*). On average, over time thrust force (T) balances the drag force (D). Newton's first law of motion indicates that due to inertia, the bird remains in uniform motion in a straight line at speed (v) as long as there are no other external forces working on it. A final assumption is that all rotational moments are compensated by the flight technique of the bird. In reality, the four forces have different points of application, moving about rather than being fixed. However, we assume that averaged over time the application centers coincide. The model regards average air mass fluxes over time periods exceeding several wing beat cycles.

A flying animal generates lift chiefly by accelerating masses of air downward by its flapping wings. The same wing beats generate thrust by accelerating air backward. Assume that the total mass of air affected by the flyer per unit distance flown is a cylinder with the diameter of the maximum wingspan *b* and a cross-sectional surface area of $(\frac{1}{2}b)^2\pi$ (Figure 4.11). Over one unit distance, the volume of this cylinder equals $(\frac{1}{2}b)^2\pi$. Hence, the affected mass of air per unit distance, m_a, equals $(\frac{1}{2}b)^2\pi\rho$, where ρ is the density of air. (This mass is expressed in kilograms in SI units). With known forward speed v, we can estimate the mass of air affected per unit time, known as mass flux, $\dot{m}$, as

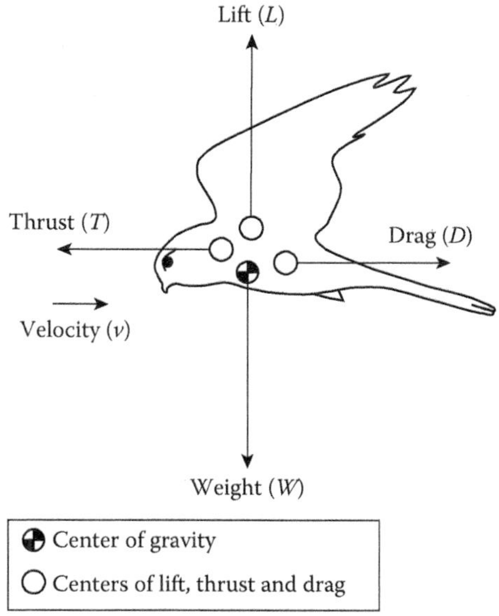

Figure 4.10. Four resultant forces in a vertical medial plane through a kestrel flying steadily at a uniform velocity versus at one height. The points of application of the forces are hypothetical. (From Videler, J.J., *Avian Flight*, Oxford University Press, Oxford, UK, 2006. With permission.)

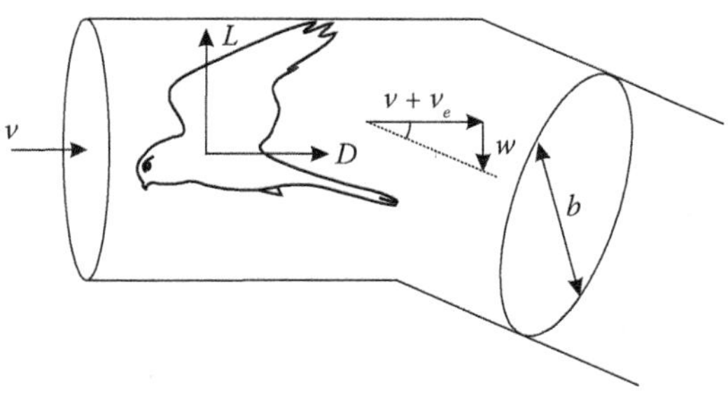

Figure 4.11. Hypothetical cylinder (diameter of the cross section b) of air deflected obliquely downward by the wing beats of a steadily flying bird. Over several wing beat cycles, L and D are the average lift and drag forces, respectively; v is the mean flight speed; $v + v_e$ is the horizontal velocity; and w the downward velocity of the accelerated air behind the bird. (From Videler, J.J., *Avian Flight*, Oxford University Press, Oxford, UK, 2006. With permission.)

$$\dot{m} = m_a v\,(\mathrm{kg\,s^{-1}}) \qquad (4.3)$$

The mass flux has both vertical and horizontal component vectors. The vertical, downward component produces lift force, as follows. The beating wings achieve an increase of the downward velocity of the air from 0 to w m s^{-1}. The downward momentum per unit time in this case is $\dot{m}w$ (N). Newton's third law tells us that the equal and opposite force is the lift force L. The vertical kinetic energy given to the air per unit time is half the downward momentum per unit time multiplied by its downward velocity squared. The induced or lift power P_L is equal to

$$P_L = \tfrac{1}{2}\dot{m}w^2\,(W) \quad (4.4)$$

Flying at a constant altitude implies that the total lift L equals the weight W. This means we can now eliminate w from the equation:

$$L = W = \dot{m}w \text{ gives:}$$
$$w = W\dot{m}^{-1}\,(m\,s^{-1}) \quad (4.5)$$

Replacing w in Equation 4.4 by Equation 4.5, we get

$$P_L = \tfrac{1}{2}\dot{m}(W\dot{m}^{-1})^2 = \tfrac{1}{2}W^2\dot{m}^{-1}$$
$$= \tfrac{1}{2}W^2(m_a v)^{-1}\,(W) \quad (4.6)$$

which reveals that the induced power used to generate lift is proportional to the weight W of the bird squared and inversely proportional to affected mass of air per unit distance m_a and the flight speed v. The wingspan, which influences or determines both m_a and W, and the vertical velocity of the air, on which W is dependent, are important factors.

Power is also required to generate enough thrust to counteract the drag.

Assume that the flying animal accelerates the air by increasing the initial velocity v with v_e (m s^{-1}), which brings the horizontal airspeed directly behind the bird at $v + v_e$ (m s^{-1}). The thrust T gained equals $\dot{m}v_e$ (N), using the mass of air affected per unit time $\dot{m}$, as in the case of generated lift discussed previously. The power P_T required to increase the kinetic energy is the total kinetic energy after the acceleration minus the kinetic energy already present due to the flight speed v:

$$P_T = \tfrac{1}{2}\dot{m}(v + v_e)^2 - \tfrac{1}{2}\dot{m}v^2$$
$$= \dot{m}vv_e + \tfrac{1}{2}\dot{m}v_e^2\,(W) \quad (4.7)$$

Drag consists of the sum of all drag forces on the moving body and beating wings of the animal. It is caused by the fact that air is pushed out of the way by the passing bird, thereby inducing dynamic pressure. According to Bernoulli's law, this pressure is proportional to $\tfrac{1}{2}\rho v^2$. The total drag force can be described as $D = \tfrac{1}{2}\rho v^2 AC_d$ (N), where A is the relevant surface area of the flyer. The drag coefficient C_d depends on the choice of the area and several unknown factors. The power required to compensate the drag equals force times velocity:

$$P_D = Dv = \tfrac{1}{2}Pv^3AC_d\,(W) \quad (4.8)$$

while flying at a uniform speed $P_D = P_T$. In that case, the total mechanical power required for flight P_{tot} is the sum of P_L and P_D.

The shape of the total power curve is U shaped (Figure 4.12) due to the contrasting relationships with velocity of the two components (Pennycuick 2008). This shape implies that there is one velocity where flight power is minimal. The minimum power speed (v_{mp}) is the point where a horizontal line just touches the P_{tot} curve. There is another optimal velocity, the maximum range speed (v_{mr}), where the tangent to the curve drawn from the origin touches the curve. It is the velocity at which the ratio of power over velocity is minimal. This ratio (W/m s^{-1}) equals the work per unit distance (J m^{-1}).

Theoretical mechanical power curves deviate from what happens in the real world because animals are not flying machines. Animal flapping flight produces both power components simultaneously, making it impossible to measure these separately.

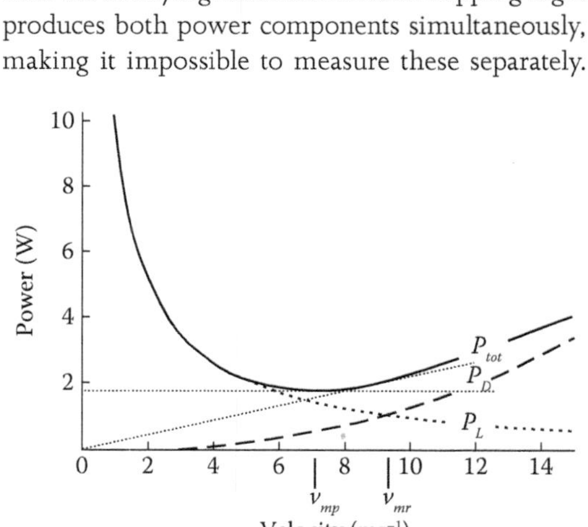

Figure 4.12. Hypothetical mechanical power curves as a function of speed. Minimum power and maximum range speeds are indicated. (From Videler, J.J., *Avian Flight*, Oxford University Press, Oxford, UK, 2006. With permission.)

In Section 4.6, we approximate how much energy animals use in real life.

4.3.2 Specific Lift-Generating Mechanisms

Under natural flight conditions, flow velocities are relatively low and air can be considered an incompressible fluid. Even the fastest wing beats do not compress the air or create voids.

Re numbers are usually larger than 1 and turbulent flow regimes are the rule.

4.3.2.1 Lift: The Wagner Effect

Wings of conventional aircraft in steady linear motion create an upward force by deflecting the horizontal oncoming airflow down. A model wing in a recirculating water tunnel seeded with tiny particles shows the flow quantitatively to fully appreciate the interaction between a conventional wing and the fluid flow (Figure 4.13). The picture shows a cross section through a transparent Perspex model of the arm wing of a northern fulmar, an oceanic bird with excellent gliding capacities. The wing section is modeled on the cross section near the end of the arm wing, just proximal of the wrist joint. The model is uniform across the entire span of the width of the water tunnel. The chord length from the leading to the trailing edge is 9.3 cm, a value that is smaller than the real chord length of 12.5 cm. The flow velocity in the water tunnel is 0.5 m s^{-1}, to match a Re number (based on the chord length) of a fulmar wing in air gliding at an airspeed of about 20 km h^{-1}. The angle of attack (between the tangent to the underside of the wing cross section and the horizontal) is about 6°. Neutrally buoyant particles are illuminated by a thin laser sheet halfway down the span of the model. The sheet shines through the Perspex model and illuminates all the particles moving around the wing section. Two successive digital images, taken 0.004 s apart, show the direction and distance covered during that time of each particle in the plane of the sheet. The two-dimensional velocity vector diagram is based on these measurements.

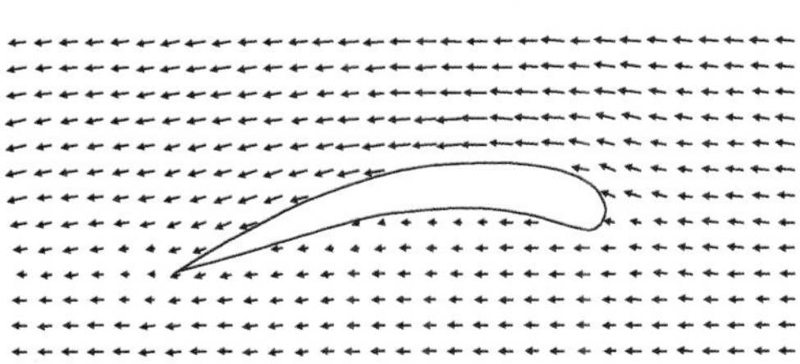

Figure 4.13. Flow pattern around the cross section through a model of the arm wing of a fulmar.

The presence of the wing model generates the vector distribution of flow direction and speed shown in the figure.

The presence of the wing causes substantial differences in velocity above and below the wing with respect to the free stream velocity far away from the model. Part of the flow moves over the bulging upper surface of the wing. Its shape constricts the free passage of the flow increasing the flow velocity and hence the dynamic pressure in the flow direction and, according to Bernoulli's law, decreases its static pressure measured perpendicular to the flow.

Under the wing, the flow velocity is reduced under the small angle of attack. The concave shape enhances the retardation of the flow because the passage widens beyond the leading edge. Unlike commercial aircraft, many bird wing cross sections have a concave lower profile, in some cases even extremely so. According to Bernoulli's law the static pressure (perpendicular to the flow) is high underneath the wing where the velocity is low. The static pressure difference under and above the wing contributes to the lifting force. However, reaction forces on the wing caused by changes in flow direction also play a substantial role. The flow anticipates the presence of the wing by partly moving upward in front of the leading edge of the wing: the up-wash. The reaction force on the wing due to the oblique up-wash pushes downward and slightly backward. Above the wing, the flow returns to horizontal over the first half of the wing chord. This change of direction is caused by a downward-directed force; the equal and opposite force on the wing is upward and much bigger than the downward force on the wing due to the up-wash. Over the rear half of the chord the flow goes downward: the net down-wash. The reaction force to this change of direction on the wing provides lifting force on the wing. The boundary layer, close to the wing, is too thin to be visualized in this velocity vector diagram, but it plays an important role. Why is the fluid following the curved upper surface of the wing? The fluid in direct contact with the wing surface has zero velocity with respect to the wing, but the velocity increases away from the surface to reach the free stream velocity. The shear between the slow flowing water close to the surface and the faster flow at some distance away from the surface gives the flow the tendency to bend to the surface.

This is known as the Coandă effect. The tendency to follow the shape of the surface causes the change in direction of the down-wash. The total lift force on the wing (and the accompanying drag force) is due to the combined effects of the shape of the wing in the horizontal flow. The rounded leading edge and the curved shape of the cross section, the angle of attack, the flow velocity, and viscous forces in the boundary layer all contribute to the flow pattern around the wing and hence to the forces generated. The net lifting result is shown by the downward flow directly behind the wing.

The interaction between wing and horizontal oncoming flow also creates a horizontal drag force. The ratio of the lift over the drag force determines the quality of the wing as a lift-generating device. The arm wing of a fulmar generates somewhere between 10 and 20 times more lift than drag.

4.3.2.2 Lift: Total Flow Pattern

Airplanes were already flying all over the world when Herbert Wagner published the theory behind the development of lift by accelerating conventional wings in 1925 (Wagner 1925). To reach the steady state of the flow pattern around a conventional wing as described above, the flow must accelerate from rest. Wagner discovered the sequence of events illustrated in the figure (Figure 4.14) that leads to the steady state where upward force is fully developed on the wing. This process takes time and is named the Wagner effect.

The figure shows how the flow pattern gradually develops around a wing cross section, with

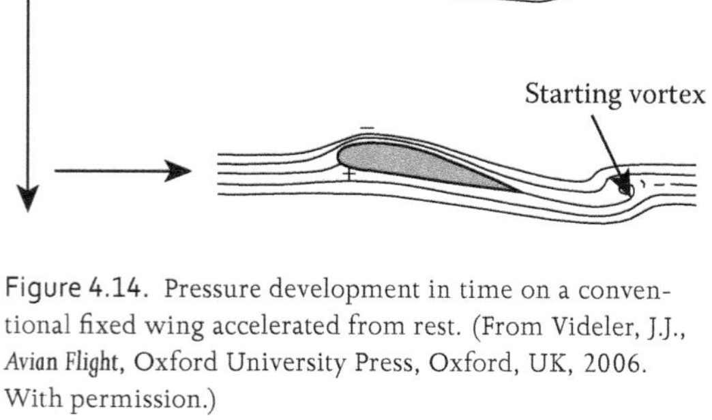

Figure 4.14. Pressure development in time on a conventional fixed wing accelerated from rest. (From Videler, J.J., *Avian Flight*, Oxford University Press, Oxford, UK, 2006. With permission.)

increasing velocity in three steps. The section has an aerodynamic profile and a small angle of attack with respect to the horizontal direction of flight. This sequence of flow pattern development happens every time an airplane starts to taxi on the runway before take-off. When the plane starts moving the air meets the wing and is forced to change direction due to the presence of the wing, the angle of attack, and the rounded leading edge. It stagnates slightly underneath the wing and most air is forced to follow a path over the upper surface of the wing. The streamlines in the picture are closer together above the wing than below, indicating that the dynamic pressure is reduced under the wing and enhanced above it. Two areas of high static pressure where the airspeed is low and two positions where the speed is high and the static pressure is low develop. Just beyond the leading edge, the pressure difference pushes the wing up, and the opposite effect can be found near the trailing edge. There is hardly any net upward force on the wing. At the sharp trailing edge, the flow pattern becomes complicated because the flow tends to move around it to the stagnation point of extremely low velocity just above the trailing edge. This induces the air to rotate counterclockwise and create a vortex, as depicted in the next picture of the cartoon. The pressure difference at the trailing edge disappears due to the development of circulation around the trailing edge. The anticlockwise-rotating vortex is still attached to the wing at this stage, but it is soon released while the velocity continues to increase. As soon as the starting vortex is shed, the flow leaves the trailing edge smoothly, the lift force on the wing is fully established, and the airplane takes off. The shedding of the starting vortex is needed to create a stable situation. This is called the Kutta condition.

In principle, the Wagner effect occurs during the start from rest during each wingstroke in flapping flight. However, experiments show that the effect is not substantial, especially not at lower Re values common for small-insect flight. Models usually ignore the effect (Sane 2003).

4.3.2.3 Leading Edge Vortex Lift

After take-off, the pressure difference between the under and upper side of the wing creates a vortex at the wing tip where air rotates upward and inward. Wing tip vortices emerge as soon as the starting

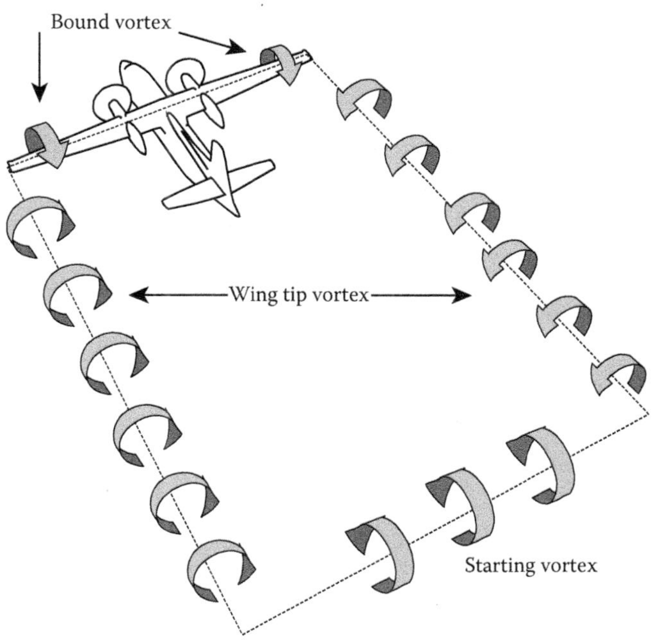

Figure 4.15. Vortices caused by a conventional Fokker 50 airplane in flight. (From Videler, J.J., *Avian Flight*, Oxford University Press, Oxford, UK, 2006. With permission.)

vortex is released and the pressure difference is established. In fact the wing tip vortices are continuous with the starting vortex and form part of an extremely long rectangular vortex "ring." (Figure 4.15). According to fluid dynamic theory, a vortex cannot end in the fluid in the direction along its center (Helmholtz's theorem). It either ends at the boundary of the fluid or forms a closed loop with itself or another vortex. The starting vortex of an aircraft forms a closed loop with the wing tip or trailing vortices and the bound vortex on the lifting wings. The bound vortex around the wings becomes visible by subtracting the average flow velocity from the local velocities around the wing. The picture of the flow around the fulmar wing can be used to verify this statement. The bound vortex is shed during landing when the speed of the aircraft drops below the value needed to maintain the pressure differences (the Wagner effect reversed).

The use of conventional lift is probably widespread in nature but precise measurements confirming this are generally lacking. The arm wings of most birds are specifically adapted to produce lift in this way, not only during gliding but also in flapping flight. Albatrosses can use attached conventional flow over the entire length of the wings because, unlike other birds, the hand wings have rounded leading edges due to increased thickness of the primary feathers. Bats and pterosaurs probably also use it, but during the downstroke only. Insects reach the Kutta condition during the mid phase of the wing strokes.

4.3.2.4 Wing-Wake Interaction

Conventional lift is reduced by increased angles of attack or if the speed drops below the value required to maintain the circulation around the wing. If the flow does not remain attached to and follow the wing surface, dramatic changes in lift forces occur. For example, large angles of attack can cause flow separation on the upper part of the wing, resulting in extreme lift reduction (stall) with dramatic consequences. A minimum velocity is needed to release the starting vortex, and a drop below this has severe consequences for the lift-generating capacity of conventional wings.

A completely different mechanism for force generation is used by delta wings (such as the Concorde). It is based on flow separation right at the leading edge of the wing. The leading edge is ideally sharp, to induce stall at even small angles of attack. The principle was known empirically before Polhamus (1971) published an analytical method to calculate lift and drag characteristics of LEVs. The picture (Figure 4.16) shows the principle of force generation by an LEV developing over a

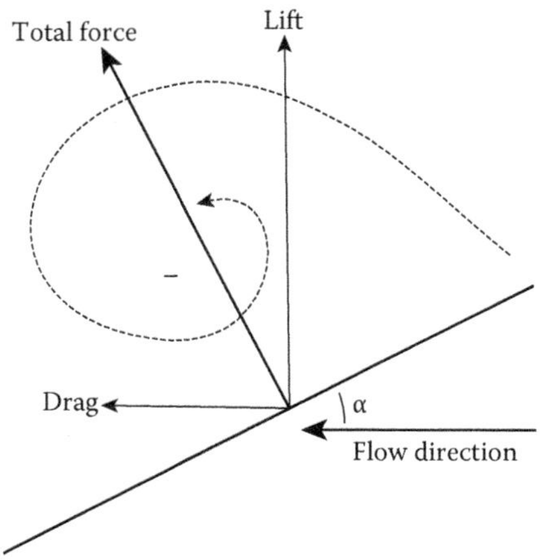

Figure 4.16. Cross section through a leading edge vortex above a flat plate.

flat plate. The flow hits the plate under a fairly large angle of attack α and separates forming a vortex over the center of the chord of the plate facilitated by the sharpness of the leading edge. The oncoming flow is slowed down underneath the foil resulting in positive static pressure. The rotating flow in the center of the vortex above the plate is fast and the static pressure is low. The resulting total aerodynamic force due to the static pressure difference is perpendicular to the plate. Note that the direction of this force is determined by the angle of attack. The total force vector can be transformed in a vertical lift force and a horizontal drag force. A smaller angle of attack diminishes the drag and increases the lift as long as a vortex develops. This system works instantaneously; there is no Wagner effect. However, the vortex keeps growing until it detaches from the surface of the plate and is shed. This causes a problem as stall is delayed at first, but it happens eventually. A new LEV then has to start growing. This means that the lift force drops to zero every time a vortex is shed. This problem can be overcome in different ways. The LEV can be transported by the oncoming flow along the swept back angles of delta wings such as those of the Concorde or the hand wings of a gliding swift and be left behind at the wing tips (Figure 4.17) (Videler et al. 2004). The LEV diameter increases along the wing as more and more air is added to the vortex. This causes the conical shape of the LEV from front to rear.

Another solution is a flapping motion of the wing, where no sweep back angle is required. During the downstroke, centrifugal forces transport the LEV in the direction of the wing tips, where shedding takes place.

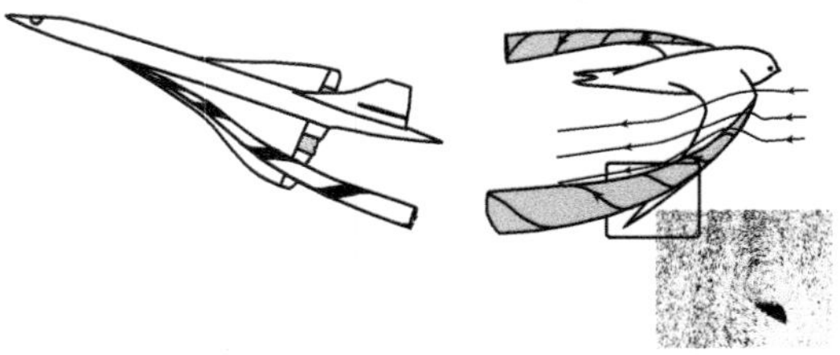

Figure 4.17. Artist impressions of leading edge vortices (LEVs) on the wings of the Concorde and the swift. The picture shows a flow visualization of the LEV above a swift wing model in a water tunnel. (Based on Videler, J.J., *Avian Flight*, Oxford University Press, Oxford, UK, 2006.)

An increasing number of biological flyers are being discovered to use LEVs: gyrating samaras (Lentink et al. 2009), insects and birds, and flapping wings of bats (Chin and Lentink 2016). During landing, when the velocity decreases to zero, most birds seem to use LEVs over the hand wings with sharp leading edges in swept-back position. The angle of attack can be used to control the amount of lift and drag produced. In insects, prominent LEVs are found stably attached to the wings during the entire stroke without shedding (Dickinson et al. 1999). The LEV rotation can be used to create a lift-enhancing effect by a special interaction between the wing and the generated wake as follows.

4.3.2.5 Clap, Fling, Peel, and Reverse Peel Mechanisms

Force measurements and flow visualization of a three-dimensional mechanical model of a fruit fly wing (scaled to the right Re) showed how the flow induced by shed stop vortices during stroke reversal can be used to generate lift (Dickinson et al. 1999). The cartoons (Figure 4.18; modified after Sane 2003) show the sequence of events during stroke reversal of an insect wing. The wing chord is drawn as a straight arrow with the arrowhead indicating the leading edge. The columns of dotted lines indicate the translation velocity and direction and hence the speed of the fluid meeting the wing. The end of the stroke is depicted in three steps on the left (a, b, and c) and the beginning of the reverse stroke on the right (d, e, and f). The white arrows indicate the direction of the movement of the chord.

a. During the entire stroke the aerodynamic force perpendicular to the wing chord caused by the LEV is directed upward and backward. Again, this can be transformed into lift and drag components.

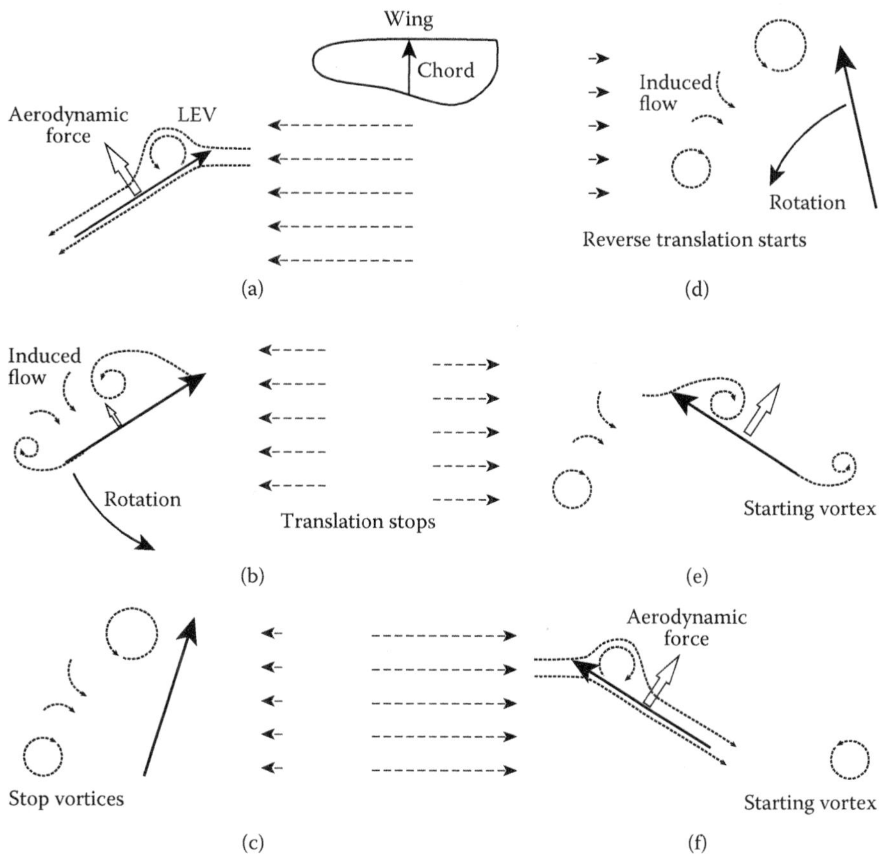

Figure 4.18. Cartoon of wing–wake interactions on insect wings (Modified after Sane, S.P., *J. Exp. Biol.*, 206, 4191–4208, 2003). The sequence of events is explained in the text.

b. The wing comes to a halt and rotates. Two stop vortices are formed: one stop vortex behind the leading edge and one stop vortex at the trailing edge. The latter stop vortex is enhanced by the rotational movement of the wing.
c. The stop vortices are shed when the wing reaches its ultimate stroke position and stops. The rotational directions of the vortices causes flow in between in the direction of the wing surface.
d. The wing now starts the reverse stroke by rotating in the new direction. It moves against the flow induced by the shed vortices.
e. The flow caused by the translation is enhanced by the induced flow that causes a large aerodynamic force very early on during the reverse stroke. The LEV develops and a starting vortex is formed and subsequently is shed by the trailing edge.
f. The reverse stroke in full swing with a stable LEV and the aerodynamic force creating lift and drag.

4.3.2.6 The Kramer Effect

Stiff wings clapping together at the end of the upstroke and flinging apart at the start of the reverse stroke (the clap and fling mechanism) were discovered as an unsteady lift-enhancing mechanism in insects (Norberg 1972; Weis-Fogh 1973; Ellington 1984). The sequence of events during the clap motion indicates how induced flow patterns relate to the resulting aerodynamic force on the wing. (The cartoon in Figure 4.19 is again based on Sane 2003.)

a. Near the end of the upstroke the leading edges of the two wings represented by the chords approach each other while the wings rotate around the lengthwise

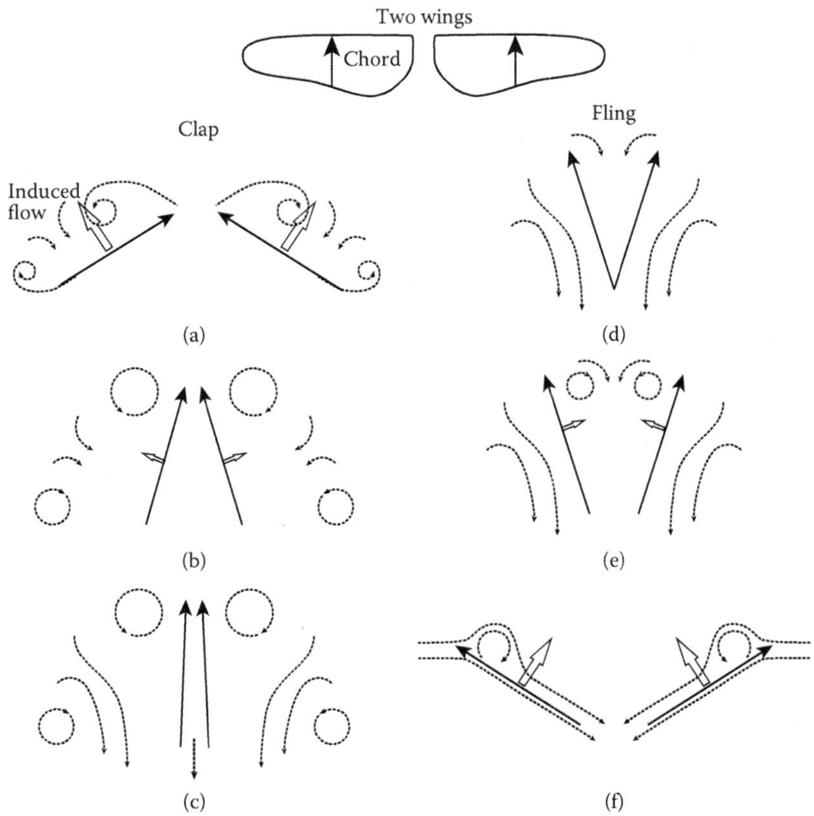

Figure 4.19. Flow patterns of the clap and fling mechanisms found in insect flight. Cartoon is based on Sane (2003) and explained in the text.

axis, accelerating the trailing edges toward each other. The LEVs continue to induce the aerodynamic force perpendicular to the wings in an oblique upward direction displayed during the stroke. Leading and trailing edge vortices create induced flow toward the wing surface.

b. Toward the end of the clap motion, stopping vortices, shed from both edges, continue to create a wake in the direction of the wings, pushing them together.

c. The induced flow dissipates in the wake, and the closing of the wedge between the wings generates a jet of air that adds to the thrust.

   Subsequent opening of the wings around the trailing edges represents another lift-generating mechanism: the fling.

d. Air rushes into the widening gap around the leading edges, inducing the buildup of LEVs.

e. Wings are separating and forces perpendicular to the wings develop due to the low pressure in LEVs.

f. The wings fling apart and perform the downstroke, each with a stable LEV responsible for the aerodynamic force in oblique upward direction. A starting vortex is also created by each wing, but these annihilate each other because they circulate in opposite directions.

The "clap and fling" mechanism works for rigid, but not necessarily for flexible wings (Miller and Peskin 2009; Percin et al. 2011). Tiny insects operating at low Re would need large forces to perform clap and fling motions due to high experienced viscosity. These can be diminished using flexible wings and a different type of movement for the wing–wing interaction, namely, the "peel and reverse peel." During a peel, the wings detach and bend starting from the leading edge and

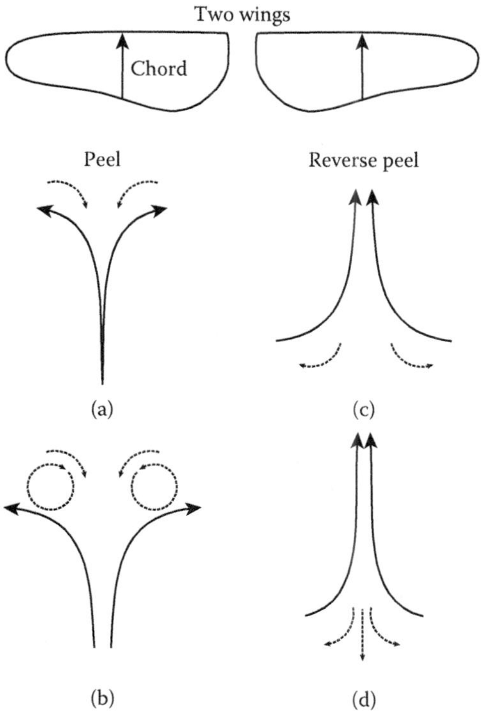

Figure 4.20. Flow induced by peel and reverse peel mechanisms. The arrowheads of the lines representing the wing cord sections indicate the leading edge. The subparts are explained in the text.

following the chord. Air rushes into the ensuing gap (Figure 4.20a). This enhances the formation of the LEVs when the wings move apart in the next stage (Figure 4.20b). The wake of the previous stroke can be more effectively captured than in the case of rigid wings. The reverse peel squeezes air backward out of the gap between the wings (Figure 4.20c and d). As there are an infinitely large number of possible variations in rigidity of wings, partial or whole, the detailed kinematics and the flow patterns are difficult to generalize. In general, the conclusion may be that wing flexibility can reduce drag forces, especially during wing–wing interaction at low Re. Lift forces are likely to be enhanced.

4.3.2.7 Added-Mass Inertia

Aerodynamic forces are generated during rotation when the stroke direction is being reversed. The air surrounding the wing is made to circulate with velocities proportional to the rotational speed (Kramer 1932; Sane 2003). The timing of the rotation determines the aerodynamic effect. A rotation preceding stroke reversal increases lift. During a delayed rotation, the leading edge moves in the new direction while rotating downward, which has a negative effect on the lift force. Rotation exactly during stroke reversal provides positive lift. Stroke reversal ideally should be executed fast because the drag during rotation increases with its duration. For efficient flight with high lift and low drag, wing rotation must occur at the stroke reversal and must be rapid (Chin and Lentink 2016).

The Kramer effect occurs most obviously during hovering flight of insects and hummingbirds, but probably also plays a role during forward flight with fast wing reversal and maneuvering flight of birds and bats.

Flapping wings not only accelerate and decelerate their own mass but also some mass of air surrounding the wings. This increases the inertia of the wings, and the reactive forces may enhance the aerodynamic forces created. Experimentally, however, it is difficult to separate the added-mass effects from other forces created by the wings. Computational studies usually estimate the magnitude of the added-mass effect by using some constant coefficients (Sane 2003).

4.4 BIODIVERSITY OF FLIGHT

Flight is being carried by air across a distance. Any object flies when vertical upward forces exerted by the air are bigger than the weight (mass times gravitational acceleration) of the object. A tiny weight in combination with a large surface area increases the ability to fly passively. Parachuting, the use of large surfaces to delay the speed of fall, is a widespread phenomenon as is gliding controlled descend. Both phenomena are considered flight; however, ballistic leaping shown by many arboreal animals is not. Dispersal through the air is a common botanical phenomenon. Entire plants, seeds, and pollen of plants can be taken by the wind and dispersed over considerable distances. Several plant seeds are morphologically adapted to delay falling speeds with astonishing efficiency (Augspurger, 1986). In the animal kingdom, many insects rely on prevailing winds for dispersion. Actively controlled flight in nature is restricted to animals. Powered flight, where beating wings are in interaction with the air, creating directional lift and thrust forces, allows animals to travel independent of the prevailing winds. This Section illustrates the diversity of flight strategies used by natural flyers.

4.4.1 Passive Displacement by the Wind

Wind is the displacement of air in time with respect to the stationary surface of the earth. Passive flyers travel in the average wind direction at or close to the velocity of the surrounding air. This way of flight is dominated by viscous forces; velocities relative to the surrounding air (airspeeds) are close to zero and body sizes are usually small. The Re values are low and well in the viscosity dominated regime. Wind distribution (anemochory) is a common way of dispersal for pollen, spores, and seeds (Hintze et al. 2013; see www.seed-dispersal.info). Distances covered can be thousands of kilometers. Plant ecologists assess wind dispersal potential by using models and a large database (Tackenberg 2003). Release height and falling velocity are considered important factors. Measurements of passively covered distances usually reveal considerable variation. Morphological traits potentially improving wind dispersal are the possession of wings, plumes, or balloon-like structures.

Pollen can also travel large distances attached to tiny wind-dispersed insects. Miniaturized insects are often important pollinators. Insects are found floating with the prevailing winds, like aerial plankton in the atmospheric boundary layer, up to 12 km in altitude. An exponential wind gradient characterizes the first few hundred meters above the Earth's surface; winds are usually steadier and stronger higher up. Close to the surface, structures may strongly influence wind speeds and turbulence and hence passive flight of biological items. The active airspeed (the velocity relative to the surrounding air) of small insects is commonly much smaller than even the lowest wind speeds. The airspeed of 60% of all insects is less than 1 m s^{-1}.

Some spiders (Aranea), spider mites (Acari), and larvae of moths (Insecta, Lepidoptera) secrete extremely fine silk lines and use these for dispersal on the wind. High temperatures and wind velocities below 3 m s^{-1} induce ballooning behavior in small spiders of less than 1 mg. From prominent positions in the vegetation, several silk threads are released by the spider to become entangled by the wind. The drag on the gossamer structure carries the spider up in the air. Darwin described the behavior of aeronautical spiders after observing them landing on the Beagle 60 M out at sea (Darwin 1845).

4.4.2 Parachuting, Gliding, Tumbling, Rotation, and Gyration of Plant Seeds

Parachutes delay the time to reach the ground after take-off from altitude. Horizontal components of wind displace the falling item during that prolonged airborne period.

Dandelion (*Taraxacum officinale*) seeds are a prime example of dispersion by parachuting. Achenes, the dispersal units, consist of a pappus disc of bristles, a long beak underneath with the actual seed dangling at the end. The surface area of the pappus disc, the length of the beak, and the mass of the seed determine the performance and static stability of the achene. In urban areas short-range dispersal is more important than in open meadows due to the lack of suitable fertile landing spots. Urban achenes drop more quickly from lower heights than those from plants living under more natural conditions in the countryside (Arathi 2012).

Samaras, winged seeds, extend the duration of the descent by gliding, tumbling, autorotation, autogyration, or by a combination of the last two mechanisms.

Gliding seeds are rare; the most prominent example is *Alsomitra macrocarpa*, a liane from tropical Asian forests. The seed is in the center of its wing near the leading edge. The mass is in the order of 200–300 mg. The span is typically about 14 cm, the area is about 60 cm^2, the leading edge is swept back some 10°, and the trailing edge is reflexed. The wing tips are slightly raised (dihedral), and cross sections have a profile that is cambered near the leading edge and flattened toward the rear (Azuma and Okuno 1987). The mean rate of descent is between 0.3 and 0.7 m s^{-1}. The lift is generated by conventional attached flow over the wing. During descent, the seeds tend to glide in phugoid motion, gaining airspeed and lift during a steep dive ending in a pitching up motion. The upward component of the oscillation ends when the speed decreases and the leading edge of the wing drops. The angle of attack of the air on the wing remains approximately constant during the repeated sequence of events.

Elongated and often twisted leaf-like seeds start to rotate around the longitudinal axis soon after being released. The rotation is induced by differential drag forces on the falling object. Maple samara seeds typically start to spin (autogyrate) within 1 m of the start of the fall. The characteristic helical motion is caused by an asymmetric mass distribution of the seed (the center of mass does not coincide with the geometric center of the seed), initially causing a tilting motion. Subsequently, aerodynamic forces with a center of pressure offset from the center of mass form a moment that causes the samara to gyrate. In the steady rotational state, aerodynamic forces nearly balance the weight of the seed (Varshney et al. 2012). In still air, the movement causes a radial air velocity distribution. The angle of attack on the wing is about 90° at the wing root and decreases to values as small as 16° toward the wing tip. Air hitting the wing with a sharp leading edge at large angles of attack separates and forms an LEV on top of the wing. On cross section, the vortex center is approximately halfway down the chord length. LEVs build up quickly and break away when they grow too large. To avoid that from happening, recall in Section 4.3 how a vortex can be displaced by the oncoming flow along a swept-back wing or by centrifugal forces. Lentink et al. (2009) showed that maple seeds create high lift by generating a compact LEV that is displaced radially by centrifugal forces to make it stable.

Combined rotation and gyration is shown by tulip tree (*Liriodendron tulipifera*) samaras (McCutchen 1977). The heavy seed is on one end of the long, leaf-like samara. During the fall at 156 cm s^{-1}, the samara drops with the seed as the lowest point rotating around the longitudinal axis while swinging around during the fall describing a helical path.

Some seeds are fitted with more than one wing. Several species have two, and the helicopter flower (*Hiptage benghalensis*) has three wings. The flight principle is probably not fundamentally different from that of single-winged seeds, but detailed research in the aerodynamics of multiple-winged seeds is not yet available.

4.4.3 Animal Flight Diversity: Gliding

Gliding involves active aerodynamic control during descend at angles less than 45°. In contrast, parachuting is strictly passive (Dudley et al. 2007).

Among insects, parachuting and directional gliding are rare. Extremely dorsoventrally flattened leaf-mimicking species of stick insects and mayflies allegedly delay falling speeds by parachuting, but there is no proof that they do so deliberately. It is more likely that camouflage is the main function of the leaf shape. Directional gliding used in intermittent bouts during flapping flight is restricted to butterflies and dragonflies.

Flying squid (Ommastrephidae) are the only other invertebrates capable of repeated flight over considerable distances. Muramatsu et al. (2013) photographed the flight behavior of schools of about 13-cm-long juveniles leaving the water in front of a vessel sailing at a speed of about 23 km h^{-1} (Figure 4.21). The average duration of the airborne periods was about 3 s, covering distances of about 30 m per flight. Launching squid break the surface using jet propulsion by contracting the mantle and ejecting water through a narrow siphon, a funnel-shaped structure connecting the mantle cavity with the exterior. Jet-propelled squid swim backward. The mantle is streamlined and ends in two terminal fins. The squid leave the water with the fins first, rolled up against the tapered end of the mantle and the arms folded tightly together. The jet continues in the air. As soon as they leave the water, the fins are unfolded to form a pair of wings at the leading part of the animal. The arms unfold and form a second wing at the rear of the backward flying squid. When reentering the water the fins are coiled back against the mantle, and the arms are stretched to maximize streamlining.

Many fish species leap into the air when chased by predators or to get rid of external parasites.

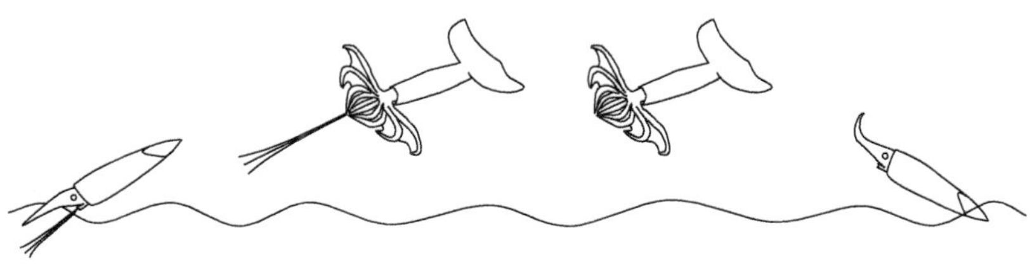

Figure 4.21. Sequence of events of a jet-propelled gliding bout of a squid. (Redrawn from Muramatsu, K., et al., *Mar. Biol.*, 160, 1171–1175, 2013.)

Marine flying fishes (Exocoetidae) are adapted to perform repeated prolonged escape glides with enlarged pectoral fins. Flight speeds between 20 and 30 m s^{-1} over up to 400 m have been measured. Some species are four winged because they use enlarged pelvic fins as well. The lower lobe of the tail fin is elongated and is used to beat the water during take-off and during touch down on wave crests to prolong the gliding distance. In fresh water, hatchetfishes (Gasteropelecidae) actually vibrate their pectoral fins during glides out of the water. They possess long, slender pectoral fins powered by large intrinsic muscles attached to a greatly expanded pectoral girdle. The body is strongly compressed to provide directional stability and low-pressure drag.

In tree frogs, gliding originated independently in two families: Hylidae and Rhacophoridae. Gliding tree frogs share elongated fingers and toes, webbed feet, and hands and skin flaps on elbows and ankles. These enable the animals to make glides at angles less than 45°, making them true gliders according to our definition. The frogs can control flight and carry out maneuvers such as banking and rolling (Emerson and Koehl 1990; McCay 2001).

Flying dragons, lizards of the genus *Draco*, extend skin folds on both sides of the body. The folds are supported by five or six extremely elongated floating ribs turning the folds in cambered wings. Escape gliding flights of around 90 m have been recorded. Several species of flying geckos (genera *Ptychozoon* and *Cosymbotus*) fly over distances of up to 60 m aided by webbed feet and skin flaps along the body and tail.

Flying snakes (*Chrysopelea paradisi*) glide to move between trees, to chase aerial prey, and to avoid predators. They are ordinary snakes with rounded bodies and no obvious further adaptation to gliding flight. In flight, they undulate the body sideways from head to tail with increasing amplitude, swimming through air like an eel in water. The horizontal distance covered during flight from 10-m height to the ground was 10 m on average, reaching an airspeed of 10 m s^{-1}, a sinking velocity of 6.4 m s^{-1}, and a horizontal speed of 8.1 m s^{-1}. The minimum recorded glide angle was only 13°. The performance of flying snakes rivals or even surpasses prominent gliders such as flying squirrels (Socha et al. 2005). An aerodynamic explanation for this mode of flight awaits discovery.

In mammals, gliding evolved independently in at least six groups. Among eutherians (placental mammals), two species of colugos (order Dermoptera), 44 species of flying squirrels, and six species of African scaly tailed squirrels (Anomaluridae) glide. Among arboreal marsupials, six species of rope dancers (*Petaurus*); the greater glider, *Petauroides volans*; and the feathertail glider, *Acrobates pygmaeus*, have gliding skills (Byrnes and Spence 2011; Jackson 1999).

The membrane (patagium) of the colugos has the largest possible extension between the tips of fingers, toes, and tail. They are the best gliding-adapted mammals, with good control over the tension in the patagium (Byrnes et al. 2008).

Table 4.1 compares gliding characteristics of mammals. The best glide ratio represents the largest horizontal distance covered per meter descend. The wing loading is the ratio of body weight over wing area.

A typical flying squirrel glide trajectory starts with a jump from a high position on a tree trunk (Figure 4.22). Along the first meters of the trajectory, acceleration is maximized due to gravity increasing glide velocity. The aerodynamic force increases with speed and results in a vertical force in equilibrium with gravity. As speed increases further, the aerodynamic force exceeds gravity, and the glide trajectory starts to curve up and the glide angle decreases. The animal slows down and the glide trajectory flattens. The aerodynamic force starts to reorient in rearward direction, and the animal slows down. Near the end, the trajectory curves increasingly upward until the animal is in vertical position when it touches the landing tree (Bahlman et al. 2012).

4.4.4 Animal Flight Diversity: Powered Flight

4.4.4.1 *Insects*

Insects have six legs and usually two pairs of wings growing from the back of the thorax.

Nobody knows how many winged insect species (Pterygota) there are, but the figure exceeds 1 million and is still growing, with new species continuously being discovered. The body mass of insects spans three orders of magnitude, from less than 1 mg up to more than a gram. Fairy flies, which measure less than 1 mm, are the smallest insects; dragonflies are among the biggest and can attain a body length of 12 cm. Size is restricted by the tracheal respiration system that limits oxygen supply and the weight of the external skeleton. The largest insects fly

TABLE 4.1
Mammalian gliding characteristics.

	Common name	Best glide ratio	Wing loading (N m^{-2})
	Marsupials		
Petaurus breviceps	Sugar glider	2.5	45–59
Petaurus norfolcensis	Squirrel glider	4.4	nd
	Eutherians		
	Rodents		
Glaucomys sabrinus	Northern flying squirrel	2.4	56–61
Petaurista leucogenys	Japanese giant flying squirrel	3.5	nd
Petaurista petaurista	Giant red flying squirrel	nd	120
Pteromys volans	Siberian flying squirrel	3.3	nd
Anomalurus derbianus	Lord Derby's anamolure	2.2	69–93
	Dermoptera		
Galeopterus variegatus	Colugo	11.3	49–71

SOURCE: Bahlman, J.W., et al., 2012.
NOTE: nd, not determined.

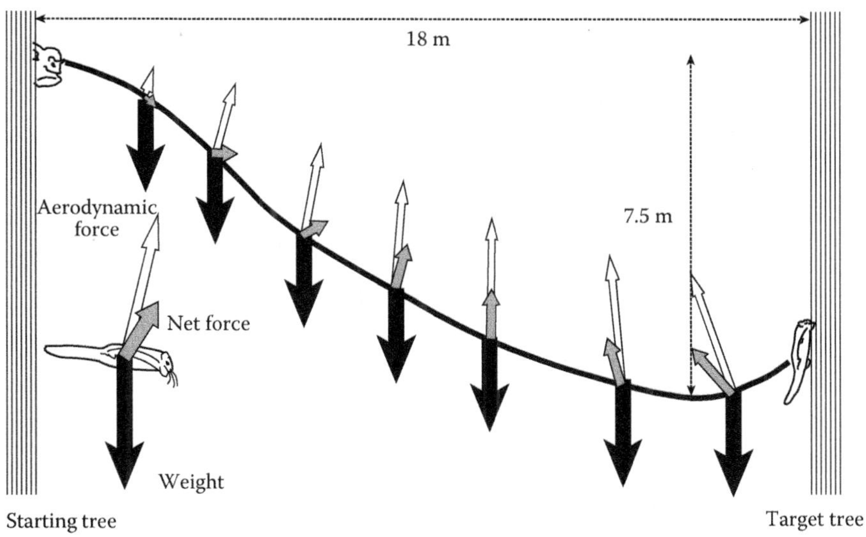

Figure 4.22. Forces on a gliding squirrel moving between trees. (Redrawn from Bahlman, J.W., et al., 2012.)

under inertia-dominated conditions with Re values exceeding 20. Details about the form and function of the flight apparatus follow in Section 4.5. The flight of only a handful species has been studied restricted to relatively large insects (locusts, dragonflies, bumble bees, butterflies, and moths). Flight performance of miniaturized insects is hardly studied, except for that of the fruit fly genus *Drosophila*. The smallest insects operate in a viscous fluid regime, with Re values around 10, which requires specially adapted wings and flight techniques.

4.4.4.2 Vertebrates

In vertebrates, wings are always modified front legs, but different modifications have arisen. In both pterosaurs and bats, the surface of the wings consists of skin folds spanning between the arm skeleton and the body and hind legs. The wing is not attached to the hind limbs in birds, where strong feathers cover and form the wing surface. The functional morphology of flight-related features of the three groups of flying vertebrates is treated in Section 4.5.

PTEROSAURS

Pterosaurs, the first flying vertebrates, originated from a nonflying ancestor in the late Triassic about 220 million years ago and disappeared abruptly during the mass extinction at the end of the Cretaceous 65 million years ago. The membrane wing was supported by skeletal elements of one elongated finger and was attached to the side of the body and the hind limbs. The wingspan of these flying reptiles varied from 20 cm to 10 m. The latter figure is based on a reconstruction starting from about half the humerus of *Hatzegopteryx*, a fragment that is 236 mm long. X-rays of a fossil made by Witmer et al. (2003) show that the part of the *Hatzegopteryx* brain responsible for coordination of movements and control (the floculus) was exceptionally large, occupying 7.5% of the brain mass (in birds it is 1–2%).

The taxonomy of pterosaurs is extremely complex; almost each find is a new species. Andress (2012) speculates that this complexity and diversity, approaching 150 recognized species, is due to early rapid adaptive radiation into the new aerial niche. Pterosaurs are divided into two major groups: Rhamphorhynchoidea, early pterosaurs possessing a long tail fitted with a plate at the end; and the later Pterodactyloidea with very short tails.

BIRDS

Birds evolved much later than pterosaurs from flightless ancestors, presumably dinosaurs, during the late Jurassic 150 million years ago. Most lineages ended in the same mass extinction event 65 million years ago, apart from a few groups that became the ancestors of modern birds. There are approximately 8500 extant species of flying birds, and each has its own specific flight characteristics. The flight of about 160 species has been studied; but of those species, only a handful have been studied in detail, including structural and flight functions as well as aerodynamics. Even so, we know more about the flight of birds than of that of the other vertebrate flyers.

BATS

Bats were the last flying vertebrates to evolve, with the earliest fossils dating from 52 million years ago. They form the single mammalian order of the Chiroptera that is subdivided into two suborders: Megachiroptera and Microchiroptera, with one common ancestor (Altringham 1996). The mega bats are usually large Old World tropical fruit-eating bats weighing up to 1500 g, but sizes of smaller species overlap with those of the micro bats. There are 186 extant species recognized. Their range is limited to Africa, Asia, and the northern part of Australia. Micro bats weighing between 1.5 and 150 g are found around the world, but not close to the poles. There are between 800 and 1000 species living today.

4.5 FUNCTIONAL MORPHOLOGY OF FLYING ANIMALS

Insects and vertebrates are the only groups of animals that conquered the air by developing structures for powered flight.

In vertebrates, flight evolved three times independently, each time by modifying the front legs to wings (Figure 4.23). The skeleton of the arm is adapted to support the wing surface in each case. The outer parts are reduced in weight to minimize inertial forces during the wing strokes. Muscles powering the wings are mainly based on the body near the center of gravity. The surfaces of bat and pterosaur wings are (were) skin flaps; those of birds are covered with feathers. The skin membrane of pterosaurs and bats stretches between the arms and the legs, involving both limbs in flight. Birds use only the arms to fly; the legs are decoupled, move independently, and can better be used for other functions.

Insect wings are not modified legs, and there are usually four wings, not two. The design of the flight apparatus is fundamentally different from those of the vertebrate flyers.

4.5.1 Birds: Quadrupeds with Feathered Front Legs

Birds evolved feathers to aid in flying. During flight the wings, which are feathered arms and hands, are beaten up and down, stretched and retracted, and rotated forward and backward. Another unique feature is the tail, which is adapted to fly. The tail can rotate and the tail feathers can be

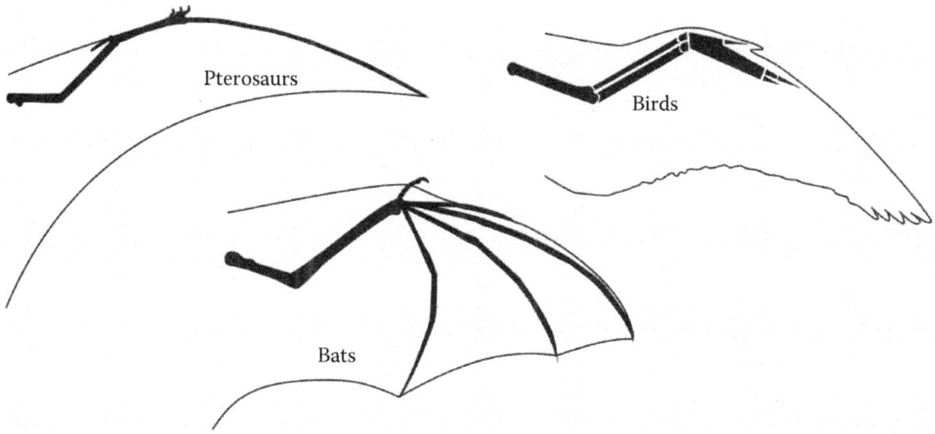

Figure 4.23. Flight evolved three times among vertebrates by modifying the front legs differently each time.

spread and closed. To understand how the building plan of the flight apparatus enables birds to fly, we need insight into the structure of the feathers, the anatomy of the wings, and the functional anatomy of the tail and other flight-related structures of the body (the following account is mainly based on Videler 2006).

4.5.1.1 Feathers

The number of feathers on a bird is species and size dependent: a swallow has about 1500 feathers and a swan 25,000 feathers. Feathers are dead epidermal structures mainly consisting of the protein keratin, like mammalian hair. They serve many functions. The insulating down of birds' under plumage helps to keep body temperature constant, bristles and filoplumes have tactile functions, and contour feathers realize a smooth airtight coverage. We are especially interested in the largest contour feathers: the primary and secondary remiges of the wings and the rectrices of the tail, because these feathers give the wings and tail their aerodynamically functional structure (Figure 4.24). The common contour feather has a shaft over the entire length and two vanes, one on either side, along the distal part. The proximal part of the shaft, the calamus, is a hollow tube that grows from a follicle in the skin. Along the shaft, where the vanes appear, the calamus becomes the rachis. The rachis is filled with spongy tissue surrounded by a stiff wall. Cross sections show that on the dorsal side, it is smooth and has a convex shape; the sides are approximately flat and the underside has a central

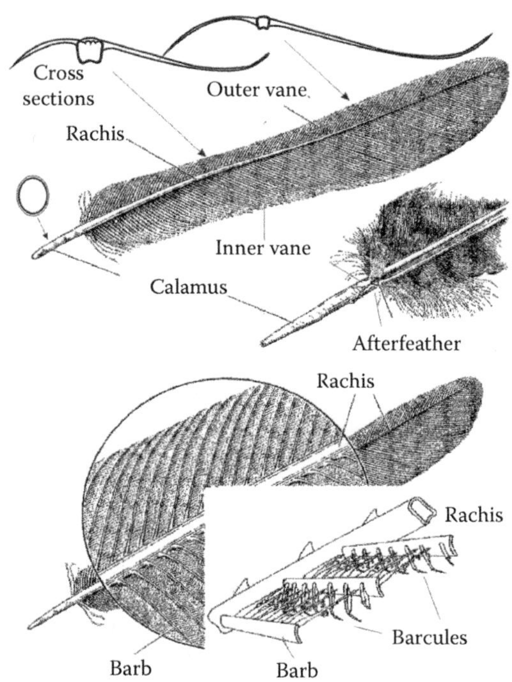

Figure 4.24. Contour feather structure. (Based on Lucas, A.M., and Stettenheim, P.R., *Avian Anatomy: Integument*, Agricultural Handbook 362, U.S. Department of Agriculture, Washington, DC, 1972.)

depression forming a groove along the rachis to the feather tip. (On the ventral side a small afterfeather can be found in the center of the rachis. It is commonly interpreted as a vestigial structure.) The vanes consist of rows of interlocking, parallel barbs implanted under an acute angle on the rachis pointing in the direction of the tip. Barbs are usually cambered on cross section, thereby making them stiffer. Each barb is fitted with rows of barbules on both sides. Proximal and distal barbules overlap and interlock using various microstructures such as flanges, hooklets, and spines. The width of the vanes is not necessarily symmetrical on both sides of the rachis. Narrow vanes are stiffer than wider vanes. Cross sections show that vanes are curved away from the rachis, either downward, upward, or downward first and then upward.

Contour feathers are light, strong, and wind tight. Microstructures of the barbules give the vanes strength and are dense enough to make them wind tight. Wear of interlocking microstructures decreases the functionality of feathers and makes periodic moulding necessary.

4.5.1.2 Internal Anatomy

The skeleton of a bird consists of the familiar components of all quadrupeds (Figure 4.25): a skull on one side of the vertebral column ending in a tail. Ribs attached to the vertebrae enclose the rib cage and are connected at the front to the breastbone (sternum). The legs are connected in the hip joints to the pelvic girdle that is firmly united with several vertebrae in the synsacrum. Each arm is joined by the shoulder joint to the pectoral girdle including the shoulder blade (scapula) and coracoid. In birds, the left and right clavicles are joined into the wishbone (furcula) and firmly attached to the distal end of the coracoid.

The shoulder joint is an approximate ball and socket joint. It faces laterally and allows the wing large up and down and fore and backward movements as well as some rotation around the length axis. The wing is supported by the humerus, the radius, and ulna and by the skeleton of the hand. The radius and ulna articulate with the humerus at the elbow and with two carpal bones (the radiale and ulnare) in the wrist joint. The wrist is a double joint, where the carpal bones articulate with the carpometacarpus (fused carpals and metacarpals) of the hand skeleton as well. Birds usually only have three digits with one or two phalanges each. The first digit forms the skeleton of the alula or bastard wing. The elbow and wrist joints allow the wing to stretch and fold. Stretching tightens two skin folds, the propatagium between the wrist and the shoulder joint and the metapatagium connecting the elbow with the trunk. The proximal condyle of the humerus articulates with the ends of the scapula and the coracoid. The coracoid is connected to the sternum, and

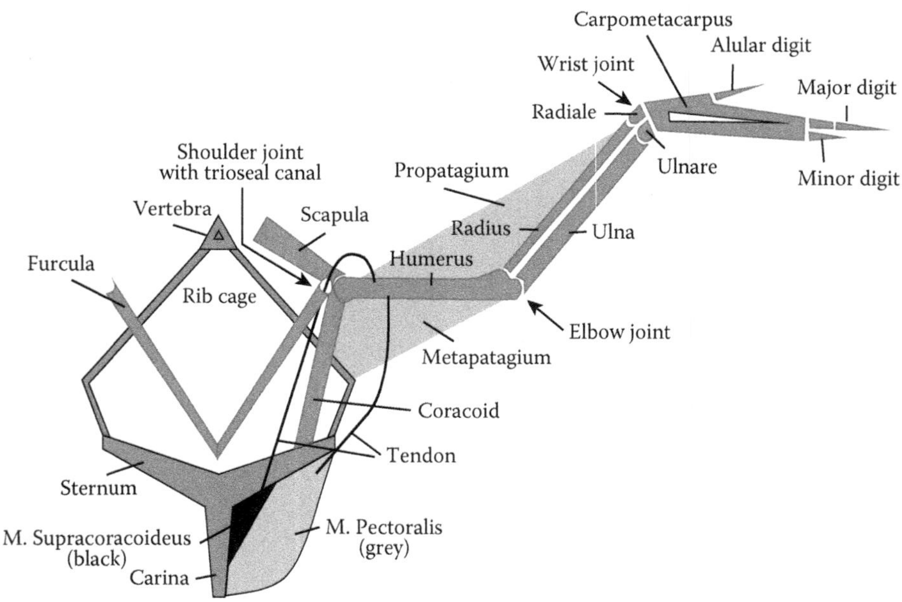

Figure 4.25. Schematic overview of the flight-related internal anatomy of the left wing, pectoral girdle, and rib cage of a generalized bird. (From Videler, J.J., *Avian Flight*, Oxford University Press, Oxford, UK, 2006. With permission.)

its length fixes the distance with the shoulder joint and counteracts the pulling forces of the main flight muscles. The sternum carries a central bony keel, the carina. Ribs, vertebral column, and sternum form a closed cage. The main flight muscles, the pectoralis and the supracoracoideus, originate on the sternum, the carina, and the coracoid. The main downstroke muscle, the pectoralis, inserts from below on an anterior crest on the humerus. It pulls the wing down and causes forward rotation (pronation) of the wing during the downstroke. The main upstroke muscle, the supracoracoideus, is situated underneath the pectoralis and forms a tendon that passes through the triosseal canal in the shoulder joint to insert from above on the upper part of the humerus. Its action lifts the wing and causes rearward rotation (supination). The canal acts as a pulley for the tendon.

4.5.1.3 Wings

Various types of contour feathers implanted in the skin of the arm allow a bird to fly. Each feather is unique and situated in a well-defined position. Feathers are arranged in tracts. Arrays of tracts are strictly organized (Lucas and Stettenheim 1972).

The wing of a barnacle goose in Figure 4.26 serves as an example. The contour feathers covering the front part of the arm wing are marginal coverts, implanted on the propatagium, the metapatagium, and on the skin covering humerus and radius and ulna. Small marginal coverts provide the rounded leading edge shown in the left cross section. Feathers of increasing size overlap each other like the tiles on a roof, with the smallest on the shoulder and largest further back. Larger marginal coverts overlap rows of secondary coverts, and these in turn overlap the greater coverts covering the implants of the secondaries of the arm wing and the primaries of the hand wing. The almost symmetrical tips of 11 secondaries form the sharp trailing edge of the arm part. Feathers of the hand wing include the alula feathers attached to the alular digit. This little winglet can be moved with respect to the rest of the wing. The most prominent feathers are the 10 primaries. By convention, number X is the most distal feather. The sharp narrow vane of that feather forms the main part of the leading edge of the hand wing. The narrow vanes of primaries VIII and IX are emarginated toward the tip. When the hand wing is fully spread emarginated feathers cause gaps forming a slotted wing.

The left profile, closest to the body of the bird, has a rounded leading edge and is extremely cambered. It differs therefore substantially from a conventional aircraft wing. The profile on the right through the primaries has a sharp, one-vane-thick leading edge and is hardly cambered. The central

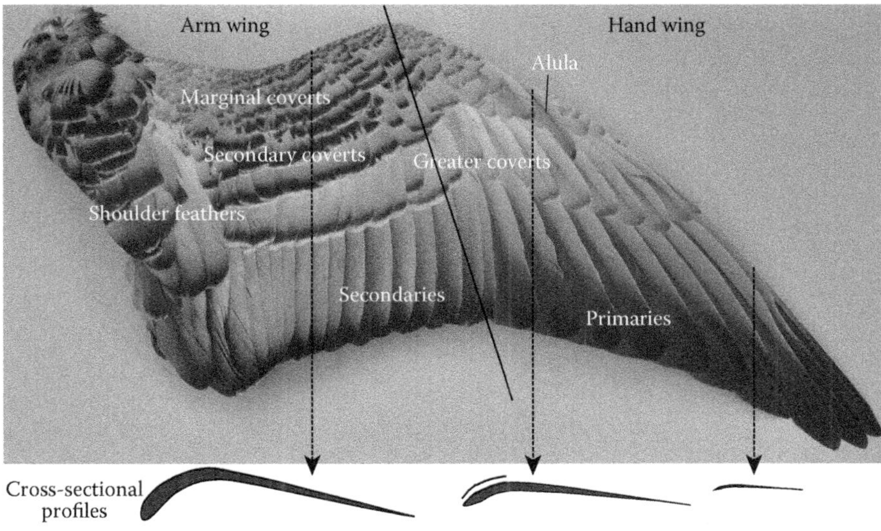

Figure 4.26. Top view of the right wing of a barnacle goose (*Branta leucopsis*) with the names of the most important feather groups and the contours of cross sections taken at three positions along the wing. The oblique line indicates the division into the arm and hand wing parts.

profile looks more like an aircraft wing, with a rounded leading edge some camber and a sharp trailing edge. The profile shows the alula in elevated position.

4.5.1.4 Wing Actions

The structure of the shoulder joint allows the following wing movements: up and down, fore and aft, and supination and pronation (backward and forward rotation along the length axis of the wing, respectively). The head of the humerus in most birds is not completely spherical, but more of an egg-shape, thereby reducing the freedom of movement. The range of possible movements is also limited by ligaments around the joint.

The mechanism moving the wing is surprisingly uniform among flying birds (Figure 4.27). The movements of the upper arm, the forearm, and the hand are flexion and extension of the wing and circumduction of the hand. The distal head of the humerus forms the elbow joint with the proximal endings of the ulna and the radius. When the wing is stretched, the shape of this joint severely limits dorsoventral rotation of the forearm with respect to the upper arm. Freedom of movement in the horizontal plane allows stretching and flexing of the forearm. During these movements, the radius shifts parallel to the ulna, inducing flexion and extension of the hand. The parallel shift has long been attributed to the shape of the distal head of the humerus. A knob on the head was thought to push the radius in an outward direction. A close examination by Vazquez (1994), however, showed that the shape of the humerus condyles in the plane of interaction with the radius and ulna were circular in flying birds. Rotation around a circular knob will not result in relative shift of the bones involved.

During wing flexion, the drawing parallel action of the radius and ulna is caused by collision of bulging muscles of the forearm and the upper arm when the elbow is flexed to angles smaller than 60°. The pressure of the abutting muscles dislocates the radius from the end condyle of the humerus and pushes it against the ulna. The shape of the facets at the position where radius and ulna meet moves the radius distally toward the wrist. During wing extension, elbow and wrist movements are also coupled. When the elbow angle widens the radius slides along the ulna because collateral ligaments attach it to the humerus. The distal end of the radius pulls via the radiale on the frontal edge of the carpometacarpus, extending it. More muscles and tendon complexes play a role in these complicated movements. The effect of the drawing parallel system on the movements of the hand becomes clear when we study the multiple joints of the wrist. Five bony elements are connected in the wrist: the radius and ulna, the radiale and the ulnare, and the hand skeleton. The shape of the connections provides and restricts the freedom of movement between the hand and the forearm. Vazquez (1994) distinguishes two distinct joints. The movement of the radius, both carpal bones, and the hand around the ulna head defines the first joint

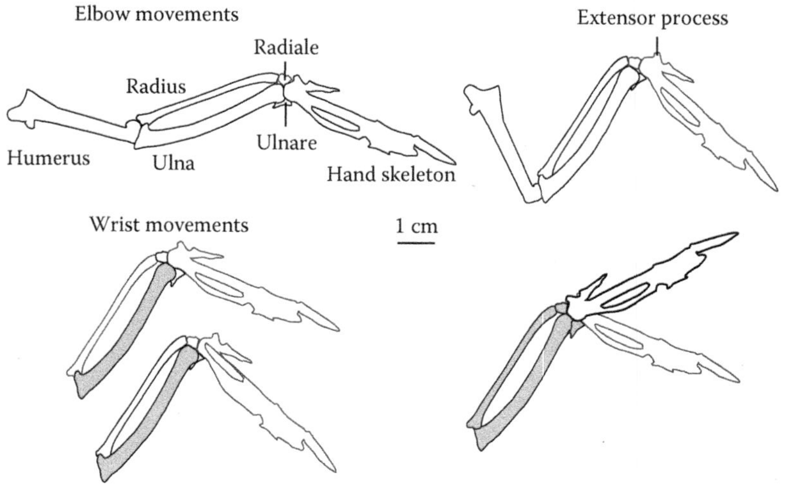

Figure 4.27. Movements of the skeleton of a pigeon wing. (Based on Vazquez, R.J., *Zoomorphology*, 114, 59–71, 1994.) (From Videler, J.J., *Avian Flight*, Oxford University Press, Oxford, UK, 2006. With permission.)

(drawing on the left). In the second joint, the hand flexes and extends with respect to the other bones (drawing on the right). During the downstroke, the plane of the hand is parallel with the plane of the wing. The wing is stretched and the hand cannot flex dorsally or ventrally about the wrist. The only movement possible is flexion in the plane of the wing. During the downstroke, the primaries of the hand wing attain the greatest vertical velocity and inflict large rotational forces on the wrist. The position of the supporting skeleton in the hand wing is close to the leading edge. The primaries form a large surface behind the skeletal support, causing a strong pronation tendency during the downstroke that results in forward rotational forces on the wrist joint. These forces are counteracted by an interlocking mechanism formed by ridges on the carpometacarpus, the ulnare, and the ulna in the joint. Supination of the hand wing around the skeleton in the leading edge is prevented by the radiale in articulation with the radius and the carpometacarpus where a ridge stops the movement. In most birds, the wrist joint can change from a stiff construction into a flexible construction during the early stages of the upstroke. This is caused by the ulnare gliding along the winding articular ridge of the ulna to its other extreme position. Due to this action, the hand can rotate over 90° with respect to the plane of the arm wing. Some birds show such a backward and upward hand flick also during vertical take-off and landing. A similar movement is also made when the bird folds its wings into the rest position. In the skeleton of the hand, the major and minor digits can be slightly curved in the plane of the hand, so movements perpendicular to that plane are limited. The alular digit is supported by one or two phalanges. The terminal one bears a claw in some orders. The phalangeal joints are saddle joints allowing 2° of freedom. The joint between the alular digit and the carpometacarpus is more complex. It allows the alula to be abducted and adducted from and to the leading edge of the wing as well as moved up and down. The joint also allows supination in the up position and pronation in the down position.

4.5.1.5 Tail

A bird's tail is supported by a few caudal vertebrae and the pygostyle, the modified coalesced last vertebrae of the vertebral column (Figure 4.28). More cranially, some caudal vertebrae are fused with the pelvic girdle in the synsacrum. The concave anterior and convex posterior globular surfaces forming the articulations between the free vertebrae allow movements in all directions. The pygostyle extends caudally in a vertical plate. Its connection with the last free vertebra is a horizontal hinge joint with a transverse hemicylindrical notch in the anterior part of the body of the pygostyle. On each side of the pygostyle, the rectricial bulbs form the seat of the large tail feathers, the rectrices. The bulbs are fibroadipose structures, partly encapsulated by muscle. Sockets on each side of the caudal vertebral column form joints in which the bulbs can move. Muscles are responsible for spreading the tail fan by pulling the calami of the rectrices together. Other caudal muscles function to hold and move the adjustable tail fan as a whole. Up to 24 rectrices in the tails of birds differ in length and shape to form an almost infinite number of tail designs. The variation is

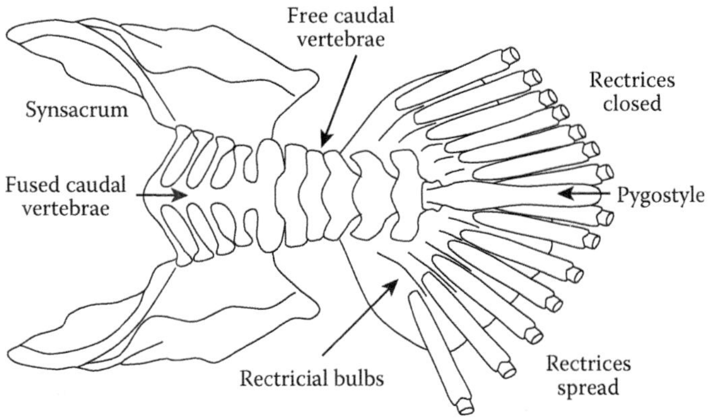

Figure 4.28. Schematic ventral view of the end of the vertebral column of the chicken.

not always related to flight; it can also be ornamental in males. Tail shapes not only vary among species and sexes but also may also change quickly in time due to spreading and closing of the tail fan. The left and right halves are usually symmetrical, but differences in the amount of spreading and tilting can create high degrees of asymmetry. The outer tail feathers can have a narrower outer and a wider inner vane, but most other feathers are left and right symmetrical. Birds usually have between 8 and 24 rectrices, but these are vestigial in grebes. In many birds, all rectrices are equal in length. This gives a folded tail a rectangular shape, and when spread the segment of a circle. In forked tails, the rectrices are increasingly shorter toward the center. Deeply forked tails have an inverse circular shape when spread. Shallow fork tails may show a straight trailing edge in the spread-out position. Some species have a wedge-shaped tail with slightly longer central feathers and shorter outer feathers. Such a tail is slender and spade shaped when spread out (the magpie is an example). Peculiar extremely elongated central feathers occur in both sexes in very distant groups. Flight-related functional explanations of tail configurations are usually fairly general, and rarely, if ever, backed up by experimental proof.

4.5.1.6 Other Flight-Related Parts of the Body

The overall shape of the body of a bird, including the head, usually approaches that of a streamlined body of rotation with a rounded leading part and a pointed trailing end. Streamlining means that the largest diameter is situated at approximately one-third of the length and that the ratio of diameter over length is about one quarter. This shape offers the smallest drag for the largest volume. Some, usually slower, species are not closely streamlined. Birds with long necks either stretch these during flight, as storks and swans do, or keep the necks folded as pelicans and herons. This influences the position of the relatively heavy head relative to the bird's center of gravity. Beaks form the leading structures in the flight direction of the flying birds. Legs and feet are generally tucked in during flight, but they are important during take-off and landing, because they can operate as airbrakes (webbed feet in particular), carry load, regulate the position of the center of gravity, and even serve as sea anchors (in Wilson's storm petrel).

4.5.1.7 Adaptations to Extreme Flight Conditions

Hummingbirds deviate so much from the basic bird pattern that their wing design deserves a separate coverage (see below). Hummingbirds are known to use a high (closely around 40 Hz) wing beat frequency for prolonged hovering in still air and for horizontal flight up to 50 km h^{-1}. The arm wings are extremely short with respect to the dominant hand wings.

On the other end of the breadth of bird flight apparatus designs, oceanic birds such as albatrosses and giant petrels are adapted to use long, slender wings to glide in the high winds of the southern oceans, migrating over thousands of kilometers at low energetic costs. The arm wings are elongated and the hand wings are relatively reduced in size.

HUMMINGBIRDS

The hummingbird configuration is described accurately by Stolpe and Zimmer (1939) and more recently by Hedrick et al. (2012). The relative dimensions of the bones of the wing skeleton are very different from those of other birds (Figure 4.29). The arm is extremely short because the humerus and radius and ulna are short and kept in a V-shaped position during flight. The angle of the V varies slightly; it cannot be greatly enlarged in a stretch because nerves and blood vessels run straight from the shoulder to the hand. The hand wing is relatively the longest found in birds. Hertel (1966) indicated that the hand wing of a hummingbird occupies 81% of the wing length against 41% in the case of a buzzard. There are only six partly overlapping secondaries in the arm; 10 long primaries form the main surface of the wings. The sternum bears a substantial carina. The main flight muscles, the pectoralis and the supracoracoideus, occupy about 27% of the body mass, with the pectoralis being only 2 times as big as the supracoracoideus (these figures are 18% and 12 times in passerines, respectively, according to Greenewalt 1975). An extremely long scapula supports the shoulder joint; it runs down the body to almost reach the pelvic girdle. Figure 4.30 shows the very short humerus with its bizarre shape. The articulating surface with the shoulder joint is not at the terminal position of the humerus, but there is a condyle at the inner side of the proximal end. The condyle in this position is a unique character of hummingbirds. The tendon of the supracoracoideus contains a sesamoid bone. It is attached to the outer part of the humerus head, runs

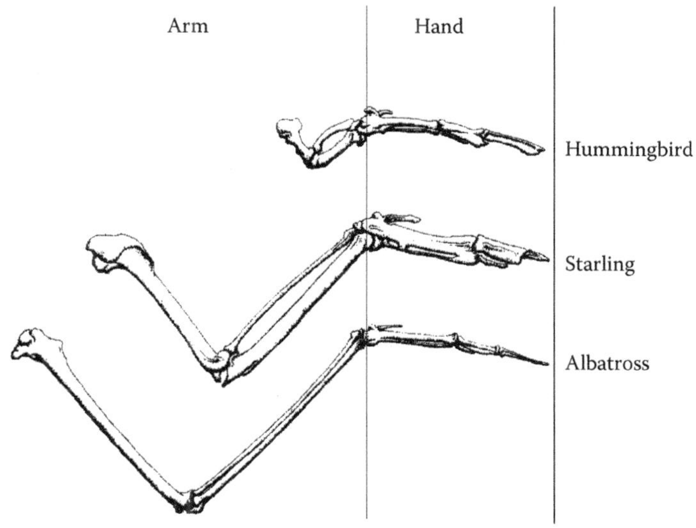

Figure 4.29. Relative dimensions of the skeleton of the forelimb of three species. The skeletons of the hand are drawn at the same length. (Based on Dial, K.P., 1992.)

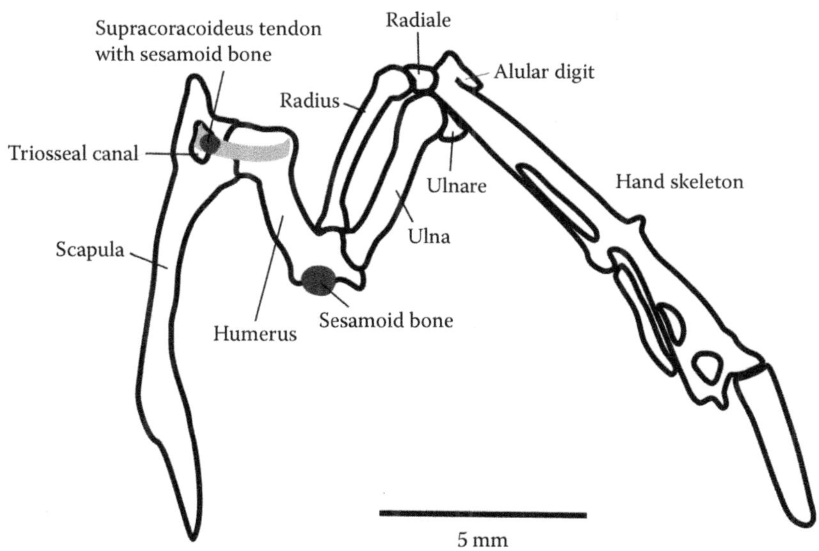

Figure 4.30. The special skeletal structures of a hummingbird wing. The drawing is based on photographs taken from an alizarin-stained skeleton of a cleared specimen. (From Videler, J.J., *Avian Flight*, Oxford University Press, Oxford, UK, 2006. With permission.)

through the triosseal canal and from there down to the muscle on the sternum. Contraction of the supracoracoideus causes adduction and rearward rotation around the length axis (supination) of the humerus. This rotation causes in fact the back (up) stroke of the wing. Pronation of the humerus by the pectoral muscle inserting on the front part of the humerus head will result in the forward (down) stroke. The unique long-axis humerus rotation converts small muscle strains into large-amplitude hand wing motion.

The elbow joint is peculiar, too, because it is obviously not designed to stretch. The muscles of the arm wing are extremely well developed and encapsulate the joint, keeping it in folded position. The extensor muscle (the scapulotriceps) has changed its function. The tendon contains a large sesamoid bone on the rear (upper) side of the elbow that determines its working direction. The sesamoid bone sits in a dent in the distal end of the humerus. Its presence causes the extensor muscle to rotate the ulna and radius backward (upward) instead of stretching the arm.

The capacity to rotate is even bigger in the complex wrist joint between the hand skeleton and the radius and ulna. The alula digit is reduced and immobile. (Hummingbirds have no alula.) The primaries are firmly attached to the bony elements, supported by cartilage and connective tissue. The pectoralis powers the hovering wing beat during the forward stroke and the supracoracoideus during the backstroke. These muscles rotate the vertical triangle formed by the V-shaped humerus and radius and ulna. The hand wing is attached to this triangle at the wrist and follows the movement. Combined rotations of radius and ulna and of the wrist joint enable the extreme rotation of the wing plane during the backstroke where the wing is used in almost upside-down position. The anterior edge of the hand wing is the leading edge throughout the entire beat cycle. Most of the hand wing has a positive angle of attack during both strokes.

During flight varying from hovering to horizontal flight at 12 m s^{-1} wing beat frequencies and wingspan hardly varies. The angles of the body and the stroke plane change with velocity as well as other kinematic parameters such as angular velocities and angles of attack. Wing beat amplitudes vary according to a U-shaped curve, being 109° during hovering, decreasing to a steady value around 100° at speeds up to 8 m s^{-1}, increasing to 126° at 12 m s^{-1} (Tobalske et al. 2007).

Albatrosses and Giant Petrels

Large oceanic birds are adapted to prolonged fast gliding in high winds. They have extremely long arm wings in common. The most extreme dynamic gliders among birds, the albatrosses and giant petrels, are believed to be capable to lock the wings in stretched position and in doing so avoid spending muscular energy to fulfill that task. Both the source of this knowledge and the mechanism behind it are difficult to track down. Hector (1894) gives, after "re-examining the wing of a large albatross in the flesh," the following record of his findings:

> The extensor muscular tendon, instead of being attached as in other birds only to a fixed process at the distal extremity of the humerus, is also attached by a subsidiary offset to a projecting patelloid bone which is articulated with the process, and thence proceeds to the radial carpal bone, and thence onward along the radial aspect of the manus, where it expands into fibrillae that embrace the quills. When the wing is fully extended the thrust of this projecting process on the elbow joint causes a slight rotation of the ulna on the humerus, so that the joint becomes locked, which renders the wing a rigid rod as far as the wrist joint. At the same time the slight play permitted by the articulation of the patelloid bone on the process allows of the transmission of the muscular pull from the shoulder to the manus without unlocking the joint.

This description does not explain clearly what is actually happening. Yudin (1957) disagrees with the idea that sesamoid bones are involved and offers an alternative explanation. The locking mechanism he describes is shown in Figure 4.31a. Tube-nosed birds have a bump in the saddle on the proximal end of the ulna. The sliding radius finds a stable position on each side of the bump. In fully flexed position, the radius has moved to the distal most point pushing via the radiale the carpometacarpus in the folded position. During extension, the radius slides along the ulna toward the humerus and needs to be pulled over the bump to reach the stable locked position of the fully stretched wing. However, X-rays I took of the elbow joint of a wandering albatross in extended and flexed positions (Figure 4.31b) suggest that the locking mechanism applies when the wing is completely folded after landing. This locking mechanism saves energy during the breeding period when the bird keeps its extremely long wings closed to the body with little muscle force. Pennycuick (1982) found a lock at the shoulder joint of albatrosses and the southern giant petrel. It consists of a fan shaped tendon running from the carina to the deltoid crest on the humerus. This tendon is superficial in the wandering albatross, the black-browed albatross, and the light-mantled albatross, and deeper inside in the southern giant petrel. By manipulating dead animals, Pennycuick found that the shoulder joint came up against a lock when raised to the horizontal position after the stretched wing had been moved forward to the fully protracted position. The lock no longer operated when the humerus was retracted a few degrees from the fully forward position or when the tendon was cut.

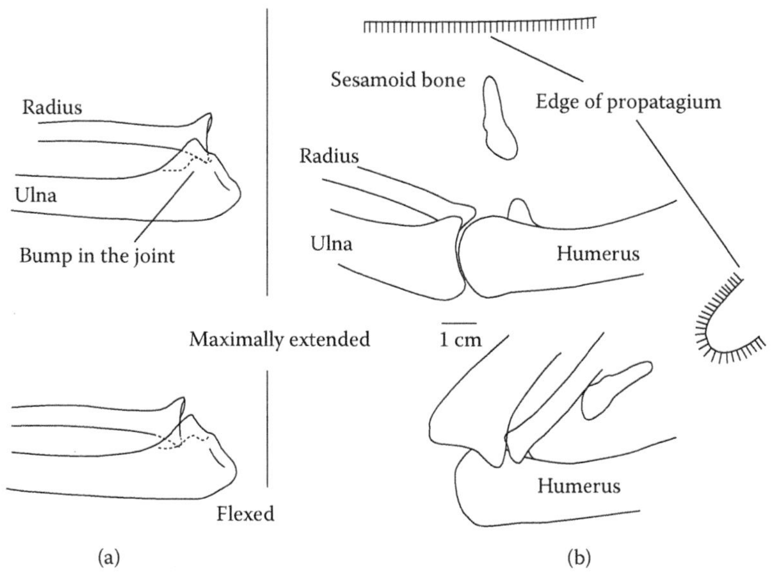

Figure 4.31. (a) The locking mechanisms in the wings of albatrosses and giant petrels according to Yudin (1957). The position of the radius with respect to the ulna in extended and flexed position and the bump in the joint are indicated. (b) Outlines of the skeleton from X-rays of the elbow joint of a wandering albatross in extended and flexed position (right wing in ventral view). The radius seems to snap in the distal-most position during extreme flexion only, securing the wing while folded. (From Videler, J.J., *Avian Flight*, Oxford University Press, Oxford, UK, 2006. With permission.)

The hand wings of albatrosses and giant petrels deviate from that of all other birds because the structure of the primary feathers is different (Boel 1929). Normally, the vanes of primaries are thinner than the shaft (see Figure 4.24). The upper surface is level with the dorsal surface of the shaft; on the lower face the shaft sticks out and forms a rim. The primaries of albatrosses and giant petrels differ from that pattern because their vanes are thick and form a smooth surface on the top and on the lower face of the feather. Stacks of these primaries form a rounded leading edge on cross section instead of a sharp leading edge as is the case in all other birds.

4.5.2 Flight with Skin Flaps: Bats

In bats the mammalian building plan is modified to make agile powered flight possible (Figure 4.32). The arms and legs support and move a large skin fold as one highly flexible wing. The feet are fitted with claws, allowing the typical hanging upside-down resting position. There is a lot of variation among the anatomies of the flight apparatuses of the micro and mega bats. The following description concentrates on the structure and function of the flight apparatus in micro bats, which have the majority in terms of number of species.

The arms are attached to the body in the shoulder joint between the humerus and the scapula. The rib cage bears a sternum that is connected to the scapula at the shoulder, joined by the curved collar bone (clavicle). The clavicles replace, in structure and function, the coracoids we saw in birds. (The coracoid bone does not exist in mammals.)

The wing skeleton consists of the humerus, a strong radius, and a reduced ulna; six carpal bones, five metacarpals, and four fingers. The first finger, the thumb, sticks outside the wing membrane and carries a functional claw. Its short metacarpal supports the wing. The other four metacarpals and the digits of fingers 2–5 are elongated and extremely slender. These and skeletal elements of the backward-pointing hind legs support the wing membrane. The second and third metacarpals and fingers are close together, strengthening the leading edge. To the rear, the femur, the tibia (and greatly reduced fibula), the calcaneum (a bony process on the heel bone) of the hind limb, and the tail support the wing membrane. All these skeletal structures form more than two dozen joints that are under more or less independent control (Schwartz et al. 2007). The wing consists of several functional parts.

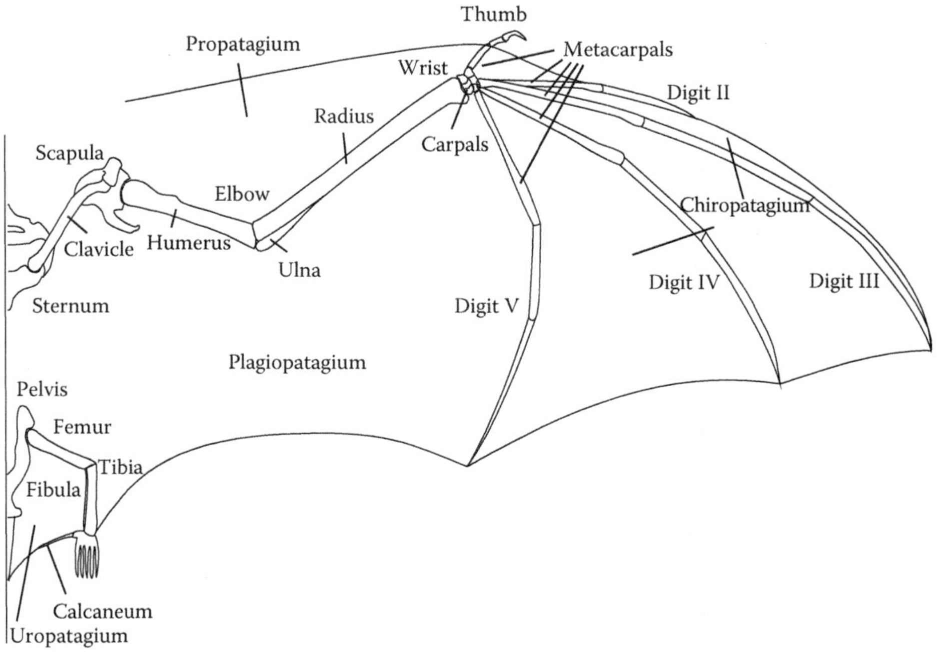

Figure 4.32. Schematic of the flight apparatus of a bat (not to scale).

The propatagium stretches from the neck to the metacarpal of the second finger, and it is supported by the short metacarpal of the thumb. A narrow membrane stretches between the second and third finger. These two flaps form the leading edge of the arm wing and can be moved up and down by the thumb, changing the camber of the wing. The main surface of the hand wing (the chiropatagium) consists of two parts separated by the bones belonging to the fourth finger. The armwing (plagiopatagium) stretches between the fifth finger and body and legs. The membrane between the tibia, the calcaneum, and the tail is the uropatagium or tailwing.

X-rays of flying bats reveal the mechanism used to beat the wings (Panyutina et al. 2013). Three joints participate in moving the humerus during the wing beat cycle caused by muscles situated on the body. About nine muscles are mainly responsible for the wing beat action (Figure 4.33). As in birds, the pectoralis is the main downstroke muscle, but the upstroke is powered completely differently.

The first joint is between the sternum and the clavicle. During a wing beat cycle, the clavicle rotates back and forth in the joint over 45°, and it swings laterally to and from the body over an angle of about 30°. The second joint between the clavicle and the scapula allows the rotation; the lateral movement of the clavicle displaces the shoulder joint. The humerus attachment in the shoulder joint provides the most degrees of freedom. The head of the humerus is either spherical or elongated, depending on the group of bats. The humerus can rotate up and down, fore and aft, and it is able to rotate

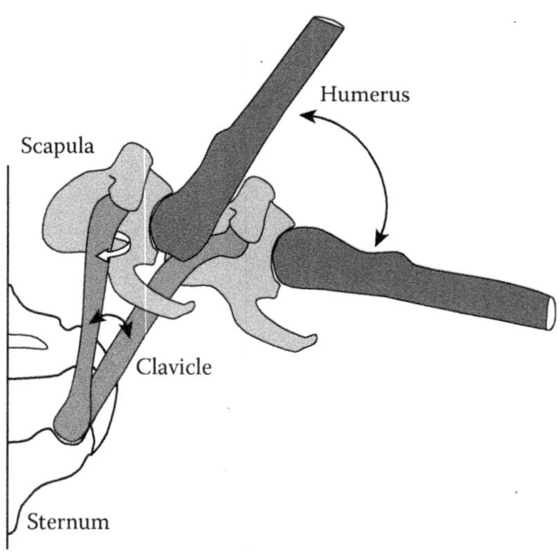

Figure 4.33. Bats: wing beat action of clavicle, scapula, and humerus. (Based on Panyutina, A.A., et al., *Anat. Rec.*, 296, 382–394, 2013.)

around its longitudinal axis. With respect to the rib cage the shoulder joint moves as a whole; viewed from the side it describes an ellipse and a circular movement in frontal view during each wing beat cycle. The wingstroke cannot be described as a simple upstroke and downstroke. The upward motion of the humerus proceeds that of the wrist; the fingers follow that movement, which finally reaches the wing tips.

Wing extension is mainly caused by the contraction of one muscle, the supraspinatus, situated on the scapula. It stretches the humerus, and this causes a chain reaction of muscles and tendons extending the elbow and spreading the hand (Neuweiler 2000). Both wings can be moved independently. All membranes are highly compliant and can be stretched and folded in flight. Muscles and elastic fibers inside the membrane maintain tension and control stiffness. The stretch and stiffness vary across the regions of the wing. The hand wing bones bend easily and can deform to measure half the resting length. Sensory hairs all over the wings report airflow conditions to the brain.

The aerodynamic characteristics of bat wings are difficult to assess due to the compliant nature of the wings. LEVs in combination with conventional flow patterns are most likely present on the beating wing, generating lift and thrust at any time of the cycle. The protruding thumb might serve as a generator of small-scale vorticity to separate steady and unsteady flow patterns. The complicated wake structure of flying bats results from the highly unsteady nature of the interaction between the membranous wing and the air in flight (Hedenström et al. 2007).

4.5.3 Flight with Skin Flaps: Pterosaurs

Pterosaurs are extinct, and the way their flight apparatus worked has to be deduced from the fossil record. This provokes a lot of disagreement among the many specialists in the field, especially so because these other-worldly flying reptiles are extremely popular. Fossilized skeletal elements are generally the only clues to deduct the way of flight as traces of the soft parts of at least 65-million-year-old remains are hard to find. Because the rare impressions of wing membranes belong to deceased specimens, they do not show the wings in flight. We can put together a pretty clear picture, though, from skeleton fossils (Figure 4.34): the wings were attached to both arms and legs, like in bats. The fossils tell us that to reduce weight, bones were hollow as they are in birds and part of an extensive air sac system used for respiration (Claessens et al. 2009). To reduce wing mass, the hand skeleton was strongly reduced in both the number and the diameter of the elements. Muscles serving the beating wings must have mostly originated on the main body to move the generally short and strong humerus from there. The rib cage was very sturdy compared to those of birds and bats. Several dorsal vertebrae were fused and their dorsal spines support a long

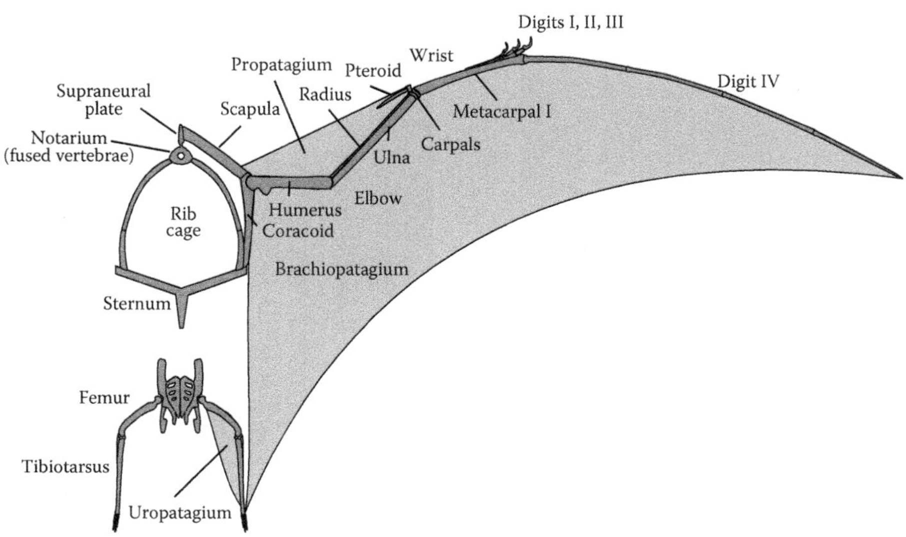

Figure 4.34. Schematic of the presumed flight apparatus of pterosaurs.

plate of bone, the supraneural plate, to which the scapula connected. The ribs were coalesced with the fused vertebrae and connected with a strongly built sternum. Coracoid and scapula formed a strong arch between the supraneural plate and the sternum. The clavicles were missing or fused with the sternum. The glenoid cavity for the articulation with the humerus was situated on the coracoid just under the connection with the scapula. The current opinion is that a strong pectoral muscle originated on the sternum and coracoid and inserted on the humerus to bring the wing down and that several muscles situated on the back were responsible for the upstroke.

The wing was supported by the humerus, radius, and ulna; carpals; metacarpals; and the four phalanges of the fourth finger. The humerus was relatively short and stout. The articulation with the glenoid cavity was saddle shaped to allow mainly up and down movements. It usually had a prominent deltopectoral crest serving as the point of attachment for the pectoral muscle. The elbow connection with the radius and ulna formed a simple hinge that could probably articulate over angles between 30° and 150°. Flexing the elbow caused the radius to slide along the ulna flexing the wrist by about 50° (Wilkinson 2008). Several carpal bones (fused or not) formed the wrist. One carpal connected to the pteroid, a bone unique to the pterosaurs. The joints of the wrist between radius and ulna and the metacarpals allowed rotational movements of the hand wing with respect to the arm. Flexion of the hand with respect to the arm over 150°–170° was possible. The first metacarpal is large with three slender metacarpals close aside, but these stretch usually not over the full length of the large metacarpal and are in that case not connected to the carpals of the wrist. The first, second, and third clawed fingers contain two, three, and four phalanges, respectively. The gigantic fourth finger usually has four phalanges. The joints between these are not very flexible. The joint between the first phalange and the metacarpal can be flexed over about 170°, allowing the wing finger to fold against the body.

There is still a lot of controversy among experts about the shape and structure of pterosaur wings. Witton (2013) reports the current state of the discussion. Recent interest in artificial aerial vehicles has increased interest in the precise configuration. The leading edge of the wing was formed by the propatagium stretching between the wrist and the shoulder. The pteroid bone supported that wing section by pointing from the wrist to the body and not straight ahead as has been suggested (Bennet 2007). This can be deduced from the fact that the pteroid articulated with the side of the preaxial carpal that would have prevented it from pointing forward. How much control the moveable pteroid gave over the leading edge is unclear. It is agreed that the main membrane stretched between the arm, the wing finger, the body, and the hind limb. Most of the discussion regards the attachment to the latter (Elgin et al. 2011). In all, 11 fossils show the connection in detail, and all suggest that the wing attached to the ankle. Another wing flap may have spanned between the legs, including the tail. But again, there is no agreement about various possible configurations. It could be a fully extended membrane stretching between the legs, including the tail and supported by an elongated fifth toe, or no membrane at all. Another possibility is that it was a small stretch of skin between the femur and the tibiotarsus not including the tail.

Imprints seem to show that the flight membranes had a complex meshwork of blood vessels, connective tissue, muscle fibers, and a layer of unique pterosaur fibers thought to be responsible for varying elastic properties in different parts of the wing. Some of these 0.2-mm-thick fibers stretch from the leading to the trailing edge. At the trailing edge, there are hints indicating that special fibrous structures served to keep the trailing edge taut. Palmer and Dyke (2012) believe that there is no need for such a structure as long as the shape of the trailing edge is concave enough to keep it tight when the wing is fully stretched.

The functional anatomy of the flight apparatus suggests that pterosaurs were presumably agile flyers that conquered most of the aerial niches that we see occupied by contemporary birds and bats. There are still many flight-related anatomical features to inspire scientists to search for aerodynamic explanations. For example, the role of the large cranial crest carried by the males of many species has been suggested to serve as a front rudder, improving flight control and turning abilities (Chatterjee and Templin 2012) during aerial courtship displays.

4.5.4 Winged Insects

Winged insects (Pterygota) are arthropods with a segmented exoskeleton mainly consisting of polysaccharide chitin fibers embedded in a protein matrix. It is a fibrous composite material that combines high strength with elastic flexibility. The insect body basically consists of three sections: head, thorax, and abdomen (Figure 4.35a). Maximum size and performance are restricted by the respiration system that is based on slow diffusion of oxygen through the body by the trachea, a simple system of thin-walled tubules. The following simplifying generalized descriptions of the flight apparatus are mainly based on Wooton (1992), Dudley (2000), and Chapman (2013).

Basically, the thorax contains three segments: prothorax, mesothorax, and metathorax (these segments can be integrated into one box). Each segment carries two legs. The prothorax is usually strongly reduced; the wings are attached to the mesothorax and metathorax, one pair of wings each. These two segments are mostly combined into the pterothorax because the two composing parts are not distinguishable. The thorax (Figure 4.35b) can be viewed as a box with sides formed by cuticular plates, the sclerites. The top plate is the notum, the bottom is the sternum, and the sides are named left and right pleuron. Anterior and posterior septa, the phragmata, form the intersegmental boundaries. Elastic elements of the thorax store and release kinetic energy during flight. Muscles inside the thorax control the wing actions.

Wing attachment to the thorax is complex and variable (Figure 4.36). A dorsoventral furrow on the outside of the pleuron indicates the position of a thick ridge on the inside. The wing is attached to the notum and hinges by using the pleural process above the ridge as fulcrum. A leg is attached below the ridge. A dorsal view on the wing attachment shows how small sclerites, the pteralia, are attached to the notum and form the start position of the wing veins. The shape, position, and articulation of the pteralia vary enormously and determine the degrees of freedom of the movements of the wing.

Insect wings are extensions of the exoskeleton and consist of a membrane formed by two

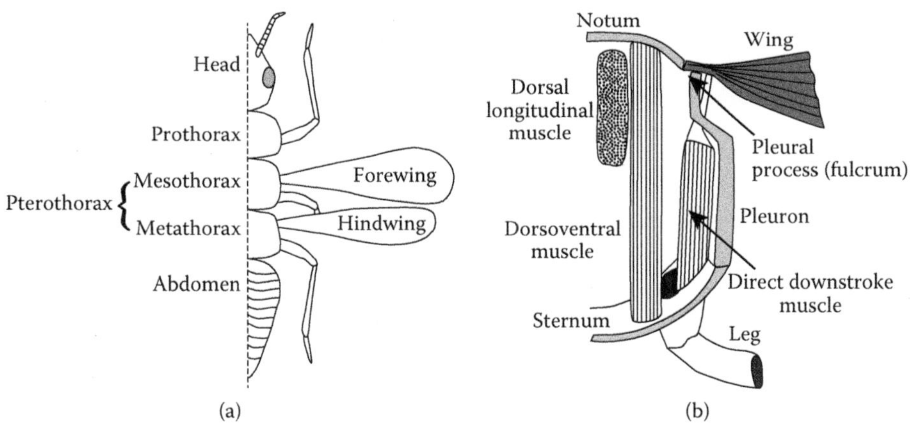

Figure 4.35. (a) Basic body plan of a winged insect. (b) Highly schematic view of insect major flight muscles and wing joint.

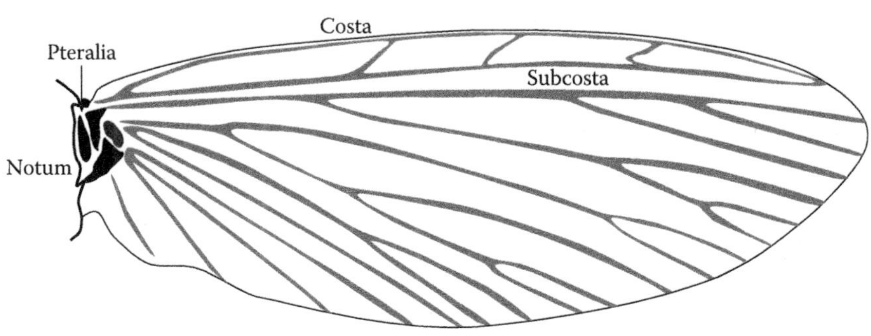

Figure 4.36. Schematic dorsal view of an insect wing.

cuticular layers supported by a meshwork of veins. There is no muscular tissue in the wings, and all the movements are induced by muscles from inside the thorax or from external forces. The vein system gives the wing rigidity and creates zones that can deform under outside pressures during flight and makes folding possible.

Veins are usually tubular with circular or oval cross sections and can be filled with hemolymph (insect blood), air-filled trachea, and nerves. Veins serve many functions. They determine the varying flexibility and stiffness of different parts of the wing and distribute blood, oxygen, and sensory information. Patterns of venation are highly specific and complex. Major longitudinal veins run from the pteralia to the outer wing margin. Bifurcation, cross-venation, and merging of veins are common features. The leading edge of the wing is usually stiffened by two thickened veins, running from the hinge with the thorax to the tip of the wing. Cross veins interconnect with longitudinal veins at junctions where they form flexible joints. The rigidity of wings can be increased by corrugation, where veins form the dorsal and ventral ridges. Flexion mainly occurs along flexion lines where veins are easily bendable due to thinner walls or smaller cross sections or by having a flexible annular structure. Folding often follows flexion lines, but can also happen along special fold lines. Both transversal and longitudinal folding is found in various insect orders. Wings are rarely flat, but they show relief by being cambered, corrugated, or both, usually to improve resistance against dorsal bending. Relief can have an important aerodynamic function. Camber can be structural and permanent or actively induced, for example, by pronation of the leading edge. Donoughe et al. (2011) found flexibility enhancing resilin and cuticular spikes in vein joints. Resilin, a flexible rubber-like protein (Weis-Fogh 1960), is the most elastic material in the world. Corrugated damselfly wings have resilin on top and underneath the vein joints of both the dorsal and ventral ridges (Figure 4.37). Resilin in joints of dragonflies is found below the joint of dorsal ridges and above it in the ventral ridges. The joints of wing veins of odonates may have spikes on the dorsal or ventral side, or both sides. Joints without spikes also occur. The function of spikes may be to stop the flexion or to adjust stiffness.

Wing motions can be up and down, and rotation occurs along the length axis for pronation (leading edge down) and supination (leading edge up) and fore and aft movements. Wing camber can be changed in some insects through muscular action from the base of a wing. The wings of some groups can also be flexed backward and extended. In some insect groups, wings can be folded using muscles inside the thorax. Muscle function in relation to the wing actions is extremely complex and mostly not fully understood. Muscles can either function

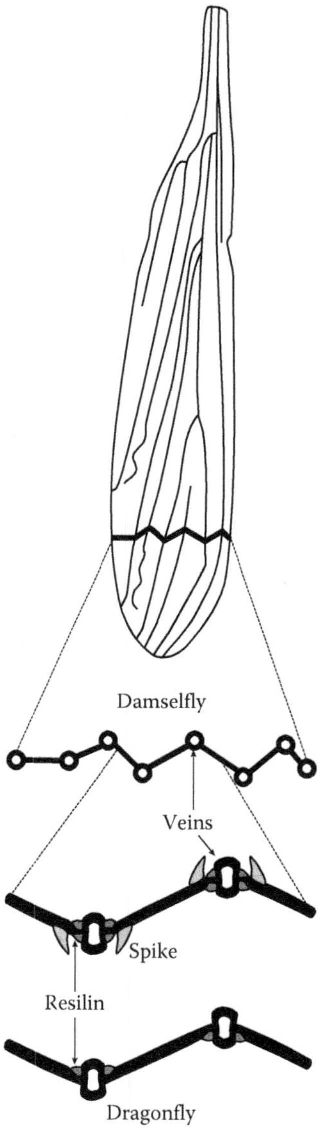

Figure 4.37. Corrugation, veins, and vein resilin and spikes of a damselfly. Dragonfly ridge and groove structure are added for comparison. (Based on Donoughe, S., et al., *J. Morphol.*, 272, 1409–1421, 2011.)

directly or indirectly. Direct acting muscles can be connected to the pteralia to perform wing actions or may serve to change the elastic behavior of the thorax or the hinge region. In all insects, indirect dorsoventral muscles between the notum and the sternum (and possibly some leg segments) depress the notum to elevate the wing during the upstroke (Figure 4.35b).

The downstroke can be realized by indirectly operating dorsal longitudinal muscles connecting the front and the rear partitioning septum of the pterothorax or by the actions of direct muscles. Contraction of the dorsal longitudinal muscles elevates the notum antagonistically with the dorsoventral muscles and thus depresses the wings using the same lever action over the pleural process. Some groups use this system exclusively to pull the wing down. Dragonflies, damselflies and cockroaches primarily use direct muscles for the downstroke. These pull outside the fulcrum of the pleuron. Other groups depress the wings by a combined action of the indirect longitudinal dorsal muscles and direct muscles.

Innervation of indirect wing muscles is asynchronous, with intermittent single nerve impulses. The contraction frequency of asynchronous muscles (also named fibrillar muscles) can approach values of 1000 Hz. The elastic thorax resonates due to the rapid action of the fibrillar muscles, and this in turn activates these muscles by stretching them. The resonance frequency of the thorax determines the wing beat frequency. The elasticity of the thorax not only activates the fibrillar flight muscles but also stores elastic energy toward the end of each half-stroke and releases it during the start of the following half-stroke. A large resilin patch in the hinge is stretched at the end of each upstroke or downstroke and releases the stored energy during the start of the next half-stroke, losing only 3% of the stored energy. This rapid repeated action is performed without fatigue even at the extremely high frequencies used by insects.

Direct downstroke muscles are commonly synchronous with one contraction after each innervation. Synchronous muscles usually operate at frequencies below 50 Hz, but they can also operate much faster at the cost of power output per unit mass. Various small muscles, operating synchronously, control rotational, fore and aft movements, and folding of the wings. Diptera for example have 18 control muscles. We have to bear in mind that every species of insect has its own specific system to operate the wing beat action and that our knowledge so far is based on a few species only.

Most insects have two pairs of wings. Odonata are able to move their forewings and hindwings independently. In other groups, the pairs of wings are coupled in some way. There are several possibilities of wing–wing coupling. Jugal coupling involves an enlarged lobe-like area near the posterior margin of the forewing that folds under the hindwing in flight. Many moths have a spine on the forward edge of the hindwing that hooks on to a loop on the underside of the forewing. Most butterflies, mayflies, and some moths show wing coupling where an area of the forewing overlaps the hindwing and pushes it down in flight. Rows of hooklets and overlapping brushes are used to connect the wings.

Insect wing surfaces can be fitted with various minute structures including tactile sensilla, colored scales, hairs, spines, and cones. Some of these microstructures may interact aerodynamically with the airflow, for example, to create microturbulence to reduce the thickness of the boundary layer and wing inertia. However, the function of many surface structures is not well understood.

The diversity of wing shapes is overwhelming. There are in fact as many different wings as there are species of flying insects. There are millions of extant species with numbers increasing as more are discovered. We only know the number of species in each order by approximation. The classification is disputed and changes continuously. A brief list of flight-related characteristics of some prominent orders follows.

Cockroaches (Blattodea, 4000 species) have leathery forewings (when present) and fan-like membranes as hindwings.

Beetles (Coleoptera, 350,000 species) normally possess thick, hardened wings that act as stabilizers in front and thin, flapping, membranous hindwings that are folded underneath the forewings when at rest. The wings are operated by indirect muscles.

Some earwigs (Dermaptera, 1200 species) are flightless. Those that do fly have two pairs of wings. The front pair is small and thick; the membranous hindwings can be folded extensively to fit under the tiny forewings.

Dipterans (Diptera, 151,000 species) have only one pair of wings: the forewings. The upstroke and downstroke are operated by indirect muscles. The miniaturized hindwings (halteres) function as gyroscopes during flight.

Strepsiptera (600 species) also only have one pair of wings, but these are the hindwings. In this group the forewings are miniaturized halteres.

Bugs including cicadas and aphids (Hemiptera, 85,000 species) have two pairs of wings: thickened front wings with membranous tips and transparent hindwings.

Sawflies, wasps, bees, and ants (Hymenoptera, 125,000 species) have two pairs of connected membranous wings, acting as one during flight operated by indirect muscles.

Scorpion flies and hanging flies (Mecoptera, 550 species) have four narrow wings with numerous cross veins.

Mayflies (Ephemeroptera, 2500 species) normally carry two pairs of densely venated wings. The hindwings are much smaller than the forewings and in some groups not present.

Termites (Isoptera, 2000 species). Some species carry two pairs of membranous wings; others are flightless.

Dragonflies and damselflies (Odonata, 5000 species) are fitted with two pairs of large, membranous wings moving independently.

Grasshoppers and relatives (Orthoptera, 13,000 species) have two pairs of wings: one pair is leathery and one pair is membranous.

Stick and leaf insects (Phasmatodea, 2600 species) lack forewings, but they have fan-shaped hindwings.

The adults of caddisflies (Trichoptera, 7100 species) are fitted with two pairs of hairy wings.

Butterflies and moths (Lepidoptera, 120,000 species) have two pairs of overlapping wings, acting as a unit during flight. They are covered by tiny, usually colored, scales.

The variation within the orders is overwhelming. Among Lepidoptera, for example, wing shapes vary from wings divided in several feathered plumes such as those of the white plume moth (*Pterophorus pentadactylus*) with forewings consisting of two plumes and hindwings of three feathered plumes, to the aerodynamically advanced wings of the hawk moths (Sphingidae). The hawk moths are among the fastest flying insects, reaching speeds of more than 50 km h^{-1}. The fastest is allegedly the horsefly *Hybomitra hinei* (Diptera), with a top speed of 145 km h^{-1}.

4.6 THE METABOLIC COST OF FLIGHT

Animals do not fly on kerosene, so what do they burn? Their engines consist of muscles moving the wings by contractions. Energy is released by degrading ATP, an energy-rich molecule with three phosphate groups, into ADP with two phosphate groups. The loss of one phosphate delivers the energy used by cross bridges to drag thick and thin muscle filaments along each other while shortening or increasing tension. Reverting ADP to ATP requires energy that comes from burning lipids (fat), proteins, or carbohydrates as fuel. This process uses oxygen and produces carbon dioxide (CO_2), water, and heat. Per unit mass the ratio of CO_2 produced over the amount of oxygen (O_2) used (the respiratory quotient [RQ]), the amount of water and heat, and the amount of energy delivered differ among fuel types.

Exchange of O_2 and CO_2 during respiration are frequently used to measure energy turnover.

Fuel	Energy kJ g^{-1}	CO_2 l g^{-1}	CO_2 kJ l^{-1}	O_2 l g^{-1}	O_2 kJ l^{-1}	RQ
Lipids (fat)	39.7	1.43	27.8	2.01	19.8	0.71
Proteins	17.8	0.70	25.4	0.95	18.7	0.74
Carbohydrates	16.7	0.80	20.9	0.80	20.9	1.00
Commonly used figures for an unknown mixture of fuel					20	0.79

NOTE: RQ = respiratory quotient.

In principle, there are several possibilities to experimentally determine the metabolic energy used in flight: directly from the amount and type of fuel used or from the heat produced, or indirectly by measuring the RQ or the amount of water produced. One problem is that it is usually not clear what type of fuel is used.

For animals that do not consume food during flight (e.g., micro bats, swallows, and swifts), measurement of loss of mass is an option to determine the metabolic cost of flight. This approach is especially relevant in cases where it is possible to establish which fuel has been burned. In addition, the duration or distance flown must be substantial to obtain figures that reach significance over the inevitable noise caused by all kinds of inaccuracies and conditions not under control of the investigators. With measurements in wind tunnels, some of the experimental control problems can be minimized. Closed wind tunnels can double as respirometers to measure oxygen consumption and CO_2 production. Birds and bats wearing masks used to collect the respiratory gases can be flown in open wind tunnels at a range of speeds to measure the metabolic costs of flight.

Heartbeat frequencies provide an indication of the amount of blood that is circulating. It should therefore be possible to use the measurement to approximate the volumes of gases exchanged per unit time. However, the relationship between heart rate and oxygen consumption has to be established for each individual case. An advantage is that in reasonably sized, free-flying animals heart rates can be monitored over long periods.

Two methods use stable (as opposed to radioactive) isotopes to estimate energy turnover: doubly labeled water (DLW) and heavy carbon (HC).

In DLW, the hydrogen molecules are replaced with deuterium, a stable isotope of hydrogen with two neutrons, and the oxygen is heavy oxygen, with 18 instead of 16 neutrons per atom. The resulting water (H_2O) is chemically identical but physically distinct. It follows the fate of water in the fundamental reaction that releases energy from fuels: carbohydrate (any fuel) + $O_2 \rightarrow CO_2 + H_2O$.

In the DLW method, the heavy water is injected in the blood circulation of the experimental animal. A first blood sample is taken as soon as the sample is mixed into the blood and gives the initial concentration of heavy water in the blood. These concentrations are then compared in a second sample, taken after a precisely monitored flight period. The deuterium gets diluted by water produced in the burning of fuel. The CO_2 produced in the equation takes one of its oxygen molecules from water in the body and heavy oxygen present in CO_2 in the blood sample betrays this. The amount of energy used can now be calculated if the RQ (the substance burned) is known or can be estimated.

The HC method requires an injection with a solution of sodium bicarbonate ($NaHCO_3$) in which the carbon is heavy carbon, with 13 neutrons instead of 12 neutrons. In the blood, this forms an equilibrium with CO_2. The ratio between the concentrations of the two isotopes can be measured in the expired CO_2 and the change of that ratio during the exercise is a measure for the CO_2 production and therefore for the energy used.

The registrations of the flight activities of the experimental animal during the period between the measurements are crucial. Flight is the most energy-expensive activity and the greater the proportion of the time spent the more accurate the estimates are. The methods can be used in the field but also under more controlled conditions in wind tunnels. The reliability of the results of flight costs studies with experimental flight duration, and numbers tested are important determinants. In that respect, the most reliable figures for birds have been measured in wind tunnels using DLW. In a wind tunnel in Lund, Sweden, four red knots flew 28 sessions of 6–10 h at a speed of 15 m s^{-1} and used about 13.5 W each (Kvist et al. 2001). Rosy starlings used 8 W while flying at 11 m s^{-1} during 27 bouts of 6 h each in the wind tunnel of the Max Planck Institute for Ornithology in Andechs, Germany (Engel et al. 2006).

4.6.1 Insects

Measuring metabolic rates during flight of insects is not an easy task because of the small sizes of the animals and the minute respiration volumes. Closed circuit wind tunnels have to be extremely small to allow the measurement of oxygen consumption, CO_2 production, or both above the noise. An advantage is that insects are easily induced to beat their wings while tethered to a fixed position. Airflow at different speeds can be applied. Tethered flight is not as artificial as it may seem: honey bees show tethered flight naturally while either ventilating the hives or during fanning behavior to disperse pheromones (Junge 2002).

Fewer than 10 studies can be found in the literature where figures are presented estimating the metabolic costs of flight in insects. All of these studies except one date from the second half of the last century. The smallest insects measured are fruit flies, black flies, and mosquitoes. They weigh in the order of a few milligrams and have flight costs of less than a milliwatt, which is equivalent to the amount of energy used by modern hearing aids. Blow flies and horse flies are about 10 times heavier and use about 10 mW, which is in the range of a laser pointer. The most frequently investigated insect is the honey bee. The body mass is around 100 mg. A worker leaves the hive to forage weighing 80 mg, using in the order of 40 mW, and returns fully loaded with pollen at 140 mg, burning 65 mW (Wolf et al. 1989). The flight speed of insects does not affect the energy use a great deal. Most of the energy is probably used to make the thorax resonate by using the main flight muscles. Smaller muscles fine-tune the wing action to adjust the speed from hovering on the spot to the maximum speed without changing the costs much. Bumble bees seem to be surprisingly cheap flyers for their body mass. There is one measurement by Ellington et al. (1990) showing that a bumble bee of 0.5 g can fly using less than 20 mW. The biggest insect measured is the locust. The metabolic rate of a 2-g animal was measured by Weis-Fogh in 1952; it used 174 mW. The oxygen consumption of a 1-g locust showed that it used 134 mW according to Snelling et al. (2012), equivalent to the maximum output of a Wi-Fi router.

4.6.2 Birds

The energy consumption figures for birds in flight have been measured much more often. All the techniques mentioned before have been used mainly during the past 40 years of the twentieth century. A few papers on this topic have been published more recently. Forward flight and hovering on the spot have been measured. Sustained flight figures cover a large range of body masses and vary from a 5.5-g hummingbird that uses 1.8 W to a 10-kg wandering albatross gliding over the southern oceans at a cheap rate of about 40 W. Selected reliable data for the most economical performance of 39 species are shown in Table 4.2 and in Figure 4.38. All birds were measured flying at the maximum range speed, covering large distances as economically as possible. The dotted line in the graph of flight cost versus body weight plotted logarithmically follows the maximum values found; only the bar-headed goose value is above that line. It probably represents a kind of limit. The slope is two thirds. If that is the case, the maximum flight costs in birds in relation to body mass follows: maximum flight costs (W) ≈ 60 mass$^{0.667}$ (kg).

The exponent suggests that the relationship is determined by some surface area, because mass is proportional to length cubed. It is not the surface area of the lungs because that increases with mass in birds. Wing surface area over a large number of species (excluding hummingbirds) has been shown to increase as mass$^{0.72}$. This number is close enough to assume that wing area is an important factor determining the cost of flight.

Some birds manage to fly at considerably lower costs in comparison to their weight than others. The lowest flight cost value is that of the 17.8-g house martin using 1 W; the maximum value of 135 W is used by the 2600-g bar-headed goose. Albatrosses are relatively cheap for their body weight because they use a dynamic gliding strategy to use the prevailing high winds typical for the southern oceans. Barn swallows, martins, and swifts are aerial feeders that minimize the cost of flight during foraging by interspersed gliding, amounting to values between one quarter and half the flying time. Wilson's storm petrel uses wind energy to fly at low costs during foraging, gliding close to the surface, and by hanging in the air as a kite on one spot by using its webbed feet dangling in the water as sea anchor. Sooty terns are also extremely aerial oceanic birds that use frequent gliding to lower costs. Thrush nightingales are long-distance migrants, but it is not clear how they minimize flight costs; natural selection has improved flight performance of long-range migrants.

4.6.3 Bats

There are 12 reports of measurements of flight costs in bats: 8 on micro bats and 4 on mega bats (Table 4.3). These reports cover the size range adequately, from the 6.7-g common pipistrelle to the black flying fox of 820 g, one of the biggest bats, with more than a meter wingspan.

The flight costs indicated here are costs at maximum range speeds, to the best approximation, and hence best performance values. These are related to body mass by a power curve suggesting the following: bat flight costs (W) $\approx$ 63 mass$^{0.82}$ (kg) (Note that the equation above for birds differs from this equation for bats because it is an indication of the maximum values for flapping flight.)

Birds and bats of similar sizes can be compared: The micro bats have an average value of 142 W kg^{-1}. Birds in the same weight category, with masses <22 g, use 157 W kg^{-1}. This indicates that small birds are less economical flyers than equal-sized bats. Mega bats achieve, on average, masses of 78 g, and use 1 W kg^{-1}, whereas birds in the same size class (from 93 to 820 g) are slightly cheaper at 76 W kg^{-1}.

TABLE 4.2
Accumulated flight cost data for birds arranged in order of body mass.

Species	Mass (g)	Flight cost (W)	Method	Speed (m s^{-1})	Duration	Gliding (%)
Green violet-ear	5.5	**1.8**	R/W	11	1–8 min	
Palestine sunbird	6.2	**1.6**	C/B		2 min	
Gould's violet-ear	8.5	**2.5**	R/W	11	1–8 min	
Pine siskin	12.5	**3.0**	G/F	15	56 min	0
Sand martin	13.7	**1.6**	D/F		12.7 h	21
Zebra finch	14.5	**2.2**	C/B		2 min	
Barn swallow	17.3	**1.3**	G/F		>2 h	
House martin	17.8	**1.0**	D/F			54
Barn swallow	19.0	**1.3**	D/F			
Barn swallow	19.0	**1.6**	D/F	11		
House martin	19.7	**1.1**	G/F		>2 h	54
Chaffinch	22.3	**4.5**	G/F	15	56 min	0
Brambling	23.2	**4.6**	G/F	16	52 min	0
Thrush nightingale	24.7	**1.7**	G/W	7.9	15 min	0
Thrush nightingale	26	**1.9**	G/W	10	12 h	0
Eurasian bullfinch	29.5	**5.6**	G/F	14	60 min	
Swainson's and hermit thrushes	30	**4.3**	D/F	13	7.7 h	
Budgerigar	35.0	**4.1**	R/W	12	0.5–2 h	
Common swift	38.9	**1.8**	G/F		>2 h	70–80
Wilson's storm petrel	42.2	**1.8**	D/F		2–4 days	
Purple martin	50	**4.1**	D/F	8	4–8.5 h	
Rosy starling	71.6	**8.0**	D/W		>6 h	
European starling	73	**9.0**	R*/W	16	>90 min	17
European starling	77	**10.5**	D/F	14	3.5 h	
European starling	89	**12**	R/W	9.9	12 min	
Red knot	128	**13.5**	D/W	15	6–10 h	
Common kestrel	180	**13.8**	G/B	9	49 days	30
Sooty tern	187	**4.8**	D/F	10	8–23 h	5–25
Common kestrel	213	**14.6**	D/F	8	2–5 h	
Common teal	237	**13.2**	G/W	11.5	15 min	0
Fish crow	275	**24.2**	R/W	11	15–20 min	
Laughing gull	277	**18.3**	R/W	12	20–30 min	
Bar-tailed godwit (♂)	282	**17.8**	G/F	16	>24 h	
Laughing gull	322	**26.3**	R/W	13	20–30 min	
Pigeon	330	**25.4**	R/W	13	>1 h	
Bar-tailed godwit (♀)	341	**24.2**	G/F	16	>24 h	
Pigeon	394	**31.9**	D/F	17	7–8 h	
Pigeon	394	**33.1**	G/F	17	7–8 h	
Pigeon	425	**34.1**	R/F	19	3 h	

(Continued)

TABLE 4.2 *(Continued)*

Accumulated flight cost data for birds arranged in order of body mass.

Species	Mass (g)	Flight cost (W)	Method	Speed ($m\ s^{-1}$)	Duration	Gliding (%)
Pigeon	442	**26.8**	R/W	10	10 min	
Chiuahuan raven	480	**32.8**	R/W	11	30 min	
Red-footed booby	1001	**24.0**	D/F		5–28 h	
Barnacle goose	2100	**102**	H/F	14-20		
Cape gannet	2507	**85**	D/F	13	14–18 h	
Cape gannet	2580	**81**	D/F		1 day	
Bar-headed goose	2600	**135**	H/F	16-21		
Laysan albatross	3064	**24**	D/F		3 days	
Northern gannet	3210	**97**	D/F		4–11 h	0
Black-browed albatross	3580	**22**	H/F	8	9 days	91
Grey-headed albatross	3707	**28**	D/F		3–4 days	97
Southern giant petrel	3900	**68**	D/F			76
Wandering albatross (♀)	7300	**31**	D/F		3–9 days	97
Wandering albatross (♂)	9310	**45**	D/F		2–8 days	97
Wandering albatross (♀)	9360	**43.8**	D/F	5	4–7 days	97
Wandering albatross (♂)	10,740	**38.1**	D/F	5	4–9 days	97

SOURCE: Sources can be found in Videler (2006), except Mullers et al. (2009) (second cape gannet) and Engel et al. (2006) (rosy starling).

NOTE: The abbreviations under the column "Method" are B = indoor flight, C = heavy carbon, D = doubly labeled water, F = free flight, G = mass loss, H = heart rate telemetry, R = respirometry (R* without mask) and W = wind tunnel.

Bold figures emphasize the actual costs.

4.6.4 Comparing Flight Costs

Is it possible to compare metabolic costs among animals, and with aircraft? An Airbus A380 with a fruit fly? Flyers of hugely different body masses can only be compared using dimensionless figures for the cost of transport (COT). To compare insects, birds, bats, and airplanes, the only way is to compare the amount of energy used per unit weight and per unit speed.

Energy used is expressed in Watts, equal to Joules per second. If we divide this figure by the weight of the flyer in Newtons multiplied by its speed in meters per second, we reach a dimensionless number ($W/N\ m\ s^{-1} = J\ s^{-1}/N\ m\ s^{-1} = J/Nm = -$). Insects like fruit flies and honey bees score COT's of around 6. There is not enough data on insects to detect a trend with body mass. Birds and bats, however, show that COT values decrease with body mass. Table 4.3 shows the values dropping from 4.13 for the 6.7-g common pipistrelle to 0.66 for the black flying fox weighing 820 g. Birds show the same trend, although the data are more scattered. If we exclude frequent gliders the correlation becomes stronger. The 25- to 26-g thrush nightingales score between 0.9 and 0.7 at speeds of 8 $m\ s^{-1}$ and 10 $m\ s^{-1}$, respectively. A 237-g common teal reaches a COT of 0.5 when flying in the same wind tunnel at a speed of 11.5 $m\ s^{-1}$. A 2.6-kg bar-headed goose, who uses continuously flapping flight, has been measured at speeds varying from 16 to 21 $m\ s^{-1}$. Its COT varies with these speeds between 0.33 and 0.25. The gliding habits of wandering albatrosses reduce their COT to around 0.1, making them the cheapest flyers in the animal kingdom. How about airplanes?

The largest commercial airliner, the Airbus A380, scores a COT of 0.1, matching the wandering albatross, assuming a transported mass of 376,000 kg and a speed of 900 $km\ h^{-1}$. The Boeing 737-400 (5000 kg, 850 $km\ h^{-1}$) flies at a COT of 0.3. The decreasing trend with mass is shown among aircraft just as well as among birds and bats.

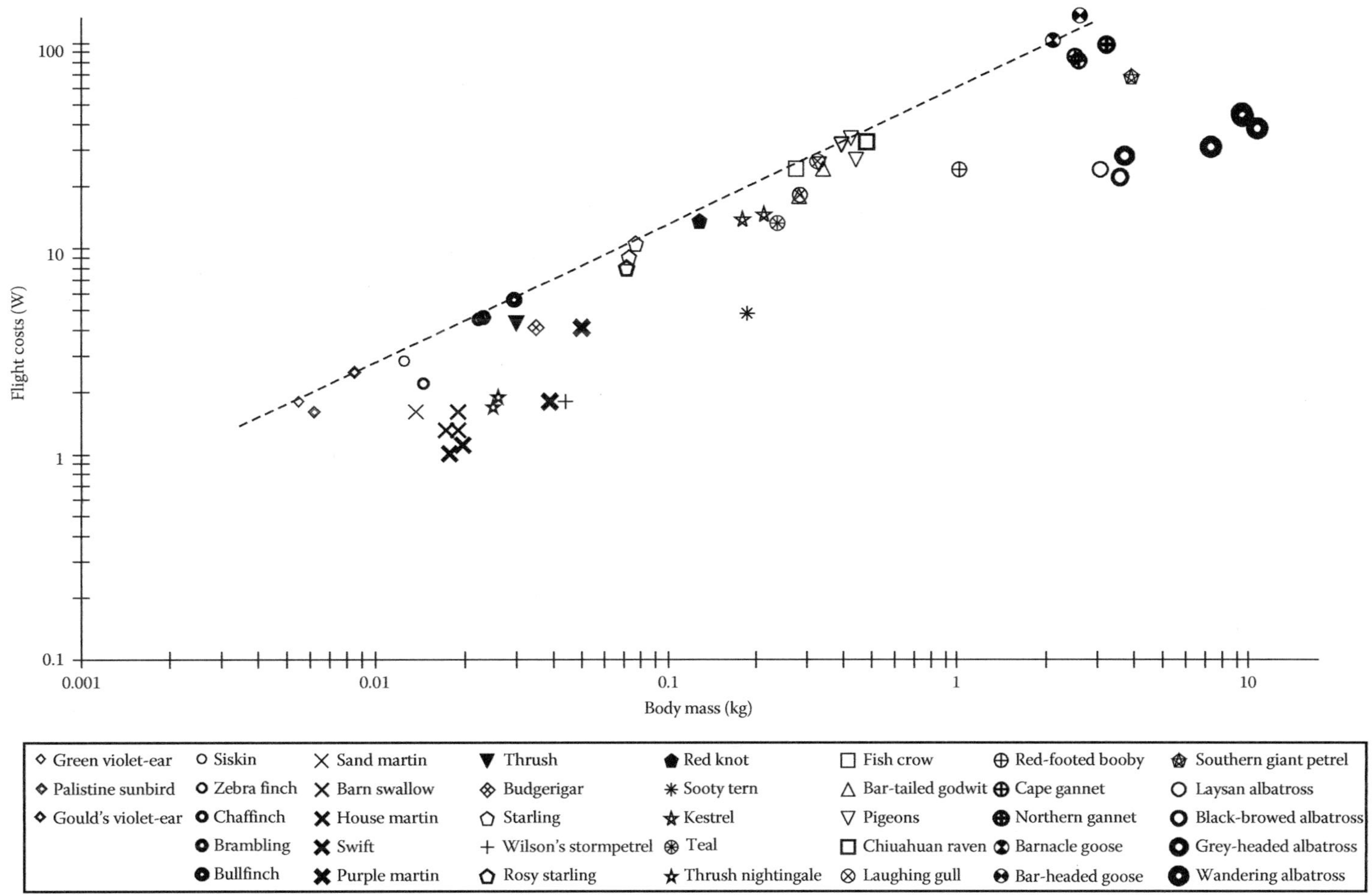

Figure 4.38. Empirically determined flight costs of birds as a function of the body mass (note the logarithmic scales). (From Videler, J.J., *Avian Flight*, Oxford University Press, Oxford, UK, 2006. With permission.)

TABLE 4.3
Empirically determined flight cost data for bats.

Species	Mass (g)	Flight costs (W)	Speed ($m\ s^{-1}$)	COT	Mass specific ($W\ kg^{-1}$)	Method	Reference
Pipistrellus pipistrellus	6.7	**1.1**	4.1	4.13	164.2	D/B	1
Plecotus auritus	7.9	**1.19**	4.3	3.60	151.0	D/B	1
Glossophaga commissarisi	8.6	**1.23**	6.5	2.24	143.0	G/W	2
Hylonycteris underwoodi	8.7	**1.31**	5.9	2.63	150.0	G/W	2
Glossophaga soricina	11.6	**1.60**	7.2	2.00	138.5	G/W	2
Carollia perspicillata	18	**2.50**	7.0	2.02	138.9	C/W	3
Leptonycteris yerbabuenae	22	**2.38**	6.0	1.84	108.2	C/W	4
Phyllostomus hastatus	93	**8.76**	8.0	1.20	94.2	R/W	5
Hypsignathus monstrosus	258	**22.94**	8.0	1.13	88.9	R/W	6
Eidolon helvum	315	**23.25**	8.0	0.89	73.8	R/W	6
Pteropus poliocephalus	574	**40.00**	8.5	0.75	69.7	R/W	6
Pteropus gouldii	820	**52.26**	8.9	0.66	63.7	R/W	7

NOTE: The abbreviations under the column "method" are B = indoor flight, C = heavy carbon, D = doubly labeled water, G = mass loss, R = respirometry and W = wind tunnel.

REFERENCES: 1: Speakman and Racey (1991) (speed: Norberg 1990); 2: Winter and von Helversen (1998) (speed: Winter 1999); 3: von Busse et al. (2013); 4: von Busse (2011); 5: Thomas (1975); 6: Carpenter (1986); 7: Thomas (1981).

4.7 EVOLUTION OF POWERED ANIMAL FLIGHT

In the animal kingdom, powered flight evolved independently at least four, but probably many more times. Each time it required the development of wings and the technique to use the wings in interaction with the air to generate lifting forces exceeding the weight of the animal. Evolution does not produce innovations from scratch as an engineer would do. Errors in the process of DNA duplication result in random, very slight changes to an existing design (Jacob 2001). The result is never the perfect solution an engineer would try to produce with the available technological means because there is no preconceived plan. Flight was not a distant goal to be reached. Unlike engineers, nature is not pressed for time in its design procedure; instead, it takes hundreds of millions of years innovating continuously through trial and error (Gregory 2008). Any innovation needs to be derived at by small steps, with each intermediate giving advantage over the previous to allow selection of the genotype.

A gradual change from terrestrial to aerial locomotion is difficult to imagine. Flying is lethally dangerous if it is not done properly and with the right equipment. How did it manage to evolve so many times?

4.7.1 Evolution by Natural Selection

About 1.5 centuries ago, two British scientists, Alfred Russel Wallace and Charles Darwin, hypothesized how organisms diversified and speculated and how natural selection drives the process of gradual changes resulting in an endless array of new species. This process we call evolution. The insight implies that all living organisms took 4 billion years to reach their current state and are still evolving. The principle is simple but often not well understood. In each generation, offspring show slight variations due to small genetic differences emerging by chance. The environment subsequently selects the individuals that are best fitted to meet the current criteria successfully and rewards these individuals with more fertile offspring. This is a gradual process where time is not a limiting factor. Innovations always start off from the structures and functions at hand. To understand the origin of species and their properties, it is necessary to gain insight into the selection criteria. Every species living or extinct is or was best adapted to the circumstances in which it lives. Adaptations are found in the smallest details. Extreme conditions often induce the weirdest adaptations. Nature has often arrived at several independent solutions for complicated technical problems. Extinct and extant species are not intermediate steps in any "right" direction. The direction of evolution is only determined by changing ecological circumstances. Every organism carries properties of the developmental history of its ancestors. To what extent organisms are adapted optimally depends on the stringency of the selection criteria and on the impact these criteria have on the ability to produce fertile offspring.

Fossils reveal the successive slight modifications that lead to the formation of the complex organs that enable flight. Fossilization is a rare process, and arrays of fossil remains showing the complete history of an unbroken chain of ancestors do not exist. Relationships among (usually) fragments of petrified animal remains are never absolutely clear and commonly sources of controversy.

Flight is a prime example of independent evolution of a complex behavior that evolved in different groups of organisms according to fundamentally different principles.

The evolution of a flight apparatus requires many parts of the body to change in concert. These parts were probably serving different functions at the beginning of the long process.

4.7.2 The Obscure Distant Past of Early Insects in Flight

Winged insects (Pterygota) radiated massively during the Carboniferous period 360–300 million years ago. Early fossils are extremely rare and show fully developed wings. It looks as if winged insects appeared abruptly, with nearly the current diversity of forms evident straight away. Noticeable is the widespread gigantism (i.e., dragonflies with a wingspan of more than 70 cm). This was made possible by higher oxygen levels, increasing the amount of oxygen that could be absorbed through tracheal diffusion.

The origin of insect wings is highly controversial. There is no agreement whether there was one wingless ancestor or whether there were more. It may be that the following hypotheses are all true, but for different lineages. There is no conclusive proof for any of these scenarios describing the transition of flightless ancestor to winged insects.

Arboreal ancestors may have developed flattened lateral lobes on the thorax or on abdominal segments to serve as gliding aids when moving from tree to tree. Many taxa do indeed possess extensions of the external skeleton. To turn these into wings, hinges would have had to develop between the lobes and the body.

Aquatic larval gills, gill covers, or both may have been preserved in the adult stage and transformed into wings above the legs in two segments. Because these structures are already hinged and moveable, it is easier to imagine how they transformed into flapping wings.

Another possibility is that legs or leg segments turned into wings. The genes involved in the developmental control of wings and legs of modern insects are the same. Numerous body segments, each fitted with pairs of legs, are part of the basic body plan of arthropods. If two segments merged with the thorax, this could provide extremities that could freely turn into wings, leaving the six legs as they are. Bear in mind that a change of form and function of appendages also turned wings into halteres to perform a gyroscopic function in true flies.

The emergence of flight is separate to the emergence of wings, although off course closely related.

We saw that tiny insects can easily be dispersed by wind. Even wingless insects can use wind to travel. To improve dispersal potential, hairy bristles or other fringed extensions on the back would produce extra viscous drag (Norberg 1972b). Beating wings may have evolved from these bristle or extensions, for more control.

It is possible that early wings, from whichever origin, did not make flapping motions but instead served as gliding devices. Gliding down from trees at reasonably flat angles to increase the distance covered and decrease the impact during landing would have high survival value.

Wings may have been originally used to row through water, and later to sail passively to eventually become movable enough to produce the small amounts of thrust in interaction with the air needed to move over the water surface. From there, it is one more step to beating wings that are strong enough to produce the lift required to make the animal fly at first aided by the wind but eventually independently. Support for this scenario comes from present-day stoneflies that show different forms of aquatic, semiaquatic, and aerial locomotion with increasing characteristic speed (Marden and Thomas 2003).

Insects with aquatic larval stages start their adult life stage by emerging from the water. Stoneflies use their wings as rowing oars and as sails to skim the water surface (Marden and Kramer, 1994).

This is regarded as one step away from developing powered flight in a safe way. A fossil with an imprint of an insect that landed on a muddy surface at some instant 308–314 million years ago provides some possible evidence (Knecht et al. 2011). The trace was preserved in fine-grain sandstone (Figure 4.39). It shows impressions of the six legs and very thin lines where the leading edges of the wings touched the mud. There are also traces of wing beats, used by the insect to escape the sticky mud. Marden (2012) reanalyzed the traces and experimented with extant stoneflies (genus *Taeniopteryx*) that have many features in common with the early insect that interacted with the mud. Wings of stoneflies are folded on top of each other and have two parallel leading edges that could touch mud in accord with the traces found. The thorax is weakly developed and not wider than the abdomen, indicating a lack of strong flight muscles (Marden and Kramer 1995).

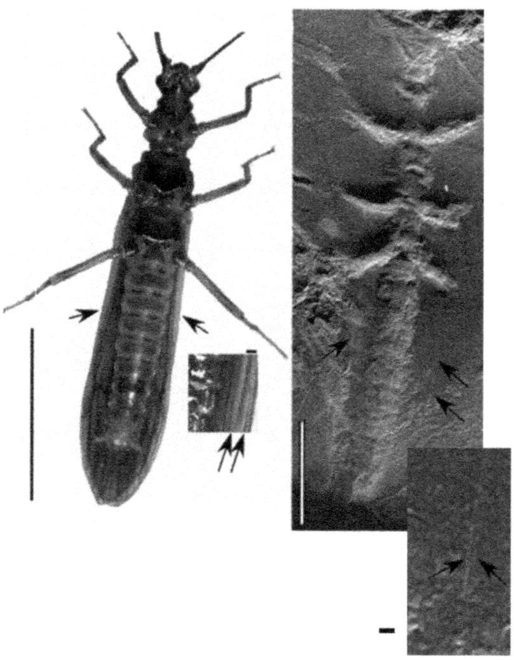

Figure 4.39. (a) Ventral view of a live *Taeniopteryx* stonefly. Arrows show the folded fore- and hindwing leading edges. Inset is a close-up of the paired wing edges. (b) Wing edge marks (at arrowheads) beside the abdomen of the trace fossil. (From Knecht, R.J., et al., *Proc. Natl. Acad. Sci. USA*, 108, 6515–6519, 2011. With permission.) The inset shows a close-up of the two fine parallel marks (at arrowheads) beside the abdomen. Scale bars: (a) 5 mm (inset 0.1 mm); (b) 10 mm (inset 0.4 mm). (After Marden, J.H., *Evolution*, 67, 274–280, 2012.)

4.7.3 The Origins of Flight on Leather Wings

About 100 million years after the first insects took flight, the next animals mastered the art of flight. More than 100 species of pterosaurs lived worldwide from 248–65 million years ago. They did not survive the mass extinction at the end of the Cretaceous. Many different flying species occurred more or less simultaneously. No fossils of intermediate stages showing some reptiles with flaps of skin that could be of use to parachute or glide have yet been discovered. Systematic relationships are difficult to establish because of the usually fragmented nature of the fossils.

Section 4.5 showed that only one finger of pterosaurs (the fourth) was enclosed in the wing membrane and three fingers were left out. The wings lengthened gradually by increasing the length of the phalanges of the fourth finger. The dimensions of the arm skeleton, humerus, and radius and ulna did not lengthen in proportion to the finger. The wingspan was determined by the sum of the length of only four phalanges of the wing finger (one phalange must have been lost at an early stage of evolution). The trailing edge of the membrane stretched out from the tip of the finger to the ankle. The leading edge was supported by a bone only known in pterosaurs, the pteroid (see Section 4.5).

The oldest fossil bat, *Icaronycteris index*, emerged 50 My, 15 million years after the last pterosaur lived. Again, no fossils of intermediate forms between early tree living ancestors and modern bats have so far been found. The remains of the first bats look very similar to extant micro bats and were obviously fully capable of flight, with membranous wings supported by all five fingers (see Section 4.5).

In summary, for both pterosaurs and bats we have no notion of how wings and flight behavior evolved during the early stages of these groups. Much more is known about the evolution of flight in birds, although the sequence of events leading to flight in that group is still enigmatic as well.

4.7.4 The Beginning of Bird Flight

4.7.4.1 *Archaeopteryx lithographica*

The debate on how flying birds evolved took off some 16 decades ago with the find of the first *Archaeopteryx* fossil in 150-million-year-old (late Jurassic) marine sediments near Solnhofen in

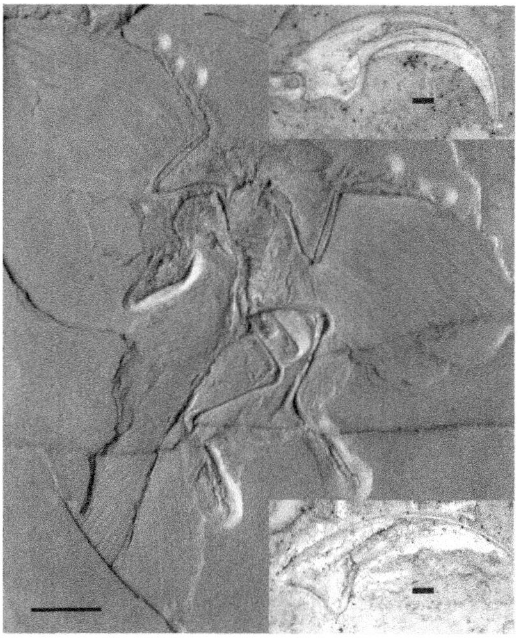

Figure 4.40. The Berlin specimen of *Archaeopteryx* (scale bar, 5 cm). The insets are the claws of the Teylers Museum specimen. The upper is the finger claw, the lower the foot claw (scale bars, 1 mm).

Bavaria, Germany (Figure 4.40). Now, 11 more or less complete fossils and one loose feather of *Archaeopteryx* have been unearthed as well as a wealth of other well-preserved specimens of assumed early birds, but the debate is still fierce and undecided.

Many Mesozoic (248–65 million years ago) bird fossils from rich deposits in China and from other parts of the world show the extensive radiation of bird-like animals during that period. Most of that fauna did not survive the mass extinction event 65 million years ago. We have to look for the ancestors of present-day birds in younger deposits, but the birds from the period after the mass extinction found so far are very similar to extant birds.

It is now almost universally accepted that birds are feathered dinosaurs. However, some believe that truly feathered animals with wings share a common ancestor with dinosaurs and should be treated as sister groups (Feduccia 2013). *Archaeopteryx* is the most ancient truly feathered species, and the circumstances under which it lived are better known than that of other Mesozoic birds. The Solnhofen quarries produce fine-grain sandstone with fossils preserved in great detail. The Solnhofen environment was a shallow tropical sea speckled with low islands comparable to the present Gulf of Cariaco in Venezuela. Fossilization in the marine sediment occurred fast, with little signs of a long decomposition phase (Buisonjé 1985; Viohl 1985). At first, *Archaeopteryx* remains were categorized as theropods, small, bipedal, lizard-hipped, carnivorous dinosaurs because some features indicate a close relationship with this group. Dinosaur features of *Archaeopteryx* are most obvious in the skeleton of the head and in that of the rear part. The skull has dinosaur characteristics, but it contained a relatively large brain. There is no bird-like beak, but there are bony jaws with simple conical teeth. The vertebral column, the ribs, belly ribs, and the long tail skeleton are theropod features. The pelvic girdle has the characteristic theropod long pubis. The pectoral girdle consists of a wishbone (furcula, the fused clavicles), shoulder blades (scapulae), and coracoids. A sternum was absent or cartilaginous in all but the seventh fossil discovered in 1992. The glenoid fossa for the attachment of the humerus at the shoulder joint of *Archaeopteryx* seems to be directed laterally, as in modern birds; scapula and coracoid meet at the joint, but there is no triosseal canal. It is not clear whether the tendon of a supracoracoid muscle (the upstroke muscle in modern birds) passed the shoulder to insert on the upper part of the humerus. There is a prominent, elongated deltopectoral crest on the anterior side of the humerus in all fossils, hinting at strong pectoral muscles despite the apparent lack of an equally strong origin for this muscle. Radius and ulna are long and slender and closely resemble the modern avian configuration. *Archaeopteryx* could fold the hand part of the wings backward along the underarm, at least in part, because the wrist contains two carpal bones that make this articulation possible. Vazquez (1992) points at slight differences between the wrist of *Archaeopteryx* and that of modern birds and argues that the *Archaeopteryx* condition does not allow full retraction of the hand part of the wing. The hand part occupies 40% of the length of the wing and is supported by three clawed fingers. The metacarpals seem to be unfused. The finger claws in all fossil remains are remarkably sharp and pointed without signs of wear, whereas the toe claws are worn (see Figure 4.40, insets). There is a remarkable difference between the last joints of the finger and the toe bones. The phalanges of the finger nails have two sockets where the distal condyle of the penultimate phalanx could fit, the equivalent joints of the foot claws show only a single socket. The wings are exceptionally large compared to a those of a bird of similar size, and it would require a lot of power to flap such heavily built wings. The careful reconstruction of Rietschel (1985) of the right wing shows that the hand part occupies about 40% of the length of the entire wing (Figure 4.41). The arm part was well cambered. Cross sections through the arm would have shown a rounded leading edge, a large cord, and a sharp trailing edge. The arm wing obviously generated lift conventionally, even at high angles of attack. The hand part consists of 12 primaries with narrow vanes, forming the leading edges of the feathers. The leading edge of the hand wing is sharp and wedged backward, and the primary shafts are slightly curved to the rear. The hand wings are in fact swept-back delta wings with sharp leading edges. At low speeds and high angles of attack, delayed dynamic stall creates LEVs above that type of wings, a configuration especially adapted to cope with angles of attack as high as 60°–70° at low velocities. *Archaeopteryx* could not flap its wings because it lacked the apparatus to do that, but it could presumably spread and close its hand wing as a fan, regulating the amount of lift required for each speed. This concept provides a clear role for the three claws on the fingers. Aircraft designed to generate lift with attached flow over the proximal wing parts as well as with separated LEVs over sharp-edged, swept-back distal wing sections often require a device to separate the two flow patterns. Wing fences and saw-toothed leading edges are installed to generate vortices that clear up the boundary layer and separate one type of flow from the next. The fingernails, situated between the arm and hand part of the wing, are leading edge protuberances possibly serving, as in aircraft, to control the formation of LEVs (Ashill et al. 1995; Barnard and Philpott 1997; Lowson and Riley 1995; Videler 2000). Leading edge claws of bats, pterosaurs, and the alula of birds may have the same function. The claws of *Archaeopteryx* probably performed that task in a dynamically optimal way. Each of the three claws could be used according to the amount of spreading of the hand wing, or even in quick succession if needed. The claws that were not in use could be kept withdrawn between the feathers. This function would explain why the claws do not show any wear and why a bistable joint formed the connection between the last digit and the rest of the finger. Because of this feature, the nails could be in two resting positions: while extended and while retracted.

The tail of *Archaeopteryx* is fundamentally different from that of all extant birds. Up in the air,

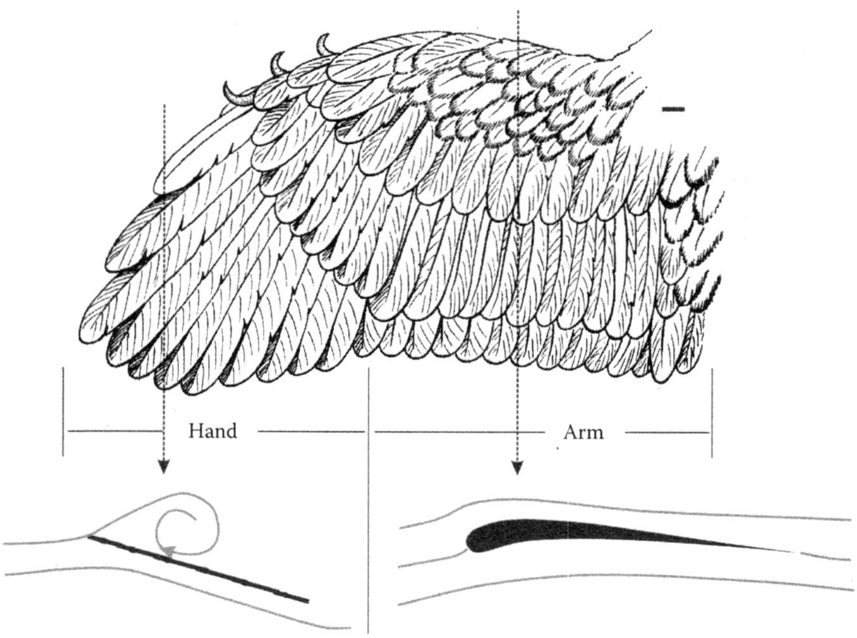

Figure 4.41. Rietschel's (1985) reconstruction of the *Archaeopteryx* wing (ventral view of the right wing). Scale bar, 1 cm. The arrows indicate the positions of the cross-sectional profiles below. Hypothetical flow patterns on the arm and hand wing are indicated. (From Videler, J.J., *Avian Flight*, Oxford University Press, Oxford, UK, 2006. With permission).

it would generate considerable amounts of both lift and drag, but the flexible structure of the bony support makes it difficult to control these forces.

The imprints of the primary feathers are one of the most striking features of *Archaeopteryx*. They stunningly resemble the primaries of modern birds consisting of a shaft with asymmetrical vanes consisting of barbs, each with numerous barbules, probably attached to each other with hooks and grooves as in extant feathers (Griffiths 1996). In summary, *Archaeopteryx* was a peculiar feathered animal with well-developed wings closely resembling chicken wings, with long legs and tail and a toothed beak. I try to reconstruct how it lived and died subsequently in this chapter, but first I consider its role in current ideas about the origin of flight in birds.

4.7.4.2 *Conflicting Views and the* Archaeopteryx *Case*

Several scenarios dominate thinking about the origin of flight in birds. The tree down (arboreal) theory predicts that birds climbed up trees by using claws on fingers and toes and that flight started by gliding down with the use of wings from the height of trees. The ground up (cursorial) hypothesis assumes that fast-running, bipedal dinosaurs developed flapping wings for stabilization, gradually offering the possibility to take off from the ground. The arboreal approach requires at least two skills: climbing trees and gliding. The nails on the hand and feet of *Archaeopteryx* are indeed suitable for climbing, but as noted, the fingernails are not worn like those of the toes in any of the specimens, indicating they had not been used for climbing. Ground-dwelling, nonclimbing birds and theropod dinosaurs have the same type of claws (Peters and Görgner 1992). The foot of *Archaeopteryx* is unsuitable for perching (Wellenhofer 1993). The wings are not optimally designed to glide either (Rayner 1985a). Soft and safe landing requires backward rotation (supination) of the wings to use them as airbrakes. In modern birds, this movement is powered by the supracoracoid muscle originating on the sternum and inserting on the top of the humerus via the triosseal canal (Poore et al. 1997). There is no evidence that *Archaeopteryx* could use this type of equipment. To quote Balda et al. (1985): "An animal the size of *Archaeopteryx* would need considerable morphological change to enter the gliding niche at sublethal speeds."

Another clinching argument against the arboreal hypothesis is the absence of trees for *Archaeopteryx* to climb.

An argument in favor of the cursorial hypothesis is that the morphology of the hind legs shows that *Archaeopteryx* was an agile, bipedal runner. Lift-generating forelimbs could have increased stability of fast-running, bipedal animals, even if these forelimbs were not yet suitable for flight (Balda et al. 1985). In opposition, Rayner (1985b) argues that the ground speed required for an animal of the size of *Archaeopteryx* would have to be in the order of 6–7 m s^{-1}, that is, about three times faster than the highest estimated running speeds of extant lizards and birds. Another argument against this hypothesis could be the following: The fitness-increasing reason to run fast would be to escape from predators or to catch fast-flying insects. In both cases, take-off would be disastrous because speed would reduce immediately due to drag not balanced by thrust.

Several scholars in the field have proposed variations of the two hypotheses. Dial (2003), for example, supported the cursorial model by showing that young partridges use wing beats to assist them during uphill running as a kind of practice for real flight.

In my opinion, *Archaeopteryx* represents a special case of a cursorial beginning of flight (Videler 2000, 2005). Like the insects discussed here, it lived a life skimming over mud flats and shallow water by using its wings in interaction with the wind to glide, with thrusting feet hitting the water or mud surface. I propose that it used a similar technique to do that as the extant Jesus Christ lizard (*Basiliscus*) (Figure 4.42). Insight in the ecology supports this view.

4.7.4.3 Archaeopteryx *Ecology*

The well-studied fossil environment around *Archaeopteryx* offers a unique opportunity to obtain insight into its life as part of an ecosystem. There were no trees in the Solnhofen environment; thus, suggesting that *Archaeopteryx* was an arboreal animal originating from a distant forested area is far-fetched. All the *Archaeopteryx* remains are embedded in marine sediment, and the most parsimonious explanation is that they are found on the spot where they died. The bodies of five

Figure 4.42. Artist impression of the Jesus Christ lizard (a) and *Archaeopteryx* (b) running over water. (From Videler, J.J., *Archaeopteryx*, 18, 27–34, 2000. With permission.)

specimens are very much intact, thereby excluding the possibility of long-distance transport from a terrestrial place of death.

What did *Archaeopteryx* eat? No stomach contents that could give an indication of the exploited food sources have been assigned. Elzanowski (2002) deduced from the size and strength of the teeth that soft-bodied arthropods made up a substantial part of the diet of *Archaeopteryx*. The contemporary water skaters (*Chresmoda obscura*) were very abundant in the area where *Archaeopteryx* lived. Their exoskeleton was probably rather thin to reduce the body weight, allowing surface tension forces to carry the approximately 5-cm-long animals. A picture emerges of *Archaeopteryx* as a highly specialized shorebird, feeding on water skaters on shallow waters and wet mud flats. The ability to run across the water surface at a speed of at least 2 m s^{-1} would fit in this scenario. Running over water is a fairly widespread technique in the animal kingdom and does not require a miracle to be performed. Basilisk lizards run over water by repeatedly hitting the water surface with their feet to support their body weight. Western Grebes

Figure 4.43. Cartoon of Western grebes (*Aechmophorus occidentalis*) running over water during courtship display. (From Videler, J.J., *Avian Flight*, Oxford University Press, Oxford, UK, 2006. With permission.)

weighing 1200 g run on water without flapping their wings to generate lift during courtship displays. Could *Archaeopteryx* have used a similar technique?

The strong *Archaeopteryx* tail, kept in a straight backward position, could have generated varying amounts of lift and drag to control and stabilize the run across the water. At the low speeds used for running, small movements would not have had disastrous effects on the control, but they could have helped to keep the animal balanced.

Archaeopteryx was not able to beat its wings forcefully and would therefore have had to keep them in gliding position during runs across the water. The wings of the London and Berlin specimens are stretched as in a gliding position.

The ecological advantages of running over water for *Archaeopteryx* could have been the possibility to escape from terrestrial predators and the exploitation of distant feeding grounds around islands in a shallow tropical sea.

Its wings were suitable to generate enough lift to take care of much of its body weight, but they would not have been able to generate thrust. *Archaeopteryx* would not have been able to fly. If *Archaeopteryx* was going to evolve into a flying animal, it would have to lose weight and it would have to develop the morphological requirements to beat its wings. That would include a proper sternum, preferably with a crest and the supracoracoideus muscle and tendon system to take care of the wing's twisting movement during the upstroke. Is there any evidence of such evolution? Of three of the *Archaeopteryx* skeletons, the position in the layers of sediment in Solnhofen is known.

The London specimen was found deepest (and is therefore oldest) and is the largest skeleton. The seventh fossil (the Munich or Solnhofer Aktien Verein specimen) was found in a layer 14.6 m higher up. Its size was about 73% of that of the London remains, and it probably had a calcified sternum. The Maxberg *Archaeopteryx* from an in-between layer 8.5 m deeper down was of an intermediate size. These facts (Wellnhofer 1993) do not prove, but support the view that *Archaeopteryx* was in the process of evolving into a flying bird. Younger fossils from the same site might provide the answer to the question of how a flying *Archaeopteryx* was built. However, by the time the *Archaeopteryx* could fly, it would no longer be running between islands, and the chances of getting killed during one of these trips would be reduced; therefore, the chances to become fossilized would be remote.

4.7.5 Radiation of Ancient Mesozoic and Modern Paleocene Birds

During the Cretaceous period, animals with bird characteristics emerged worldwide with fossil finds from Spain, Patagonia, and China. Specimens of a small theropod dinosaur named *Sinusauropteryx* from northeastern China are considered of extreme interest because the entire skin is covered with a layer of presumed feather-like structures. Feduccia (2013) agrees with Lingham-Soliar (2003) and believes that these structures have nothing to do with feathers, but instead are remains of collagenous skin fibers. Other finds of small dinosaurs clearly have feathers, but they are interpreted as secondarily flightless early birds. Rich fossil beds in China contain many birds that lived in a freshwater and woodland environment. The great diversity shows little coherence, but some specimens are interesting regarding flight abilities.

In 1915, Beebe predicted the existence of flight feathers on the hind legs of bird ancestors, based on the development of dove feathers. Until now, 11 fossils have been discovered in China with that configuration (Zheng et al. 2013). Both front and rear legs carry pennaceous feathers; the clawed toes and fingers are left free. These are assumed to have been arboreal animals capable of controlled gliding, but not powered, flight.

Several other specimens do show characteristics associated with flapping flight. *Jeholornis prima* was an arboreal animal with large clawed feed suitable for perching on branches. It had a well-built sternum and large deltoid crests on the humerus, but it lacked a triosseal channel (Zhou and Zhang 2002). Large numbers of *Confuciusornis sanctus* have been found. It was a 25-cm-tall bird with a flat sternum, only a few tail vertebrae, and a pygostyle. Its lifestyle was probably comparable with that of parrots. *Liaoningornis longidigitris* from the same deposits is the first ancient bird found with a keeled sternum. *Protopteryx fengningensis*, about the size of a starling, had more features of modern birds: a sternum with a crest and lateral processes, elongated coracoids, a prominent clavicula, a triosseal channel, and an alula.

The fossil of the first bird with an alula as well as a clear triosseal channel was found in Spain. *Eoalulavis hoyasi* had features of a small wader, with a wingspan of 17 cm. It was probably the best flyer of the Mesozoic birds that appeared on earth 45 million years after *Archaeopteryx*.

Only a few flying aquatic birds managed to escape from the mass extinction event 65 million years ago. These birds are believed to have been the ancestors of birds such as ducks, loons, plovers, and tube-noses. The passerines, the most numerous group of birds, started to diversify between 30 million years ago and 10 million years ago, leaving very few fossils.

REFERENCES

Alexander, D.E. 2015. *On the Wing: Insects, Pterosaurs, Birds, Bats and the Evolution of Animal Flight*. Oxford University Press, Oxford, UK.

Altringham, J.D. 1996. *Bats, Biology and Behaviour*. Oxford University Press, Oxford, UK.

Anderson, J.D., Jr. 1997. *A History of Aerodynamics and Its Impact on Flying Machines*. Cambridge University Press, Cambridge, UK.

Andress, B. 2012. The early evolutionary history and adaptive radiation of the Pterosauria. *Acta Geologica Sinica* (English edition) 86: 1356–1365.

Arathi, H.S. 2012. A comparison of dispersal traits in dandelions growing in urban landscape and open meadows. *J Plant Stud* 1: 40–46.

Ashill, P.R., G.L. Riddle, and M.J. Stanley. 1995. Separation control on highly-swept wings with fixed or variable camber. *Aeronaut J.* 99: 317–327.

Augspurger, C.K. 1986. Morphology and dispersal potential of wind-dispersed diasporas of neotropical trees. *Am J Bot* 73: 353–363.

Azuma, A., and Y. Okuno. 1987. Flight of a samara *Alsomitra macrocarpa*. *J Theor Biol* 129: 263–274.

Bahlman, J.W., J.W. Schwartz, D.K. Riskin, and K.S. Breuer. 2012. Glide performance and aerodynamics of non-equilibrium glides in northern flying squirrels (*Glaucomys sabrinus*). *J Roy Soc Interface* 10: 20120794.

Balda, R.P., G. Caple, and W.R. Willis. 1985. Comparison of the gliding to flapping sequence with the flapping to gliding sequence. In: *The Beginnings of Birds* (M.K. Hecht et al., eds.), pp. 267–277. Freunde des Jura-Museums, Eichstätt.

Barnard, R.H., and D.R. Philpott. 1997. *Aircraft Flight*, 2nd edn. Longman, Harlow, 378 pp.

Barnes, J. 1991. *The Complete Works of Aristotle*, 4th edn. Princeton University Press, Princeton, NJ.

Beebe, C.W. 1915. A tetrapteryx stage in the ancestry of birds. *Zoologica* 2: 39–52.

Bennet, S.C. 2007. Articulation and function of the pteroid bone of pterosaurs. *J Vert Paleontol* 27: 881–891.

Bernoulli, D. 1738. *Hydrodynamica*. JR Dulsecker, Strasbourg.

Boel, M. 1929. Scientific studies of natural flight. *Trans Am Soc Mech Eng* 51: 217–242.

Borelli, A. 1680. *De motu animalium*. Angeli Bernabo, Rome.

Buisonjé, P.H. 1985. Climatological conditions during deposition of the Solnhofen limestones. In: *The Beginnings of Birds* (M.K. Hecht et al., eds.), pp. 45–65. Freunde des Jura-Museums, Eichstätt.

Byrnes, G., N. T.-L. Lim and A. J. Spence. 2008. Take-off and landing kinetics of a free-ranging gliding mammal, the Malayan colugo (Galeopterus variegatus). *Proc. R. Soc. B* 275: 1007–1013.

Byrnes, G., and A.J. Spence. 2011. Ecological and biomechanical insights into the evolution of gliding in mammals. *Integr Comp Biol* 51: 991–1001.

Carpenter, R.E. 1986. Flight physiology of intermediate sized fruit bats (Pteropodidae). *J Exp Biol* 120: 79–103.

Chapman, R.F. 2013. *The Insects: Structure and Function*, 5th edn. (S.J. Simpson and A.E. Douglas, eds.). Cambridge University Press, Cambridge, UK.

Chatterjee, S., and R.J. Templin. 2012. The flight dynamics of *Tapejara*, a pterosaur from the early Cretaceous of Brazil with a large cranial crest. *Acta Geologica Sinica* (English edition) 86: 1377–1388.

Chin, D.D., and D. Lentink. 2016. Flapping wing aerodynamics: From insects to vertebrates. *J Exp Biol* 219: 920–932.

Claessens, L.P.A.M., P.M. O'Connor, and D.M. Unwin. 2009. Respiratory evolution facilitated the origin of pterosaur flight and aerial gigantism. *PLoS One* 4: e4497.

Dale Guthrie, R. 2005. *The Nature of Paleolithic Art*. University of Chicago Press, Chicago, IL.

Darwin, C.R. 1845. *The Voyage of the Beagle* (reprint 1967). JM Dent, London.

Dial, K.P. 1992. Avian forelimb muscles and nonsteady flight: Can birds fly without using the muscles in their wings? *Auk* 109: 874–885.

Dial, K.P. 2003. Wing-assisted incline running and the evolution of flight. *Science* 299: 402–404. doi: 10.1126/science.1078237.

Dickinson, M.H., F.O. Lehmann, and S.P. Sane. 1999. Wing rotation and the aerodynamic basis of insect flight. *Science* 284: 1954–1960.

Donoughe, S., J.D. Crall, R.A. Merz, and S. Combes. 2011. Resilin in dragonfly and damselfly wings: Its implications for wing flexibility. *J Morphol* 272: 1409–1421.

Dudley, R. 2000. *The Biomechanics of Insect Flight*. Princeton University Press, Princeton, NJ.

Dudley, R., G. Byrnes, S.P. Yanoviak, B. Borell, R.M. Brown, and J.A. McGuire. 2007. Gliding and the functional origins of flight: Biomechanical novelty or necessity? *Ann Rev Ecol Syst* 38: 179–201.

Elgin, R.S., D.W.E. Hone, and E. Frey. 2011. The extent of the pterosaur flight membrane. *Acta Palaeontol Polonica* 56: 99–111.

Ellington, C.P. 1984. The aerodynamics of hovering insect flight: IV Aerodynamic mechanisms. *Philos Trans Roy Soc London B* 305: 79–113.

Ellington, C.P., K.E. Machin, and T.M. Casey. 1990. Oxygen consumption of bumble bees in forward flight. *Nature* 347: 472–473.

Elzanowski, A. 2002. Archaeopterygidae (upper Jurassic of Germany). In: *Mesozoic Birds: Above the Heads of Dinosaurs*. (L.M. Chiappe and L.M. Witmer, eds.), pp. 129–159. University of California Press, Berkeley, CA.

Emerson, S.B., and M.A.R. Koehl. 1990. The interaction of behavioral and morphological change in the evolution of a novel locomotor type: 'Flying' frogs. *Evolution* 44: 1931–1946.

Engel, S., H. Biebach, and G.H. Visser. 2006. Metabolic costs of avian flight in relation to flight velocity: A study in Rose-Coloured Starlings (*Sturnus roseus* Linnaeus). *J Comp Physiol B* 176: 415–427.

Fahy, F. 2009. *Air: The Excellent Canopy*. Horwood, Chichester, UK.

Feduccia, A. 2013. Bird origins anew. *Auk* 130: 1–12.

Gibbs-Smith, C.H. 1962. *Sir George Cayley's Aeronautics 1796–1855*. Her Majesty's Stationery Office, London, UK.

Greenewalt, C.H. 1975. The flight of birds. *Trans Am Philos Soc* 65: 1–67.

Gregory, T.R. 2008. The evolution of complex organs. *Evol Educ Outreach* 1: 358–389.

Griffiths, P.J. 1996. The isolated *Archaeopteryx* feather. *Archaeopteryx* 14: 1–26.

Hector, J. 1894. On the anatomy of flight of certain birds. *Trans NZ Inst.*

Hedenström, A., L.C. Johansson, M. Wolf, R. von Busse, Y. Winter, and G.R. Spedding. 2007. Bat flight generates complex aerodynamic tracks. *Science* 316: 894–897.

Hedrick, T.L., B.W. Tobalske, I.G. Ros, D.R. Warrick, and A.A. Biewener. 2012. Morphological and kinematic basis of the hummingbird flight stroke: Scaling of flight muscle transmission ratio. *R Soc London B* 279: 1986–1992.

Hertel, H. 1966. *Structure form and Movement.* Van Rostrand-Reinhold, New York.

Hintze, C., F. Heydel, C. Hoppe, S. Cunze, A. König, and O. Tackenberg. 2013. D^3: The dispersal and diaspore database—Baseline data and statistics on seed dispersal. *Perspect Plant Ecol Evol Syst* 15: 180–192.

Houlihan, P.F. 1997. *The Animal World of the Pharaohs.* Thames and Hudson, London, UK.

Jackson, S.M. 1999. Glide angle in the genus *Petaurus* and a review of gliding in mammals. *Mammal Rev* 30: 9–30.

Jacob, F. 2001. Complexity and tinkering. *Ann NY Acad Sci* 929: 71–3.

Junge, M. 2002. *Kinematische und strömungsmechanische Untersuchungen zum Fächelverhalten der Honigbiene (Apis mellifera L.).* Dissertation, Universität des Saarlandes, Saarbrücken.

Knecht, R.J., M.S. Engel, and J.S. Benner. 2011. Late Carboniferous paleo-ichnology reveals the oldest full-body impression of a flying insect. *Proc Natl Acad Sci USA* 108: 6515–6519.

Kramer, M. 1932. Die Zunahme des Maximalauftriebes von Tragflügeln bei plötzlicher Anstellwinkelvergrösserung (Boeneffekt). *Z. Flugtech. Motorluftschiff.* 23: 185–189.

Kvist, A., Å. Lindström, M. Green, T. Piersma, and G.H. Visser. 2001. Carrying large fuel loads during sustained bird flight is cheaper than expected. *Nature* 413: 730–732.

Lentink, D., W.B. Dickson, J.L. van Leeuwen, and M.H. Dickinson. 2009. Leading-edge vortices elevate lift of auto-rotating plant seeds. *Science* 324: 438–440.

Lighthill, J. 1990. *An Informal Introduction to Theoretical Fluid Mechanics.* Oxford University Press, Oxford.

Lilienthal, O. 1889. *Der Vogelflug als Grundlage der Fliegekunst*, 3rd edn. Oldenbourg, Munchen.

Lingham-Soliar, T. 2003. Evolution of birds: Ichthyosaur integumental fibers conform to dromaeosaur protofeathers. *Naturwissenschaften* 90: 428–432.

Lowson, M.V., and A.J. Riley. 1995. Breakdown control by delta wing geometry. *J Aircraft* 32: 832–838.

Lucas, A.M., and P.R. Stettenheim. 1972. *Avian Anatomy: Integument. Agricultural Handbook* 362, U.S. Department of Agriculture, Washington, DC.

Maquet, P. 1989. *On the Movements of Animals.* Springer-Verlag, Berlin.

Marden, J.H. 2012. Reanalysis and experimental evidence indicate that the earliest trace fossil of a winged insect was a surface-skimming neopteran. *Evolution* 67: 274–280.

Marden, J.H., and M.G. Kramer. 1994. Surface-skimming stoneflies: A possible intermediate stage in insect flight evolution. *Science* 266: 427–430.

Marden, J.H., and M.G. Kramer. 1995. Locomotor performance of insects with rudimentary wings. *Nature* 377: 332–334.

Marden, J.H., and M.A. Thomas. 2003. Rowing locomotion by a stonefly that possesses the ancestral pterygote condition of co-occurring wings and abdominal gills. *Biol J Linn Soc* 79: 341–349.

Marey, E.J. 1890. *Le vol des oiseaux.* Masson, Paris, France.

Marinoni, A. 1976. *Il codice sul volo degli uccelli.* Giunti-Barbèra, Firenze, Italy.

McCay, M.G. 2001. Aerodynamic stability and maneuverability of the gliding frog *Polypedates dennysi. J Exp Biol* 204: 2817–2826.

McCutchen, C.W. 1977. The spinning rotation of ash and tulip tree samaras. *Science* 197: 691–692.

Miller, L.A., and C.S. Peskin. 2009. Flexible clap and fling in tiny insect flight. *J. Exp. Biol.* 212: 3076–3090.

Mullers, R.H.E., R.E. Navarro, S. Daan, J.M. Tinbergen, and H.A.J. Meijer. 2009. Energetic costs of foraging in breeding Cape gannets *Morus capensis. Mar Ecol Progr Ser* 393: 161–171.

Muramatsu, K., J. Yamamoto, T. Abe, K. Sekiguchi, N. Hoshi, and Y. Sakurai. 2013. Oceanic squid do fly. *Mar Biol* 160: 1171–1175.

Neuweiler, G. 2000. *The Biology of Bats.* Oxford University Press, Oxford, UK.

Newton, I. 1686. *Philosophiae Naturalis Principia Mathematica.* Cambridge, UK.

Norberg, R.Å. 1972a. Flight characteristics of two plume moths, *Alucita pentadactyla* L. and *Orneodes hexadactyla* L. (Microlepidoptera). *Zool Scripta* 1: 241–246.

Norberg, R.Å. 1972b. Evolution of flight in insects. *Zoolo Scripta* 1: 247–250.

Norberg, U.M. 1990. *Flight.* Springer-Verlag, Berlin, Germany.

Palmer, C., and G. Dyke. 2012. Constraints on the wing morphology of pterosaurs. *Proc Roy Soc B Biol Sci* 279: 1218–1224.

Panyutina, A.A., A.N. Kutznetsov, and L.P. Korzun. 2013. Kinematics of chiropteran shoulder girdle in flight. *Anat Rec* 296: 382–394.

Pennycuick, C.J. 1982. The flight of petrels and albatrosses (Procellariiformes) observed in South Georgia and its vicinity. *Philos Trans R Soc Lond B* 300: 75–106.

Pennycuick, C.J. 2008. *Modelling the Flying Bird*. Academic Press, San Diego, CA.

Percin, M., Y. Hu, B.W. van Oudheusden, B. Remes and F. Scarano. 2011. Wing flexibility effects in clap-and-fling. *Int. J. of Micro Air Vehicles* 3(4): 217–227.

Peters, D.S., and E. Görgner. 1992. A comparative study on the claws of *Archaeopteryx*. In: *Papers Avian Paleontology* (K.E. Campbell, ed.). Natural History Museum Los Angeles, Los Angeles, CA.

Polhamus, E.C. 1971. Predictions of vortex-lift characteristics by leading edge suction analogy. *J Aircraft* 8: 193–199.

Poore, S.O., A. Sánchez-Haiman, and G.E. Goslow. 1997. Wing upstroke and the evolution of flapping flight. *Nature* 387: 799–802.

Rayner, J.M.V. 1985a. Mechanical and ecological constraints on flight evolution. In: *The Beginnings of Birds* (M. Hecht et al., eds.), pp. 279–288. Freunde des Jura Museums, Eichstätt.

Rayner, J.M.V. 1985b. Cursorial gliding in Proto-birds, an expanded version of a discussion contribution. In: *The Beginnings of Birds* (M.K. Hecht et al., eds.), pp. 289–292. Freunde des Jura-Museums, Eichstätt.

Reynolds, O. 1883. An experimental investigation of the circumstances which determine whether the motion of water shall be direct or sinuous, and of the law of resistance in parallel channels. *Philos Trans Roy Soc Lond* 174: 935–982.

Rietschel, S. 1985. Feathers and wings of *Archaeopteryx*, and the question of her flight ability. In: *The Beginnings of Birds* (M.K. Hecht, J.H. Ostrom, G. Viohl and P. Wellnhofer, eds.), pp. 251–260. Freunde des Jura-Museums, Eichstätt.

Sambursky, S. 1987. *The Physical World of Late Antiquity*, 1st paperback ed. Princeton University Press, Princeton, NJ.

Sane, S.P. 2003. The aerodynamics of insect flight. *J Exp Biol* 206: 4191–4208.

Schwartz, S.M., J. Iriarte-Diaz, D.K. Riskin, A. Song, X. Tian, D.J. Willis, et al. 2007. Wing structure and the aerodynamic basis of flight in bats. 45th AIAA Aerospace Sciences Meeting and Exhibit, 8–11 January 2007, Reno, Nevada. *Science Series* 36: 29–37.

Slijper, E.J. 1950. *De vliegkunst in het dierenrijk*. E.J. Brill, Leiden, The Netherlands.

Snelling, E.P., R.S. Seymour, P.G.D. Matthews, and C.R. White. 2012. Maximum metabolic rate, relative lift, wing-beat frequency and stroke amplitude during tethered flight in the adult locust *Locusta migratoria*. *J Exp Biol*. 215: 317–3323.

Socha, J.J., T. O'Dempsey, and M. LaBarbera. 2005. A 3-D kinematic analysis of gliding in a flying snake, *Chrysopelea paradisi*. *J Exp Biol* 208: 1817–1833.

Speakman, J.R. and P.A. Racey. 1991. No cost of echolocation for bats in flight. *Nature* 350: 421–423.

Stolpe, M., and K. Zimmer. 1939. Der Schwirrflug des Kolibri im Zeitlupenfilm. *J für Ornithologie* 87: 136–155.

Sunada, S., and C.P. Ellington. 2000. Approximated added-mass method for estimating induced power for flapping flight. *AIAA J* 38: 1313–1321.

Tackenberg, O. 2003. Modeling long-distance dispersal of plant diasporas by wind. *Ecol Monogr* 73: 173–189.

Thomas, S.P. 1975. Metabolism during flight in two species of bats, *Phyllostomus hastatus* and *Pteropus gouldii*. *J Exp Biol* 63: 273–293.

Thomas, S.P. 1981. Ventilation and oxygen extraction in the bat *Pteropus gouldii* during rest and steady flight. *J Exp Biol* 94: 231–250.

Tobalske, B.W., D.R. Warrick, C.J. Clark, D.R. Powers, T.L. Hedrick, G.A. Hyder, and A.A. Biewener 2007. Three-dimensional kinematics of hummingbird flight. *J Exp Biol* 210: 2368–2382.

Varshney, K., S. Chang, and Z.J. Wang. 2012. The kinematics of falling maple seeds and the initial transition to a helical motion. *Nonlinearity* 25: C1–C8.

Vazquez, R.J. 1992. Functional osteology of the avian wrist and the evolution of flapping flight. *J Morphol* 211: 259–268.

Vazquez, R.J. 1994. The automating skeletal and muscular mechanisms of the avian wing (Aves), *Zoomorphology* 114: 59–71.

Videler, J.J. 2000. *Archaeopteryx*: A dinosaur running over water? *Archaeopteryx* 18: 27–34.

Videler, J.J. 2005. How *Archaeopteryx* could run over water *Archaeopteryx* 23: 23–32.

Videler, J.J. 2006. *Avian Flight*. Oxford University Press, Oxford, UK.

Videler, J.J., E.J. Stamhuis, and G.D.E. Povel. 2004. Leading edge vortex lifts swifts. *Science* 306: 1960–1962.

Viohl, G. 1985. Geology of the Solnhofen lithographic limestones and the habitat of *Archaeopteryx*. In: *The Beginnings of Birds* (M.K. Hecht et al., eds.), pp. 31–44. Freunde des Jura-Museums, Eichstätt.

von Busse, R,. S.M. Schwartz, and C.C. Voigt. 2013. Flight metabolism in relation to speed in Chiroptera: Testing the U-shape paradigm in the short-tailed fruit bat *Carollia perspicillata*. *J Exp Biol* 216: 2073–2080.

von Busse, J.R.S. 2011. The trinity of energy conversion–Kinematics, aerodynamics and energetics of the lesser long-nosed bat (*Leptonycteris yerbabuenae*). PhD dissertation, Humbold University, Berlin, Germany Shaker Verlag: Aachen.

Wagner, H. 1925. Über die Entstehung des dynamischen Auftriebes von Tragflügeln. *Zeits für Angewandte Mathematik und Mechanik* 5: 17–35.

Weis-Fogh, T. 1952. Fat combustion and metabolic rate of flying locusts (*Schistocerca gregaria* Forskål). *Phios Trans Roy Soc Lond* B237: 1–36.

Weis-Fogh, T. 1960. A rubber-like protein in insect cuticle. *J Exp Biol* 37: 889–907.

Weis-Fogh, T. 1973. Quick estimates of flight fitness in hovering animals including novel mechanisms for lift production. *J Exp Biol* 59: 169–230.

Wellnhofer, P. 1993. Das siebte Exemplar von *Archaeopteryx* aus den Solnhofer Schichten. *Archaeopteryx* 11: 1–47.

Wilkinson, M.T. 2008. Three dimensional geometry of a pterosaur wing skeleton, and its implications for aerial and terrestrial locomotion. *Zool J Linnean Soc London* 154: 27–69.

Winter, Y. 1999. Flight speed and body mass of nectar-feeding bats (Glossophaginae) during foraging. *J Exp Biol* 202: 1917–1930.

Winter, Y., and O.V. Helversen. 1998. The energy cost of flight: Do small bats fly more cheaply than birds? *J Comp Physiol B* 168: 105–111.

Witmer, L.M., S. Chatterjee, J. Franszosa, and T. Rowe. 2003. Neuroanatomy of flying reptiles and implications for flight, posture and behaviour. *Nature* 425: 950–953.

Witton, M.P. 2013. *Pterosaurs: Natural History, Evolution, Anatomy*. Princeton University Press, Princeton, NJ.

Wolf, T.J., P. Schmid-Hempel, C.P. Ellington, and R.D. Stevenson. 1989. Physiological correlates of foraging efforts in honey-bees: Oxygen consumption and nectar load. *Func Ecol* 3: 417–424.

Wooton, R. 1992. Functional morphology of insect wings. *Ann Rev Ent* 37: 113–140.

Yudin, K.A. 1957. *O nyekotoryikh prisposobityel'nyikh osobyennostyakh kryila trubkonosyikh ptits (otryad Tubinares)*. [On certain adaptive properties of the wing in birds of the order Tubinares.] *Zoologiceskij Zurnal* 36: 1859–1873.

Zheng, X., Z. Zhou, X. Wang, F. Zhang, X. Zhang, Y. Wang, G. Wei, S. Wang, and X. Xu. 2013. Hind wings in basal birds and the evolution of leg feathers. *Science* 339: 1309–1312. doi: 10.1126/science.1228753.

Zhou, Z., and L. Hou. 2002. The discovery and study of Mesozoic birds in China. In: *Mesozoic Birds: Above the Heads of Dinosaurs* (L.M. Chiappe and L.M. Witmer, eds.), pp. 160–183. University of California Press, Berkeley, CA.

CHAPTER FIVE

Terrestrial Locomotion

REINHARD BLICKHAN

5.1 INTRODUCTION

This chapter is about metazoan animal locomotion on land. The animals discussed spend at least significant parts of their respective life histories moving around either in or on the ground or on trees and other vegetation. It considers functional morphology; biomechanics; kinematics; other aspects of bioengineering; and, briefly, robotics. Like the preceding chapters on swimming and flying, this chapter begins with a summary of important points and issues that are the broad background for the detailed, more specialized discussions making up most of the chapter. More general points and issues relating to basic concepts and to biological evolution are discussed in Chapters 1 and 2.

5.1.1 Invasions of the Land

Evolutionary transitions from aquatic environments and lifestyles to terrestrial environments and lifestyles were difficult and complex for those relatively few animal lineages that were ultimately successful in invading the land. Only 9 of the 32 presently recognized phyla include lineages that made the transition. Successful persisting invasions became possible only after many major changes occurred, both in the animals themselves and in their environments.

The very different evolutionary trajectories followed by the nine successful groups have been described in detail by multiple authors. Some lineages seem to have made the transition relatively quickly (in geological terms), whereas others took many millions of years. Important relevant books and review articles include Little (1990), Gordon and Olson (1995), Clack (2006), Long and Gordon (2004), and Carroll (2016).

Here I outline and summarize only the major issues and factors involved in the overall process. As summarized in Chapter 2, multiple lines of evidence and databases provide the supporting evidence for the histories. The evidence is direct (trace and actual fossils), contextual (taphonomy), indirect (molecular phylogenies and studies of embryonic development of living forms), and inferential (reverse engineering from studies of living forms and modeling of fossil forms). Figure 2.2 and Table 2.2 summarize the phylogenetic relationships and the taxonomy of the major groups discussed in this chapter.

Assuming that life on Earth will continue to have a long-term future, there is food for thought in considering whether invasions of the land by diverse animal groups are either still ongoing or perhaps are only in early stages. It is unlikely that they have either stopped or ended.

5.1.2 The Phylogeny of Invasions

The nine phyla including lineages that are presently terrestrial may be divided into two broad groups:

1. Animals that are mostly small that, when free-living (meaning they are not parasites), spend their lives burrowing in generally moist soils, leaf litter, organic

detritus, or a combination of these habitats. These lineages are called cryptic.

2. Larger free-living animals that are most often active and surface dwelling.

The phyla including forms that are mostly to entirely cryptic include the Tardigrada (water bears), Nemertini (ribbon worms), Nematoda (roundworms), and Nematomorpha (hair worms). The nematodes are one of the most diverse and numerous of the phyla, including many forms that are terrestrial and cryptic. The other three phyla mentioned are not very diverse, and each includes only a few cryptic forms.

The remaining five phyla are the Mollusca (some gastropod snails are terrestrial), Annelida (some oligochaete earthworms and some leeches are terrestrial), Onychophora (a small entirely terrestrial phylum related to arthropods), Arthropoda (the major terrestrial lineages include the myriapods [centipedes and millipedes; living forms almost all terrestrial], chelicerates [scorpions, spiders, and related groups; living forms almost all terrestrial], insects [almost all terrestrial], and crustaceans [important sublineages having living terrestrial representatives include isopods, amphipods, crayfish, and crabs]), and Vertebrata (aside from the fishes, some of which are surprisingly terrestrial, all major lineages are partly or dominantly terrestrial).

5.1.3 Environmental Factors and Timing

Animals could not successfully persist in terrestrial environments until a wide range of abiotic conditions became tolerable for them. An incomplete list of these conditions includes air temperatures, ultraviolet radiation fluxes from the sun, atmospheric oxygen and carbon dioxide partial pressures, and partial pressures of toxic gases (e.g., methane, hydrogen sulfide).

The single essential biotic condition that then also had to be met was the availability of adequate supplies of food. Without sources of food away from the waters' edge excursions onto land could not last long. The geographic distribution and timing of successful invasions of the land by plants were major factors affecting what animals could do.

Plant invasions initially occurred along shorelines, ranging from open oceanic coasts to protected bays, estuaries, shores of rivers and streams, and shores of ponds and lakes. The cryptic groups probably were the first animals to move ashore because they were largely shielded from the harsh abiotic environmental conditions by their burrowing habits. The more surface-dwelling larger forms followed, possibly herbivores first and carnivores soon thereafter.

All of these conditions were first met on geographically meaningful scales for geologically significant periods of time beginning about 450 million years ago (mega-annums) (My), during the later Ordovician Period. Most of the major animal invasions of the land began then. As mentioned, the evolutionary histories of various groups were diverse and followed widely different physical pathways and time lines. Some pulmonate gastropod mollusks apparently began to be terrestrial much later, during the Mesozoic Era. The invasion of the land by the vertebrate tetrapod lineage (our ancestors) mostly occurred over about 25 My during the late Devonian and into the Carboniferous periods. Secondary reinvasions by vertebrates, back into aquatic habitats, such as the transitions that led to seals and whales, happened at various times much more recently.

Although largely speculative, it is worth asking in broad terms why animals invaded the land at all. The reasons probably involved a combination of evolutionary pushes and pulls. The pushes probably included the presence in aquatic environments of stringent competition for resources plus an abundance of predators, parasites, and diseases. The pulls probably included the absence of these factors plus the opening up of a vast array of empty ecological niches in huge expanses of biologically unoccupied territory. It may be a cliché, but nature abhors a vacuum.

5.1.4 Anatomy, Functional Morphology, Biomechanics, and Physiology

In addition to the environmental factors just mentioned, there are two features of terrestrial habitats that were major selective factors that affected all surface-dwelling animals: the force of gravity and limitations on water availability. These two features together required large numbers of evolutionary changes in all lineages with respect to anatomy, functional morphology (Hildebrand et al. 1985), biomechanics (Alexander 2006; Biewener 2003), and physiology.

Gravity plays relatively minor to no roles in the daily lives of most aquatic animals. Most swimming animals are close to or at neutral buoyancy as a result of a wide array of structures and mechanisms that balance sinking forces against flotation forces. Those forces are generated by structures either denser than water (e.g., skeletons) or less dense than water (e.g., gas-filled bladders). Metabolic energy is always needed to maintain the structural integrity and functionality of the structures, but those energy requirements are relatively small compared to the overall metabolic rates of the entire animal.

Animals on land, away from the water, must deal with gravity continuously. Maintenance of posture and any movements resulting in positional changes in either height or direction require substantial amounts of antigravity energy. These energy demands use substantial fractions of overall metabolic rates. Resisting gravitational forces also required the evolution of stronger skeletons (both internal and external), different amounts and arrangements of muscles, changes in limb positions and sizes, and an array of coordinated changes in nervous system structure and function.

The water in which aquatic animals live also creates a variety of physiological problems for them. Major sets of those problems relate to water balance and ionic regulation. The specifics of those problems and their multiple solutions are not relevant here. What is relevant is that, despite those problems, water is always available for the physiological needs of aquatic animals.

Water is not always available to terrestrial animals. Indeed, in many situations liquid water may be absent or only present in short supply. Thus lineages invading the land necessarily had to evolve arrays of structures and processes that contributed to abilities to conserve and efficiently use water. These changes had to coevolve with the antigravity changes occurring so that, at all stages, the animals involved were functional, internally coordinated entities.

This suite of stringent requirements and complex structural and functional changes probably account for the substantial amounts of geological time associated with almost all invasions of the land. They also probably explain why so few animal phyla have successfully completed those invasions.

An intriguing aspect of the changes made by the major terrestrial lineages relates to the extents to which nature has conserved the basic body plans (bauplans) of the various groups from their aquatic origins through to their terrestrial successes today. Terrestrial groups of snails, annelids, and arthropods are anatomically surprisingly similar to their aquatic forebears. Their locomotor mechanisms and controls are broadly similar as well. Terrestrial vertebrates are mostly very different anatomically from the sarcopterygian fishes from which they derived, but they all retain large numbers of structures and functions that had their origins in the aquatic realm (Shubin 2008).

Evolutionary biologists often describe such situations as examples of preadaptations. Preadaptation is defined as a circumstance in which nature has readjusted and often repurposed important structures and functions first evolved in animals in environments different from those in which they are later used by the descendants of the animals in which they arose. Note also that the variety of independently evolved patterns of adaptation to life on land just discussed is a group of convergent homoplasies.

5.1.5 Solutions for Terrestrial Locomotion

This chapter is primarily devoted to legged locomotion of macroscopic animals on level ground. This introduction highlights the vast abundance and variety of ingenious investigations nature has undertaken to permit animals to locomote in the terrestrial environment. At the end, I briefly mention bioinspired and biomimetic robotic studies.

There are two primary categories of body plans related to locomotor modes: animals without limbs and animals with limbs, including as described here, many transitional solutions.

5.1.5.1 Locomotion without Limbs

The four most diverse terrestrial invertebrate groups lacking limbs are the flatworms, nematodes, snails (with and without shells), and annelids (oligochaetes). Land-dwelling flatworms creep or glide on their ventral surfaces. Terrestrial nematodes and worms are almost all cryptic burrowers that spend most of their lives underground or in leaf litter. They use several different locomotor modes to move around (e.g., sinusoidal wriggling, peristaltic contractions of their body walls). The leaf- or ribbon-shaped Plathyhelmithes (flatworms) have unsegmented bodies. Outer circular, inner longitudinal, and in addition dorsoventral (parenchymal) layers of muscle fibers allow shape modifications. Further inside, cell layers reinforced by collagen fibers serve as a soft skeleton. The animals are able to swim, crawl, and dig. Nematodes (roundworms) possess below their rater stiff cuticle tubes with only longitudinal muscles. The muscle cells are subdivided in active and passive segments. The muscles act against a hydroskeleton composed of the strong cuticle opposing the highly pressurized body cavity (pseudohemocoel). With this system, the animals can perform righting movements, flip from side to side, or undulate. Annelids such as the earthworm also have a pressurized body cavity (coelom) and are segmented animals. They possess a collagen-reinforced cuticle. Each ring segment entails at least ring and longitudinal muscles attached to the separating septa (not in leeches), thereby allowing each ring to lengthen and shorten for use in peristaltic burrowing and creeping (Zimmermann et al. 2009). Leeches can use an inchworm mode of locomotion.

Some modes of locomotion of these cryptic animals are not very far from those observed in legged locomotion. In fact, we must keep in mind that also for tetrapods (see below) and for humans the body or trunk plays an important role with respect to force and movement generation. Many annelids have stiff bristles (chetae) of chitin attached to their skin to provide traction, and in polychaetes these chetae are attached to parapodia. As extensions of the body wall the parapodia and the chetae can be moved by muscles, and the animals can use them for walking patterns and for burrowing. Depending on substrate and size, burrowing may consist of wedging (crack propagation), fluidizing, or removing small pieces of solids. Tardigrada, or water bears, are tiny animals that live in rather moist environments and move at a maximum speed of about 0.3 body length per second (Greven and Schüttler 2001). They possess eight limb-like clawed stubs that can be shortened by muscles against internal pressure (extension) and are used to cling on surfaces and for propulsion. In fact, they can display slow, but quite regular, walking movements.

For crawling and creeping locomotion, I must refer to holometabolous insects, that is, those animals that, like transformers, completely change their general appearance and their modes of locomotion from worm-like crawling to exceptional walking and flying. They all have six articulated legs like the adult animal, the imago. Some of them, the typical caterpillars of butterflies and moths (Lepidoptera), have in addition up to four pairs of prolegs on the abdomen that are tipped with hooks and an anal pair of prolegs. These prolegs are not articulated but instead are driven by muscles with the hydraulics as an antagonist. Some caterpillars move using the typical inchworm mode of locomotion (geometrids), whereas others use a forward-progressing, shortening wave supported by the prolegs in which the gut proceeds the forward movement of the legs (Simon et al. 2010), a feature only possible with open coelom. It is obvious that the sack-like abdomen of caterpillar (scarabaeiform) larva is not suitable for active propulsion but instead must be dragged using the six articulated legs.

Returning to "real" soft-bodied animals a relatively few land snails and slugs are cryptic, but most are not. Like most other gastropods, they crawl or creep on large, flat feet. For those hauling heavy shells, loading of the soft bodies may not be that different under water or on land, and attaching to surfaces in a wave-swept environment is not possible without generating considerable forces. Snails move by muscular waves within their feet or by extension and contraction of the whole foot. Now the foot comprises a hydroskeleton in which the oblique muscles slightly lift segments from the ground, resulting in a wavy contact pattern traveling in the direction of locomotion (contraction wave) or retrograde (extension wave in marine snails). In the direct mode, the lifted and contracted wave sections are shifted anterior and extend at the front, and there they form a new extended segment traveling caudal with progression (Section 5.8.1). Snails are able to creep across smooth surfaces tilted in any direction. Here the animals take advantage of adhesive and rheological properties of mucus, allowing differential friction. In static segments (interwave region) the few tenths of micrometers thick mucus sticks the animal to the wall. In stressed segments, that is, in those wave segments in which the muscles shear and pull the mantel off the surface, the mucus strained above yield strength flows. This is reported to be the most expensive mode of locomotion and above all not a very good strategy to escape quickly (Lauga et al. 2008). Nevertheless, each gardener is able to report on the success of these soft-bodied animals.

Three groups of living vertebrates lack limbs. All caecilian amphibians (gymnophionans) and all snakes are limbless. The caecilians are mostly cryptic burrowers, although a few forms are aquatic. Caecilian burrowing biomechanics are, to our knowledge, undescribed. The snakes use a wide variety of locomotor modes (Jayne 1986), depending upon both the substrates over which they are moving and the speeds at which they wish to move (see below). There are also a relatively few groups of lizards that are limbless or nearly so. Many of those forms are cryptic burrowers.

Let us turn to the various styles of locomotion used by snakes. In lateral undulation, a caudal traveling wave is used to push against substrate this mode is also rather useful during swimming. Sidewinding (Edwards 1985) represents an undulatory gait used on slippery sand requiring less metabolic cost than legged locomotion (Secor et al. 1992). The snake places its head into the sand in the direction of motion. Then the lifted segments following caudally are placed behind the head in to the sand at about 45° to the movement direction and the head searches for the next static support. The caudally traveling wave assures that continuously all segments of the snake push to the ground as support leaving a trace of the stretched out animal while the snake moves diagonal to it. The discrete placement of the head determines the interval of these oblique snake tracings ending in the movement direction with the tail. During concertina locomotion, the animal creeps in tunnels by anchoring with strongly bent segments. The anchor travels caudally. Rectilinear locomotion is mainly used by large snakes. Here segments with erected scales travel caudally, and the relative movement of body and scales provides propulsion (Lissmann 1950). Finally, in slide pushing the animal's weight-supporting segments alternate with the sliding segments and simultaneous bending allows progression.

As undulatory swimmers, eels are well adapted to creep across wet meadows for a mile, similar to snakes. When burrowing, even animals with legs may behave like snakes. For example, the sandfish lizard actually "swims" rather quickly through desert sands (Maladen et al. 2009; Crofts and Summers 2011).

5.1.5.2 Legged Locomotion

We have seen already several examples in which soft-bodied invertebrates approach legged-like locomotion. Raising a stiff or flexible trunk with strut-like appendages at least partially above the ground offers new possibilities for locomotion. By using pectoral fins, many amphibious fish approach limbed locomotion. At slow speeds, they tend to "walk" by using their paired pectoral and pelvic fins (Gordon et al. 1978; Nyakatura 2016). For higher speeds, they tend to jump, usually quite accurately in terms of direction (Ashley-Ross et al. 2014). Some climb steep slopes or vertical surfaces by using wriggling body motions. The animals spend substantial periods of time out of water. Many of them are active, for example, pursuing food items or defending territories.

When legs appear, the lateral flexible bodies of their precursors do not hamper legged locomotion. The flexible salamanders also retained their ability for undulatory swimming and in addition their ancient sprawled posture. Due to neuromechanical interactions, bending of the trunk switches from a standing wave reinforced by leg placement and enhancing step length while walking on land to a caudally traveling wave while swimming (Frolich and Biewener 1992; Ijspeert et al. 2007). Similarly, crocodiles are excellent swimmers and also decent walkers, and some are even able to gallop.

Concentrating on legs, the number used varies widely between groups. Some animals are almost entirely bipedal (humans; most kangaroos; most birds, most of the time) or may be facultatively bipedal (many other groups of primates and some lizards when running). Most tetrapod groups are quadrupedal. Insects have six walking legs. Some insect species (e.g., mantises and mosquitos) are functionally tetrapedal. Chelicerates such as crabs have 8 legs and myriapods have many more—up to more than 100 in some millipedes. Terrestrial crustaceans have 10 walking legs. Predictably, this diversity in both bauplans and numbers of limbs has led to great diversity in locomotor modes. Speed over the ground plays a major role in determining the mode used by animals under given sets of circumstances: legged animals all have different gaits that they use for moving slowly, moderately fast, or very fast. For bipedal species, we distinguish walking, running, and hopping and for quadrupeds slow metachronal walking, trotting, pacing, pronking, and various galloping gaits. These classifications originally were historical and based on experience. The first, still largely valid scientific criterion is purely kinematic and based on phase relationships of leg placement (Hildebrand 1965; Hildebrand 1977; Biknevicius and Reilly 2006). This definition favors the idea that different gaits come about by central limb coordination. It ignores that many gaits are expressions of resonant or attractive modes of operation of dynamical systems; the latter actually represents the major theme of the treatment here.

Terrestrial locomotion in animals is not restricted to flat surfaces, and it is impossible to restrict it to steady state. To cope with bumpy surfaces, steep slopes, walls, and trees, locomotion involves jumping, climbing, and arboreal locomotion. Excellent jumpers can be found, for example, among arthropods, amphibians, and mammals (Section 5.7). Climbing up and descending down slopes and steeper surfaces—up to vertical—requires the ability to attach or to hang on, as mentioned for snails. Spiders (Kesel et al. 2004) and insects such as ants and bees have specialized leg tips that allow them to adhere to various surfaces and easily move across walls and roofs (Federle and Clemente 2008). In vertebrates, many species possess claws and sometimes spectacular specialized toes, allowing them to cover smooth surfaces (e.g., tree frogs, Barnes et al. 2009b) or ceilings (e.g., geckoes; Autumn et al. 2006). Mammals—look at your palm or some claws—possess some, but much less spectacular, adaptations and are able to climb on trees or rocks (Cartmill 1985). Large gaps can be bridged by reaching out. However, even small rats are not able to climb across smooth ceilings.

Talking about reaching out brachiation represents a special type of arboreal locomotion unique to primate vertebrates in which the animals swing like pendulums from branch to branch. Our children love to do this on playgrounds. This enforced special adaptations in the motion system (Usherwood and Bertram 2003; Ahlborn et al. 2006).

Finally, the ability to get started or to stop, to accelerate or to decelerate, to change directions more or less rapidly (maneuverability), are major features of any type of locomotion and provide a rich landscape for adaptation or specialization.

This leads us to locomotory versatility. Most species are able to explore a wide range of locomotory modes. Humans are very familiar with walking and running. We are able to hop, but not like a kangaroo. However, in contrast to the kangaroos, we are also good climbers and decent brachiators. Compared to snakes, we are bad creepers, but we are able to creep. Humans certainly walk within air, which is a fluid, and we experience minor static lift, but we are no balloons and also do not use air for propulsion. There is a large gray zone in which the realms overlap. Think of running ostriches or cockroaches (Full and Koehl 1993) supported by lift, or a rather heavy crab walking across the sea floor. Moreover, many species are familiar with several realms. A bird is able to fly and to run; we are able to swim; and more so, snakes are excellent swimmers that can be rather annoying in a natural pond. Besides the elegance of animal locomotion, it is this versatility that allows animals to cope with a complex and fast-changing environment that still is most discouraging for robot engineers.

In many cases, evolution has found different solutions for the same task. Some of the different solutions are enforced and also limited by the bauplan or by the materials involved. There is a whole world of rather intriguing perspectives of locomotion remaining to be revealed. Now, taking a closer biomechanical look at a selected aspect, namely terrestrial legged locomotion, we actually ignore the fact that animals are not just walking machines and have their own phylogenetic history. Nevertheless, let us dare to start with a basic, and as such, a rather simplifying treatment.

5.1.6 Basics of Locomotion on Level Ground

During terrestrial locomotion on level ground, the weight of the animal is largely supported by the solid substrate, in contrast to swimming (see Chapter 3) and flying (see Chapter 4) wherein the body is supported by the surrounding fluid. In a steady-state, loss-free situation, locomotion would only require such a support. The system must be able to generate a vertical component of the ground reaction force that, on average, equals body weight mg. Free-of-cost wheels provide for support. As mentioned, neglecting losses due to drag generated by the air, rolling, and internal friction (partly induced by the support) forces would be only necessary for acceleration and deceleration (for a general comparative review, see LaBarbera 1983). This is certainly also valid for a single wheel. There is no change in potential mgh and kinetic energy $\frac{m}{2}v^2$, that is, total mechanical energy is constant.

The rolling egg has frequently been used as a metaphor for walking; just replace the round wheel by a wheel with an elliptic rim. As it rolls, the shaft is lifted periodically. Here, too, the total mechanical energy is constant. The cyclic increase in potential energy is compensated by a decrease in kinetic energy and vice versa, similar to observations for walking (see below). This assumes zero rolling friction, a goal that can be approximated in reality. Zero rolling friction is, for example, related to the lack of hysteresis of the contact material and requires a smooth surface. As soon as the surface becomes bumpy, the egg experiences impacts, causing losses. In contrast, sufficient static friction is in this case necessary to prevent the egg from gliding and to guarantee rolling.

Returning to the wheel, if we take losses into account, it is necessary to compensate those losses and propulsive force components parallel to the substratum. In the case of unsteady situations, such as jumping, it is necessary to generate additional forces for acceleration and deceleration. Similar to the supporting forces, these forces must be generated by interaction with the substratum. Forces perpendicular as well as parallel to the substratum are also necessary in a loss-free climbing situation (tilted substrate).

In general, the forces onto the substrate necessary to move the center of gravity (COG) are generated by movements of body parts relative to the COG. In a stationary situation or while animals try to cover longer distances, these forces must be applied and generated cyclically. Examples are our legs or the scales of certain snakes (Lissmann 1950). These movements and the support against gravity are directly or indirectly powered by muscles. Internal support of muscle action is provided by skeletons (e.g., hydrostats, rigid skeletons). Humans or animals (as a machine) must fight losses occurring while interacting with their environment and also are forced to compensate for internal losses. These losses are related to the mechanical properties of the mechanical constituents and also indirectly to external forces.

Legged systems consist of a large central body (head, trunk, tail or an abdomen) and attached legs. Contact to the substrate is provided by the distal tip of the legs, sometimes supported by other appendages. Neglecting drag external forces necessary for support, propulsion, or acceleration are mostly transmitted by the legs.

5.2 PRINCIPLES OF LEGGED LOCOMOTION

On bumpy ground, locomotion with compliant legs offers an advantage compared to locomotion with wheels. The most important difference compared to wheels is that legs provide for discontinuous support. Legs, supporting the body weight, with leg tips moving with the ground (or, in other words, being positioned at zero relative speed with respect to the ground), function like wheels (Figure 5.1). For real legs, the movement goals of constant horizontal speed and zero vertical oscillations are difficult to achieve and rather expensive. Issues contributing to these expenses are distal masses, hip torques necessary to support the body weight without redirecting the ground reaction force, and the enormous control effort to match speed at touchdown. Multiple legs sharing body weight (Section 5.5.5) and oscillations of a heavy trunk (Section 5.2.6) might help to reduce this cost.

Considering these arguments, it seems that having wheels instead of legs would be a good idea. However, this is not the case. Our wheeled vehicles perform economically only within an environment created artificially (e.g., streets, highways, rails). Animals may not have developed wheels because (among other reasons), on average, they evolved within a bumpy environment

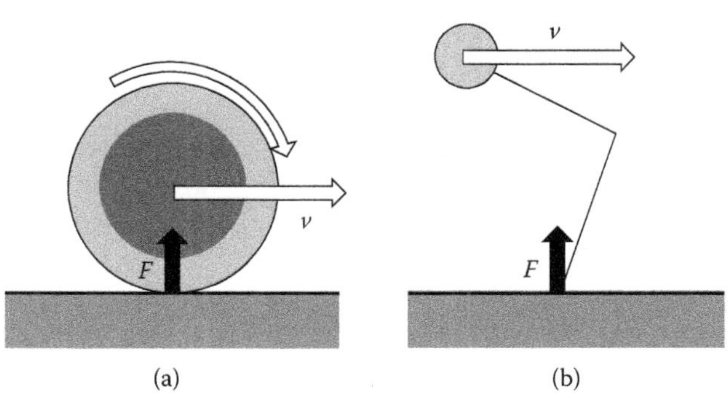

Figure 5.1. (a) A wheel supports weight and in the ideal case its rim touches tangentially the ground. (b) This can be achieved from a legged system, but it requires compliant legs, a suitable moment in the hip, and the tangential placement of the leg tip at zero relative speed with respect to the ground.

lacking highways. In such an environment, rolling friction increases dramatically and wheeled locomotion loses its advantage and may even be enforced to stop. There are some exceptions in the animal kingdom, such as desert spiders (Armstrong 1990; Rechenberg 2009) and mantis shrimps (Full et al. 1993), that seem to mimic wheeled locomotion under certain conditions. In these cases, the whole animal mimics a wheel with rather adverse consequences with respect to exteroception. The impact of the rimmed wheel on bumpy ground is similar to that of a spoke of the spoked rimless wheel resulting in a redirection of the ground reaction force and in deceleration and acceleration. Legs of animals and walking machines have the advantage that contact with the ground can be made at rather discrete points and by careful stepping and active adaptation-effective surface roughness as seen by the COG can even be reduced. Furthermore, while climbing, the footholds provided by a bumpy ground can be used to prevent sliding down an oblique surface.

Nevertheless, the simplest and most frequently cited model for walking is the rimless wheel that is similar to the idea of a "compass gait" (e.g., Borelli 1680). The rimmed wheel is replaced by a series of spokes touching the ground one after the other. The impact at touchdown of the next spoke results in losses that must be compensated to achieve a steady-state condition (e.g., by rolling down a slope). It can be avoided by an adapted convex shape of the virtual rim or, similar to the case of the rolling egg, by a continuous lengthening and shortening of the leg length with zero radial touchdown velocity (see above).

The evoked images of the rimmed and the rimless wheel inexplicably entail specific mechanical properties. Besides of important geometric details and those of action to be specified, they imply mechanical properties such as infinite stiffness of the spoke, the rim, or both. On contact, the spoke or the rim must build up force to carry the load. The time course of this buildup depends on the relative speed and mechanical properties of the colliding objects. The force generated by a springy contact depends on compression and correspondingly starts with zero force and then builds up with a force in dependence to its stiffness. A given speed of compression results in a higher rate of force development for a system with a higher stiffness. The assumption of infinite stiffness results in an infinitely fast rise of the reaction force. (In an undamped system, it would provoke high-frequency oscillations.) Low stiffness implies a slow buildup of force. It requires time and displacement until the leg is able to take up a certain load. (For completeness, consider parallel viscous effects. In this case, force builds up instantaneously if there is a relative speed of the surfaces at touchdown.) We see that the implicit assumption of an infinite stiffness represents a limit with strong implications. In fact, animals seem to control the touchdown process and the leg's response to avoid such adverse conditions.

Another issue is the detail of action. If we assume contact at a single point without adhesion, which is not the case for all animals, the tips of the legs or the feet will not allow transmission of torques perpendicular to the plane of movement. However, depending on the torques at the hip, that is, the joint coupling to the body, the direction of the ground reaction force can be modulated. In this way, it is possible to control the angular momentum of the system with a single leg. In particular, generation of only vertical forces omitting acceleration and deceleration requires the generation of tangential force components perpendicular to the leg axis. In a first step, hip torques are neglected and the assumption is made that axial forces acting at a single spot—the center of pressure—point toward the center of mass. Moreover, let us start with the description of running. In contrast to walking, this gait is not complicated by the issue of double support.

5.2.1 The Spring-Mass Model for Walking and Running

The simplest model describing very closely the dynamics of the center of mass during bipedal walking and running is the spring–mass model. Neither during walking nor during running do the legs of animals and humans represent stiff poles or spokes; they are compliant (although leg stiffness may be high at low walking speeds). This results in the gradual buildup and decay of forces. The compliance can be taken as being largely achieved actively by suitable activation of musculature. The use of natural modes of a simple system may save energy and may facilitate control. The expert reader may miss the inverted pendulum, so it further below.

Ignoring all details embedded in the movement of appendages, the principal dynamics of a system is expressed by the time course of the ground reaction force. The latter determines the acceleration of the center of mass and thus with given initial conditions the time course of its velocity and displacement. Due to single (velocity) and double (displacement) integration, many details are smoothened when calculating displacements. Thus, the time course of the center of mass is rather insensitive with respect to details in the time course of ground reaction force: higher frequencies f are suppressed proportional to f^{-2}. (Note: For frequencies below 1 Hz, the argument is inverse.) While investigating posture, it may be advisable to monitor positions. Correspondingly, the calculation of the principal energetics of the center of mass, depending on velocity and vertical displacement, is rather insensitive to detailed assumptions on the time course of force generation as long as the net momentum generated by each leg is similar to the leg observed (see below). However, the task of the legs and actuators strongly depends on the detailed force pattern.

In the first step to become familiar with the peculiarities of the model and some general properties, I restrict discussion to the one-dimensional case of bouncing at the spot. Humans use this mode of "locomotion" during rope skipping, and some species, for example, cranes (*Grus grus*), use it during courtship. In fact, an engineer designing a fast machine may be well advised to start with hopping.

5.2.2 Hopping on the Spot

Introducing the spring–mass model, which I later discuss as suitable to describe bipedal walking and running and some other bouncing gates, it seems to be useful to revert to the one-dimensional spring–mass model. Most students are familiar with this model; it is relevant for swinging and hopping on the spot, and it has both during-contact and during-flight simple analytic solutions wherein general aspects of its dynamics (e.g., force patterns) and relations, such as the interaction of ground and flight phase, can be made transparent. Also, it can be used to describe hopping at the spot.

We are all familiar with the suspended undamped spring–mass model or the simple harmonic oscillator. In fact, by rotating it upside down, having the point mass placed on the mass-less spring (Figure 5.2) and ignoring the balance issue, the equations and its solutions are the same. The point mass represents the body and the mass-less spring the leg(s). But before I go into mathematical details, think about the coarse behavior of the system. The mass placed on top of the spring will compress it proportional to its weight. As soon as the spring is deflected, it will start to oscillate around the resting value. The natural cycle frequency ω will be determined by its properties (mass m; stiffness k), but the amplitude of the oscillation will depend on the initial condition, that is, how much and how fast it has been deflected at the beginning.

If our system should mimic a hopper, it is confronted with boundary conditions. One such condition is that the mass should not pass the ground. This limits the maximum compliance of the system. Moreover, the spring should not stick to the substrate (no adhesion), that is, it should not be able to generate tensile forces, or it should detach as soon as the length exceeds its unloaded length. Once detached, the system should behave ballistically; it has an aerial or flight phase. The ballistic behavior determines the initial condition at its next contact. If we do not want it to take off, we should be more careful with the initial conditions. We can imagine that oscillations without take-off are related to walking, whereas those with take-off are related to saltatory gaits. Once the system takes off, the vertical take-off velocity determines the time the system spends in the air. We realize that as long as the system remains attached, its frequency of oscillation is well determined. Once it is allowed to take off, there is a difference between the cycle frequency of the system and the natural frequency of the oscillator.

Now have a look at the details. During flight—if there is a flight phase—the system behaves ballistically, that is, $m\ddot{y} + F_{gravity} = 0$, or

$$\ddot{y} + g = 0 \tag{5.1}$$

where m is mass; y is vertical position, with the double dot indicating acceleration; $F_{gravity}$ is the weight of the mass; and g is gravitational acceleration. Assuming the initial conditions of $y(t = 0) = y_0$ and $\dot{y}(t = 0) = v_0$, then the well-known solutions are

$$y = -\frac{g}{2}t^2 + v_0 t + y_0 \tag{5.2}$$

with

$$\dot{y} = -gt + v_0 \tag{5.3}$$

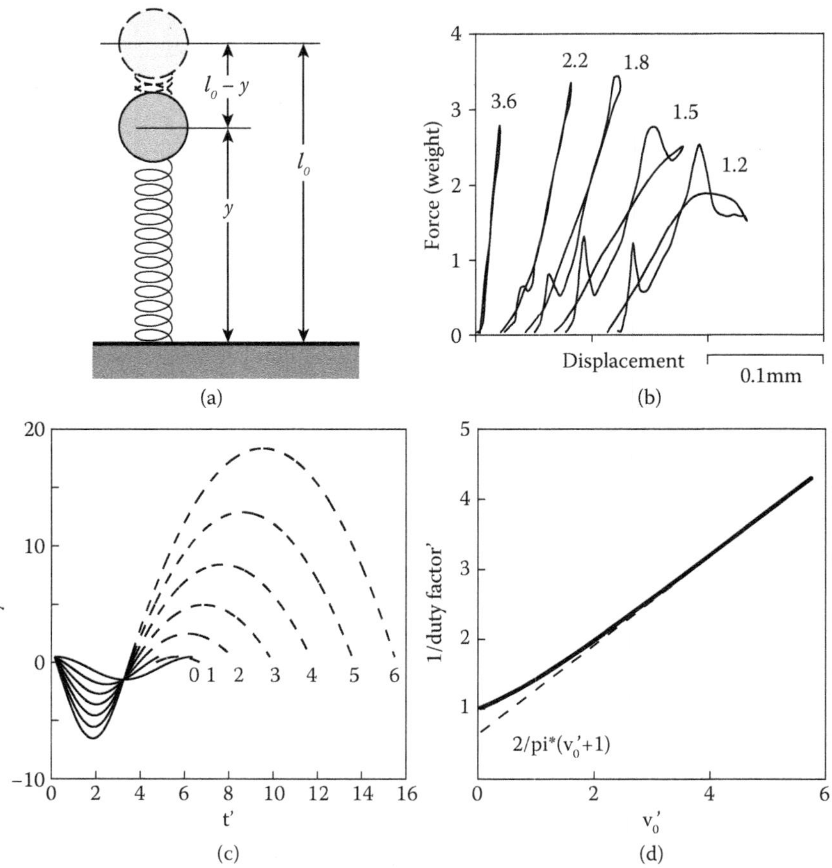

Figure 5.2. (a) Spring-mass model. Dashed: initial condition at t = 0. $l_0 = y_0$ may correspond to the unloaded spring length l_u. (b) Vertical ground reaction force in dependence of displacement of the center of mass during free human hopping on the spot at given frequency. For hopping at and above the preferred frequency of about 2 Hz (1.8 Hz, 2.2 Hz, 2.8 Hz), the characteristics is spring like. For lower frequencies (1.5 Hz, 1.2 Hz), the characteristics displays significant deviations. The line with bumps represents loading the smooth line unloading. (After Farley, C.T., et al., *J. Appl. Physiol.*, 71, 2127–2132, 1991.) (c) Time course of the dimensionless deflection y' of the spring–mass model in dependence of dimensionless time t' for different dimensionless landing velocities v_0' (0 … 6). Solid line: contact, dashed line: flight. (d) Inverse duty factor in dependence of dimensionless landing velocity v_0' (bold line). The dashed line and the inserted formula represent an approximation for the inverse duty factor on the dimensionless landing speed for speeds above 4.

where t is time. Assuming symmetry (equal heights at take-off and landing), the time of the aerial phase amounts to $t_a = \frac{2v_0}{g}$.

The equilibrium of forces involves during contact weight ($F_{gravity}$), force generated by the spring (F_{spring}), and inertial force are $m\ddot{y} + F_{spring} + F_{gravity} = 0$

Formulated within the coordinates given in Figure 5.2a,

$$m\ddot{y} + k(y - l_u) + mg = 0 \text{ or}$$
$$\ddot{y} + \frac{k}{m}(y - l_u) + g = 0 \tag{5.4}$$

where k is spring stiffness. The homogeneous equation is $\ddot{y} + \frac{k}{m}y = 0$ and the inhomogeneous equation is

$$\ddot{y} + y\frac{k}{m} = \frac{k}{m}l_u - g \tag{5.5}$$

The general solution for the homogeneous equation y_h entails the natural circle frequency ω_n with $y_h = a\sin(\omega_n t) + b\cos(\omega_n t)$, with $\omega_n = \sqrt{\frac{k}{m}}$. A special solution of the inhomogeneous equation is $y_{inh} = l_u - \frac{g}{\omega_n^2}$. The general solution as a linear combination of the homogeneous and the inhomogeneous equations results in

$$y = a\sin(\omega_n t) + b\cos(\omega_n t) + l_u - \frac{g}{\omega_n^2} \quad (5.6)$$

that yields for the velocity

$$\dot{y} = a\omega_n \cos(\omega_n t) - b\omega_n \sin(\omega_n t)$$

The solution with coefficients a and b complying with the initial conditions $y(t = 0) = y_0$ and $\dot{y}\ (t = 0) = v_0$ is

$$y = \frac{v_0}{\omega_n}\sin(\omega_n t) + \left(\frac{g}{\omega_n^2} + y_0 - l_u\right) \cos(\omega_n t) + l_u - \frac{g}{\omega_n^2} \quad (5.7)$$

and for the velocity

$$\dot{y} = v_0 \cos(\omega_n t) - \left(\frac{g}{\omega_n^2} + y_0 - l_u\right)\omega_n \sin(\omega_n t)$$

Now consider the special case where the initial length of the spring equals its unloaded length ($y_0 = l_0 = l_u$; Figure 5.2a), then both equations simplify further:

$$y = \frac{v_0}{\omega_n}\sin(\omega_n t) + \frac{g}{\omega_n^2}\cos(\omega_n t) + l_0 - \frac{g}{\omega_n^2} \quad (5.8)$$

$$\dot{y} = v_n \cos(\omega_n t) - \frac{g}{\omega_n}\sin(\omega_n t) \quad (5.9)$$

Now we can see a sinusoidal oscillation occurs around an equilibrium value that is the length of the unloaded spring $l_u = l_0$ reduced by the compression due to the weight of the system $\frac{g}{\omega_n^2} = \frac{gm}{k}$. Let us assume the system starts without flight phase with zero velocity ($v_0 = 0$). Under this condition, the first term vanishes in both equations and the amplitude of the full-wave oscillation is determined by the compliance related to body weight. For long flight phases, the first term dominates and the time course of displacement and hence of the ground reaction force approaches a sinus half wave. The magnitude of the contact time t_c is determined by the period $T_n = \frac{1}{2\pi}\sqrt{\frac{m}{k}}$ of the harmonic oscillator. Its actual value depends on the landing velocity or the height of the aerial phase and ranges from the full period to half of it:

$$\frac{T_n}{2} < t_c \leq T_n \quad (5.10)$$

The sum of the flight and contact time, or the period of the movement cycle, corresponds to the period or a full-wave oscillation for the limit of a vanishing flight phase and approaches the flight phase (plus half the period of the oscillator) for high jumps or high energies.

The duty factor is defined as the ratio of the contact time to the period of the movement cycle $DF = \frac{t_c}{t_a + t_c} = \frac{t_c}{T_{step}}$. Its limits are approached for the conditions of vanishing flight time $DF = 1$ and 0 for large flight times. Note for the limit of an infinitely stiff spring, the latter condition is always fulfilled, and deflection due to gravity can be neglected. In reality, typical duty factors are around 0.5.

As pointed out, the model is adequate for hopping, less valid for running, and even less valid for walking (see below).

Dimensionless formulation: Let us consider a dimensionless formulation achieved by the following substitutions: $t' = \omega_n t$, $y' = \frac{y\omega_n^2}{g}$. Time is now given in units of the reciprocal circular frequency, and displacements are given in units of the compression achieved by loading the spring with body weight. This results in $\dot{y}' = \frac{\dot{y}\omega_n}{g}$ and $\ddot{y}' = \frac{\ddot{y}}{g}$. It is reasonable to present accelerations in units of gravity (g) and the forces in units of body weight. (It should be noted that the apostrophe here does not denote a spatial derivative.)

The equations in Equation 5.4 are now given as follows:

$$\ddot{y}' + \left(y' - l_0'\right) + 1 = 0 \quad (5.11)$$

with the general solutions

$$y' = v_0'\sin(t') + \cos(t') + l_0' - 1 \quad (5.12)$$

$$\dot{y}' = v_0'\cos(t') - \sin(t') \quad (5.13)$$

The dimensionless velocity corresponds to the Groucho number introduced by McMahon et al. (1987). With a factor of $\frac{1}{2\pi}$, it normalizes the velocity to that achieved while falling during the period of the system. The ratio of the dimensionless speed $\dot{y}$ and distance y' corresponds to the inverse of the Strouhal number $\sigma = \frac{\omega y}{\dot{y}}$. It represents the dimensionless progression (in units of compression of the spring loaded by body weight; see above) within a period of the oscillator divided by 2π (circle period). Dimensionless stiffness is 1. For a given length of the system, the vertical displacement, that is, the dynamic behavior of the system and, for example, the duty factor merely depend on the dimensionless landing velocity.

It should be noted that guided by stiff-legged walking, for example, Alexander (1991) introduced the Froude number $v'_{FR} = \frac{v}{\sqrt{gl_0}}$ as a dimensionless velocity

$$v'_{FR} = \frac{v}{\sqrt{gl_u}} = \frac{v'}{\sqrt{k'}} \tag{5.14}$$

with

$$k' = k\frac{l_0}{mg} \tag{5.15}$$

The Froude speed has the advantage that running speed can be scaled just knowing leg length, whereas the Groucho speed requires knowledge of leg stiffness. Systems of similar size running at similar Froude numbers may run at different Groucho speeds in case leg stiffness differs.

Observations: Now let us compare theory with some experimental observations. Up to now, hopping on the spot represents an important experimental paradigm in the research of the biomechanics and motor control of locomotion. It is dynamic, entails flight and contact phases, and is experimentally simple (on the spot). Ground reaction forces can be registered with force plates, and the displacement of the COG can be determined by integrating the forces, providing the initial conditions are available. The latter can be facilitated by starting the experiments while standing on the plate—initial velocity zero, initial height of the COG can be guessed based on anthropomorphic data (e.g., Biewener 1992; Zatsiorsky 2002).

One of the most intriguing findings is that human hoppers prefer a frequency of 2 hops per s or 120 hops per min (e.g., Farley et al. 1991). In these hops the leg displays a rather spring-like behavior (Figure 5.2b). This, not by chance, corresponds to the musical tempo of an allegro and is the dominating rhythm in modern music. Yes, music is related to hopping, and as movements of the body and its entities must deal with mechanics, much of the control, that is, of the output of our brain, as well as the internal imagination of the controls, should be adapted to mechanical requirements.

During fast locomotion, many features of hopping are conserved. The bouncing locomotion of a kangaroo is also termed "hopping." Obviously, now the spring legs move backward during contact (retraction) and forward in the flight phase (protraction). The loading and unloading of the spring during contact must now be tuned to leg retraction. This key issue cannot be ignored or neglected. For example, reducing the kangaroos hop to the simple vertical bounce described above leads to wrong leg stiffness. Furthermore, the nonlinear interaction of horizontal and vertical components is necessary for self-stable modes of locomotion (Section 5.6) that may help us to understand gaits dominated by the energetics of the center of mass.

5.2.3 Running and Running-Like Gaits: Two-Dimensional Model

Many of the features described above remain valid when adding a second dimension (two-dimensional model [2D]) to the system; however, the extension also results in new features. Let us first consider what may be an obvious benefit of such an attempt.

The rimless wheel was already introduced. Now, imagine the spokes being elastic and bouncing forward (avoiding double support); this now seems to be close to running. In an alternating sequence, the legs are placed obliquely to the ground and then are compressed and extended to generate the take-off. It can be envisioned that a similar series of events can describe hopping forward in a human or a kangaroo. Even a similarity to a horse seems to be obvious. It is much more surprising that such a mechanism also describes "running" in cockroaches, where not the single legs, but the summed action of all legs, can be described by a single substitute spring. The insight that locomotion can be understood by similarity with respect to dynamics across species and several gaits has largely altered our view and treatment of human and animal locomotion.

Insertion: This concept evolved on the bases of experiments in the 1960s by G. Cavagna and coworkers. They introduced "force plates as ergometers" (1975) and used a force-plate track to calculate fluctuations in velocities and displacements of the center of mass of a variety of runners, hoppers, and trotters (Cavagna et al. 1977).

The flight phase is still ballistic, with the equations for the vertical component as given above 5.2, but with v_0 being replaced by v_{TOy} and y_0 by y_{TO}. The index *TO* refers to take-off. For the independent horizontal component,

$$\ddot{x} = 0$$

and

$$x = v_{TOx}t + x_{TO}$$

with

$$\dot{x} = v_{TOx}$$

We now address the contact period. Using the Cartesian formulation, we now have to describe the horizontal x and vertical y position of the COG. Similarly, the compression of the spring $\Delta l = l_0 - l$ can be subdivided into the respective components $\Delta x = x_0 - x$, $\Delta y = y_0 - y$. Again, the spring is assumed to be unloaded ($l_0 = l_u$) at the beginning of contact (Figure 5.3a). The forces in the respective directions are proportional to these compressions. Using the intercept theorems result in

$$\ddot{x} - \frac{k}{m}\Delta x = \ddot{x} - \frac{k}{m}(x_0 - x) = \ddot{x} - \frac{k}{m}\left(x\frac{l_0}{l} - x\right)$$
$$= \ddot{x} - \frac{k}{m}x\left(\frac{l_0 - l}{l}\right) = 0$$

or explicitly

$$\ddot{x} - \frac{k}{m}x\left(\frac{l_0}{\sqrt{x^2 + y^2}} - 1\right) = 0 \qquad (5.16)$$

Similarly, we obtain for the vertical component

$$\ddot{y} - \frac{k}{m}y\left(\frac{l_0}{\sqrt{x^2 + y^2}} - 1\right) + g = 0 \qquad (5.17)$$

Using Lagrange's equations of the second kind, the equations can be obtained with the kinetic energy $E_{kin} = \frac{m}{2}(\dot{x}^2 + \dot{y}^2)$, the gravitational potential energy as $E_{pot} = -gy$, and the potential energy stored in the spring as $E_{spring} = \frac{1}{2}k\left(l_0 - \sqrt{x^2 + y^2}\right)^2$. Note that the sum of the horizontal and vertical kinetic energy interact with the potential energy stored in the spring and its components and with the gravitational energy in the dynamic situation.

Using the normalization stated above, the dimensionless equations are

$$\ddot{x}' - x'\left(\frac{l_0'}{\sqrt{x'^2 + y'^2}} - 1\right) = 0,$$
$$\ddot{y}' - y'\left(\frac{l_0'}{\sqrt{x'^2 + y'^2}} - 1\right) + 1 = 0 \qquad (5.18)$$

This represents a coupled system of nonlinear differential equations of second order.

Only approximate analytical solutions are available. Solutions can be obtained numerically using standard algorithms (e.g., Runge–Kutta method).

The system inherits its basic components from the one-dimensional system. Characterizing parameters are the mass m, spring stiffness k, resting length $l_u = l_0$, and the gravitational acceleration g; or in the compact dimensionless formulation, a single parameter, the relative leg length l_0'. Complying with intuition the length of the leg spring relative to its compression caused by the body weight must be similar to obtain similar behavior for similar dimensionless initial conditions.

Let us now consider the space of solutions. Here it is helpful to limit considerations to single loading-unloading cycles, that is, the system should leave the ground when forces drop to zero or the spring approaches its rest length. However, it should be noted that in situations where animals can stick to the ground, as observed in many

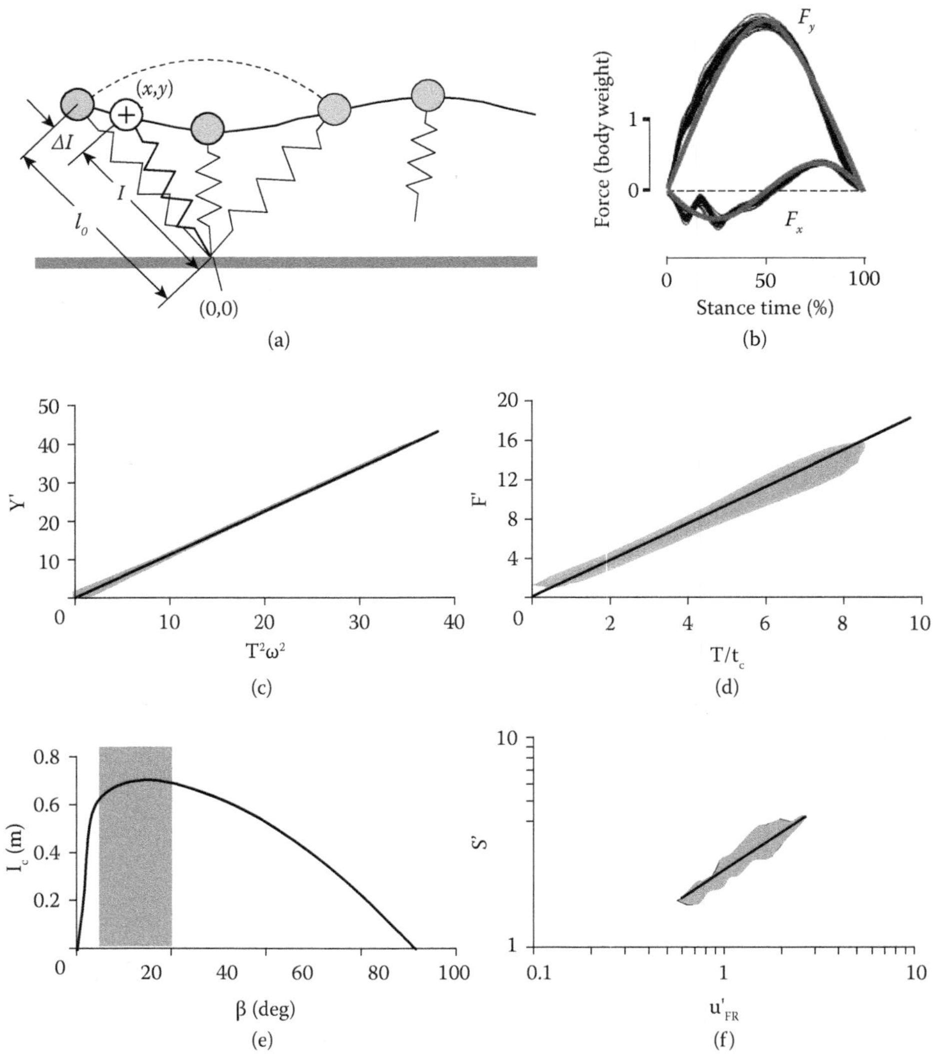

Figure 5.3. The planar spring-mass model (a, b) and some general properties supported by observations (c through f: shaded). (a) The planar spring-mass model. Spring compression Δl proportional to the axial force can be decomposed into its components Δx and Δy. (b) Measured (group of thin tracings; speed 4 m s^{-1}) and predicted (bold line) tracings of the horizontal F_x and vertical F_y component of the ground reaction force. (After Geyer, H., et al., *Proc. Roy. Soc. B Biol. Sci.*, 273, 2861–2867, 2006, fig. 7.) (c) The total dimensionless vertical excursion of the system is proportional to T^2 as would be expected in a purely ballistic situation. (d) Similarly, the weight specific vertical reaction force $F' = F/mg$ is linearly related to the inverse of the duty factor. The higher the relative flight phase, the higher the force). (e) Humans and animals use flat touchdown angles β (the angle of the velocity of the center of mass with respect to the horizontal at touchdown) to maximize contact length l_c (the distance covered by the center of mass in the contact phase). [(c through e): From Blickhan, R., The spring-mass model for running and hopping. *Journal of Biomechanics*, 22, 1224–1225, Copyright 1989. figs. 6,7. With permission from Elsevier.] (f) The length specific stride length $s' = s/l_0$ increases proportional to ${u'_{FR}}^{0.6}$. (After McMahon, T.A., and Cheng, G.C., The mechanics of running: How does stiffness couple with speed? *Journal of Biomechanics*, 23(Suppl 1), 73, Copyright 1990, fig. 11. With permission from Elsevier.)

insects and climbing vertebrates, different solutions might be relevant.

As for the one-dimensional case, system behavior strongly depends on the velocity vector of the point mass at touchdown. It also depends on the orientation of the spring with respect to the ground. This orientation can be described by the touchdown angle $\alpha_0 = \alpha_{TD}$. It is obvious for a touchdown velocity aligned with the orientation of the spring, the system just behaves similar to the one-dimensional system: the mass is reverted to where it comes from (neglecting gravity). In animals, such rebound situations are of significance when reverting at sight of a danger. If the touchdown velocity is not aligned, the system either rotates forward or backward. The focus here is on the former solution. For a steady-state situation, the task now becomes critical. Unloading of the spring must be synchronized with rotation. For a given system and a given velocity at touchdown, there is only a certain angle of attack that allows for symmetrical bounces. [For such solutions (1-periodical), each step remains the same; however, the solution can also be 2 or n-periodical repeating their pattern after n steps.] Such symmetrical solutions can be found by shooting. It should be noted that similar to the one-dimensional case in principle depending on the value of the velocity at touchdown, we can observe bounces with high or low duty-factor. However, as the radial and tangential components of speed alter with rotation dependencies are not simple. Furthermore, the period of oscillation of the system is affected by the rotation around the foothold, the latter unloading the spring due to centrifugal forces.

It may be pointed out that like most animals the human runner prefers solutions with rather flat angles of the touchdown velocity which maximize contact length. Considering only solutions within this range leads to some simple relations (Figure 5.3) well supported by experimental data.

It is important to realize that the stiffness of the spring differs from vertical stiffness. The latter is determined by the vertical excursion which is always smaller than spring compression. For Froude running velocities u'_{FR} above 2, the vertical stiffness can be approximated (McMahon and Cheng 1990). Ignoring deceleration during contact speed can be approximated by

$$u = \frac{l_c}{t_c} = \frac{2\, l_0 \sin \varphi_0}{q\, T_{n\ vert}} \tag{5.19}$$

where $\varphi_0 = \varphi_{TD}$ represents the touchdown angle with respect to the vertical and the contact time t_c is assumed to be represented as a fixed fraction q of the period of the vertical $T_{n\,vert}$ harmonic oscillator with $T_{n\ vert} = \frac{1}{2\pi}\sqrt{\frac{m}{k_{vert}}}$. Introducing dimensionless quantities (Equation 5.14) and (Equation 5.15)

$$k'_{vert} = \pi q^2 \frac{u'^2_{FR}}{\sin^2 \varphi_0} = \pi q^2 \frac{u'^2}{k' \sin^2 \varphi_0} \tag{5.20}$$

As the quantities k', q, and φ_0 only slightly depend on running speed, the normalized vertical stiffness is dominated by the quadratic dependency on running speed.

An approximate solution for the system of coupled equations has been obtained using the radial distance of the point mass to the point of rotation and the rotation angle as coordinates (derivation and details see Geyer et al. 2005). With the neglect of the angle dependency of the gravity term the equations transform into a central force system which can be integrated analytically for small compressions. The solution can be written as

$$r(t) = l_0 - \frac{|\dot{r}_{TD}|}{\hat{\omega}_n} \sin \hat{\omega}_n t + \frac{\dot{\varphi}_{TD} l_0 - g}{\hat{\omega}_n^2}\left(1 - \cos \hat{\omega}_n t\right) \tag{5.21}$$

$$\varphi(t) = \pi - \alpha_0 + \left(1 - 2\frac{\dot{\varphi}_{TD^2} - g/l_0}{\hat{\omega}_n^2}\dot{\varphi}_{TD} t\right) + \frac{2\dot{\varphi}_{TD}}{\hat{\omega}_n}\left[\frac{\dot{\varphi}_{TD^2} - g/l_0}{\hat{\omega}_n^2}\sin \hat{\omega}_n t + \frac{|\dot{r}_{TD}|}{\hat{\omega}_n l_0}\left(1 - \cos \hat{\omega}_n t\right)\right] \tag{5.22}$$

with $\hat{\omega}_n = \sqrt{\frac{k}{m} + 3\dot{\varphi}_{TD}^2}$. The centrifugal forces increase the effective circular frequency $\hat{\omega}_n$ of the system. It is nicely seen that the radial motion corresponds to the motion of a one-dimensional spring-mass model (see above) depending on the radial $\dot{r}_{TD}$ and angular $\dot{\varphi}_{TD}$ components of the touchdown velocity with the latter diminishing gravitational acceleration due to centrifugal acceleration. The angular motion has a linear characteristic modulated by trigonometric functions. The centrifugal forces increase the effective circular frequency $\hat{\omega}_n$ of the system. The contact time resolves to

$$t_c = \frac{1}{\hat{\omega}_n}\left[\pi + 2\arctan\left(\frac{g - l_0\, \dot{\varphi}_{TD}^2}{|\dot{r}_{TD}|\hat{\omega}_n}\right)\right] \tag{5.23}$$

For the centrifugal term compensating gravitational acceleration, the system approaches the "infinite stiff" behavior in with $t_c = \frac{\pi}{\hat{\omega}_n} = \frac{1}{2}\hat{T}_n$ for actual finite stiffness.

The forward component modifies the behavior of the system and alters the space of suitable solutions. We here concentrated on symmetrical solutions in which touchdown mirrors the take-off. In animals, asymmetry is rather the rule even during steady-state locomotion. Moreover, locomotion is mostly unsteady (Section 2.6 for the extreme case of jumping). Some energetic conservative aspects of such unsteady cases can be still well understood based on this model others demand sinks and source of energy. Sticking at this simplifying level, the question arises whether combinations of springs could also help to shed some light on the dynamics of special gates observed in animal kingdom. The simplest enhancement is represented by a combination of two springs and this leads us to walking. Human walking is characterized by phases in which two legs have ground contact simultaneously (double support). Multilegged support is typical for walking in general (Section 2.5). However, multiple supports can also be observed in different running gaits. Next, I concentrate on symmetric walking as the simplest and most basic example.

5.2.4 Walking: The Influence of Double Support

We traditionally call a slow-legged gait a walk. However, we can obviously run on the spot, so speed does not seem to be a good criterion but we just used the flight phase to distinguish between both gaits. With the understanding of the principal dynamics of gaits, dynamic criteria moved into the foreground. It could be shown that in a variety of animal walkers lifting the center of mass at midstance is typical leading to the concept of the stiff-legged inverted pendulum as a model (Section 5.2.5) It is assumed that the COM of the walker vaults over the supporting stiff leg during the single-support phase. In contrast, running is characterized by lowering the center of gravity at midstance. Following the same nomenclature, a compliant inverted pendulum as described above would be an adequate model. In fact, following this criterion, Groucho running (McMahon et al. 1987), named after the famous American comedian (Groucho Marx), is possible at low speeds and without flight phases. Looking at reality and across a complete cycle, there is no stiff-legged walk. While walking with crutches, one step is supported by the stiff crutch connected to a compliant shoulder and the next step is supported by a compliant leg. Yes indeed, the COG tends to be lifted but exchange between potential and kinetic energy (Section 5.5) never comes close to 100%. Real legs are in general compliant during walking pushing us back to a not too compliant spring–mass model—but now with double support and without flight phases.

In order to describe the system during double support, another parameter such as the distance x_{ds} between the two legs must be specified to describe the system (Figure 5.4). The governing equations are now

$$\ddot{x} - x\frac{k}{m}\left(\frac{l_0}{\sqrt{x^2+y^2}} - 1\right) + (x_{ds} - x)\frac{k}{m}\left(\frac{l_0}{\sqrt{(x_{ds}-x)^2+y^2}} - 1\right) = 0 \tag{5.24}$$

$$\ddot{y} - y\frac{k}{m}\left(\frac{l_0}{\sqrt{x^2+y^2}} - 1\right) - y\frac{k}{m}\left(\frac{l_0}{\sqrt{(x_{ds}-x)^2+y^2}} - 1\right) + g = 0 \tag{5.25}$$

What is the new basic feature introduced by the double contact? Acceleration of the point mass is now determined by the sum of forces exerted by both legs. Forces below body weight generated with the individual leg may add up for the two legs to forces above weight. Furthermore, if we assume that each leg is only able to generate axial forces generating reactions at their tips pointing toward the COG, the net reaction force has a line of action somewhere between the legs far from the points of support. In fact, in principle assuming telescope-springs with bending strength the two forces at the leg tips are not necessarily aligned to the legs. Even for a system with a point mass, these torques could be balanced by the counteracting leg. In the case of the single-leg contact, counteracting torques are missing. The point mass has zero moment of inertia. An extended body with inertia would be able to produce a dynamic counteracting torque. We here assume axial or radial forces of the legs. Similarly, it is here assumed that the two legs have similar properties (length, stiffness). It has to be pointed out that system energy is still constant depending on initial conditions. There are no losses due to collisions or other factors.

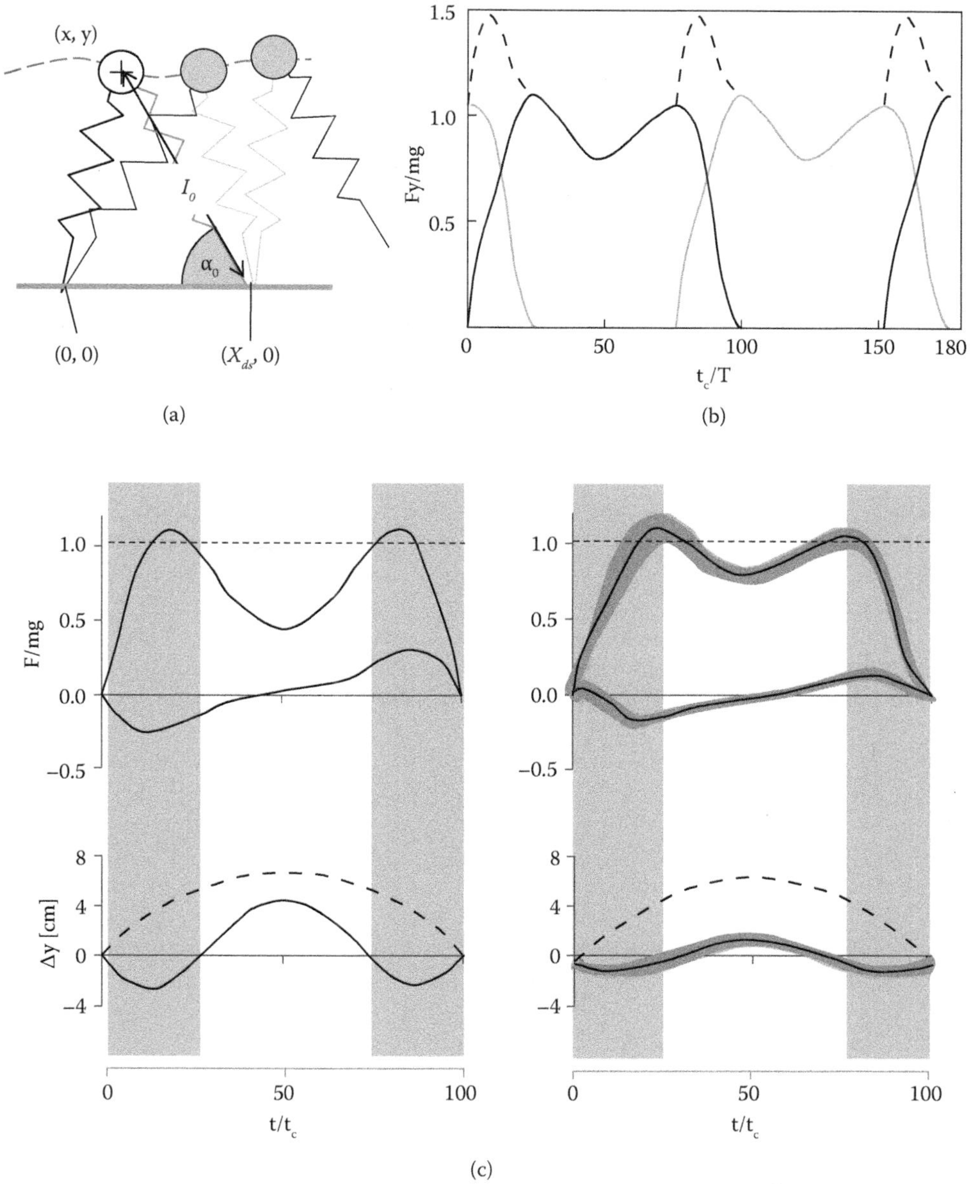

Figure 5.4. Model and patterns of forces and displacements. (a) During double support, the mass is supported by two springs. The contralateral leg (gray) is placed at x_{ds}. (b) Dashed line: Superposition of the vertical ground reaction forces of two consecutive steps (data from experiment in c). t/t_c: time relative to contact t_c of a single leg in percent. (c) Above: Pattern of the components F_x, F_y of the ground reaction forces. Below: Patterns of the displacement. Arc for a stiff leg: dashed line. Left: Simulated, rather dynamic, walking mode. Right: Experimental observations; shaded: range of variation while walking on a treadmill. (After Geyer, H., et al., *Proc. Roy. Soc. B Biol. Sci.*, 273, 2863, 2006, fig. 3.) Stippled: double support.

During a movement cycle, the system switches from a system with one leg at the ground described by the equations above (Equation 5.18) to a system with two legs at the ground and again to the system with one leg at the ground.

Numerical solutions again very well mimic the dynamic footprint of the system the time course of the ground reaction force. Moreover, for the same system energy steady state solutions are observed differing with respect to parameter combinations and system dynamics. Steady state solutions can be asymmetric with respect to the pattern of the ground reaction force. Similar patterns with increased system energy may describe saltatory gaits with two legs in succession in ground contact, such as skipping.

As observed in nature, the center of mass is lifted at midstance, but does not reach the elevation for stiff-legged walking. Indeed, compliant walking can be understood as a measure to reduce fluctuations of potential energy, it provides smoother movement. Animals prefer a smooth ride and try to reduce the impacts (see below).

It may be helpful to visualize another important feature. By focusing at the force patterns of a single leg we miss the fact, that acceleration of the center of gravity is determined by the sum of forces. Mean vertical ground reaction force must equal body weight during steady-state locomotion. At midstance, a single leg produces a force below body weight. Simultaneously, the body is lifted to its maximum height. In the midst of the double support, the combination of the two legs provide a force above body weight and the center of mass reaches its lowest point. Now, we realize that the system does not behave very different from a system with flight phases: The point mass is not lifted with a stiff leg to its maximum height but it is thrown by the momentum generated during double support (Figure 5.4).

From another point of view, we may consider the motion of the system as a transversal oscillation of a system with variable stiffness. During midstance, the system oscillates close to the 1D-situation we lined out at the beginning. Then, the two springs interact generating a net spring with variable stiffness and wandering foothold. At the angles or distances d observed, this stiffness is larger than the value of a single spring. The net spring stiffens during double support (doubling at maximum if both springs would operate in parallel). The system oscillates with varying vertical stiffness covering one cycle during half a stride. We now can imagine that such a system may have oscillatory modes with higher frequencies (Geyer et al. 2006), which Andre Seyfarth indeed observed while his child ran on the instrumented treadmill. By enhancing our perspective further, we may now evaluate force traces of animals walking and running with a number of legs simultaneously on the ground rather differently (Figure 5.5a). With corresponding assumptions, the model is easily extended considering a moving "effective footprint" generated by rolling over the foot (Bullimore and Burn 2006) or by the interaction of multiple feet (Srinivasan and Holmes 2008). By including deviations from symmetry and oscillatory modes as well as hip torques, more accurate descriptions are possible.

An extension of such a model has been used to investigate the half-bound of a pika (Hackert et al. 2006). Besides the sequential touchdown of the two front-legs, a rotational spring was used to mimic the crucial action of the spine. Only solutions with a stepping pattern similar to experimental observations—trailing contact, trailing and leading contact, leading contact, and aerial phase—were considered. The animal mass was condensed within a point pass. Forces exerted by the hindleg were assumed to act only during extension and to point toward the animal's center of gravity. The rotational spring is assumed to be preloaded at touchdown. The work produced by extension of the hind leg is dissipated at the front leg by changing stiffness after maximum compression. These assumptions follow the findings identified by experiments. To mimic nature, stiffness and positions must be adjusted carefully.

Compliant action of legs seems to be essential in animal locomotion. Many engineers avoid compliancy in order to avoid undesired oscillations. Animals in turn seem to take advantage of oscillatory modes (Section 5.6). We also have seen that inclusion of the simultaneous action of different legs has largely enhanced the spectrum of different gaits which can be described. Nevertheless, since the 1960s, the inverted pendulum dominates our dynamical image of walking. It qualitatively describes the exchange between potential and kinetic energy of the center of mass. Frequently ignored, it also introduces losses at touchdown, an issue further explored by modern collision models (Section 5.5.5).

5.2.5 Stiff-Legged Walking

We have seen how we by combining springs can easily proceed to describe a variety of gaits. Nevertheless, let us return to studies on the rigid inverted pendulum, a model representing a pacemaker with respect to our understanding of gaits. Let us first have a glimpse at the spoked wheel, and assume that each spoke is replaced by a (massless) leg. We see easily that true wheel-like locomotion could be generated by just binding a piece of the rim to the legs (e.g., McGeer 1990). Because the swing leg is assumed to be massless, it takes no effort to move it to the desired angle. Conveniently, there is neither a vertical

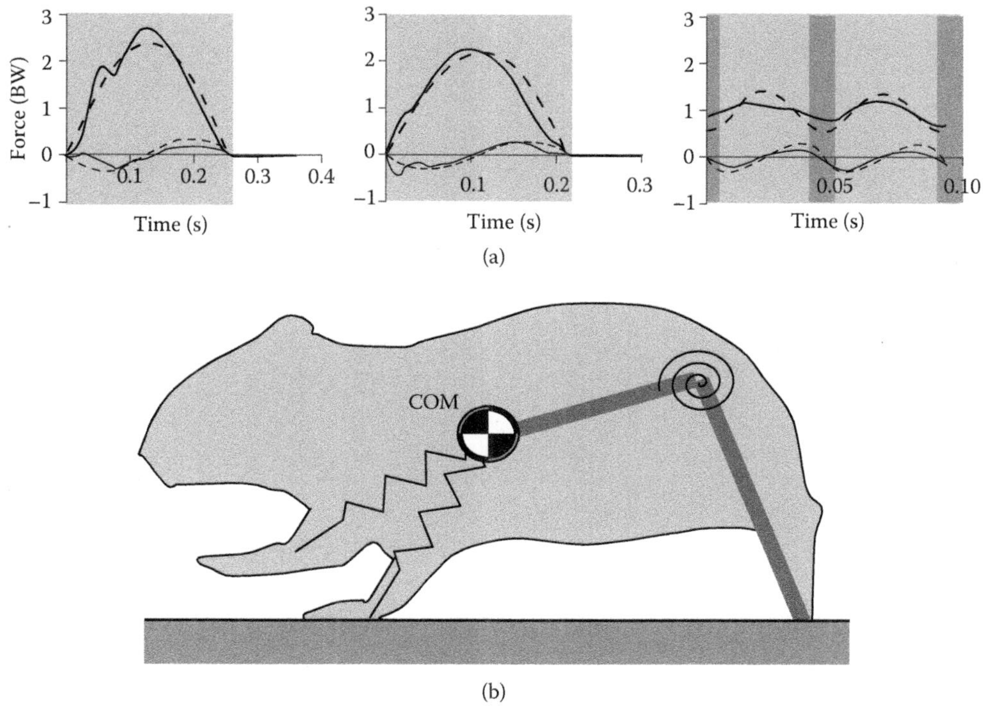

Figure 5.5. (a) Ground reaction forces in body weight (vertical component: fat lines; front-aft component: thin lines) measured (line) and simulated (dashed line) using the spring-mass model. The description for the human runner (left) and the horse (middle) is more precise than for the cockroach (right). Stippled: contact phase; dark stippled area in the cockroach: double support. (After Srinivasan, M., and Holmes, P., How well can spring-mass-like telescoping leg models fit multi-pedal sagittal-plane locomotion data? *Journal of Theoretical Biology*, 255, 5, Copyright (2008), fig. 4. With permission from Elsevier.) (b) The model for the galloping pica mimics the compliant, sequential decelerating action of the front legs and the accelerating action of the spine driving the stiff hind legs. (After Hackert, R., et al., Mechanical self-stabilization, a working hypothesis for the study of the evolution of body proportions in terrestrial mammals? *Comptes Rendus Palevol*, 5, 545, Copyright (2006) fig. 1. With permission from Elsevier.)

displacement nor a fluctuation in vertical speed. In fact, rolling over feet is what we mostly prefer to do during walking. The strategy cheaply reduces vertical oscillations as long as our feet mimic the correct radius of curvature. However, our feet are too short, and a complete avoidance of oscillations based on stiff-legged walking is impossible. The question arises why we do not have sufficient long feet? We return to this question later when we talk about the subdivision into segments. Here we just argue that evolutionarily relevant environments may be rather bumpy and that short feet and elastic reactions (see above) are thus advantageous. Shortening the feet has the consequence that after midstance, the walker falls on its stiff leg. Vertical excursions and oscillations with respect to horizontal speed are the consequence. Even a very careful placement of the leg cannot provide a pure tangential touchdown velocity of the leg tip, and a pure rolling egg mechanism is impossible. (However, we could use a flexible or an adjustable leg to achieve this.) At the instant of touchdown, the radial speed of the runner is stopped instantly (depending on the viscoelastic properties of the leg and the ground; Figure 5.6a). This impulse leads to an energy loss. The ratio of the momentum p (note point mass sitting on massless legs) after (+) and before (−) the impact amounts to

$$\frac{p_+}{p_-} = -\cos(2\alpha_0);\ \frac{E_{kin^+}}{E_{kin^-}} = \cos^2(2\alpha_0) \quad (5.26)$$

The losses vanish for angles of attack of 90°. All kinetic energy is lost for the angle of attack α_0 of 45°. Strolling down a shallow slope is sufficient to compensate this loss. We see below how this approach can be used to asses cost of locomotion in various gaits. Collisional models have been

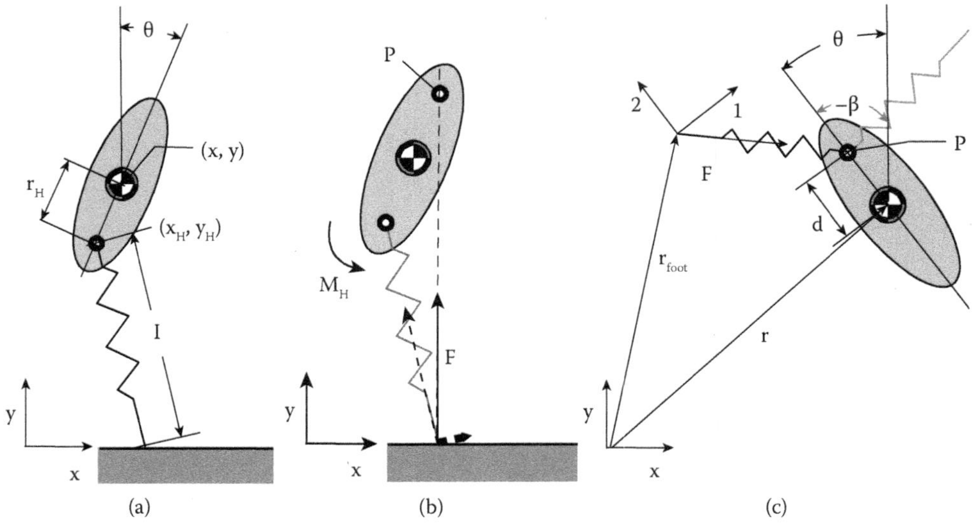

Figure 5.6. (a) Stiff-legged walker. The flash symbolizes the collision implicating losses. (b) Description of quadrupedal walking using a compound spring-mass model. (c) The ratio (CR, crosses) between the observed vertical oscillations of the center of mass and those predicted by the stiff-legged model indicate considerable system compliance. Energy recovery (ER, squares) is extremely high. (After Usherwood, J.R., et al., *J. Exp. Biol.*, 210, 538, 2007, fig. 4.)

developed in many variations and in different contexts, starting with the rimless wheel (Bekker 1956; Margeria 1976) up to studies on passive walkers (Garcia et al. 1998) and more recent studies (Kuo et al. 2005).

Under the assumption that a quadruped walks just like two rigidly coupled bipeds (four-bar linkage system), the approach can be expanded to walking of a dog (Usherwood et al. 2007). The comparison between data obtained from force-plate measurements and model predictions clearly demonstrate that during walking legs in dogs are compliant. The deflections of the center of mass observed divided by those predicted by the stiff-legged model (CR) are between 0.2 and 0.6 (Figure 5.6c). The energy recovery (Section 5.5.2) approaches, on average, a remarkable value of 0.8. In this example, collision loss largely depends on phase between the legs. Even though the contralateral legs are assumed to be out of phase, the state of the hind limbs influences the collision of the front limbs and vice versa. Also, load distribution becomes important. The mass distribution in dogs results in the fact that the losses in the hind limb contribute only about one-third of the losses that are, in turn, rather unaffected by the instant at which the energy is replenished. In contrast, the losses of the front leg dominate and react rather sensitively to replenishment timing. The minimum in the range of phases used in the real dogs is reached when energy is replenished directly after forefoot contact. Within the range of phases observed, both tangential and radial replenishment at the hind limb would be effective.

Simplifying our mental image of dynamical systems can enhance transparency. We see below (Sections 5.3.2 and 5.5.5) how the concept of collisions can be used to explain multilegged locomotion, and how different amounts of elastic restitution can be taken into account. As long as we are aware of neglected issues, such approaches are helpful. One issue not elaborated on dynamically is the distribution of legs at an extended trunk. Moreover, we ignored the mass distribution of the trunk or its moment of inertia. Some studies give us a glimpse for the wide range of possible solutions and answers. Next, I discuss the modifications introduced by the trunk during human running and how closely they are related to the planar dynamics of a running cockroach.

5.2.6 Dealing with a Heavy Trunk

If the body of the runner is not reduced to a point mass, the moment of inertia of the body can be used as counterweight while generating torques in the hip. In this situation, reaction forces at the leg deviate from axial alignment even while the system is supported by a single leg. As long as the center of mass is located at the hip, torque generation and trunk rotation are affected by the radial shortening of the legs in that this affects the lever for the hip actuator. As soon as the center of mass moves away from the hip, the radial forces on the leg influence rotation and the situation becomes more complex. Again, initial conditions exist in which the rotating and counter-rotating radial forces lead to a steady-state situation.

It is important to realize that in such a system the rotation of the trunk feeds back to the loading of the leg. Thus, controlling the position of the trunk applying a hip torque affects loading of the leg. This is also true if several legs are in contact with the ground as, for example, during bipedal walking or during quadrupedal galloping. In fact, it is proposed for walking (Maus et al. 2008, 2010) that leaning forward may be used to control acceleration and slope negotiation by using the position of the trunk.

To describe the dynamics of the system (Figure 5.7a), the locus of the hip (x_H, y_H) must be distinguished from the locus (x, y) of the COG, with

$$x_H = x + r_H \sin\theta \text{ and } y_H = y - r_H \cos\theta \quad (5.27)$$

In analogy to the equations above 5.18, listing the accelerations generated by the elastic radial component (parallel to the leg, first addend in the equation below), by the tangential component (perpendicular to the leg, second addend) produced by the hip torque M_H, and by the component due to gravity gives

$$\ddot{x} = \frac{k}{m} x_H \left(\frac{l_0}{\sqrt{x_H^2 + y_H^2}} - 1 \right) + \frac{M_H}{m} y_H \frac{1}{\left(x_H^2 + y_H^2\right)} \quad (5.28)$$

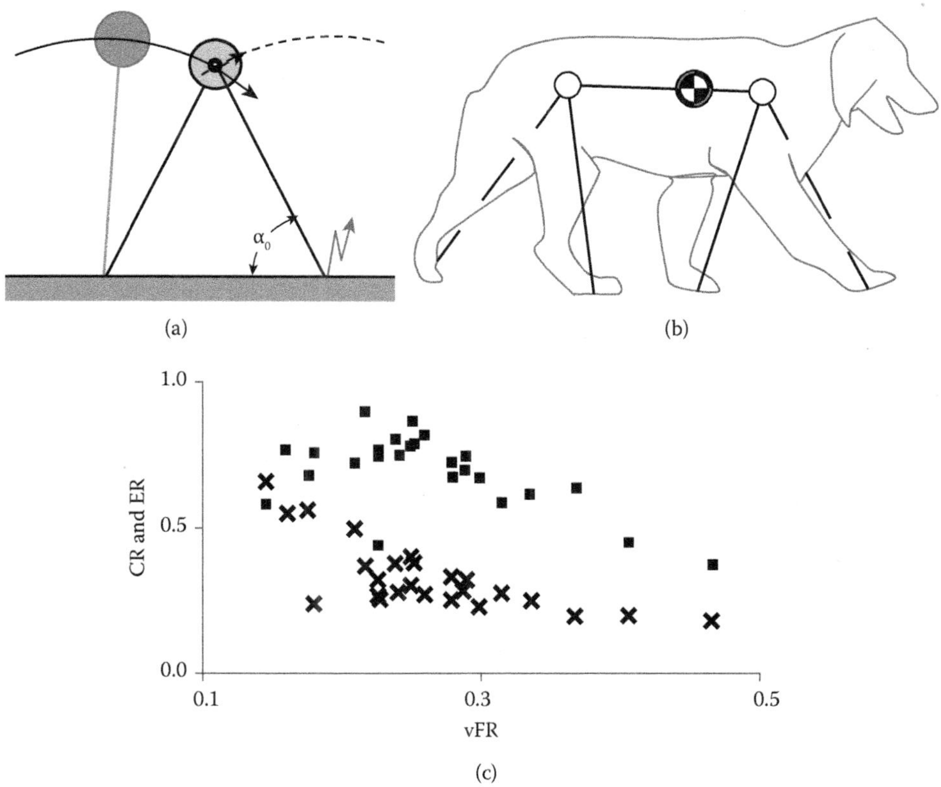

Figure 5.7. (a) Model of a bipedal runner including a heavy trunk. The origin of the system of coordinates is located at the foot. (The second leg is omitted. After Maus, M., et al., *Advances in Mobile Robotics*, pp. 623–629, Coimbra, Portugal, 2008.) (b) A hip torque generating a reaction force F pointing toward a point P above the center of gravity stabilizes the system. The virtual leg describing the net action of the tripod in action is connected via a peg at P to the animal's body. (c) Model of a running insect. The contralateral acting force of the following tripod is placed at an angle of $-\beta$ to the animal. (After Schmitt, J., and Holmes, P., *Biol. Cybernetics*, 83, 501–515, 2000.)

$$\ddot{y} = \frac{k}{m} y_H \left(\frac{l_0}{\sqrt{x_H^2 + y_H^2}} - 1 \right) - \frac{M_H}{m} x_H \frac{1}{\left(x_H^2 + y_H^2\right)} - g \tag{5.29}$$

For zero torque in the hip, the second summand vanishes. Moments generated at the hip strongly affect the movement of the center of mass.

In addition, we must take care for the balance of the moment that considers the inertia of the body I and again the contributions to torque with respect to the center of mass generated by the radial and tangential components

$$\ddot{\theta} = \frac{k}{I} \left(\frac{l_0}{\sqrt{x_H^2 + y_H^2}} - 1 \right) \left((y_H - y) x_H - (x_H - x) y_H \right) + \frac{M_H}{I} \left(\frac{x \cdot x_H + y \cdot y_H}{\left(x_H^2 + y_H^2\right)} \right) \tag{5.30}$$

If the hip is located at the COG, the radial contribution to torque described by the first addend vanishes, and the tangential component merely depends on the moment generated at the hip. For a center of mass located above the hip, the radial loaded leg generates torques with alternating signs. This can result in stable solutions without the application of hip moments (Maus et al. 2010). The model can be expanded to treat the influence of more legs, that is, to investigate different gaits such as walking, trotting, and galloping.

It should be noted that this system and the describing equations are very similar to one of the models used to describe stability in cockroaches (Schmitt and Holmes 2000). In the following experimental evidence, the action of the cockroach with its environment in the horizontal plane is described by a pair of alternating "virtual" legs fixed to a heavy body (Figure 5.7c). A slightly different description is used including global and body-fixed Cartesian coordinates, a "peg in a slot," allowing to shift the hip with respect to the long axis of the body. Because the forces act in the horizontal plane, there is no gravity term:

$$m\ddot{r} = \mathbf{F}, \quad \mathbf{I}\ddot{\theta} = \left(\mathbf{r}_{foot} - \mathbf{r}\right) \times \mathbf{F} \tag{5.31}$$

with the locus of the foot r_{foot} (1 ft at a time), the locus of the COG r, the force exerted by the spring **F**, the moment of inertia of the body **I**, and the orientation of the body with respect to the y-axis θ. The forces of the legs push the body forward, as well as from one side to the other, and rotate the animal. Similarly to the human case described above, a peg in front of the COG helps to stabilize the system.

The coupling to a heavy trunk influences operation of the legs in all gaits, and neglect is only suitable in special cases. This is even more emphasized for systems in which the trunk is flexible. Flexible trunks are abundant in animal kingdom. In some cases (e.g., snakes), this is all that is left, and frequently it largely determines leg placement (e.g., salamanders). But this listing indicates a rather kinematic perspective. In the following sections, I treat the trunk as a mechanical system possessing its own modes of oscillations and thus possibly dominating the rhythm of locomotion and the emergence of gaits. In terms of mathematical description, this issue requires a short excursion into beam theory. Describing the dynamic action of compliant systems represents an approach that seems promising in both research on animals and in robotics.

5.2.7 Trunk as Beam Bending the Spine

So far we have reduced the trunk of animals to a point mass or a rigid body. The dynamics of locomotion is dominated by the more or less elastic collisions of the legs and their coupling introduced by the trunk. However, especially during galloping in many vertebrates the trunk is rather flexible. This flexibility has the effect of increasing stride length (Fischer and Witte 2007) and thus enhances speed of locomotion (think of the hare or the cheetah). Humans also retained this flexibility. Besides of bending, torsion of the trunk does play a role. There are different ways to introduce the corresponding degrees of freedom, one is to subdivide the trunk into segments coupled by visco-elastic elements (Section 5.2.4) and or muscles another is to introduce continuum elements such as flexible bars. Let us here turn to the bar element.

The response of the trunk of quadrupeds as a bar depends on its loading and/or clamping both are related to the coupling of the trunk to the legs (Alexander 1988). Combining bending

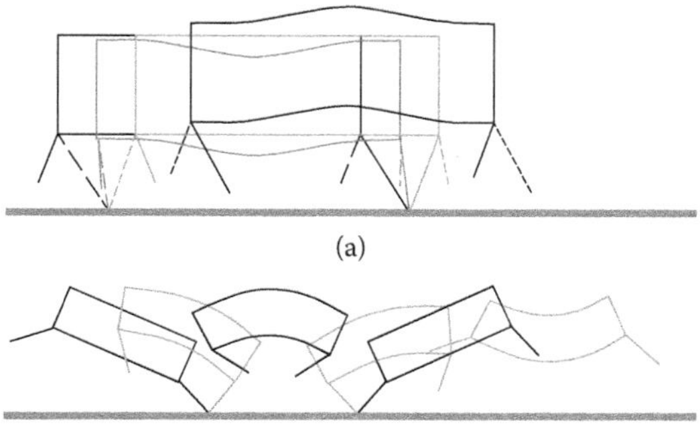

Figure 5.8. Bending modes of the trunk during trotting (a, half stride) and during galloping (b, stride). During the trot, the action of the legs clamps the trunk, thereby limiting its bending. The bending mode during galloping as indicated by the boundary conditions (see text) differs from the mode depicted. The model animals proceed from left to the right.

with rocking motion leads to the typical pattern observed for galloping mammals which in principle would allow for energy storage in the trunk (Figure 5.8). Deceleration at the front legs induces bending of the trunk and the stored energy is released in the next phase resulting in accelerative forces transferred by the hind-limbs and leads to propulsion and to protraction of the forelimbs.

Such a sequence offers the advantage of additional elastic storage. Research on tissue providing such a storage in the immediate vicinity of the spine confirmed adaptive flexibility being much higher in a seal than in a monkey (Gal 1993a,b), however, compared to the animals size, the stored energy e.g. in the jaguars spine seems to be negligible. However, we must recall, that our leg joints without muscle activation are rather flexible too and there is only a minor amount of energy stored. These changes with activation of the muscles in series. In any case, the quasi-elastic operation of the leg dominates its principal dynamics. If this is also valid for the trunk the issue seems to be worthwhile to be revisited. Especially at high speed the flexible trunk not only enhances stride length but also saves energy (Seok et al. 2015).

Modeling the trunk as a system with a single link may not reasonably describe the prevailing bending modes which may be important in stabilizing gaits and allowing for energy storage. The modes significant for vertebrate locomotion will not only include bending in the sagittal but also in the frontal plane and torsion (O'Reilly et al. 2000; Witte et al. 2004) and both are also significant in human locomotion.

The flexible trunk enhances principal compliance. This may not be relevant for a small insect, but it is rather helpful for large vertebrates. And as we have seen the flexible trunk and its bending modes can serve as a clock determining the rhythm and modes of locomotion. A classical pace maker is the suspended pendulum of the swing leg. As we neglected the mass of the legs we did not stumble across this issue so far and we turn to it in the following chapter. We will see that suitable subdivision into segments and mass distribution does represent an important ingredient and should not be ignored in legged machines.

5.3 SWING LEGS: PENDULUMS AND PASSIVE WALKERS

So far, we considered only the mass of the trunk either as point mass or as a distributed mass; we neglected the mass of the legs. This is an idealization that works decently when treating the contact phase that is dominated by redirecting the falling body mass. Movement of massless legs during the swing phase can be set arbitrarily and do not require any force or energy. This arbitrariness is strongly reduced when including mass.

5.3.1 Walking: The Swinging Pendulum as a Pacemaker

If we assume that the swinging leg is suspended at the hip that, in turn, moves at a given attitude with fixed speed and we ignore friction, the swinging leg is able to swing like a suspended pendulum. The frequency f of a mathematical pendulum consisting of a point mass and swinging with sufficiently small amplitudes is given by

$$f = \frac{1}{2\pi}\sqrt{\frac{g}{l}} \tag{5.32}$$

with the gravitational acceleration g and the length of the pendulum l.

If the leg is allowed to swing freely, the swing time will set the pace. This seems to be the case during walking, but not in fast gaits such as running. There acceleration and deceleration to provide the rotation of the swing leg is enforced by muscles (and to an unknown extent by visco-elastic tissue).

The equation above must be modified for a physical pendulum where mass distribution must be taken into account. In this case, the body is assumed to be rigid with arbitrary mass distribution.

$$f = \frac{1}{2\pi}\sqrt{\frac{mgl_{CM}}{I}} \qquad (5.33)$$

with the moment of inertia I of the rotating body with respect to its center of mass, the mass m of the body (here, the mass of the leg), and the distance l_{CM} of the center of mass of the body with respect to the axis of rotation. In fact, the frequency can be used to calculate the moment of inertia of bodies under investigation (e.g., Dowling et al. 2006).

Frequently, the center of gyration or reduced length of the pendulum is used to describe the net effect of mass distribution. It represents the distance from the axis where the total mass contracted to a point mass would give the same frequency as the rigid body described above

$$l_{gyr} = \frac{I}{ml_{CM}};\ f = \frac{1}{2\pi}\sqrt{\frac{g}{l_{gyr}}} \qquad (5.34)$$

In general, the natural legs are not rigid pendulums. They are flexible and shorten during protraction. Reasons for this are clearance with respect to ground—becoming especially important on rough ground—and reduction of the moment of inertia to adjust the period of the swing or to reduce forces necessary to drive the pendulum. It should be noted that the effect on period is not dramatic. Halving the distance of all mass points halves the length of gyration and reduces the period by a factor of $1/\sqrt{2}$. In contrast, it strongly reduces the moments necessary to rotate the legs by a factor of 1/4. Making the legs slender, for example, by halving the mass halves the moments.

The fact that the swing leg sets the pace during walking has received much attention. Besides of a number of attempts to describe observations in nature, it led to numerous numerical and technical models called passive walkers consisting of stiff supporting legs combined with freely moving swinging legs. The simplest system approaching the behavior of a wheel is the "synthetic wheel" (McGeer 1990), consisting of two alternating spokes of small mass compared to the body mass placed on the hip with sufficiently long rim segments attached to it (Figure 5.9). Under such conditions, it is possible to neglect acceleration and deceleration of the systems center of mass induced by the swinging leg, and we can treat the system like a physical pendulum suspended at the hip and allowed to swing freely (neglecting also shuffling). To mimic wheeled motion, the swing leg leaves the ground with a rotational velocity $\dot{\varphi}_{mean}$ given by the forward speed of the system $v = \dot{\varphi}_{mean} l$, that is, the leg will for some time continue to swing back, and it should have the same speed when taking over after touchdown, that is, it should retract before touchdown to match forward speed. Given the swing amplitude φ_{swing} step duration T_{step} amounts to

$$T_{step} = \frac{2\varphi_{swing}}{\dot{\varphi}_{mean}} \qquad (5.35)$$

The swinging leg pendulum has a sinusoidal motion with

$$\varphi = \varphi_{swing}\cos\omega_l t + \frac{\dot{\varphi}_{mean}}{\omega_l}\sin\omega_l t \qquad (5.36)$$

For $\omega_l T_{step} = 4.058$, the swing leg has again the rotational velocity of $\dot{\varphi}_{swing}$. This implies that the step duration amounts to 0.646 the swing period T_l. In other words, the step duration that is half the stride period exceeds the half-period of the pendulum by a factor of 1.29. It should be noted that in the region where the small angle solution of the pendulum is valid, the pendulum period does not depend on deflection amplitude. Within a limited range (for human-sized walkers <3 m/s), this mechanism allows walking at different speeds by changing the amplitude of the pendulum.

With more heavy (human-like) legs, the acceleration of the body mass due to the swinging legs cannot be ignored, and vice versa the induced movement of the center of mass may pump the motion of the swing leg. This may compensate to some extend the shortening of the swing half-time with respect to the step duration (Verdaasdonk et al. 2009).

Despite of the numerical difference between the period of the stepping cycle and the suspended passively swinging limb, it is evident that in this model both periods are strongly coupled. The period of the swinging leg determines

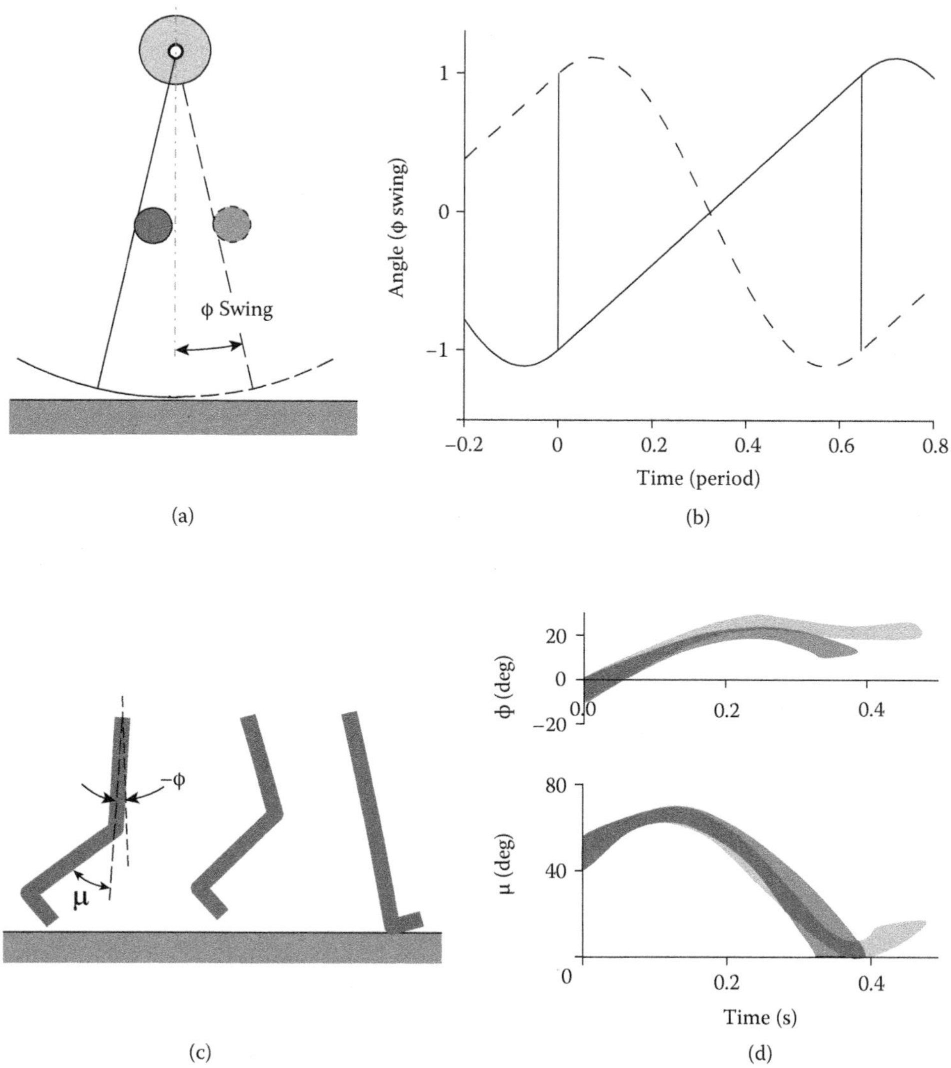

Figure 5.9. Swing legs as pendulums. (a) Simple model of a wheel-like bipedal walker with a swing leg (synthetic wheel). (b) For each leg, sinusoidal oscillations (swing) alternate with periods in which the rotation angle changes at a constant rate. Angle: Deflection angle. (After McGeer, T., *Int. J. Robotics Res.*, 9, 66, 1990, fig. 5. With permission from SAGE.) (c) Double-segmented pendulum. (d) Simulations (transparent dark shading) based on the double pendulum starting with initial conditions as measured, guided at the hip, and using individual anthropometric data revealed rather similar time courses of the hip and knee angle as measured light shading. (After Selles et al. Copyright 2001, fig. 3, p. 1173. With permission from Elsevier.)

stepping frequency. Moreover, the synthetic wheel does not require muscle work. Nature does not seem to use the synthetic wheel. We have noted that our legs are heavy. Our feet or the feet of some animals can be used as curved "rim segments"; however, they are too short to support the center of mass throughout the gait cycle. Correspondingly, we must consider either elasticity (see above) or collision. So far, no investigations are available including an elastic stance leg combined with a passive swing leg. However, a long line of research exists taking the aspect of collisions into account, which is addressed below.

Let us first turn to a different issue: clearance. When we walk, we move our toes just millimeters above the ground, but we do not touch it. Only if we are lazy or tired do we start to shuffle. We can achieve this by raising our hips in the frontal plane (e.g., Inman et al. 1981) or by bending our legs. Assuming that efficient walking may be a passive process, the question arises whether a passive mode does exist providing the correct motion. This is indeed the case. Let us consider a double pendulum consisting of a thigh and a shank with a foot rigidly attached at 90° to the shank with human dimensions (length, inertia). We suspend it by a pin-joint to the hip and prescribe the movement of this joint and the initial conditions of the segments according to experimental data. Then, the double pendulum behaves very similar, but not identically, to the real leg (Selles et al. 2001). In experimental investigations using technical models, McGeer (1990) could show that to provide clearance, the exact leg properties, especially the length distribution, must comply with observations in nature, but the reason seems to rest upon the double support phase. Let me cite McGeer (l: length of the leg): "Natural proportions are 0.46 l in the thigh and 0.54 l in the shank (Chapman and Caldwell 1983), as opposed to the 'obvious' choice of 50/50 which would maximize swing foot clearance per unit knee flexion. The flaw in the 'obvious' reasoning is that rotations induced at support transfer lead to substantially deeper flexion, and to better foot clearance, if the knee is placed higher on the leg. Unfortunately we learned this the hard way, after building the legs for a 50/50 model." In contrast to the stiff-legged models that lock the knees close to touchdown, real legs slightly bent their knees, indicating quasi-elastic behavior (Figure 5.9b). Retraction before touchdown and after lift-off is not visible in the tracings documented by Selles et al. (2001).

Recalling that support and rebound are already demanding functions for a leg, it is intriguing to observe the delicate adjustment of properties to the swing-phase. Principally, the action of the swing leg must somehow fit to the dynamics of the stance leg or the rest of the system. This interdependency is treated extensively in the investigation of passive walkers.

5.3.2 The Advent of Passive Walkers: Combining Dynamics of Stance and Swing

Now that we know that the passive swinging leg mimics the general features observed in nature, we may return to the rhythm of walking. In their pioneering work, Mochon and McMahon (1980) (see also McMahon 1984) developed a model for "ballistic walking" that couples the swing leg to the stiff stance leg, allowing for interaction (Figure 5.10a). The stance foot was ignored. In these simulations, arbitrary initial conditions were allowed that did not drive the system off the ground, produced clearance of the swinging leg at the beginning of the simulation, and caused the leg to extend when the leg hits the ground. Similar to the simple model of the synthetic walker (see above), the range of solutions displayed step times largely independent of step length. Furthermore, the half-period of the compound pendulum representing the swing leg was clearly longer than the step time observed for larger persons and found in the simulations (factor of ~1.4). This shortening is contributed to interaction of the stance and swing leg (Figure 5.10b). Carefully matched experiments and modeling using the ballistic walking model, including as an enhancement foot extension and initial conditions from experimental observations, result in swing times clearly below observations (factor of ~0.8; Selles et al. 2001).

The ballistic walker only considers the action of the model during the swing phase. It ignores the double support phase and the aspect of collision (see above). These have been taken into account by later models (e.g., McGeer 1990; Garcia et al. 1998). The aspect of load transfer mentioned reasoning about the subdivision of the walker's leg (segmentation) is only described in the context of a treatment merging all phases. It seems to be especially tricky to build a walker that is able to walk slowly at near zero slope conditions. The condition is that the center of mass of the leg is located on the line connecting hip and the contact point to guaranty close to static stability. Furthermore, the COG of the shank-foot segment should be located at the parallel line through the knee (Garcia et al. 2000). The ballistic walkers introduced by McGeer (1990) and followed up, for example, by Garcia et al. (2000) that included walkers with stiff swing legs and those with a knee joint (Figure 5.10c and d), largely inherits the properties of the synthetic wheel (see above). To compensate for the impact losses (see above), these walkers must stroll down a gentle slope. And, in fact, they do so in the simulation as well as in reality.

The equations of motion can be found in McGeer (1990). Garcia et al. (2000) describes the modeling process as follows: "The motions of the walkers are governed by standard rigid

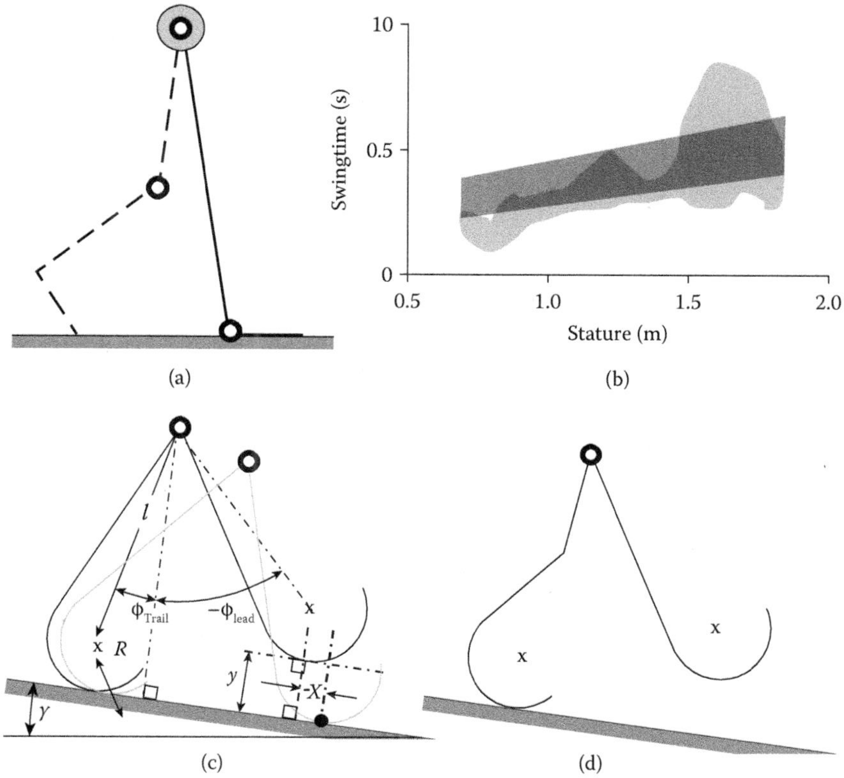

Figure 5.10. (a) Model of the ballistic walker. The legs possess anthropomorphic inertia. The small circles mark the joints active in the swing leg (dashed) and the stance leg (full). (b) Swing time versus stature for the simulation (dark transparent) and after experimental data (light shading). (Redrawn after Mochon, S., and McMahon, T.A., Ballistic walking: An improved model, *Mathematical Biosciences*, 52, 247, Copyright 1980, fig. 5. With permission from Elsevier). (c) Stiff-legged dynamic walker with curved feet. (d) Walker with knees. [(c) and (d): After Garcia, E., and de Santos, P.G., *Robotica*, 23, 13–20, 2000.]

body dynamics. In our way of formulating these, all of the equations described below are based on angular momentum balance about various points. A walking step starts right after heel strike, just when the old stance leg (which is now the swing leg) starts swinging. Equations of motion for the "three-link mode" are integrated forward in time until knee strike is detected. Assuming no rebound (perfectly plastic knee collision), jump conditions then determine the post–knee strike state. Equations of motion for the "two-link mode" are now integrated forward in time until heel strike is detected. Assuming no rebound (perfectly plastic heel collision) and no ground impulse at the trailing foot, jump conditions are used to determine the post–heel strike state from the pre–heel strike state. This completes one walking step. The double support is in this model instantaneous. We note that solutions exist deviating from those observed in human locomotion.

Like in the synthetic wheel, cadence remains nearly invariant with speed, and different speeds are accommodated by enhancing deflection or stride length. It should be noted that similar to the synthetic wheel, the system has two solutions: one described above with a long swing duration matching the speed at touchdown and a short one in which only angular symmetry is obtained. The latter adds the cost of decelerating the swing leg with the impact at touchdown. However, its step duration is shorter as well as a given speed leg excursion. Considering experimental evidence, it may well be closer to reality. The measurements with technical models seem to support this notion (observed short period is ~1.1-fold and for the long period ~1.3-fold the value observed for the human walker). We have seen above while discussing the ballistic walker that introducing a short foot and a knee has the advantage of automatic clearance during protraction. By considering the stance phase, a longer, curved foot would be of advantage because it would reduce the collision. The length of our foot might represent a compromise, minimizing total costs including both swing and stance aspects. It is important to note that introduction of a knee introduces asymmetry (McGeer 1990). The rolling contact of an asymmetric foot facilitates extending moments at the knee generated by the ground reaction force. It is obvious that human walkers possess a foot that can be flat to allow for standing and is adjustable compliant to provide for curvature during walking. Furthermore, the heel is in the aft of the ankle joint; thus, the impact at heel-strike can generate adverse moments flexing the knee, which is desirable for a compliant leg (see above). Nevertheless, by comparing real anatomy with idealizations via numerical and technical models, we are able to better understand the functions in nature.

With increasing slope, the passive walker increases speed until becoming instable. As pointed out above, the walker increases speed prominently by increasing the angles of excursion. Without detailed calculations, energy balance arguments can be used to derive the dependencies verified in experiments (here again, we follow Garcia et al. 2000). For simplicity, let us return to the simple model without knees (Figure 5.10a).

The collision at heel contact is assumed to be completely plastic. The velocity of the foot parallel ($\dot{x}$) and perpendicular ($\dot{y}$) to the surface can be approximated as

$$\dot{x} = R\left(\dot{\varphi}_{trail} - \dot{\varphi}_{lead}\right) + l\ \cos\varphi_{trail}\ \left(\dot{\varphi}_{trail} - \dot{\varphi}_{lead}\right) \quad (5.37)$$

$$\dot{y} = -l\ \sin\varphi_{trail}\ \left(\dot{\varphi}_{trail} + \dot{\varphi}_{lead}\right) \quad (5.38)$$

with R being the radius of the foot, l being the fixed distance between the hip and the center of the foot arch, $\varphi_{trail,\,lead}$ being the deflection of the trailing and leading leg, and γ being the slope angle. We have seen above that the excursion increases with the slope, that is, $\varphi_{trail} \sim \gamma^p$, with some $p > 0$. Let us find the suitable exponent p. As the period remains largely unaffected by the slope: $\varphi_{st} \sim \gamma^p$. For small angles and shallow slopes, $\dot{y} \sim \varphi_{trail}\dot{\varphi}_{trail}$ and $\dot{x} \sim \dot{\varphi}_{trail}$. For the same condition, the dissipated energy E_{loss} can be approximated as

$$E_{loss} = \frac{1}{2} m_{body}\dot{y}^2 + \frac{1}{2} m_{leg}\dot{x}^2 \quad (5.39)$$

The potential energy per step available due to the slope can be taken as $E_{pot} \sim step\ length \cdot slope \sim \gamma^{p+1}$ with $p > 0$. For the mode in which the horizontal velocity of the leg at touchdown $\dot{x}$ is zero $E_{loss} \sim \gamma^{4p}$ or for the equilibrium between kinetic and potential energy: $p + 1 = 4p$ or $p = \frac{1}{3}$. For the other, short mode where the leg shuffles on the ground after reaching the correct excursion $E_{loss} \sim \gamma^{2p}$ hence $p + 1 = 2p$ or $p = 1$. From this, the scaling of the power ($P = mgv \sin\gamma$) can be derived.

One of the advantages of basic models is that they can reveal principle strategies. Considering the losses observed, the question arises what might be intelligent measures for compensation. McGeer (1990) identified a mixture of radial and tangential measures. The radial measure consists of extension and push of the trailing leg and the tangential of adjusting the position of a trunk with inertia requiring a net moment at the hip that must be balanced by corresponding ground reaction forces. McGeer (1990) also shows that for each degree of descent the human walker has to lean back by 7° to compensate. The experimental biped "Dynamite" required slopes of several degrees to walk.

Passive behavior of the system largely characterizes human walking, and this may be true for many large species. This is increasingly explored in legged robots. In fast gaits such as running, it is generally assumed that leg retraction is largely actively driven. Nevertheless, control may mostly take advantage of the leg segmentation to facilitate control and to increase efficiency. In the following section, some first steps are described, and we use the opportunity to introduce the vectorized formulation of a suspended compound pendulum. We return to this formulation in Section 5.7.4.

5.3.3 The Swing Leg during Running: Approximations and Vectorized Formulation

As mentioned above, interaction between a springy contact and a pendulum flight phase is unexplored so far. Corresponding studies require consideration of trunk inertia (see above). In the only study available (Knuesel et al. 2005), dynamical interaction is omitted and the suspension point is not accelerated. The mathematical pendulum inherits the angle and angular velocity of the spring at take-off (the spring is able to actively lengthen beyond l_0), and at touchdown the swing angle corresponds to the angle of attack of the leg spring. Similar to the findings for the synthetic wheel, stable solutions of such a system were found for angles of attack decreasing with increasing speed (Figure 5.11a).

Undoubtedly, during running there are active contributions to leg swing. Nevertheless, the preferred frequency used by joggers correlates highly with the natural frequency of the compound pendulum (Huat et al. 2004). This is an example how small-amplitude description can be used to calculate frequencies (Figure 5.11b; compare the frequency of bending modes of a flexible trunk in Section 5.2.7).

Following Huat et al. (2004), the equation of motion of the thigh segment of the compound pendulum moving around the hip joint is

$$I_A\ddot{\varphi}_1 = -m_1 g\, r_1 \sin\varphi_1 - m_2 g\, l_1 \sin\varphi_1 + F_{x,B} l_1 \cos\varphi_1 \quad (5.40)$$

Rearranging results for small angles,

$$F_{x,B} = \left(\frac{I_A}{l_1}\right)\ddot{\varphi}_1 + \left(\frac{r_1}{l_1}m_1 + m_2\right)g\varphi_1 \quad (5.41)$$

F_{hor} is determined by the acceleration of the mass of the distal segment. For small angles, this results in

$$F_{x,B} = -m_2\ddot{x} = -m_2\left(r_2\ddot{\varphi}_2 + l_1\ddot{\varphi}_1\right) \quad (5.42)$$

Combining the last equation results in

$$\left(\frac{I_A}{l_1} + m_2 l_1\right)\ddot{\varphi}_1 + m_2 r_2\ddot{\varphi}_2 + \left(\frac{r_1}{l_1}m_1 + m_2\right)g\,\varphi_1 = 0 \quad (5.43)$$

with $I_A = I_1 + m_1 r_1^2$.

The equation of motion for the distal segment reads as

$$I_2\ddot{\varphi}_2 = F_{x,B} r_2 \cos\varphi_2 - m_2 g\, r_2 \sin\varphi_2 \quad (5.44)$$

With $I_B = I_2 + m_2 r_2^2$, this results in

$$F_{x,B} = \left(\frac{I_B}{r_2}\right)\ddot{\varphi}_2 - m_2 r_2\ddot{\varphi}_2 + m_2 g\varphi_2 \quad (5.45)$$

Combining Equations 5.45 and 5.42

$$m_2 l_1\ddot{\varphi}_1 + \left(\frac{I_B}{r_2}\right)\ddot{\varphi}_2 + m_2 g\varphi_2 = 0 \quad (5.46)$$

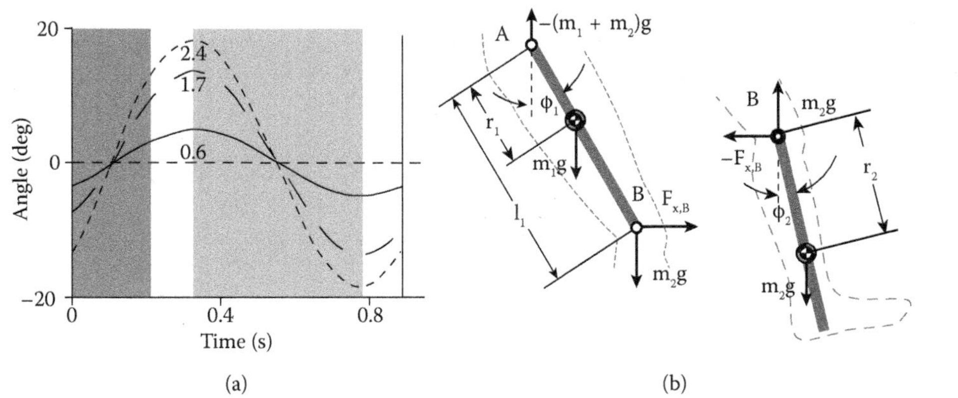

Figure 5.11. (a) Angular excursion of the swing pendulum depending on horizontal take-off speed (m/s). Dark shaded: stance, rest flight; light shaded: retraction; white: phases of retraction during flight; vertical line at the right marks the touchdown; length of the pendulum: 0.2 m. (After Knuesel, H., et al., Influence of swing leg movement on running stability. *Human Movement Science*, 24, 536, Copyright 2005. With permission from Elsevier.) (b) Free body diagram of a double compound pendulum for approximate (small angles) calculation of the pendulum frequency. (After Huat, O.J., et al., *Int. J. Comput. Appl. Technol.*, 21, 48, 2004, fig. 1.)

Assuming sinusoidal oscillations with the same frequency and phase, $\varphi_1 = \varphi_{1,max}\sin(\omega t + \phi)$ and $\varphi_2 = \varphi_{2,max}\sin(\omega t + \phi)$ lead to a linear system of equations for $\varphi_{1,\,max}$ and $\varphi_{2,\,max}$:

$$\begin{bmatrix} \left(\frac{I_A}{l_1} + m_2 l_1\right)\omega^2 - \left(\frac{r_1}{l_1} m_1 + m_2\right) g & m_2 r_2 \omega^2 \\ m_2 l_1 \omega^2 & \left(\frac{I_B}{r_2}\right)\omega^2 + m_2 g \end{bmatrix}$$

$$\mathrm{X}\begin{bmatrix} \varphi_{1,max} \\ \varphi_{2,max} \end{bmatrix} = \begin{bmatrix} 0 \\ 0 \end{bmatrix}. \tag{5.47}$$

The condition for nontrivial solutions that the determinant of the coefficient matrix is zero results in a quadratic equation (not listed) for ω^2 whose roots give the estimates for ω. The regression ($N = 5$; $r = 0.94$) between the computed (with segment mass and inertia after Winter [2009], and the calculated stride frequency shows that the model predicts lower values [the excluded outlier has been an obese subject; see Huat et al. [2004]]):

$$f_{osc} = 0.75\, f_{pref} + 0.29 \tag{5.48}$$

The reader should be aware that this procedure does really represent a small-angle approximation. It ignores all terms of higher order including angles and its derivatives, and it is assumed that $\sin\varphi = \varphi$ and $\cos\varphi = 1$. It also ignores, by assuming synchronous motion of the two segments, the complex and fascinating motions of such a pendulum. Nevertheless, the results described above are intriguing.

The complete equations can be derived from Lagrange equations (e.g., Zimmermann et al. 2009) and are

$$\begin{aligned}\left(m_1 r_1^2 + m_2 l_1^2 + I_1\right)\ddot{\varphi}_1 + \cos\left(\varphi_1 - \varphi_2\right) m_2 l_1 r_2 \ddot{\varphi}_2 \\ = -\sin\left(\varphi_1 - \varphi_2\right) m_2 l_1 r_2 \dot{\varphi}_2^2 \\ -\left(m_1 r_1 + m_2 l_1\right) g \sin\varphi_1 + M_1 - M_2\end{aligned} \tag{5.49}$$

$$\begin{aligned}\cos\left(\varphi_1 - \varphi_2\right) m_2 l_1 r_2 \ddot{\varphi}_1 + \left(m_2 r_2^2 + I_2\right)\ddot{\varphi}_2 + \\ = \sin\left(\varphi_1 - \varphi_2\right) m_2 l_1 r_2 \dot{\varphi}_1^2 - m_2 r_2 g \sin\varphi_2 + M_2\end{aligned} \tag{5.50}$$

where M_1 and M_2 represent additional torque that may act at the hip or the knee. With respect to generalization (Section 5.7.4 on jumping), it is convenient to introduce a formulation in matrix form

$$\mathbf{A}(\varphi)\ddot{\varphi} = \mathbf{B}(\varphi)\dot{\varphi}^2 + \mathbf{C}(\varphi) + \mathbf{T}(\varphi, \dot{\varphi}), \tag{5.51}$$

or

$$\begin{aligned}&\begin{pmatrix} m_1 r_1^2 + m_2 l_1^2 + I_1 & \cos(\varphi_1 - \varphi_2) m_2 l_1 r_2 \\ \cos(\varphi_1 - \varphi_2) m_2 l_1 r_2 & (m_2 r_2^2 + I_2) \end{pmatrix}\begin{pmatrix} \ddot{\varphi}_1 \\ \ddot{\varphi}_2 \end{pmatrix} \\ &= \begin{pmatrix} 0 & -\sin(\varphi_1 - \varphi_2) m_2 l_1 r_2 \\ \sin(\varphi_1 - \varphi_2) m_2 l_1 r_2 & 0 \end{pmatrix}\begin{pmatrix} \dot{\varphi}_1^2 \\ \dot{\varphi}_2^2 \end{pmatrix} \\ &+ \begin{pmatrix} (m_1 r_1 + m_2 l_1) g \sin\varphi_1 \\ -m_2 r_2 g \sin\varphi_2 \end{pmatrix} + \begin{pmatrix} M_1 - M_2 \\ M_2 \end{pmatrix}\end{aligned} \tag{5.52}$$

The action of compound pendulums undergoing extreme flexion during the swing phase as observed during fast locomotion is currently under investigation. There the issue of intelligent control exploring the properties of an actively driven pendulum may move into the foreground. Small flexion of the compound pendulum as observed during locomotion at moderate speed facilitates clearance of the foot and improves the prediction of the protraction period. In robots, passive swinging is mostly avoided or even impossible due to the hardware used. Before considering the strength of a simple spring in explaining the action of a leg during locomotion, considering the subdivision of stance legs into segments seems to be a nuisance. In nature, this subdivision seems to be an evolutionary load resulting in the fact that most legged systems consist of several segments and are driven by antagonistic muscles and actuated telescopes or springs are not available. However, it may well turn out that here segmentation (here and in the following used as a technical term) may offer helpful nonlinearities and asymmetries. Nevertheless, once gifted with a segmented leg the question arises how to deal with the segmentation.

5.4 SEGMENTED LEGS DURING STANCE

We have learned that well-balanced segmentation in a two-segmented anthropomorphic leg can help to provide ground clearance and load uptake for ballistic walkers (Section 5.3). This is related to the significance of proper layout when exploring the whip effect for example during throws or kicks (e.g., Putnam 1993). Segmentation allows to actively fold the leg during the retraction, thereby, in turn, helping to reduce the moment of inertia that is especially significant during fast gaits. It also facilitates obstacle clearance in jumps. Because all appendages have the task to reach locations within the environment (e.g., legs finding footholds in a clotted environment or "arms" reaching for prey), the range of movements of the leg covered also represents an import biomechanical aspect strongly influenced by segmentation. As to my knowledge, general performance criteria including those for the work space developed for robot manipulators (e.g., van den Doel and Pai 1996) have not been applied to natural arms and legs so far. Let us here turn to the significance of the three-segmented leg that plays a major role as a constructional principle in mammalian legs.

In artwork, robot design, and the minds of scientists from antiquity to recent times, the following image prevails: "And a viviparous quadruped bends his limbs in opposite directions to a man's, and in opposite directions to one another; for he has his forelegs bent convexly, his hind legs concavely" (Aristoteles, 350 BC in Nussbaum 1986). This concept ignores the true three-segmental nature of the mammalian legs and their orientation (Figure 5.12a). The major reasons for this misconception are that the most proximal segment is frequently hidden under fur and voluminous muscle packages and that the functional role of the scapula as a third segment of the front leg has been discovered only recently (rev. Fischer and Blickhan 2006). In the light of these investigations, the geometry of the front legs mirrors that of the hind legs. The thigh is equivalent to the shoulder blade (scapula), the shank to the upper arm, the foot to the lower arm, and the toes to the hand. We first ignore the foot (fourth segment) and discuss this issue below. In addition, if we conceive the whole animal as a linkage system and do not ignore bending of the trunk (Section 5.2.7), the latter also contributes to an extending and shortening segmented system that we now would hesitate to call a "leg."

First, let us address the question of whether some aspects of segmentation can be interpreted as a measure to reduce the cost to compensate the loads generated by the ground reaction force.

5.4.1 The Economy of Segmentation

Ignoring segment mass, cost of work, rotational velocity, and the different levers of the muscles at the two joints, we assume here that joint moment determines the cost of a configuration and we search for three-segmented configurations that minimizes such a cost function (Gunther et al. 2004). In a two-segmented leg, the angle of the knee joint is—besides of its mirror—unambiguously determined by leg length. It is obvious that a given ground reaction force pointing toward the hip generates moments in the knee that increase with leg shortening. For a given reaction force, the moment at the joint is determined by the distance of the line of action from the joint h. Shortening a two-segmented leg with segments of equal length starting with an extended configuration causes a monotonous increase of the lever until the leg length (or the inscribed joint angle) vanishes and the lever reaches a value of 0.5 l_{max}:

$$h = \frac{1}{2}\sqrt{{l_{max}}^2 - l^2} \qquad (5.53)$$

If we start with a configuration in which one segment is shorter than the other, for example, $l_2 < \frac{l_0}{2} < l_1$ with $l_1 + l_2 = l_{max}$, then the leg is not able to shorten completely (the work space is reduced) and the maximum of $h = l_2$ is reached at an intermediate position when the segment is horizontal

$$l(h) = \sqrt{l_1^2 + l_2^2 - 2h^2 \mp 2\sqrt{\left(l_1^2 - h^2\right)\left(l_2^2 - h^2\right)}} \qquad (5.54)$$

In the case of unbalanced segment lengths, it is better to arrange the shorter segment distally because otherwise the "knee" will hit the ground and limit excursion. At this point, it should be noted that we ignore the internal leverage and the

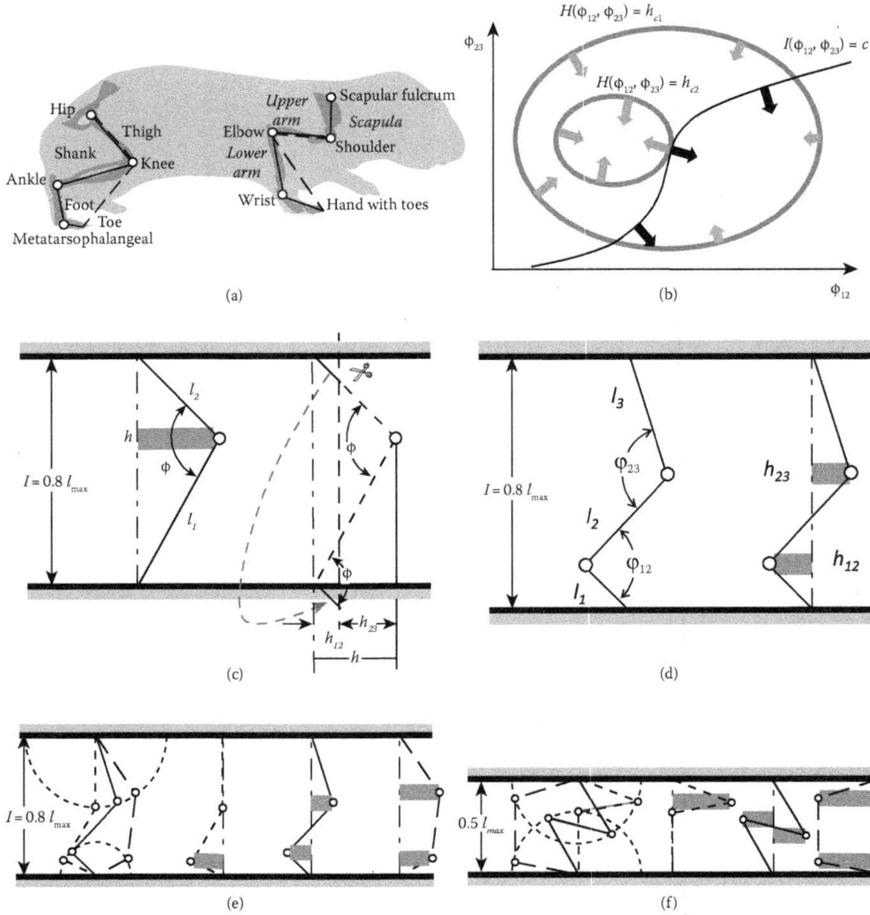

Figure 5.12. (a) The two-segmented schematization ignoring the shoulder plate as a functional segment (bold dashed lines) leads to the antique concept of opposing leg orientation (see also below). Detailed high-speed X-ray analysis revealed the functionally rather similar zigzag configuration of the mammalian front and hind leg. (After Fischer, M.S., and Blickhan, R., *J. Exp. Zool. A Comp. Exp. Biol.*, 305, 935–952, 2006.) Nevertheless, some asymmetry remains and the functionality of the orientation is still the issue of current research. Black lines with circles: segments and joints deduced; black and normal text: joints; gray italics: segments. (b) Optimization of H under the constraint of l. The arrows indicate the gradients. (c) Two-segmented leg (left) shortened to 0.8 of its maximally extended length (l_{max}). On the right: Cutting a piece off the outer segment and shifting it to the other end results in an equivalent three-segmented leg. (d) Three-segmented leg (segmentation 2:2:1). (e) For given shortening ($0.2\ l_{max}$) and segment subdivision (2:2:1), the open degree of freedom of the system allows an infinite variety of configurations. The $h_{23} = 0$ configuration offers minimal moments. (f) For strong shortening ($0.5\ l_{max}$), a segment subdivision of (1:1:1) offers slightly lower moments in the z-configuration. Dotted: h_{23} or $h_{12} = 0$, that is, the configuration at the border between z- and c-mode; normal line: best symmetric z-configuration; dashed line: worst c-configuration.

geometry function that has a singularity for the extended leg (Section 5.6.5). The height of this triangle as a function of the knee angle φ can be calculated using the partial derivative $h = \frac{\partial l}{\partial \varphi}$, to repeat this: the height (our cost function) is equal to the decrease in length with changing knee angle ($F\ \delta l = M\ \delta\varphi = F\ h\ \delta\varphi$ or $\delta l = h\ \delta\varphi$). For an extended leg, the change in leg length is small for a given change in the knee angle and it is maximum for the configuration where h, the distance with knee and line of action, reaches its maximum value:

$$l(\varphi) = \sqrt{l_1^2 + l_2^2 - 2 l_1 l_2 \cos\varphi} \qquad (5.55)$$

results in

$$h = \frac{\partial l}{\partial \varphi} = \frac{l_1 l_2}{l} \sin\varphi \qquad (5.56)$$

With this background, it is possible to discuss the impact of the leg configuration on leg stiffness k_{leg} or its inverse leg compliance c_{leg}. Compliance is defined by the ratio of the change in displacement divided by the change in force. Let us assume a force F causing finite changes in the leg angle $\Delta\varphi$

$$c_{leg} = \frac{\Delta l}{F} = \frac{\Delta\varphi}{F}h. \tag{5.57}$$

For equal segment lengths $l_1 = l_2 = l_s$, compliance is zero for the extended leg ($h = 0$) and $c_{leg} = \Delta\varphi\frac{l_s}{F}$ for the completely flexed configuration.

Let us now turn to the three-segmented configuration $l_{max} = l_1 + l_2 + l_3$ with the included angles φ_{12} (between l_1 and l_2) and φ_{23} (between l_2 and l_3). Now, with fixed segment lengths the same leg length can be achieved by infinite combinations of the two angles, that is, the segments can move. The three-segmented leg has 2 degrees of freedom; only one is fixed by the leg length. Using the cost function of minimizing the sum of the distances ($H = |h_{12}| + |h_{23}|$) of the joints from the line of action of the ground reaction force, we introduce a condition diminishing the freedom of the system.

Before going into detail, let us first address two basic configurations that we term the z-mode (or zigzag mode) and the c-mode (or bow mode). They are distinguished by the fact that the two joints connecting the three segments are either arranged each on one side of the line of action (z-mode) or both on one side of the line of action (c-mode). The bow mode seems to allow maximum levers and thus should be most expensive (so far without proof). Based on the reasoning presented for the two-segment model, the bow mode should be the most compliant configuration. In contrast, in the best z-configurations the same ground reaction force can be produced with lower joint moments. If moments count for cost, it is better to use the z-mode. Independent of cost, the configuration allows the same joint moments to generate higher reaction forces, a widespread requirement in legged systems. Think about jumping, or pushing.

Now, what is the best z-configuration? By probing with a compass you can find out that it should be the arrangement in which the long outer segment is parallel to the line of action. Then, the moment at one joint vanishes and only the moment at the joint adjacent to the shorter outer segment determines the cost. (This is also true for the case where all segments or the outer segments have the same length.) Intuitively, we may not like this situation marking the boundary between the c- and the z-modes, and we will see further below why we might—at least a little bit—stick away from it. But if we look on our leg configuration during walking and running, we realize that we are not far from this limit. We might intuitively assume that the z-mode configuration with equal angles $\varphi = \varphi_{12} = \varphi_{23}$ should be the configuration with minimum joint moments. But this is not the case. It is actually in our terms the most expensive z-configuration, but not as expensive as the extreme c-mode configurations. The fact that the situation with angular symmetry does not represent the cheapest configuration offers us another answer to a central question. It tells us that the three-segmented leg in an optimized z-configuration requires less moment to produce force than a two-segmented configuration with the same total length l_{max}. Subdividing the leg in more segments saves energy! (I assume, but this remains to be shown, that this is true for systems with more segments, and it would be worthwhile to check mathematically for the conditions.) Why should we care for two-segmented configurations, shouldn't they be obsolete? The problem introduced by the third segment is stability and control of the additional degree of freedom (see below).

Let us outline the path for analytical and numerical considerations and describe some results. Similar to the geometric situation in the two-segmented leg, we can calculate the heights (levers of the reaction force) with respect to the joints by using the partial derivatives of the leg length. Leg length is given as

$$l(\varphi_{12}, \varphi_{23}) = \sqrt{\begin{array}{l} l_1^2 + l_2^2 + l_3^2 - 2l_1l_2\cos\varphi_{12} - 2l_2l_3\cos\varphi_{23} \\ +2l_1l_3\cos(\varphi_{12} - \varphi_{23}) \end{array}} \tag{5.58}$$

The heights are given as

$$\begin{pmatrix} h_{12}(\varphi_{12},\varphi_{23}) \\ h_{23}(\varphi_{12},\varphi_{23}) \end{pmatrix} = \nabla_\varphi l = \begin{pmatrix} \dfrac{\partial l}{\varphi_{12}} \\ \dfrac{\partial l}{\varphi_{23}} \end{pmatrix}$$

$$= \frac{l_1 l_3}{l} \begin{pmatrix} \dfrac{l_2}{l_3}\sin\varphi_{12} - \sin(\varphi_{12}-\varphi_{23}) \\ \dfrac{l_2}{l_1}\sin\varphi_{23} + \sin(\varphi_{12}-\varphi_{23}) \end{pmatrix} \tag{5.59}$$

We normalize length now to the maximum length l_{max}, that is, $l_1 + l_2 + l_3 = 1$, and choose to characterize the length distribution by l_2 and the ratio $\frac{l_1}{l_3}$. We search for a minimum of our cost function $H\left(l_2, \frac{l_1}{l_3}, \varphi_{12}, \varphi_{23}\right) = |h_{12}| + |h_{23}|$ under the constraint of a given leg length $l\left(l_2, \frac{l_1}{l_3}, \varphi_{12}, \varphi_{23}\right) = const.$ Now, we must find the minimum in H. However, it should be the minimum for solutions that also fulfill the equations defining the constraint, that is, find the minimum of H while you walk along the path defined by $l = const$. At the site of the extremum, the tangents of H and l must be parallel, a requirement fulfilled when the gradient vectors are collinear (same orientation but different length; this ambiguity is expressed introducing the scalar factor γ (Figure 5.12b). Using the Nabla Operator (see Equation 5.59) to describe the gradient results in the equations

$$\nabla_\varphi H = \gamma \nabla_\varphi l \tag{5.60}$$

Eliminating γ from the two equations leads to a single equation:

$$G\left(l_2, \frac{l_1}{l_3}, \varphi_{12}, \varphi_{23}\right) = 0 \tag{5.61}$$

The details of the results of the optimization process are rather complex and so far they concentrate on z-configurations. The considerations are limited to results that take the heel height into account and do not allow joints and the heel to dig into the surface. Here only some major results are reported. First, the calculations support (with a minor exception for very short legs) the fact that independent of leg length and segment configuration, the symmetrical situations are the most expensive situations and the configurations with one joint on the line of action are the cheapest. If the lengths of the outer segments are different, then it is of advantage to align the long segment with the line of action. With concentration on those solutions, the cost increases with shortening until the middle segment is perpendicular to the leg axis (horizontal for a leg arranged vertically to the ground). It is obvious that the range of such economic leg shortenings is limited by the length of the short segment and the ground condition. Looking at different segment distributions $\left(l_2, \frac{l_1}{l_3}\right)$, this implies that the range of economic shortening is largest for the case of three segments with the same length $\left(l_2 = \frac{1}{3}, \frac{l_1}{l_3} = 1\right)$. The length of the short segment determines the range in which cheap shortening can be achieved. Design restricts the range of economic leg shortening. The strongest gain of the aligned solution compared to the worst-case symmetric solution (two-segment solution, see Figure 5.12c) is achieved for $l_2 \approx 0.45$ and $\frac{l_1}{l_3} \approx 0.4$, resulting in $l_3 = 0.4$ and $l_1 = 0.15$. In this area, the unbalanced outer segments also gain most compared to solutions with equal outer segment length. It is not bad to have a foot. The layout of segment lengths and their orientation in humans guarantee maximum savings (in the currency of static joint moments). This advantage is achieved for a slightly shortened leg (~6 %, i.e., $l = 0.94$). The situation may also be relevant for large birds, but this time the role of the foot may be taken by the thigh. For systems in a crouched posture, where pushing and acceleration with large elongations are demanded, the situation changes. Optimum solutions can be obtained by strong folding. However, this folding is limited by the ground, the heel, and anatomical conditions. Here having equal segment length offers an advantage. For an equally segmented 1:1:1 leg at $l = 0.5$, the $h_{23} = 0$ condition is not better than the condition with symmetric angles $\varphi_{12} = \varphi_{23}$. In fact, the roughly horizontal orientation of the middle segment largely diminishes differences between the total moment or H observed under angle configurations in the vicinity of that horizontal arrangement. Further shortening the leg makes the symmetrical solution even better than the $h_{23} = 0$ solution.

Based on these considerations, we can understand the segmented organization of the human leg as being a global optimum for any nearly extended leg, exploited, for example, during human (slow) running or walking. In contrast, we see that small mammals and probably also most insects have legs that are optimized to produce high forces combined with a maximal range of leg extensions. Therefore, an initially crouched posture seems to meet this combination optimally (Section 5.7). The reader might object that the considerations are rather simplifying. They indeed are. However, the effect of the muscle levers in determining the effective mechanical advantage (Biewener et al. 2004) would modulate the results, but not alter the basics. The influence of angular rates counters that of the moment and might have a stronger effect. Furthermore, stability requirements (Section 5.4.4) may conflict with optimal H-configurations in extended legs.

Here the focus is on legs that were driven without or with negligible hip torque, that is, the line of action of the ground reaction force runs through the hip. For the human leg, this represents a decent approximation. Especially for small quadrupeds and insects, this is not the case. We consider this point in the following section being well aware that we are still far away from a general picture enclosing all important geometries.

5.4.2 Leg Axis and Line of Action

The investigation outlined above assumes alignment of ground reaction force and leg axis (defined by a line connecting the point of pressure or the distal tip of the limb with the hip joint). This does represent a decent approximation for some gaits such as two-legged running; however, it neglects strong deviations observed in some other gaits and especially in small animals (e.g., small mammals and insects). In general, such a design increases moments because now the moments at the hip must be considered. (A detailed mapping is not available.) The situation changes however when more than one leg is in contact with the ground in an instant. For example, legs may push against each other as can be seen in insects (Full et al. 1991) and in small galloping mammals (Hackert et al. 2006). Cockroaches run using an alternating tripod. At midstance, the front legs push against hind legs and the left leg(s) against right leg(s). This seems like a waste of muscle force. Yet, a calculation of the muscle forces at the femur-tibia joint and the coxa-trochanter joint, as examples, revealed that taking the sprawled position as a constraint, the observed direction of the reaction vectors actually minimized the moments and, more sharply, the sum of the muscle forces (Figure 5.13). It should be noted that in the situation of many-legged locomotion, the minimization of moments might be one of several criteria determining load distribution and that, in general, all actively controlled joints must be taken into account (Günther and Weihmann 2012).

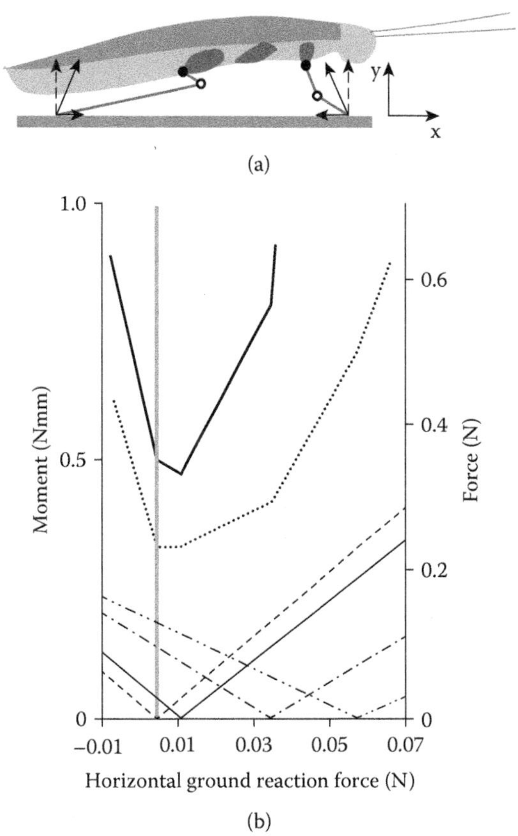

Figure 5.13. (a) Laterad view of a cockroach with the ground reaction forces (bold arrows) at the first and at the third leg supporting the animal vertically (each one-third the body weight; the second leg omitted) and with compensating horizontal components. Filled circle: coxa-trochanter (C-T) joints; open circle: femur-tibia (F-T) joint. (b) Moments at the depicted joints (line: leg 1 C-T joint; dashed: leg 1 F-T-joint; dash-dot-line: leg 3 C-T-joint; dash-double dot: leg 3 F-T-joint), total moment (dotted line) and estimated muscle force (upper bold line) in dependence of the magnitude of the horizontal force. Shaded bold vertical line: horizontal force measured. (After Full, R.J., et al., *J. Exp. Biol.*, 158, 387, fig. 10.)

Kuznetsov (1995) used the work approach to illuminate segment orientation in crouched trisegmented legs for the situation that the reaction force is oriented vertically. The work approach directly addresses energetics of locomotion (Section 5.5). In the presented case, it assumes similar costs for work and for energy absorbed, and it ignores the cost of isometric force production that, as we have seen in the quasi-static approach above, has been rather successful in explaining the human segmentation and the orientation of ground reaction forces in cockroaches. Nevertheless, it is rather illustrative and provides a basis for future research. As mentioned in Section 5.2.1, moving the center of mass at steady speed (here in direction of x) does not require any network. The supporting force is perpendicular to the motion:

$$W = \oint_{s=0}^{s=b} \mathbf{F} \circ ds = \int_{x=0}^{x=b} F_x dx + \int_{y=h}^{y=h} F_y dy = 0 \tag{5.62}$$

where x and y represent the coordinates in horizontal and vertical direction, respectively. Moreover, legs that shorten and lengthen without accelerating or decelerating the center of mass do not produce "external" work (Section 5.5.2). But one could conceive that motors shortening and lengthening the leg and generating the moments at the hip might need energy proportional to these contributions to work. (This is equivalent of using $b_1 = b_2$ in Equation [5.90]). The work for axial elongation of such a leg from midstance to lift-off must take the leg lengthening into account and the projection of the force to the leg axis.

$$W_{ax} = \int_{l(x=0)}^{l(x=b)} F_l dl = \int_{l(x=0)}^{l(x=b)} F \cos\varphi_2(l)\, dl \tag{5.63}$$

$$= \int_{l(x=0)}^{l(x=b)} F \frac{h}{l}\, dl = F\, h \int_{l(x=0)}^{l(x=b)} \frac{1}{l}\, dl = F\, h\, \ln l \Big|_{l(x=0)}^{l(x=b)}$$
$$= \frac{1}{2} F\, h\, \ln\left(1 + \frac{b^2}{h^2}\right). \tag{5.64}$$

This axial (or radial) work must be counter balanced by an equal amount of tangential energy absorbed ("negative work"; W_{tan}) in the hip joint as total work is zero. If we count (section 5.5.4) the absolute values for cost, this doubles the estimate above. This is still valid when we introduce a segmented leg (omitting moments at the distal leg tip). Here we may assume that the knee provides the work for leg lengthening. In other words, the jointed system can be described by the following equations that can be easily expanded to a three-segmented system (index 2: proximal joint, index 1: knee; Figure 5.14)

$$dW_{tot} = Fx_2 d\varphi_2 + Fx_1 d\varphi_1 \tag{5.65}$$

The constant horizontal distance covered during locomotion at constant speed v within a time increment amounts to

$$-v\ dt = h d\varphi_2 + y_1 d\varphi_1 \tag{5.66}$$

and a cost function can be formulated as the sum of the absolute values of the work per force (Section 5.4.2):

$$C = |x_2 d\varphi_2| + |x_1 d\varphi_1| \tag{5.67}$$

As mentioned above, the configuration of the two-segmented leg is completely determined for the presumed height and translation. This is different for the three-segmented leg. Here the same total leg length can be realized by different angle configurations. During translation, these angle configurations, in turn, can do work against each other, thus adding cost (in our currency). Unfortunately, Kuznetsov (1995) did not map the configurations, but he worked out principles (and geometrical criteria on the basis of the resulting equations not worked out here). A simple way to reduce work is to lock a joint, leaving an effective two-segmented leg, but with a more favorable instant geometry where the work is done by joints close to the support (Figure 5.14). If at any instant one joint does work against the others, then it is useful to lock the very joint doing negative work. This prescription results in a motion sequence resembling observations and an activation sequence for the clamping muscles.

The mathematical treatment of the three-segmented leg represents a direct transfer of the procedure lined out above. In analogy to Equation 5.67, the angle contributions $d\varphi_1$, $d\varphi_2$, $d\varphi_3$ can be used to rewrite the cost function in terms of the change of one angle, for example, $d\varphi_3$. This equation has three coefficients describing the slope of the dependencies

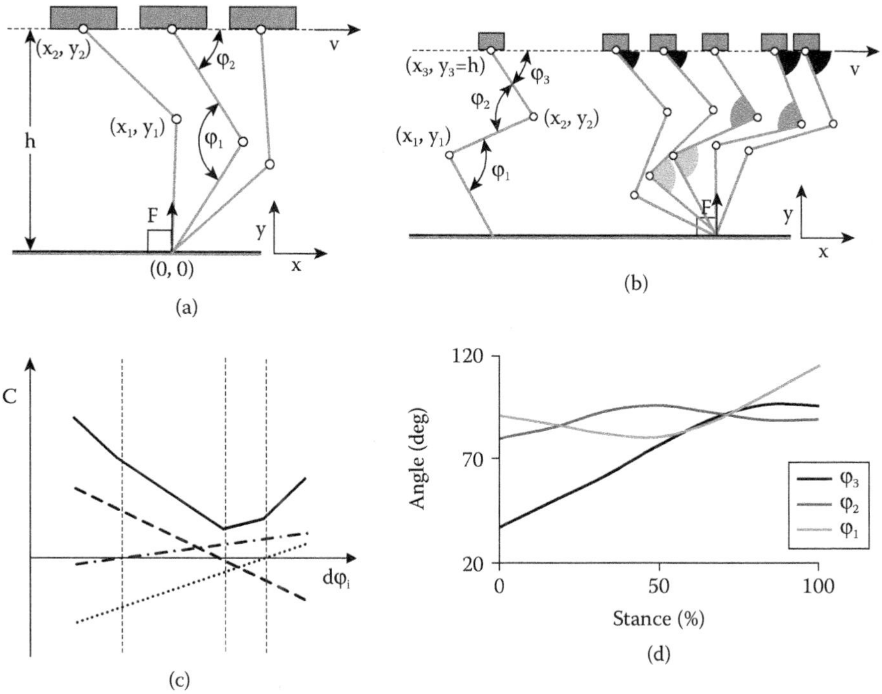

Figure 5.14. Joint work during crouching. (a) The vanishing work to move the hip horizontally at constant speed can be subdivided in two compensating contributions at the knee (joint 1) and the hip (joint 2) for the two-segmented leg. (b) The three-segmented leg (1: ankle; 2: knee; 3: hip) has 1 degree of freedom to minimize the sum of the absolute contributions to total work at each instant. (c) The linear functions (stippled, dashed, and dot-dashed lines) describing the contributions to the work compensate to zero at al increments. The sum of the absolute values shows a minimum that is reached for a proper configuration in which the contribution of one joint is zero, that is, the joint is clamped. The joint that is suitably clamped [shaded angles in (b)] changes with progression. (After Kuznetsov, A.N., Energetical profit of the third segment in parasagittal legs. *Journal of Theoretical Biology*, 172, 101, fig. 2, Copyright 1995. With permission from Elsevier.) (d) Change of the corresponding angles in the forelimb of a pika.

with respect to $d\varphi_3$. The coefficients depend on the instantaneous position of the joint:

$$C = \left|x_3 d\varphi_3\right| + \left|\left[\frac{x_2\left(x_3y_1 - x_1h\right)}{\left(x_1y_2 - x_2y_1\right)}\right]d\varphi_3 - \left[\frac{vx_2x_1}{\left(x_1y_2 - x_2y_1\right)}\right]\right|$$
$$+ \left|\left[\frac{x_1(x_3y_2 - x_2h)}{(x_2y_1 - x_1y_2)}\right]d\varphi_3 - \left[\frac{vx_1x_1}{(x_1y_2 - x_2y_1)}\right]\right| \quad (5.68)$$

Two of the three coefficients have signs opposing the third coefficient, indicating that the joints work against each other. The minimum is achieved for configurations in which the third counteracting joint does zero work, that is, the corresponding joint is locked (Figure 5.14).

Biarticular muscles are helpful at instances where neighboring joints must be blocked in the z-configurations. It is also remarkable that the cost calculated like this is less than half the cost of a two-segmented system. In conclusion, in very crouched positions and in the energetic costly situation of creeping, the introduction of a third segment can save energy. Investigations to reveal possible advantages of biological legs may also be relevant while constructing legged robots. At the outset, it was mentioned that segmented legs introduce a geometric asymmetry. What does it matter whether a leg is oriented in the one or the other direction?

5.4.3 Orientation of the Leg: Backward and Forward

Based on quasi-static considerations such as outlined above, the usefulness of counteracting reaction forces combined with leg geometry and orientation is obvious. Two-segmented legs oriented with elbows pointing toward each other offer an advantage compared to the situation where the two joints point in the same direction for the case that the reaction forces point toward the center of mass. Decelerating front legs and accelerating hind legs are substantiated in many experimental investigations on small mammals. However, the direction of such forces also has a dynamic effect: they reduce pitching. We must be careful not to overuse arguments based on the antique misconception of two-segmented front- and hind legs with joints pointing against each other in quadrupeds (compare to the beginning of this chapter). Still, there may well be a significant remaining asymmetry between the front- and hind legs. Besides, dynamic arguments (Lee and Meek 2005; Figure 5.15) illuminate important consequences of differences in morphology of the hind and front legs in quadrupedal galloping. A two-segmented springy leg with inertia actuated at the hip has directional properties. In such a case, the orientation of the leg influences dynamics. In this investigation, the legs of a model dog consist of a springy distal segment that is connected via another (knee-) spring with a proximal segment. Depending on the stiffness distribution, the vector of the reaction force at touchdown aligns differently: in the case of a connecting (knee-) spring with zero stiffness the reaction force aligns to the distal segment, in the case of a connecting (knee-) spring with finite stiffness combined with zero torque at the hip the vector of the reaction force points toward the hip. Legs were actively retracted and protracted, that is, the motors at the hip were able to produce moments. Translating this pattern to a human leg would produce a righting moment pattern for the trunk (Section 5.6).

The orientation question is still an important issue that awaits to be addressed more thoroughly. Different issues such as righting moments for the knee and righting of the trunk during bipedal locomotion and the automatic induction of moments to the trunk as well as reduction of pitching all may contribute to configurations found in animals. Ignoring such interdependencies may hamper performance in robots.

Due to inherent nonlinearities, the dynamic relationships can turn out to be rather complex. This can be observed already at the leg level itself. The seemingly simple attempt to build a leg spring by combining two torsional springs in a three-segmented leg turns out to be delicate. The demand for robust solutions helps us to understand configurations identified in animal kingdom.

5.4.4 Flipping Geometries

The introduction of the third segment or the second joint results in an additional degree of freedom that is not fixed by external boundary conditions (indeterminacy). We can use this degree of freedom to take advantage of the new geometrical conditions. However, solutions seemingly advantageous with respect to energy or cost may be unstable with respect to their structure, that is, they may collapse or alter dramatically under minor changes of the loading condition (Blickhan et al. 2007; Seyfarth et al. 2001).

The most prominent example in engineering is the Euler buckling of an axially loaded column. For a two-segmented system coupled by a torsional spring at the knee joint, a corresponding buckling would also occur in the case of axial loading if the nominal angle of the spring is 180°. In the extended position, the joint could flip to both sides and the extended leg would be unstable.

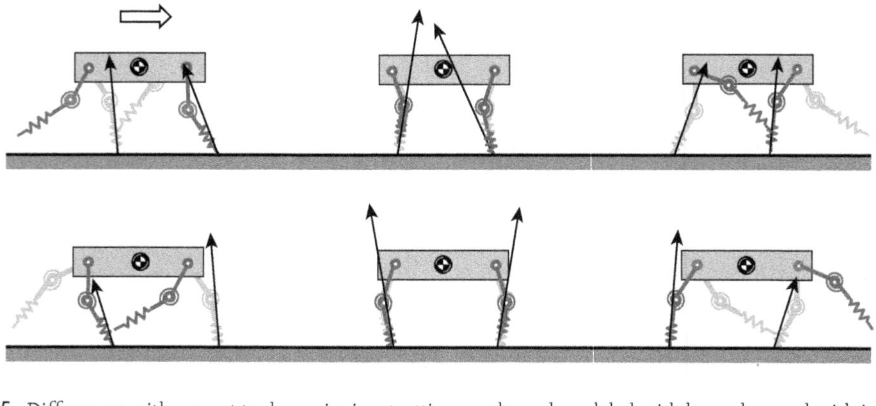

Figure 5.15. Differences with respect to dynamics in a trotting quadruped modeled with heavy legs and with inward (above) and outward elbows and knees depicted (from left to right) at 25%, 50%, and 75% of contact. The hip joint is actuated. (After Lee, D.V., and Meek, S.G., *Proc. Roy. Soc. B Biol. Sci.*, 272, 567–572, fig. 3, 2005.)

To address stability in a three-segmented leg, it is also necessary to introduce joint properties. Again, for simplicity we assume that joint moment is generated by linear rotational springs. We neglect segment inertia and assume that loading is merely produced by a point mass at the proximal tip, that is, neither at the "hip" nor at a "toe" moment is introduced and the reaction force points from the "toe" to the "hip" (axial or radial loading, see Section 5.7). Different instabilities lurk: the system might collapse completely under a given load, or it may suddenly switch from one internal configuration to the other. The latter is the issue of this section.

Numbering the segments from toe to the hip (Figure 5.16a) and neglecting inertia and neglecting the moments at the toes ($\mathbf{M}_{01} = 0$) and the hip ($\mathbf{M}_{34} = 0$), the equilibrium equations for the segments (based on free body diagrams) are

$$\begin{aligned} -\mathbf{M}_{12} - \mathbf{l}_1 \times \mathbf{F} &= 0 \\ \mathbf{M}_{12} - \mathbf{M}_{23} - \mathbf{l}_2 \times \mathbf{F} &= 0 \\ \mathbf{M}_{23} - \mathbf{l}_3 \times \mathbf{F} &= 0 \end{aligned} \tag{5.69}$$

with $\mathbf{M}_{21} = -\mathbf{M}_{12}$ and $\mathbf{M}_{32} = -\mathbf{M}_{23}$ and due to the lacking inertia of the segments $-\mathbf{F}_{21} = \mathbf{F}_{12} = -\mathbf{F}_{32} = \mathbf{F}_{23} = -\mathbf{F}_{h3} = \mathbf{F}$.

In addition, the leg length can be expressed by: $\mathbf{l} = \mathbf{l}_1 + \mathbf{l}_2 + \mathbf{l}_3$.

Because this represents a planar model with h_{12} and h_{23} being the positive distances of the joint from the line of action:

$$M_{12} = h_{12}F;\ M_{12} - M_{23} = (h_{12} + h_{23})F;\ M_{23} = -h_{23}F \tag{5.70}$$

The equilibrium is simply achieved for

$$\frac{M_{23}}{h_{23}} = -\frac{M_{12}}{h_{12}} \text{ or } = M_{12}h_{23} + M_{23}h_{12} = 0 \tag{5.71}$$

This condition can be reformulated using $\varphi_{12},\varphi_{23}$. The equations for the leg length $l(l_{1..3}, \varphi_{12},\varphi_{23})$ (Equation 5.58) and the joint distances $h_{12}(l_{1..3},\varphi_{12},\varphi_{23})$ (Equation 5.59) from the line of action haven been already been presented above. The leg force can be written as

$$F = \frac{l\left(\frac{M_{12}}{l_1}\sin\varphi_{23} + \frac{M_{12}+M_{23}}{l_2}\sin(\varphi_{12}-\varphi_{23}) - \frac{M_{23}}{l_3}\sin\varphi_{12}\right)}{l_1\left(\cos\varphi_{23} - \cos(2\varphi_{12}-\varphi_{23})\right) + 2l_2\sin\varphi_{12}\sin\varphi_{23} + l_3\left(\cos\varphi_{12} - \cos(\varphi_{12}-2\varphi_{23})\right) - 2\frac{l_1l_3}{l_2}\sin^2(\varphi_{12}-\varphi_{23})} \tag{5.72}$$

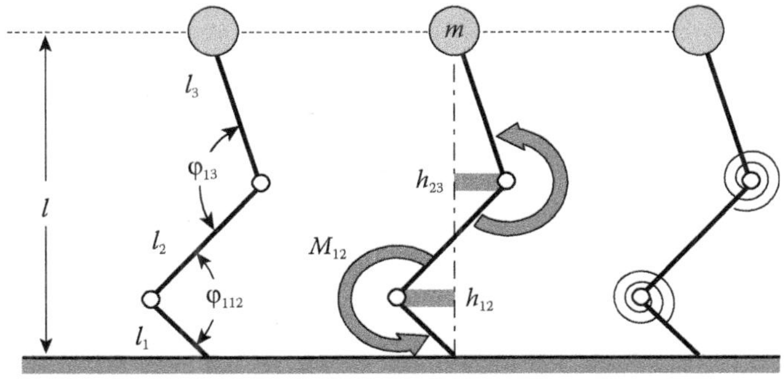

(a)

Figure 5.16. (a) Three-segmented leg (cf. Figure 5.12) now including moments and gravity acting at a point mass. Left: angles and segments; middle: moments $\mathbf{M}_{ij}$ and moment arms $\mathbf{h}_{ij}$ of the axial load, not depicted: moments at the toes ($\mathbf{M}_{01}$) and the hip ($\mathbf{M}_{34}$); right: torsional springs generating moments proportional to deflection. (b through e) Equilibria (bold black lines) in the configuration space (ankle angle φ_{12}, knee angle φ_{23}) for different segmentations (n:n:n) and nominal values (φ^0, fat crosses) and linear springs ($\nu = 1$). Crossings of those lines mark bifurcation points. Initial conditions slightly shifted from the nominal conditions (± 1°) lead during loading to adjacent asymptotic configurations described by the thin lines. Gray shading (scaled to minimum and maximum): joint work $\boldsymbol{W}_M$ being zero at the nominal condition and increasing outward; contour-lines with numbers: leg length $l = const$—note that the levels of work are in general not parallel to the contour-lines for constant leg length, that is, the energy of the leg changes along the contour-lines for leg length resulting in convex (stable) and concave (unstable, dark hatched areas) solutions; dash-dot-line: $h_{12} = 0$; dashed line: $h_{23} = 0$. *(Continued)*

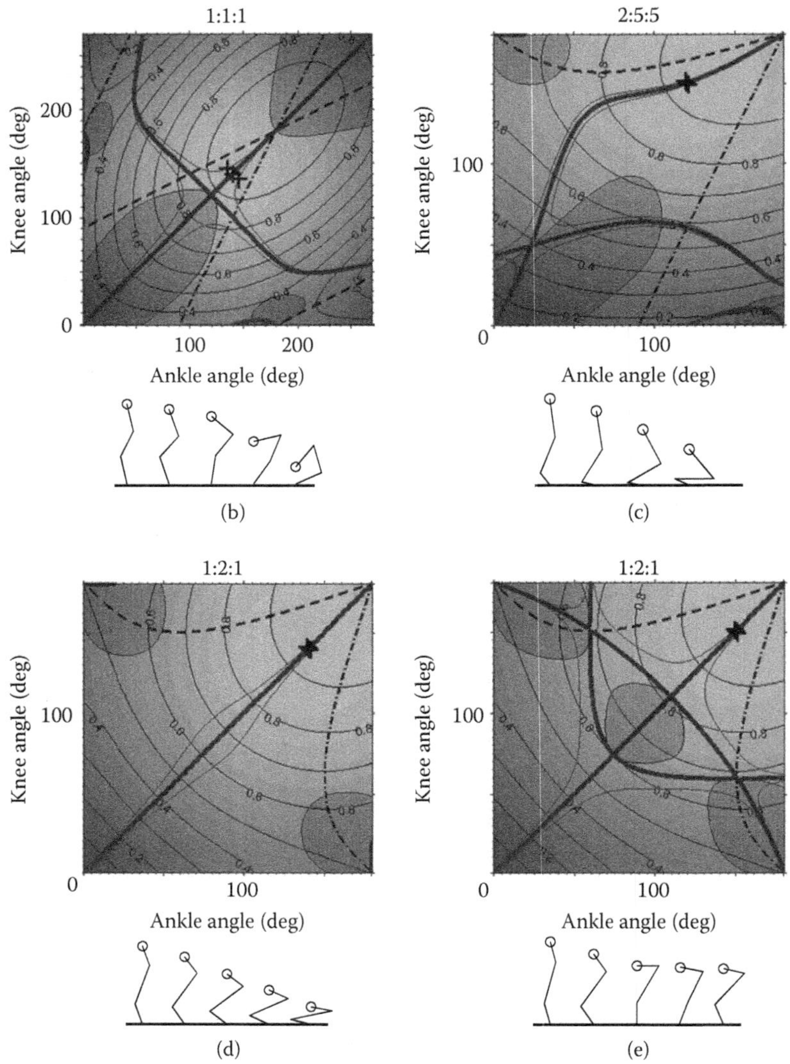

Figure 5.16. (Continued) (b) Segmentation (1:1:1), nominal angle (140, 140); bifurcation at (120, 120); stick figure from left to right: nominal conf. ($l = 0.94$), bifurcation ($l = 0.88$); - lower branch, crossing the $h_{12} = 0$ border and flipping into the c-mode: $l = 0.8$, 0.5 and 0.2. (c) Human-like segmentation (2:5:5), nominal angle (120, 150); the bifurcation (24, 50) is clearly shifted down to $l = 0.39$; stick figure $l = 0.93$, 0.8, 0.6, 0.39; following the left stable branch after bifurcation the knee will hit the ground before the border of the configuration space ($\varphi_{12} = 0$). (d) Long intermediate segment (1:2:1), nominal angle (140, 140): The bifurcation is shifted beyond the considered configuration space; stick figures: $l = 0.94$, 0.8, 0.6, 0.4, 0.2. (e) Starting with the same segmentation as in (d), but with a slightly altered nominal angle (150, 150) reveals a complicated landscape including new bifurcation sites; stick figures: $l = 0.97$, 0.81 (first bifurcation), branch to the right - 0.68 (second bifurcation) at the $h_{12} = 0$ border, following the stable branch to right (increasing ankle angle) into the c-mode ($l = 0.66$) or returning as in the last configuration ($l = 0.65$) to the stable z-mode (decreasing ankle angle).

We mentioned above that a mechanical equilibrium does not guarantee stability. Stability of a configuration is achieved if for the same leg length neighboring configurations with respect to $\varphi = (\varphi_{12}, \varphi_{23})$ have higher system energies. Thus, we have to search for local minima in $E(\varphi_{12}, \varphi_{23})$ for given leg length and for situations obeying the equilibrium condition. How could we identify local minima? Like in calculus, where we identify a minimum in $f(x)$ by the first derivative being zero $f'(x) = 0$, and the second derivative being positive $f''(x) > 0$, we can define corresponding conditions in the 2D configuration space. The first condition that the gradient of the energy is zero for a given length $\nabla_{\varphi}E(\varphi) = 0$ is identical to our equilibrium condition provided that our actuators are conservative:

$$\delta E = \mathbf{M} \circ \delta\varphi + \mathbf{F} \circ \delta \mathbf{l} = 0 \quad (5.73)$$

The gradient results in

$$\nabla_{\varphi}E(\varphi) = \begin{pmatrix} -M_{12} \\ M_{23} \end{pmatrix} + F(\varphi)\nabla_{\delta}l(\varphi)$$
$$= \begin{pmatrix} -M_{12} \\ M_{23} \end{pmatrix} + F(\varphi)\begin{pmatrix} h_{12}(\varphi) \\ h_{23}(\varphi) \end{pmatrix} = 0 \quad (5.74)$$

A stable local minimum requires that the energy increases in the immediate environment of the equilibrium point, any deflection from this point will drive it back. This condition of a convex environment is fulfilled if the Hessian matrix entailing the second partial derivatives with respect to energy is positive definite. In the 2D case of a positive definite Hessian, the eigenvectors $e_{1,2}$ define the axis of an ellipse characterizing the convex trough in the environment and the eigenvalues $w_{1,2}$ their length $\sim\left(\frac{1}{\sqrt{w_1}}, \frac{1}{\sqrt{w_2}}\right)$. The Hessian $\mathbf{H}$ of the energy is (W_M denotes work done by joint moments)

$$\mathbf{H} = \mathbf{H}_{WM} + F\,\mathbf{H}_l$$

$$= \begin{pmatrix} \frac{\partial^2 W_M}{\partial\varphi_{12}\,\partial\varphi_{12}} & \frac{\partial^2 W_M}{\partial\varphi_{12}\,\partial\varphi_{23}} \\ \frac{\partial^2 W_M}{\partial\varphi_{23}\,\partial\varphi_{12}} & \frac{\partial^2 W_M}{\partial\varphi_{23}\,\partial\varphi_{23}} \end{pmatrix} + F\begin{pmatrix} \frac{\partial^2 l}{\partial\varphi_{12}\,\partial\varphi_{12}} & \frac{\partial^2 l}{\partial\varphi_{12}\,\partial\varphi_{23}} \\ \frac{\partial^2 l}{\partial\varphi_{23}\,\partial\varphi_{12}} & \frac{\partial^2 l}{\partial\varphi_{23}\,\partial\varphi_{23}} \end{pmatrix}$$

$$= \begin{pmatrix} -\frac{\partial M_{12}}{\partial\varphi_{12}} & -\frac{\partial M_{12}}{\partial\varphi_{23}} \\ \frac{\partial M_{23}}{\partial\varphi_{12}} & \frac{\partial M_{23}}{\partial\varphi_{23}} \end{pmatrix} + F\begin{pmatrix} -\frac{\partial h_{12}}{\partial\varphi_{12}} & -\frac{\partial h_{12}}{\partial\varphi_{23}} \\ \frac{\partial h_{23}}{\partial\varphi_{12}} & \frac{\partial h_{23}}{\partial\varphi_{23}} \end{pmatrix} \quad (5.75)$$

Because we are interested in convex solutions for given leg length $l = const$, we merely investigate the convexity in the direction perpendicular to $\nabla_{\varphi}l$ or in the direction of $\mathbf{t}_l = \begin{pmatrix} -h_{23} \\ h_{12} \end{pmatrix} = t_1\mathbf{e}_1 + t_2\mathbf{e}_2$, that is, along the lines defining $l = const$.

Two kinds of transition from stability to instability are observed (Figure 5.16): a bifurcation of (Equation 5.71) solutions (torque equilibrium) require $det(\mathbf{H}) = 0$, and $\mathbf{t}_l$ is an eigenvector to the vanishing value of H. In such a case the trough degenerates and in the direction of the $l = const$ line the altering configuration in φ does not alter the system energy.

The solution aligns with a $l = const$ line, that is, in the direction of this line despite of changing leg angles the mechanical equilibrium is maintained. That is, on this solution path there are at least two different angle combinations resulting in the same leg length. Beyond this point of alignment, the solution path on which mechanical equilibrium is fulfilled runs on a rise (maximum) in the energy landscape, that is, it is unstable.

To maintain a conservative system, the joint moments are assumed to be generated by rotational springs. For human running this represents a reasonable approximation (Gunther and Blickhan 2002):

$$M_{12} = c_{12}\left(\varphi_{12}^0 - \varphi_{12}\right)^v$$
$$M_{23} = c_{23}\left(\varphi_{23}^0 - \varphi_{23}\right)^v \quad (5.76)$$

with the nominal angles φ^0_{12} and φ^0_{23} and $\varphi_{ij} < \varphi^0_{ij}$, the rotational stiffnesses c_{12} and c_{23}, and the exponent v. For the particular case of symmetrical loading with $\varphi^0_{12} = \varphi^0_{23}$ and $\varphi_{12} = \varphi_{23}$, the equilibrium condition requires

$$\frac{c_{12}}{l_1} = \frac{c_{23}}{l_3} \quad (5.77)$$

The equilibrium depends on the ratio between the stiffness and the length of the adjacent outer segments. It does not depend on the length of the middle segment.

It is always possible to operate the three-segmented leg with symmetrical joint configurations until a bifurcation occurs. Both remaining odd branches are possible if they allow further leg shortening. The $h_{ij} = 0$ conditions are very attractive (Figure 5.16). A transition from the z- into the c-mode requires to cross such a condition. The working range can be extended by bypassing the bifurcation using suitable nominal angles and segment ratios. With increasing nominal angle(s) (and thus leg length), a limited symmetrical working range (shortening from extended leg to the bifurcation point while fulfilling stiffness equilibrium) occurs for any leg in which the middle segment length l_1 exceeds 1/2. For increasing degrees of joint moment nonlinearity, this critical length $l_{2,crit}$ even decreases ($\nu = 1$: $l_{2,crit} = 0.41$, $\nu = 2$: $l_{2,crit} = 0.46$, $\nu = 3$: $l_{2,crit} = 1/3$); however, the corresponding critical upper nominal angle also rises up to full extension (180°) for $\nu = 3$. For lower nonlinearities, this limit in safe leg extension is still 171° for $\nu = 2$, but it falls down to 136° for $\nu = 1$. Up to three bifurcations emerge for $l_2 < 1/2$ (just two in Figure 5.12e because $l_2 = 1/2$). All in all, the stable working range can be increased by nonlinear joint springs that partly compensate the nonlinearity of the resulting force–length characteristic, as well as by coupling both joints with biarticular springs.

We illustrated that the distribution of lengths, the initial configuration, and the angle dependencies of the rotational springs all contribute to robust spring-like behavior of the segmented leg. Despite the fact that in general the joints of animal legs are not driven by springs, orientation and segmentation of legs follow the predicted rules surprisingly close. This may be partly due to the convergent pressure of economy (Section 5.4.1). It does not come as a surprise that energetics does represent an important issue both during evolution and in the construction of legged machines. In animals, the issue of energetics of locomotion is related to mechanics, but also to physiology. We only touch the latter aspect.

5.5 ENERGY: MECHANICS AND COST OF LOCOMOTION

5.5.1 Cost of Locomotion and Mechanics

So far, we have considered the principal dynamics of gaits. In physics and engineering, energy represents an important quantity that, due to its conservation, in many cases facilitates mathematical treatment. In motion systems, energy frequently (but not always) represents an important optimization criterion. Especially today, we would like to move from A to B with a minimum cost of transport (CT), the ratio between the consumed energy and the distance covered. This is also true in living motion systems. While driving a car, we know the CT rather well. When we walk, it seems to be a little more difficult to give decent numbers. In this book, physiology is not addressed; however, when working out models describing the general (bio)mechanics of animal locomotion the question arises whether our knowledge about the mechanics of the system can help to assess the cost of locomotion. We can alter the perspective and ask, if we assume that cost plays an important role (undoubtedly a decent assumption), to which extent can the pressure to minimize cost explain the motion and dynamics observed. If you take into account that we not only may fatigue earlier but also that the body has to carry around a part of the offshore drilling system as well as the refinery as a payload, and how consuming it is to distribute the fuel within the body, then you realize that minimizing cost is important in general, even for the case that you may sit around most of the time and need only from occasionally to run to catch a flight.

Looking at the cost of steady-state locomotion in very basic physical terms, it is obvious that we neither lift nor accelerate the system. Steady-state locomotion should be free; however, we know that this is not the case, neither while driving a car nor for us. A vehicle has to work against friction (e.g., drag, rolling friction of the wheels, friction of the bearings), and, in general, it produces some heat loss that we must compensate. What happens if humans walk and run, or animals gallop or creep? Where are the losses? As in a vehicle, losses can occur internally and externally due to interaction with the environment.

The "simple" models described above can give us already many answers. We still call them simple as they are strongly simplifying compared to full-blown multisegmental models entailing "all" participating tissues. We already met external losses in the treatment of stiff-legged models. However, reconsidering all approaches you may argue that there is no necessity of stiff-legged locomotion: you could use springs instead. Spring–mass systems are conservative, and once they are accelerated to the envisioned speed there is no need for additional supply. We have seen that we indeed do mimic the dynamics of such a system, and the reason may well be that there is an energy advantage. However, we also know very well that the legs of animals do not represent simple springs. Nature has invented tissues with remarkable spring-like behavior (Sections 5.7 and 5.8), and they are used in structures for energy storage. They usually come along with minor viscous losses that must be compensated during cyclic locomotion. Furthermore, deceleration of distal masses inevitable in real legs implemented in animals or robots leads to inevitable losses (Section 5.7.8). Importantly, versatility of the leg in the context of a variety of behaviors requires the possibility of pure acceleration or pure braking. Nature has developed legs complying perhaps with some focus with all demands. This led to the development of springs in series with the muscle as an actuator, with the spring in parallel being the exception. Correspondingly, forces exerted by tendons functioning as springs are most frequently taken up by muscles in series. This force generation demands metabolic energy. Muscles do not posses clutches, and even when no length change takes place the molecular force generators cycle to maintain the force. Correspondingly, despite the quasi-elastic operation of the leg at the principal level, this maintenance is not free of cost. Thus far, it is not possible to exactly estimate all costs of muscular activity during locomotion. Possible are intelligent approximations based on the knowledge of the dynamics of the systems, assuming certain muscle activities and related metabolic costs.

There are two approaches related to the principal level that we considered in Sections 5.2–5.4: one approach is related to the fluctuations of the total mechanical energy of the center of mass and emerged from experimental research; the another approach is related to the collisions and is concept based. The idea that the cost of locomotion is related to the fluctuations of the total energy of the center of mass composed of kinetic and potential energy has been put forward by Cavagna et al. (1963) and resulted in the development of experimental equipment ("Force Platforms as Ergometers," Cavagna 1975) and has triggered the development of simple mechanical models (Section 5.2). Well aware of the fact that some of the fluctuations in energy may be powered by energy exchange, it is assumed that all positive increments in energy (see Equation 5.81) must be powered by musculature (Cavagna et al. 1977).

If exchange of energy of the center of mass (elastic, potential, and kinetic energies) is possible, then this is an overestimate. If absorbing energy contributes to the cost—remember the muscles in series of tendons or think about legs with only minor energy storage in tendons—then considering only positive energy increments, that is, muscle work, may be an underestimate. Keeping this in mind, calculating the positive increments has been an accepted standard.

What would be the consequences of such an approach, considering the background of our models described above? The fluctuations of energy entail contact and aerial phases of the system. Ignoring drag, there are no losses due to interaction with the environment during the flight phase, and the legs do no work to accelerate or to lift the center of mass. This is different during contact where energy fluctuations may be powered by the legs and their muscles within. With speed of locomotion, the amount of the forward component of the kinetic energy increases, dominating in bouncing gaits. Correspondingly, it inevitably dominates the fluctuations. In contrast, during slow locomotion, the forward kinetic energy is of the same order or lower than the magnitude of the other components. Assuming a flight height of half the leg length l, the energy necessary to power it (cf. Bejan and Marden 2006)

$$E_{pot} = mg\frac{l}{2} \tag{5.78}$$

is the same as forward kinetic energy at which an inverted pendulum would take off due to centrifugal forces. Exchange between kinetic and potential energy is necessarily more important at low than at high speeds of locomotion where elastic storage increasingly comes into play.

Experimental estimates of the energy fluctuations of the center of mass normalized to body weight and distance covered (mechanical CT) based on force-plate records resulted in conservative estimate across a wide range of animals (Figure 5.17b). For running gaits such as running, trotting, and hopping, with a principal dynamics very similar to the spring–mass system (Section 5.2), this constant mechanical energy together with knowledge about footfall patterns can be used to obtain principal leg properties (Blickhan and Full 1993). The result of such calculations is that constant specific CT can be explained by a size-independent relative leg loading and leg stiffness. Such a dynamic similarity has consequences on leg design that have been described previously (Bullimore and Burn 2004; Alexander 2005). This excursion on scaling demonstrates that information on scaling or differences in species require additional information about the point of operation of the systems elaborated on above. Studies on load carrying (Heglund et al. 1995) represent another good example. Normal subjects largely

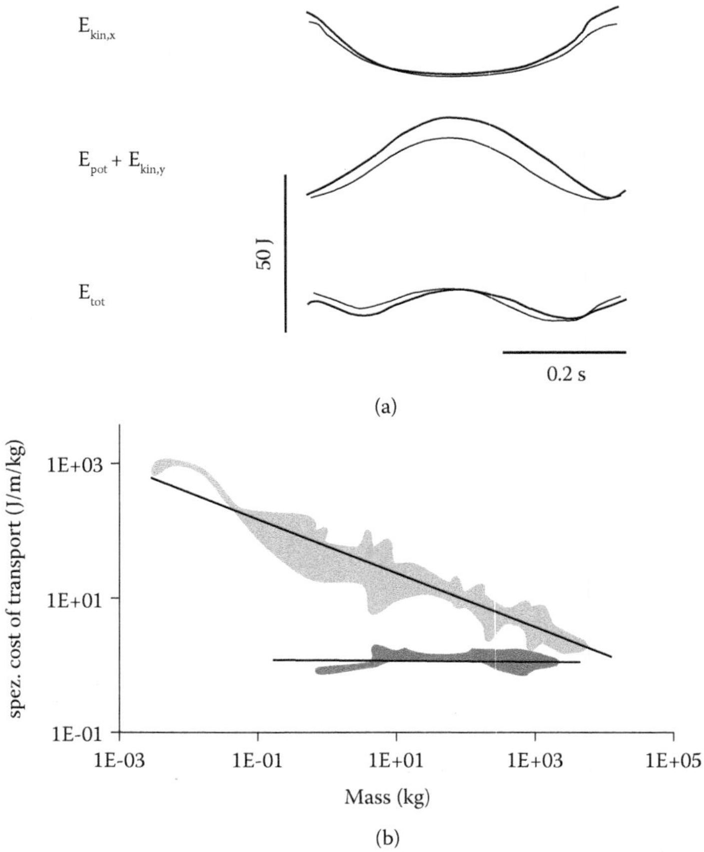

Figure 5.17. (a) Tracings of horizontal, vertical, and total energy of the center of mass for an African woman walking at a speed of about 2 m/s without load (thin lines) and with a load of 30% body weight (bold lines). The fluctuations of the center of mass (sum of the positive increments) are not significantly affected. (After Heglund, N.C., et al., Energy-saving gait mechanics with head-supported loads, *Nature*, 375, 52–54, fig. 1, p. 63, Macmillan Publishers Ltd., copyright 1995.) (b) Specific cost of transport as determined from metabolic measurements (upper group with measurements including arthropods, birds, and mammals) and from the fluctuations of energy of the center of mass. Whereas the metabolic cost strongly depends on mass, the mechanical cost of transport remains surprisingly constant. (After Full, R.J., *Energy Transformations in Cells and Organisms*, p. 180, Thieme, Stuttgart, 1989, fig. 4.)

retain their pattern of locomotion while carrying weights; however, due to the added weight, the energy fluctuations of the center of mass increase proportionally and indeed so does the measured metabolic cost. African women are able to carry loads up to 20% of body weight without additional costs and weights up to 70% of body weight on their heads much more economically than untrained people using their heads or backpacks. Measuring the energetics of the center of mass with force plates revealed that they indeed are better in taking advantage of the exchange of potential and kinetic energy (Figure 5.17a).

The "load carrying" of obese people seems to follow the same rules (Browning and Kram 2009). The total mechanical energy seems to be largely unaffected, despite of deviations of the movements (e.g., lateral sway) and so are specific (cost/body mass) metabolic costs.

Despite of the difficulty to relate it to metabolic cost, investigating the energetics of the center of mass leads to basic insights. It helped to focus on principal dynamics and it addresses a dominating cost factor. However, it neglects the complicated interaction within a segmented body. This issue is addressed in the next section.

5.5.2 Internal and External Work

Concentration on the COG has the advantage of direct comparability of predictions based on simple models and observed principal dynamics.

As pointed out above, matching principal dynamics is essential in describing the mechanics of locomotion. However, it is well known that for fast gaits, "internal work" thought of as work to change the form of the linkage system (not meaning to deform segments), including to accelerate the appendages against the center of mass, contribute strongly to metabolic cost. Experimentally, such estimates are possible based on kinematic measurements and the knowledge of the moment of inertia of the body segments. Again, such estimates largely depend on assumptions with respect to energy exchange (kinetic, potential, and elastic energies) that are possible within appendages, between the body and appendages, and on this path between appendages (see below). The total mechanical energy of the linkage system can be calculated by summing up the energy of all components in here the 2D version:

$$E_{tot} = \sum_i \left(m_i \frac{\dot{x}_{CM,i}^2 + \dot{y}_{CM,i}^2}{2} + I_{CM,i} \frac{\dot{\phi}_i^2}{2} + m_i g y_{CM,i} \right) \tag{5.79}$$

where the index CM, i denotes the center of mass of the ith segment. The external energy as the sum of kinetic and potential energies of the center of mass of the total system (see above) can be altered by external forces:

$$E_{ext} = m_{tot} \frac{\dot{x}_{CM}^2 + \dot{y}_{CM}^2}{2} + m_{tot} g y_{CM} \tag{5.80}$$

It does not show the fluctuations of the energy of the constituents. Imagine a system consisting of two segments of equal mass connected by a spring in the flight phase. The vibration energy of the segment is not visible in the energetics of the center of mass. However, we could describe it separately with coordinates relative to the center of mass:

$$E_{int} = \sum_i \left(m_i \frac{\dot{x}_{CM,i-CM}^2 + \dot{y}_{CM,i-CM}^2}{2} + I_{CM,i} \frac{\dot{\phi}_i^2}{2} + m_i g y_{CM,i-CM} \right) \tag{5.81}$$

This mechanical law dates back to J.S. Koenig (1712–1757). The kinetic energy of a system of point masses can be written as (planar case) follow:

$$\begin{aligned} E_{kin\ tot} &= \frac{1}{2} \sum_i m_i \left(\left(\dot{x}_{CM} + \dot{x}_{CM,i-CM} \right)^2 + \left(\dot{y}_{CM} + \dot{y}_{CM,i-CM} \right)^2 \right) \\ &= \frac{1}{2} \left(\sum_i m_i \right) \left(\dot{x}_{CM}^2 + \dot{y}_{CM}^2 \right) + \left(\sum_i m_i \dot{x}_{CM,i-CM} \right) \dot{x}_{CM} \\ &\quad + \left(\sum_i m_i \dot{y}_{CM,i-CM} \right) \dot{y}_{CM} \\ &\quad + \frac{1}{2} \sum_i m_i \left(\dot{x}_{CM,i-CM} + \dot{y}_{CM,i-CM} \right)^2 \\ &= \frac{1}{2} m_{tot} \left(\dot{x}_{CM}^2 + \dot{y}_{CM}^2 \right) + \frac{1}{2} \sum_i m_i \left(\dot{x}_{CM,i-CM} + \dot{y}_{CM,i-CM} \right)^2 \end{aligned} \tag{5.82}$$

as $\sum_i m_i \dot{x}_{CM,i-CM} = \sum_i m_i \dot{y}_{CM,i-CM} = 0$. It is also valid for continuous deformable systems and for the special case of segmented systems.

Although the concept of internal energy entails relevant information adding up external and internal energy to obtain total energy, it can be prone to errors, especially if we try to use the fluctuations to approach muscle work (cf. Aleshinsky 1986a, 1986b).

As outlined above, work done by the musculature is frequently approximated by the sum of positive increments (+) for one stride (Cavagna et al. 1963; Cavagna and Kaneko 1977; Winter and Robertson 1978; Saibene and Minetti 2003; Ortega and Farley 2007):

$$W = \int_0^T \left| \frac{dE}{dt} \right|^+ dt \tag{5.83}$$

It is now crucial whether the increments are summed up after summation of the contributions from each segment or before (think of two components out of phase). We should be careful in using them to describe the metabolic cost of locomotion.

Exchange and transfer: Here some general remarks on energy exchange and transfer are provided. As pointed out above, introducing the concept of energy exchange during walking and running has largely promoted our understanding of locomotion. The introduction of energy recovery by Cavagna et al. (1977) has been a useful concept and is still used in animal research:

$$ER\ [\%] = 100\frac{\Delta E_{ext,pot+} + \Delta E_{ext,kin+} - \Delta E_{ext,tot+}}{\Delta E_{ext,pot+} + \Delta E_{ext,kin+}} \quad (5.84)$$

with ΔE_+ indicating the sum of the positive increments in one stride. Recovery should be 100% for a perfect pendulum (the typical value achieved during walking is 65%, cf. Section 5.2) and 0% for the pure elastic collision with complete recovery. Introducing a segmented system, such an exchange is possible for each segment. In addition, energy can be transferred from one segment to the other, or not. The transfer can be due to the constraints inherent to the linkage system, for example, acceleration of the suspension of a pendulum due to inertia results in angular acceleration, that is, the translational work (e.g., from an adjacent segment) generates rotational energy. The reaction forces generated at the collision of a rotating stick with the ground lift the center of mass, that is, internal energy is transferred into external work. There is exchange and transfer of energy without muscle activity. Adding muscle forces obviously alters the coupling between segments. In addition to the constraint forces of the linkage system, there are now moments generated by the muscles. In all cases, transfer and exchange occur across couplings, that is, the linkages or the muscle attachments. Muscles crossing several joints directly couple the corresponding links. Via those couplings, transfer can occur across many segments, very simply: flexing the ankle joint lifts the whole body.

By calculating the energies at the joints, the transport of power can be observed and offers important general insights. Inverse dynamics gives the net forces (neglecting cocontraction or work against elastic tissues) and the net torques at the joints. Note that despite the same magnitude, the torque has opposite signs at the adjacent joints. Power at each segment can be calculated by multiplying the contact force F with the velocity vector v at the joint j and the joint torque M with the rotation velocity ω of the adjacent segment. Whereas the translational power P_{trans} is the same at the adjacent leg segments, the same torque results in different power P_{rot} at the upper (+) and the lower (–) segment:

$$P_{trans,j} = \mathbf{F}_j \mathbf{v}_j \text{ and } P_{rot,j-} = \mathbf{M}_j \boldsymbol{\omega}_{j-} \text{ and } P_{rot,j+} = \mathbf{M}_j \boldsymbol{\omega}_{j+} \quad (5.85)$$

From segment to segment, it can be seen whether power is added or absorbed. This helps to interpret the functional mechanics of the system. In the example below (after Blickhan and Full 1992) (Figure 5.18a), in the last phases of push-off of walker a high power of 533 W is generated at the ankle joint flexing the foot. This flexion results in raising the shank and the muscles in addition to rotating the shank clockwise. This is enhanced by corresponding moments around the knee joint. Up to the hip, almost all power is lost and only 23 W is transferred to the trunk. At toe-off the flow already reverses its direction. The hip and knee flexors accelerate the distal segments.

When analyzing the movements of animals, we have tools to investigate local changes in mechanical energy, can judge their contribution to total energetics, and can detect strategies of energy flow. It seems that during evolution and during ontogeny, animals have learned to take advantage of complex dynamic interactions to save energy, and it seems to be worthwhile to transfer corresponding strategies into robotics. However, compared to the net behavior of the linkage system it is much more difficult to reasonably predict the actual contribution of muscle work or of muscle action. We address some of the related issues in Sections 5.5.3, 5.5.4, and 5.7.3–5.7.5.

5.5.3 Muscle Work versus Energy Fluctuations

Simulations based on full-blown planar muscle–skeletal models clarify this issue (e.g., Sasaki et al. 2009). The forward model of a human walker (at a treadmill speed of 1.2 m/s) has 13 degrees of freedom (joints: toe, mid-foot, ankle, knee, hip) including 25 hill-type muscles per leg, with 13 functional groups (muscles within group receive the same activation pattern), viscoelastic passive torques, tendons, and elaborate feet with 38 viscoelastic elements to describe the contact. A suitable activation of the musculature has been calculated by minimizing the difference between observed and simulated kinematics and ground reaction force (Figure 5.18b; Section 5.7). Subjected to constraints, the time course of the resulting stimulation is, with minor exceptions, similar to that observed by surface electromyography. The footprints of dynamics, the ground reaction forces, have some minor glitches; however, considering that no wobbling masses (Section 5.7) and no three-dimensional (3D) hip

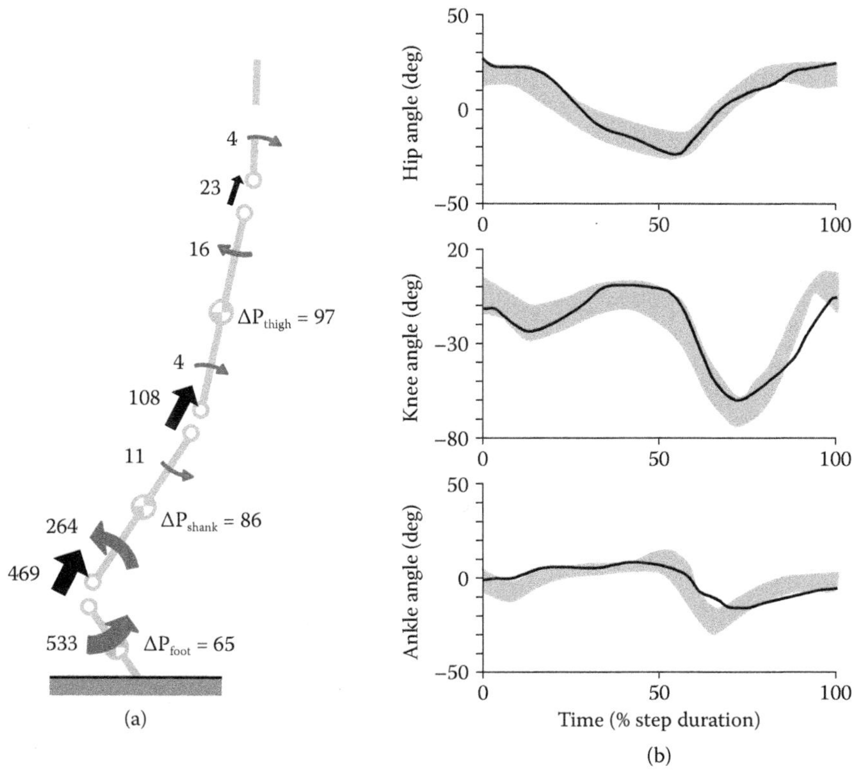

Figure 5.18. (a) Power flow (Watts) in the human leg during walking at push-off. (With kind permission from Winter, D.A., and Robertson, D.G., Springer Science+Business Media: *Biol. Cybernetics*, Joint torque and energy patterns in normal gait, 29, 1978, 141, fig. 5.) Vectors indicate the direction of the force vectors. (b) Measured (gray shade) and predicted (bold line) time course of (a) the hip, knee, and ankle angles. The time course of the ground reaction force (not depicted). (After Sasaki, K., Neptune, R.R., and Kautz, S.A., *J. Exp. Biol.*, 212, 740, fig. 2, 2009.)

movements were considered, the result is convincing and should give a good estimate of the different work estimates.

The net joint and muscle work must be sufficiently high to compensate for the losses in the shoe element. The sum of external and internal work is about two thirds the muscle fiber work. The high internal work (see also Cavagna and Kaneko 1977) is overestimated because the calculation does not use the relative movements with respect to the COG. External work alone amounts to only one-third of the muscle fiber work. In total, 148 J of joint work with intercompensation due to biarticular muscles strongly underestimates muscle work. In this case for corresponding joints, power absorbed at one joint was allowed to cancel power generated at the next joint. For comparison, in investigations on the cat, energy transfer accounted to about 15% to the joint work (Prilutsky et al. 1996). The difference between joint work (254 J) and muscle–tendon work (274 J) is due to cocontraction of antagonists. As desired, the sum of net muscle joint work and net passive joint work equals net joint work. A detailed listing of the different work components can be found in Sasaki et al. (2009: Table 1, p. 741).

As expected, coarse measures such as external (73 J) and internal (102 J) work strongly underestimate the mechanical work done by the muscle fibers (256 J). Nevertheless, we see below that they illuminate important aspects of the energetics of locomotion.

5.5.4 Muscles, Springs, and Collisions

The treatment above uses experimental data to study the energetics of locomotion. Even the forward muscle–skeleton model described in section 5.7.4 is used to closely mimic observations (kinematics and muscle activity) and from that point to disclose the different work estimates (muscle work vs. joint, external, and internal work). From the shortcomings revealed by this elaborate model, it can be concluded that costs determined by strongly simplifying approaches only relying on the dynamics of the center of mass or neglecting the issue of segmental exchange and cocontraction of antagonists must be even more off the real values. In contrast, even on the simple level, energy as an optimization criterion could reveal basic patterns prevailing in a complex system. The evolutionary pressure to reduce costs acts at all levels of the system. Here are some simple questions: Does the cost function help to decide between the different models? Does the emerging dynamics strongly depend on the cost function or on the model type? These seem to be trivial questions as, for example, spring–mass model is energetically conservative and the stiff-legged model does not come around its collisions. However, having pure springs would be hampering in many behavioral situations and the springs are only images describing the compliant leg. And we did already ask when and how is it suitable to include actuation?

Let us start again at a rather simple level (Figure 5.19). Prescribing the pattern of the ground reaction force and a stepping pattern is equivalent to introducing actuators in a leg that guarantee accelerations of the center of mass similar to those observed in nature. By allowing for variation of such a pattern, it is possible to find out what is special about it. In particular, it can be used to investigate whether for the observed patterns of the energetics of the center of mass are minimal. Alexander (1992) and Minetti and Alexander (1997) considered a two-term cosine series to describe force patterns and then investigated what particular force pattern is energetically optimal at various speeds. This is rather close to our hypothesis (Section 5.2) that walking and running represent different modes of oscillation of a (hybrid nonlinear) compliant system. For example, the specific vertical force is described as

$$F_y = \frac{3\pi}{4\,DF\left(3+q\right)}\left\{\cos\left(\pi t/t_c\right) - q\cos\left(3\pi t/t_c\right)\right\} \tag{5.86}$$

for $-\frac{t_c}{2} \le t \le \frac{t_c}{2}$ and for q typically in the region of $-0.1 \le q \le 0.4$, where DF is the duty factor. The horizontal component can be given by

$$F_{x,\text{left}} = F_y\frac{x}{y} \text{ and } F_{x,\text{right}} = F_y\frac{x - D/2}{y} \tag{5.87}$$

where x, y are the coordinates of the center of mass and D is the stride length. The forces are assumed to be aligned with the hip. Applying these forces on a system with telescope or monoarticular legs with leg masses fixed at a specific relative distance r on to the telescope results in acceleration of the system. From the latter, the velocities and displacements and energies can be calculated by integration. Problematic is the prescription of the angular speed of the legs as being constant during contact and sinusoidal during swing (Section 5.2). This approximation allows for estimation of the internal work, but it is not dynamically correct.

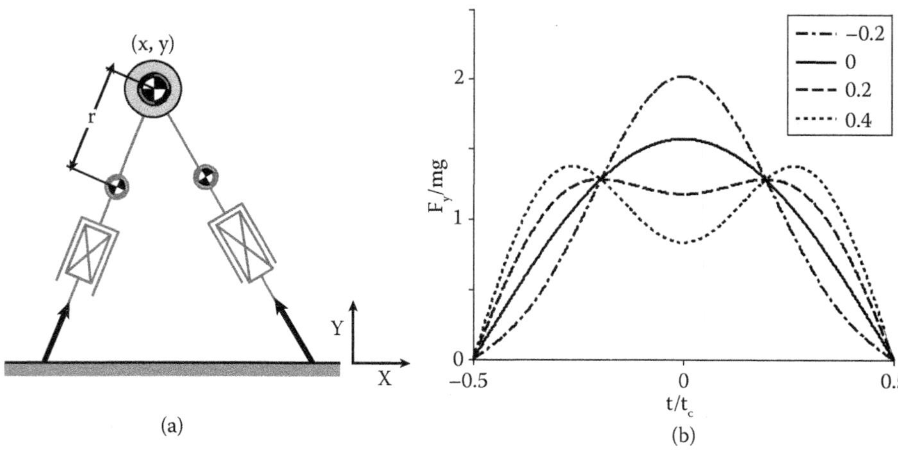

Figure 5.19. (a) Model with two actuated telescope legs. (b) Prescribed vertical force (Equation 5.86) for different values of q (legend).

Calculating the cost based on the concept of external and internal work (Alexander 1992) yields a value that can be mapped in dependency of dimensionless speed, stride length, duty factor, and the shape factor. Shifting the landscape by assuming that a substantial fraction of energy is stored in elastic elements results in optima at the combinations of the experimentally observed speeds and stride lengths. The transition between walking and running is correctly predicted.

A deeper and more general insight is possible by using energy as an optimization goal and tuning the forward dynamics of a simple system (Section 5.2.3). Srinivasan (2010) considered the models in Figure 5.20, allowing the force generated by the actuator to vary more generally than in Alexander (1992), but as in Alexander's work, attempting to find what leg forces minimize the energetic cost for given locomotion constraints. By combining the actuator with springs in series, and by considering muscles with different costs in the concentric, eccentric, and isometric realms, it is possible to judge to what extent these ingredients are essential and alter the energetics of the system.

The equations of motion as formulated for the spring mass system (see Equations 5.24 and 5.25) can be easily generalized for arbitrary ground reaction forces. Introducing the dimensionless quantities: $x = \hat{x} / l_0$; $y = \hat{x} / l_0$; $F = \hat{F}/mg$

$$\ddot{x} = \frac{F_1(x - x_{c,1})}{l_1} + \frac{F_2(x - x_{c,2})}{l_2} \text{ and } \ddot{y} = -1 + \frac{F_1 y}{l_1} + \frac{F_2 y}{l_2}$$

$$l_1 = \sqrt{(x - x_{c,1})^2 + y^2} \text{ and } l_2 = \sqrt{(x - x_{c,2})^2 + y^2} \tag{5.88}$$

$x_{c,1}$ and $x_{c,2}$ are the positions of the two feet in contact with the ground. The equations simplify when only one leg or neither leg (ballistic flight) touches the ground. The unspecified force allows to implement different models (Figure 5.20). Time is given in units of the stride period $T = s'/v_{FR}$, s': dimensionless stride length $[l_0]$; v_{FR}: dimensionless speed (cf. Section 5.2.3) $[\sqrt{gl_0}]$. Searching in a space of speeds and stride lengths close to human values ($s' = 2.5\ v_{FR}^{0.6}$; Section 5.2.3), force patterns that minimize a cost function during stance are sought. To describe muscle force, length dependencies and activation are ignored, that is, development of force only depends on the force–velocity relation (simplified Hill-type model). The quality of the results does not seem to depend on the detailed hyperbolic equations and the calculations; it can be even simplified by linear relationships (Figure 5.20B). In all models, constraints on force and force rate are implemented, avoiding singularities (e.g., stiff collisions, see below) and providing more realistic simulations.

Resting costs are ignored. Different cost functions are used and not all are illustrated here. One cost is just minimizing total work (see above):

$$C_1 = \int_{t=0}^{T} F\left([v_m]^+ - [v_m]^-\right) dt \tag{5.89}$$

Another cost is oriented at standard approaches to quantify metabolic cost in muscle–skeleton models. The bilinear function with a smaller slope b_2 for eccentric than for concentric b_1 contractions (Figure 5.20B) approximates complex dependencies (e.g., Woledge et al. 1985) and simplified versions (Minetti and Alexander 1997), assuming no cost for isometric contractions:

$$C_2 = \int_{t=0}^{T} F\left(b_1 [v_m]^+ + b_2 [v_m]^-\right) dt \tag{5.90}$$

where the index m refers to muscle speed, but the criterion can also refer to the actuator speed. Isometric costs are entailed in the cost function

$$C_3 = \int_{t=0}^{T} |F|^2\, dt \tag{5.91}$$

that is frequently used in optimizing dynamic simulations in muscle–skeletal systems. Also, the robustness of the qualitative results with respect to combinations of these cost functions is investigated:

$$C_4 = \lambda C_2 + (1 - \lambda) C_3 \tag{5.92}$$

with $0 < \lambda < 1$.

The landscape of solutions is fairly large and not completely mapped. General features such as the gait structure, the walk–run transition, symmetries were insensitive to the specific cost function. Using numerical optimization, Srinivasan (2010) obtained the actuator forces and body trajectories that minimized the above mentioned cost functions at various speeds and step lengths. The landscape of possible calculations is fairly large,

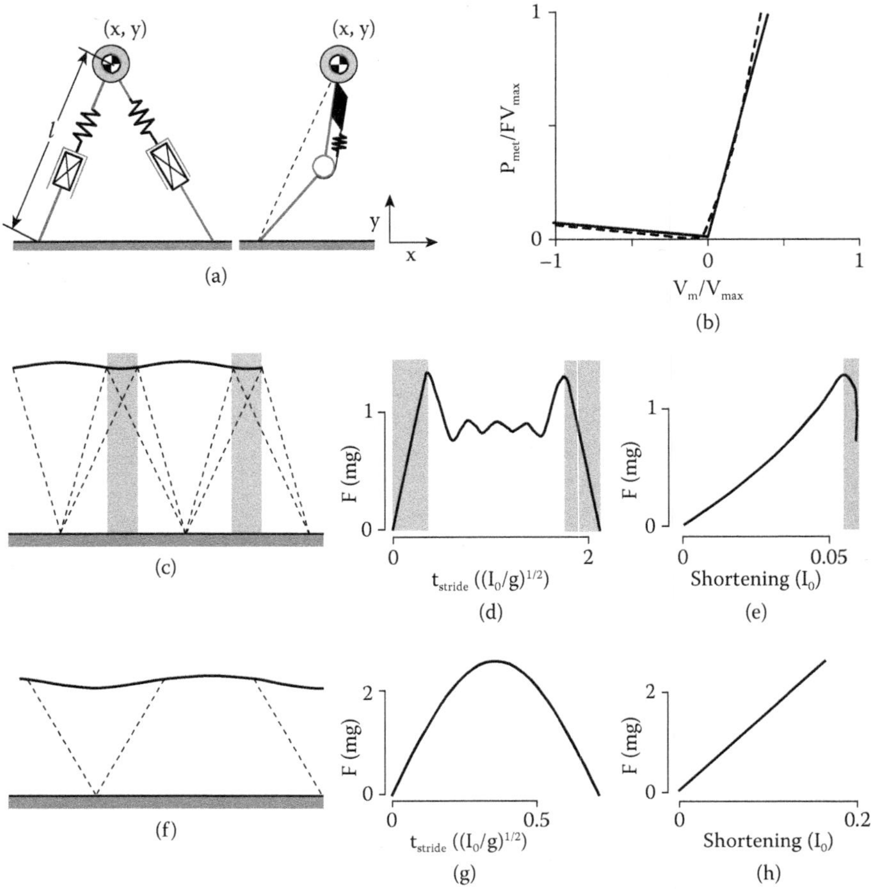

Figure. 5.20. (a) Investigated legs have telescope actuators combined with a spring (left) or are assumed to be bisegmental driven by a thigh muscle connected with a tendon to the shank (right). The compliance of the elastic elements has been varied in a wide range. (b) Close to the experimental metabolic cost of muscle work (dashed line; ATPase activity) given by Minetti and Alexander (1997), the cost is approximated by a bilinear function (normal line; Equation 5.90). (After Srinivasan, M., *J. Roy. Soc. Interface*, 8, 74–98, 2010) (c–h) Walking (upper row, c–e) and running (lower row, f–h) as simulated using telescopic actuator and a spring in series. Upper row: Walking of a pure telescope (spring compliance zero) minimizing the cost of work and limiting the force rate ($-4 \leq \dot{F} \leq 4$) for $v_{FR} = 0.39$; $s' = 1.38$. (c, f) Path of the center of mass; (d, g) Leg force; (e, h) Shortening of the spring. Shaded: periods of double support. Lower row: Running for a telescopic actuator with a spring in series for $\lambda = 0.9$, $v_{FR} = 1$, and $s' = 2.4$. Despite the contribution of cost for force development, actuator work is still minimized to almost zero. Decreasing λ to 0.75, that is, increasing the cost related to force generation results in significant impulsive contributions. (After Srinivasan, M., *J. Roy. Soc. Interface*, 8, 74–98, 2010. c, d, e: fig. 9, p. 11; f, g, h: fig. 16, p. 14. With permission of the Royal Society.)

and not all possible calculations were performed. Nevertheless, several observations could be made about the structure of these optimal gaits. It was found that general features such as the gait structure, the walk–run transition, symmetries were insensitive to the specific cost function.

Specifically, in all cases a rather stiff-legged walk was predicted at slow speeds and a "compliant" run at high speeds. Maximum leg length was observed at the beginning and the end of the stance phase. At least for systems that are constrained by force properties and facing a significant cost, walk–run transition occurs at about the same speed. There is an energetic advantage for a muscle–skeletal system to behave like a compliant inverted pendulum.

Walking with horizontal hip trajectory is always expensive (see the basic considerations above). Using positive work as an estimate for cost makes impulsive redirection the cheapest (bang–bang control) as long as the spring in series is stiff (Figure 5.20). Telescope-like work occurred around the collision, that is, pushing with the leading leg at touchdown and the trailing leg at push-off. Heel strike (Section 5.7.8) may be interpreted as a cheap way for the animals to perform "negative" work. It absorbs the necessary energy instantaneously, and after the initial push compensating for the losses, energy remains constant throughout the contact. Restrictions in peak force and force rates as well as the introduction of springs in series result in distributing the "collision" work across a longer contact segments. The muscle-cost function C_2 enforced more natural force patterns and smooth transitions of the center of mass trajectory during double support. Again considering the treatment in Section 5.2.4, the springy biped shows oscillations if not tuned, but otherwise rather natural displacement and force patterns. The tuned spring is optimal for $\lambda = 1$, where only work determines the cost. Importantly, however, it is not optimal for low λ, that is, in the case where the cost of force generation strongly contributes to the cost. To explore this observation further, we have to be careful as muscles can entail a considerable amount of parallel elasticity (e.g., titin, connective tissue). However, purely springy legs hamper versatility, for example, the spring is good for running but not for stopping. Introducing the muscle in series to actuate the spring introduces the cost of generating force. Now, the use of collisions (such as a heel strike, Section 5.7.8) represents an intelligent measure to increase economy.

Spring-like running gaits even emerged for the systems without springs. The leg yields at touchdown with a sinusoidal length pattern even for the cost function C_1. Introducing the force–velocity curve and the same cost function results in a force and length asymmetry (quite similar to nature). With a spring in series, the most economic gait is the gait with an almost isometric operating actuator, even though isometric force may be loaded with a substantial penalty (e.g., $\lambda = 0.90$; Figure 5.20) because during walking high penalties introduces significant deviations from the sinusoidal pattern. For high spring stiffness where the spring cannot save much work, the resulting force–length curve of the leg is still rather spring-like. Minimizing work results in an optimum despite substantial isometric force cost, that is, cost may be largely determined by force, whereas the optimum is largely explained by work. If cost purely depends on force, then there is no advantage of using springs. However, as soon as work counts, the spring in series helps to save energy.

The inclusion of muscle-related cost factors explained the question to which extent the principal behavior of the system is determined because the system is muscle driven and to which extent this performance would prevail in the presence of other actuators. It seems that using electromechanical drives would not strongly alter general patterns and insights. Knowing about the strength of the not-too-complicated compliant descriptions, including dynamic interactions and modes of oscillation, the question arises as to why we should care about even more reduced modeling as that used in collision models. We will see that they rather elegantly provide insight in gaits and their energetics without the necessity to numerically solve systems of differential equations.

5.5.5 The Elegance of Using Collisions to Examine Gaits

We have seen that using the mechanical energy fluctuations, the extrapolated metabolic costs, or both can be used to decide between models and to elucidate the mechanics of gaits. The extension to more legs and different gaits is the ongoing focus. We have learned that despite its rough dynamics, the simple collision model captures essential features of bipedal gaits (see above and Section 5.7.8). Ruina et al. (2005) used the model to expand on more general issues such as double support and multiple collisions as are observed in fast gaits such as the gallop. The assumption that the cost of locomotion is largely related to the cost of redirection is reasonable. Multiple contacts alter the pathway of the center of mass and thus the cost.

The restriction to shallow angle collisions (glancing) resulting in small deflection angles facilitates the estimates. In this case, speed changes parallel to the (assumed) fixed leg can be approximated as $\sim v \sin\beta \approx v\beta$, and changes in speed of locomotion are small $v^+ \approx v^- \approx v$. Then, the energy change amounts to $\Delta E = -E_a + E_g = m\frac{v^2}{2}\left(-\beta^- + \beta^+\right)$ or if we use the momentum balance $m(v^+ - v^-) = \mathbf{P}^*$ and the fact that $\beta = \beta^- + \beta^+$ then

$$\Delta E = m\frac{v^2}{2}\left(-\left(\beta^-\right)^2 + \left(\beta^+\right)^2\right) \tag{5.93}$$

The standard coefficient of restitution is defined as

$$e = \frac{v\ \sin\beta^{+}}{v\ \sin\beta^{-}} \approx \frac{\beta^{+}}{\beta^{-}} \tag{5.94}$$

Because actuators are involved, this ratio ranges from $0 \leq e \leq \infty$. To avoid infinity, a different coefficient is suggested:

$$e_g = \frac{v^{+} - v^{-}}{v^{+} + v^{-}} = \frac{e-1}{e+1} \approx \frac{\beta^{+} - \beta^{-}}{\beta} \tag{5.95}$$

completely inelastic collision: $e_g = -1$; $e = 0$; $\beta^{+} = 0$,

perfect quasi-elastic collision: $e_g = 0$; $e = 1$; $\beta^{+} = \beta^{-}$,

only generating: $e_g = 1$; $e = \infty$; $\beta^{-} = 0$.

For intermediate cases $\beta^{+} = (1+e_g)\frac{\beta}{2}$ and $\beta^{-} = (1-e_g)\frac{\beta}{2}$. From this

$$\Delta E = e_g \frac{m}{2} v^2 \beta^2 \tag{5.96}$$

It should be stressed that e_g is a number specifying the style of the rebound (Figure 5.21a). The different deflection types have different net energy balances ΔE that are not affected by the amount of energy stored. The latter is quantified by r as a measure for the fraction of absorbed energy E_a stored and used for the generated energy E_g. A quasi-elastic collision may be obtained with $r = 0$, that is, without storing elastic energy (see above)! e_g affects kinematics and the mechanical energy of the point mass r affects only metabolic energy. In the following, energetic cost is taken as the minimum values possible during a gait cycle. Taking elastic energy storage into account results in

$$\Delta E = (E_g - rE_a) - (E_a - rE_a) \tag{5.97}$$

For the case that $(E_g - rE_a) < (E_a - rE_a)$ or $e_g < 0$, the latter term (negative muscle work) may represent a lower bound for metabolic cost; in case that $(E_g - rE_a) > (E_a - rE_a)$ or $e_g > 0$, the first term (positive muscle work) may represent a lower bound. Using the equations above, it can be inferred that

$$\frac{E_m}{b} = \frac{mv^2\beta^2}{8}(1-r)(1-e_g)^2 \quad \text{for } e_g < 0$$

and

$$\frac{E_m}{b} = \frac{mv^2\beta^2}{8}(1-r)(1-e_g)^2 + e_g \quad \text{for } e_g < 0 \tag{5.98}$$

where E_m represents the metabolic energy and b is a constant. Using this approach, multiple collisions can be taken into account by summing up the angle and energy changes.

In a quasi-elastic collision ($e_g = 0$) with energy stored ($r = 0$), the cost amounts to

$$E_m = b\frac{mv^2\beta^2}{8} \tag{5.99}$$

and the deflection angle is $\frac{\beta}{2}$. If the collision is completely reabsorbing ($e_g = -1$, $r = 0$), cost amounts to

$$E_m = b\frac{mv^2\beta^2}{2} \tag{5.100}$$

and the deflection angle is β. Even with no elastic recovery, it is beneficial to use a quasi- or pseudo-elastic collision. We see that this approach is supported by the more detailed calculations outlined above.

In similar terms, dividing a single collision into two subcollisions halves the individual subcollision velocities and halves the total energy cost $(\frac{1}{2^2} + \frac{1}{2^2} = \frac{1}{2})$. This can be easily expanded to n collisions: subdividing a single collision into n subcollisions reduces the cost by a factor of $\frac{1}{n}$. Sequencing collisions have an energetic advantage. For infinite collisions, cost vanishes as the glancing angle vanishes. This is a powerful general insight.

Now we can use the collisions to estimate the CT here defined by the dimensionless quantity as CT′:

$$CT' = b\frac{W_{loss}}{mgvT} \tag{5.101}$$

with v being the animal's speed and T being the stride period. It can be estimated by counting and specifying the collisions across a given distance. Because the collisions are infinitesimally short, the angle determining the time of flight also determines the stride period. In this approximation, the time of flight amounts to $T = \frac{\beta v}{g}$ and the distance covered is $s = vT$. Assuming $e_g \leq 0$ and introducing the Froude-related (also based on d and not the usual l) number $\hat{F} = \frac{2gd}{v^2}$, we have

$$CT' = b\frac{E_m}{mgvT} = b(1-r)(1-e_g)^2\frac{\hat{F}}{16} \quad (5.102)$$

Using perfect symmetry and no elastic storage ($e_g = 0$, $r = 0$) results in

$$CT' = b\frac{E_m}{mgvT} = b\frac{\hat{F}}{16} \quad (5.103)$$

With complete storage $E_m = 0$.

For the runner with complete losses ($e_g = -1$, $r = 0$) the CT amounts to

$$CT' = b\frac{\hat{F}}{4} \quad (5.104)$$

This seems to be a big difference, however, to run using hip torques is much more expensive. Taking as an example a galloping horse using three subcollisions, this reduces cost compared to the pronk to one-third (e.g., $e_g = 0$):

$$CT' = b(1-r)\frac{\hat{F}}{48} \quad (5.105)$$

Galloping does not seem to be bad.

Ruina et al. (2005) also used a hodograph to illustrate different strategies of double support during walking by using the collision model (Figure 5.21b). Rolling over a long, curved foot (synthetic wheel) does represent the most economical situation. Cost increases for the case that the situation is approximated by push-off and heel strike at one-quarter and three-quarter points, respectively, using feet. A push-off preceding heel strike is better than simultaneous acting leading and trailing legs, and even more than the reverse sequence: heel strike before push-off. It can be seen how sequencing of several legs and feet can be used to reduce costs of collision.

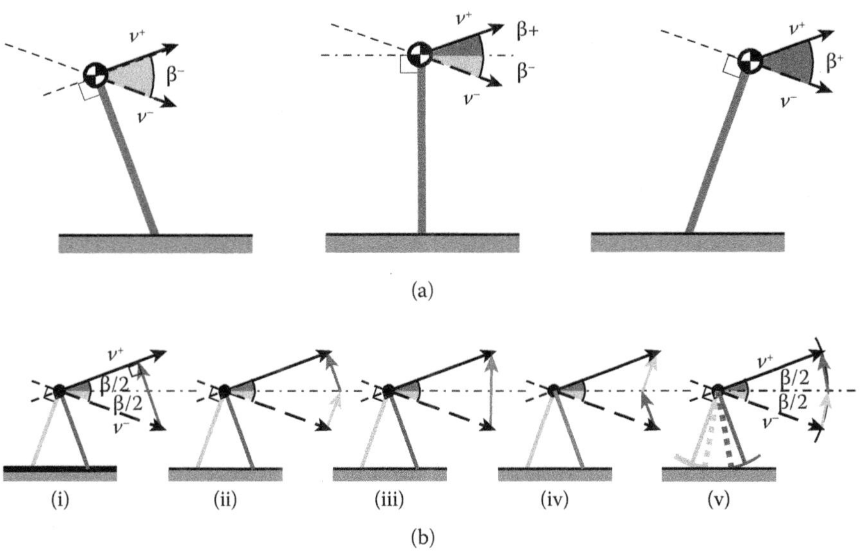

Figure 5.21. (a) Glancing conditions exclusively absorbing (left), quasi-elastic (middle), and generating (right). (b) Hodographs for collision models for step-to-step transition during walking. (i) Only leading stance leg contributes (passive walker). (ii) Heel strike of the leading leg is completed before the trailing leg pushes off (slight improvement). (iii) Simultaneous horizontally compensating impulses. (iv) The trailing leg axially compensates the axial losses of the previously collided leading leg. (v) Reduction of collision angle to $\pm\frac{\beta}{4}$ by using heel strike and toe-off with a foot. This uses collisions to closely mimic the synthetic wheel (circle, minimum cost; Section 5.3). (After Ruina, A., et al., A collisional model of the energetic cost of support work qualitatively explains leg sequencing in walking and galloping, pseudo-elastic leg behavior in running and the walk-to-run transition, *Journal of Theoretical Biology*, 237, Copyright 2005, a: fig. 2, p. 6, b: fig. 5, p. 15, with permission from Elsevier.)

In general, the cost during walking for different scenarios can be described as

$$CT' = mb\ \Lambda\ (1-r)\frac{d_{step}{}^2 v^2}{l^2} \tag{5.106}$$

with d_{step} being the step length, l being the leg length, and zero foot length ($d_{foot} = 0$). To take a convex foot into account, these values must be multiplied by $\frac{\left(d_{step} - d_{foot}\right)^2}{d_{step}{}^2}$ for the first four cases:

Figure 5.21b	Λ	Situations
i	1	Passive dynamic walking
ii	$\frac{3}{4}$	Heel strike before push-off
iii	$\frac{1}{2}$	Simultaneous push-off and heel strike
iv	$\frac{1}{4}$	Push-off before heel strike
v	$\frac{1}{8}$	Toe and heel at ¼ and ¾ points

The general insights gained by abstraction should remain to be valid independent of distribution of redirection over time and of the use of elastic storage to replenish losses.

The cost of locomotion is both important for animals as well as for legged machines. Correspondingly, we find that we can interpret many features of animal locomotion before this background, and because batteries or combustion machines are heavy, we can learn from animals how to reduce cost. In contrast, for example, when facing a predator, energy may not be the key issue any more (Section 5.7). And we have seen above when discussing segmentation (Section 5.4.4) that robust performance or insensitivity to disturbances may also be important. This idea led to a rather fruitful research line indicating that stable operation of a system might be an important factor influencing also the construction of locomotor systems.

5.6 STABILITY OF LOCOMOTION

Stability is the ability of a disturbed system to return to its origin, a certain state, or to reach an envisioned goal. The goal may be to, for example, maintain stationary locomotion (speed, direction, or both), to just to hold the beer mug quiet, or to securely catch prey. Disturbances are abundant in motion systems. The most obvious disturbances are introduced by moving across rough ground, the standard situation in nature, or it may occur in the course of interaction with other subjects. These disturbances are external. However, the reaction to internal disturbance may be more important, because any genetic deviation, deviations introduced by exercise, aging processes, injuries, or stochastic outputs of neurons can be considered as disturbances. Dripping over your own feet may just reflect a response to an occasional error in the motor program. Even without disturbances, it is by no means trivial to generate a program allowing for steady-state locomotion if the system itself has properties in which small deviations of the neuronal program may lead to large deviations in system behavior. Such a system does not need any "disturbance" to trip over its feet! Classically, we would call the field of cybernetics for help and force the system to return to its goal by using neuronal feedback. From the theory of dynamic systems, we know that attractivity can be an inherent mechanical system property.

One may argue that it is only of academic interest whether a system shows attractive behavior. This would be the case if neuronal control was physiologically cheap and always readily available, but it is not. The control of the numerous degrees of freedom of the system is energetically expensive and the challenge to organize the control of such a system in the short time demanded quickly overloads even large processors. In the following, the requirements on neuronal control are reduced by an intelligent construction of the mechanical system (Blickhan et al. 2007; Dickinson et al. 2000; Hogan and Sternad 2013).

It is important to keep other aspects of stability in mind, for example, maximizing stability can be a bad choice. We are able to alter movement goals quickly. For example, think of a hare darting from side to side. In this case, it must be at least possible to quickly switch between states of high stability. Maneuverability is inversely proportional to stability. Although this inverse property is obvious, so far we are still on the verge of a thorough understanding.

Following the history of investigations, we start with the simples aspect, static stability, being well aware that animals during locomotion tend explore their possibilities far beyond static stable situations. Below, we further see that they have learned to take advantage of stable modes in highly dynamic situations. This treatment directs us into the world of dynamic systems. Engineers are, in general, well aware of the complex behavior of such systems, and we show that it seems to be rather worthwhile to consider corresponding issues in the design of legged machines.

5.6.1 Static Stability

Static stability in locomotion is fulfilled as long as the vertical projection of the COG is located within the polygon of support. The system loses its balance as soon as the COG transcends the margin of the polygon. Such a polygon of support degenerates to a line or a point with only two or one support points, respectively. Therefore, we need at least three points of support to prevent a static system from falling. In such a case, small disturbances are counteracted by the torques generated by the ground reaction forces of the legs with respect to the COG. It corresponds to our experience that a polygon of support that is large compared to the height h of a system provides high stability. For standing humans, the ratio of the smallest distance of the polygon edge (stability margin) compared to the height is unfavorable (~0.1); for a standing pig, it is a bit better (~0.4); and for an ant it is quite good (~5). During locomotion, it is necessary to periodically replace the foothold. A two-legged system would only have a chance to maintain static stability for step lengths of less than the size of the feet. This should be much easier for a quadruped. But even there, leg placement and timing are critical. The quadruped can afford to lift a leg, and it is still supported by three legs (tripod). Intermittently, to guaranty stability while the COG is pushed across the border of the former tripod, it should use four legs. Leg placement is not possible in zero time. McGhee and Frank (1968) demonstrated that of the six permutations possible moving one foot after the other to move in a sequence of tripods, only three of them allowed static stable transitions (Figure 5.22).

Using an iterative approach (ignoring inertia), they were further able to specify the longitudinal stability margin, that is, the minimum longitudinal distance between the center of mass and the edges of the polygon of support, for four-legged strict forward locomotion. For each leg, the specified parameters consisted of duty factor, the proximal and lateral foot position relative to step length, and their phase relative to the first leg. The optimum, that is, maximum stability margins given relative to the spacing of the legs for regular gaits (equal duty factors for all legs and equal spacing of front and hind legs) are 1 4 2 3 (Figure 5.22): $DF-\frac{3}{4}$; 1 3 4 2: $\frac{3}{2}DF-\frac{5}{4}$; 1 2 4 3: $\frac{3}{2}DF-\frac{5}{4}$. The longitudinal stability margin of 1 4 2 3 is always above that observed for the other stable walks. Moreover, to be stable the duty factor must be above $\frac{3}{4}=0.75$ for the sequence 1 4 2 3 and even above $\frac{10}{12}=0.83$ for the other two gaits. The demand of static stability strongly restricts footfall patterns and speed.

This concept has been extended and generalized by Song and Waldron (1987). The first basic aspect is that increasing the number of legs ($n \geq 4$, n: *even*) allows static stability at a lower duty factor $\left(\frac{3}{n} \leq DF < 1\right)$. An insect with six legs can obtain static stability for $DF = 0.5$, a spider for 0.375, and a centipede can keep its legs most of the time in the air without losing static stability. For static stable wave gaits, that is, gaits in which legs on each side are placed with constant phase shift, the phase φ_i of the legs for machines with n legs (n: even) are defined as

$$\phi_{2m+1} = fract(m \cdot DF),\ m = 1, 2, \ldots, \frac{n}{2}-1 \text{ and } \frac{3}{n} \leq DF < 1 \tag{5.107}$$

where $fract(X)$ is the fractional part of the real number X, and m denotes the successive legs after leg 1 (for numbering, see Figure 5.22; odd indices, left legs; even indices, right legs; Table 5.1).

For such gaits with $DF > 0.5$ and at least $n = 4$, the longitudinal stability margin (S; Figure 5.22d through f) amounts to

if $\frac{1}{2} \leq DF$, or if $DF > \frac{2}{3}$ and $S_R \leq S_{Rb}$:

$$S_1 = \left(\frac{n}{4}-1\right)S_P + \left(1-\frac{3}{4DF}\right)S_R \tag{5.108}$$

if $DF > \frac{2}{3}$ and $S_R > S_{Rb}$:

$$S_2 = \left(\frac{n}{4}-\frac{1}{2}\right)S_P + \left(\frac{1}{4DF}-\frac{1}{2}\right)S_R \tag{5.109}$$

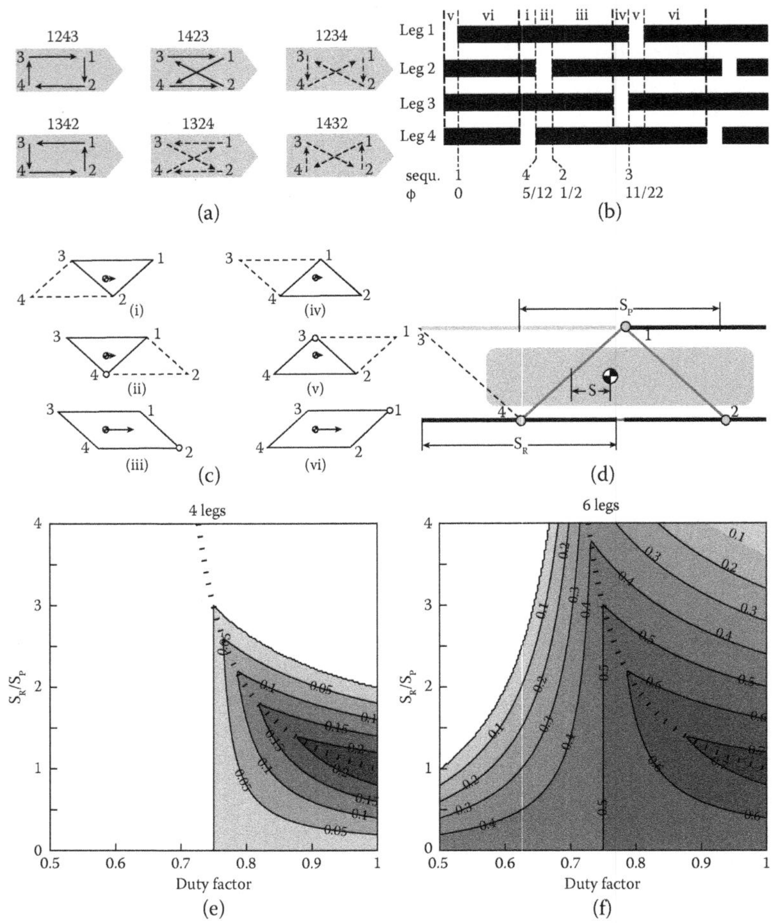

Figure 5.22. Stable quadrupedal crawl sequences. (a) Crawl sequences moving from left to right (shaded background). Arrows indicate placement of next leg. 1: left leading leg; 3: left trailing leg; 2: right leading leg; 4: right trailing leg. The sequences with dashed arrows are unstable. (b) Stepping pattern for the most stable crawl. Bold lines: contact of the respective leg; $DF = 11/12$; sequence: 1423; phases see bottom row; roman numbers on top refer to polygons of support depicted in (c). (c) Polygons of support. The arrows indicate direction and amount of movement using this support. Circles: new leg placed; dashed: leg lifted off. (d) Distances used to calculate the longitudinal stability margin S (the smaller of the distances to the polygon of support here depicted for instant iv) (cf. c). The bold gray and black horizontal lines mark the longitudinal leg tip translation with respect to the body (hatched). (e, f) Longitudinal stability margins calculated after Song and Waldron (1987). The relative distance of the center of support S_P is assumed to be 1. Altering support distance S_P does not affect the general pattern only the magnitude of the stability margin increases in proportion. In the white area the margin becomes negative. The dotted line marks the condition $S_R = S_{Rb}$ and delineates a stability crest. (a and c) (After McGhee, R.B., and Frank, A.A., *Math. Biosci.* 3, 331–351, 1968.)

TABLE 5.1

Phases (legs on the left side; Equation 5.107) and duty factors of static stable wave gaits in dependence of leg number.

n	φ_3	φ_5	φ_7	DF
4	DF			$\geq \frac{3}{4}$
6	DF	fract(2 DF)		$\geq \frac{1}{2}$
8	DF	fract(2 DF)	fract(3 DF)	$\geq \frac{3}{8}$

with $S_{Rb} = \frac{DF}{3DF - 2} S_P$, where S_R is longitudinal leg tip translation during support/stride length, S_P is distance of the centers of the supports/stride length, S is longitudinal stability margin normalized to S_P, and (for additional information) is speed $v = S_R/DF$.

The condition $S_P = S_R = DF$ leads to the following much simpler equation (Ting et al. 1994)

$$S = \frac{n}{4} DF - \frac{3}{4}. \quad (5.110)$$

The longitudinal stability margin increases with the number of legs. The factor preceding the translation S_P is always positive and increases with the number of legs. Correspondingly, and in agreement with intuition, increasing the distance between the centers of support always increases static stability and more so with an increasing number of legs. The second addend is more complex. It changes sign in dependence of the duty factor, and it now depends on its magnitude compared to the first addend whether there remains a positive stability margin. Due to the changing sign of the second addend a statement such as stability increases with tip translation is not possible. The addend becomes negative for $DF < 3/4$ in S_1 and in S_2 for $DF > 1/2$.

It is obvious that not only the distance from the polygon edge but also the weight of the object contributes to stability. An energetic static stability margin normalized to the weight of the system (NESM; Garcia et al. 2005) uses the increase in potential energy that is necessary to tilt the system (assumed to be rigid) around the polygon edges i:

$$S_{NESM} = \min_{(i)}\left(\frac{mgh_i}{mg}\right) \text{with } h_i = |\mathbf{r}_i|(1-\cos\phi)\cos\psi \tag{5.111}$$

and $|r_i|$ is distance between the COG for the untilted object (vector perpendicular to the tilt axis), ψ is the angle between r_i and the vertical, φ is the angle between the distance vector from the edge to the COG while untilted and the same vector at the instant of tipover, and h_i is the height the COG is lifted during the tilt from standing to tipover. The edges can be oblique with respect to the horizontal plane. (For expansion to compliant robots, see Nagy et al. 1994).

It is worthwhile to consider static stability and the concept of static stability during slow locomotion, that is, within a realm of speeds where accelerations are expected to be small compared to gravity and in which there is sufficient time to place the next supporting leg while moving within the polygon of support. In many cases, such a restriction can be acceptable and animals are using corresponding gaits and strategies. However, they also learned to expand the realm of stable locomotion to higher speeds.

5.6.2 Dynamic Postural Stability

Several definitions tried to overcome the static considerations outlined above and take the dynamics of the system into account. One fundamental concept is the concept of the zero moment point (ZMP; Vukobratovic et al. 1990). This concept is used, for example, in the control of the famous humanoid robot "ASIMO" (Honda, Minato, Japan). I am not aware of a zoological study based on this approach. Let us return to the original definition: "ZMP is defined as that point on the ground at which net moment of the inertial forces and gravity forces has no component along the horizontal axes" (Vukobratovic and Borovac 2004).

The distance of the ZMP from the polygon of support can be taken as a stability measure. The system is stable if the point falls within the polygon of support, and it is tilted around its edge as soon as it falls outside.

Vukobratovic and Borovac (2004) developed this concept for bipedal walkers with sufficient large feet. With one foot at the flat floor, the following equations are valid for the dynamic equilibrium (Figure 5.23):

$$F_R = -m_{foot}\, g - F_{ank} \tag{5.112}$$

$$\mathbf{r_R} \times \mathbf{F_R} = -\mathbf{r}_{CM,\,foot} \times \mathrm{m}_{foot}\, \mathbf{g} - \mathbf{M}_{ankle} - \mathbf{M}_{COP} - \mathbf{r}_{ankle} \times \mathbf{F}_{ankle} \tag{5.113}$$

where $\mathbf{F}_R$ is the ground reaction force; $\mathbf{F}_{ankle}$ is the force at the ankle joint; $\mathbf{m}_{foot}$ is the mass of the foot segment; $r_{R,\,CM,\,ankle}$ are the position vectors of the ground reaction force (center of pressure), the foot's COG, and the ankle joint, respectively; $\mathbf{M}_{ankle}$ is the torque at the ankle joint; and $\mathbf{M}_{COP}$ is the torque with respect to the center of pressure. Because friction is only able to generate torques around the z-axis $\mathbf{M}_{COP} = (0\ \ 0\ \ M_{COP})$. The equilibrium with respect to the z-component must be guaranteed by suitable movements and is not treated in depth here. Therefore, it is possible to restrict the considerations to the horizontal components that results in

$$(\mathbf{r}_R \times \mathbf{F}_R)^{hor} = -\mathbf{r}_{CM,\,foot} \times \mathrm{m}_{foot}\, \mathbf{g} - \mathbf{M}_{ankle}^{\;hor} - (\mathbf{r}_{ankle} \times \mathbf{F}_{ankle})^{hor} \tag{5.114}$$

As mentioned above, as soon as r_R, as determined by the right side of this equation, falls

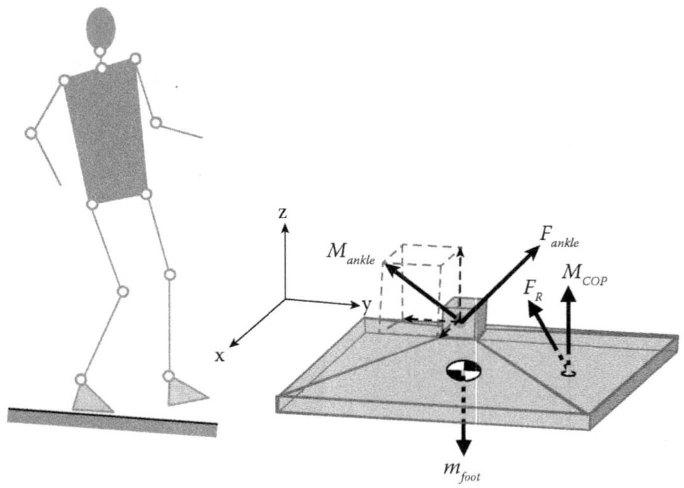

Figure 5.23. Forces and moments at the foot of a walker (see text). (After Vukobratovic, M., and Borovac, B., Int. J. Humanoid. Rob., 1, 160, fig. 1, 2004. With permission from World Scientific.)

outside of the support polygon, the foot would be tilted and the system becomes unstable. In the case of a multilegged system, the right terms might be simply replaced by the horizontal component of the dynamic moment introduced by the inertia of the system

$$(\mathbf{r}_R \times \mathbf{F}_R)^{hor} = \left(\mathbf{r}_R \times (-m\mathbf{g} + m\mathbf{a}_{CM})\right)^{hor}$$
$$= -\mathbf{r}_{CM} \times m\mathbf{g} + (\mathbf{r}_{CM} \times m\mathbf{a}_{CM})^{hor} + \dot{\mathbf{L}}_{CM}^{hor} \quad (5.115)$$

with $\mathbf{r}_{CM}$ being the position vector of the center of mass of the system, $\mathbf{a}_{CM}$: being the acceleration of the system; and $\dot{\mathbf{L}}_{CM}$ being the time derivative of the angular momentum of the system. In bipedal walkers, stability estimates require the calculation and prediction of the ground reaction forces, centers of pressure, positions of the edges of the feet, position of the center of mass, and horizontal transversal and angular accelerations of the center of mass. In a system with high degrees of freedom, this represents a considerable task. Investigations on human walkers reveal that the position the ZMP is within the supporting polygon of the feet (Mrozowski et al. 2007).

A normalized dynamic energy-based stability margin closely related to the static energy stability margin described above (NESM) and the zero momentum point (ZMP) has been introduced by Garcia and Santos (2005). Again, the critical situation is reached as soon as the forces and moments generated by the ground reaction force are not able to compensate disturbances such as generated by external forces (e.g., accidents, manipulations) or placing legs. Critical tumbling is reached as soon as the torques and moments generated by the ground reaction forces with respect to any edge of the support polygon vanish. In addition, Garcia and Santos (2005) assume that the kinetic energy becomes zero in this critical situation. Knowing the tumbling edge from the polygon of support, which may have a slope, the difference between the potential energy at the instance of the disturbance to that in the critical situation as well as the kinetic tilt energy can be estimated. The changes in potential energy ΔV are composed of the contributions by weight ΔV_G, reaction force ΔV_F, and ground reaction force ΔV_M:

$$\Delta V = \Delta V_G + \Delta V_F + \Delta V_M \quad (5.116)$$

The work against gravity amounts to $\Delta V_G = mgh$. The work ΔV_F by the reaction force $\mathbf{F}_R$ without the contribution of gravity with respect to the edge i ($\mathbf{e}_i$ is the unit vector in the direction of the polygon segment in counterclockwise direction)

$$\Delta V_F = \int_{\theta_1}^{\theta_2} \left((\mathbf{F}_R - m\mathbf{g}) \times \mathbf{r}_R \right) \circ \mathbf{e}_i \, d\theta \quad (5.117)$$

and the work of moments induced by manipulation moments ($\mathbf{M}_M$, e.g., grip) and inertia $\mathbf{M}_I$

$$\Delta V_M = \int_{\theta_1}^{\theta_2} (\mathbf{M}_M - \mathbf{M}_I) \circ \mathbf{e}_i d\theta \qquad (5.118)$$

The kinetic energy can be assessed from the actual momentum L_i with respect to the polygon edge,

$$E_{kin} = \frac{1}{2} I_i \omega_i^2 \qquad (5.119)$$

with $\omega_i = \frac{L_i}{I_i}$ and $L_i = (r_{cog} \times m\, v_{CG}) \circ e_i$.

It is obvious that these concepts, although dynamic, ignore the adaptive deformation of the system and especially the possibilities to stabilize the system by proper placement of the feet in the next step. Stability is assumed to be achieved by torques produced by contact points available and not by future contact points. In fact, dynamic stabilization by means of the next step is more the rule than the exception. It is necessarily the only possibility in gaits with flight phases such as running, trotting, and galloping. However, it also occurs during sufficiently fast walks and in quadrupedal walking and in hexapods. A classical study is the work of Jayes and Alexander (1980) on the locomotion of turtles. A naïve observer would assume that these slow-moving animals use always static equilibrium. However, the stepping pattern clearly indicated phases with only two leg tips on the ground (Figure 5.24).

Falling from short legs on an armored belly might not be that bad, but twice each cycle? Well, the animals do, in fact, not touch the ground. Support of the center of mass is not a problem with at least two legs in contact with the ground. Because the rate of force development is limited, we may expect that sometimes the vertical force drops below weight. However, more critical is tilting of the carapace around the axis through the two supporting points. The experiments show roll, pitch, (and yaw) movements with amplitudes of just about 5°, 10°, and (5°), respectively. The amount of roll and pitch depends on the resulting moments exerted by the reaction forces and the moment of inertia of the bodies. By using a forward simulation describing the time course of the ground reaction force, including the phase relation, between the legs by using a Fourier series, Jayes and Alexander (1980; Section 5.5) investigated the degree of roll and pitch depending on the phase of the feet.

In the following, the basic approach in this pioneering study is described. Jayes and Alexander (1980) restrict this in a first attempt to an even series as the basic pattern of the ground reaction force F exerted by each foot can be approximated by a series of cosine half-waves with a period of 2 DF/f, where DF is duty factor, f is stride frequency, and C is maximum amplitude of F

$$F(t) = C\cos(\pi f t / DF) \qquad (5.120)$$

interrupted by short intervals of zero force where the leg is protracted. The even series for this pattern is

$$F(t) = \frac{a_0}{2} + \sum_{k=1}^{\infty} a_n \cos(2\pi f n t) \text{ with}$$

$$a_n = C\; 2f \int_{-DF/2f}^{+DF/2f} \cos(\pi f t / DT)\cos(2\pi n f t)\, dt.$$

yielding

$$F(t) = \frac{2\,C\,DF}{\pi}\left\{1 + \sum_{n=1}^{\infty} \frac{\cos(\pi n\, DF)\cos(2\pi n f t)}{1 - 4n^2 DF^2}\right\} \qquad (5.121)$$

An equal weight distribution is assumed that implies that during a stride, each foot has to carry a quarter of the body weight. This can be used to determine C. The phase of each foot (e.g., δ) can be easily implemented: $\cos(2\pi n f t)$ is replaced by $\cos(2\pi n f t - \delta)$. The total force can be calculated by adding the contributions of each foot. The calculations of the moments neglect the minor contributions of the horizontal force components. The levers with respect to the COG are calculated assuming symmetrical foot placements, taking the dimensions of the turtle into account and a constant speed v such that the longitudinal displacement of the feet $x = v\,t = \frac{s\,f}{DF}$, where s represents the step length. The moments normalized by the respective moments of inertia are integrated to yield the rotation angles. [In the integration, the authors nicely use the fact that for $y = \cos(at)$ the

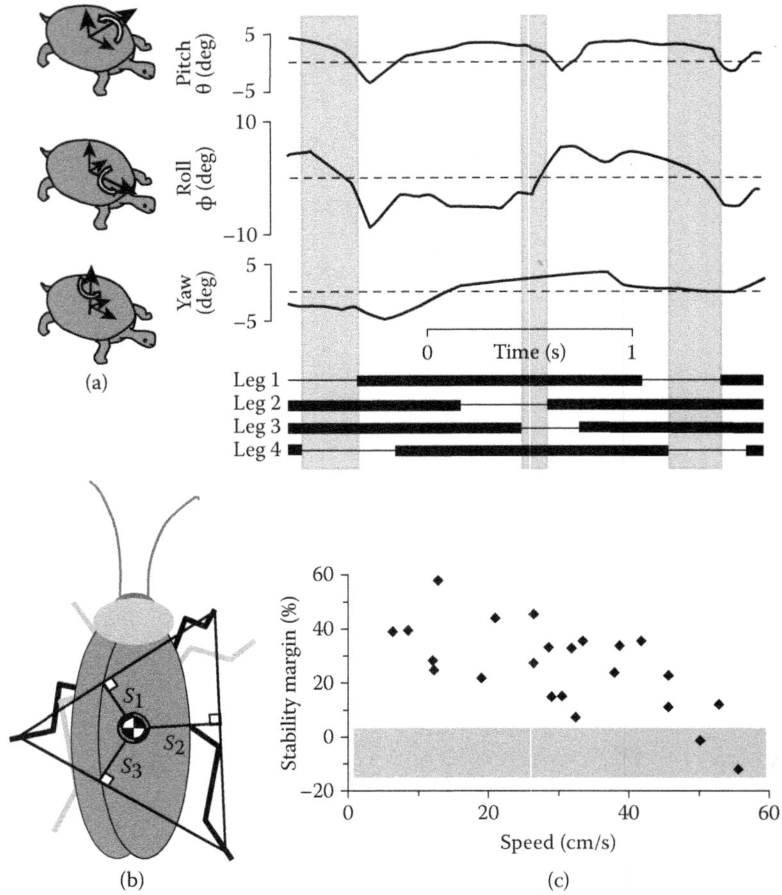

Figure 5.24. Rotation of a fast-walking turtle (a) Gray shaded areas where only two legs are supporting. Despite of using an alternating tripod with increasing speed cockroaches run statically instable. (After Jayes, A.S., and Alexander, R.M., The gaits of chelonians: Walking techniques for very low speeds. *J. Zool.* 1980. 191. 353–378, fig. 6, Copyright Wiley-VCH Verlag GmbH & Co. KGaA.) (b) Stability margins S_i with respect to the three edges of the polygon defined by the supporting tripod (black legs). (c) Stability margin in percent of maximum possible for polygon as observed during locomotion in dependence of speed. Gray shaded: negative area defining instability. (After Ting, L.H., et al., *J. Exp. Biol.*, 197, 251–269, B: fig. 1, p. 255, c: fig. 7, p. 260, 1994.)

second derivative is $\ddot{y} = -\frac{1}{a^2}\cos(at)$, or vice versa $y = -a^2\ddot{y}$.]

Jayes and Alexander (1980) found the smallest value for role and pitch for a trot-like step pattern, the gait that is used by the animals. Given a duty factor of 0.83, several static stable gaits would have been possible (see above). However, the turtle's ability to generate forces quickly is rather limited (albeit saving energy), and the other gaits would have resulted in much stronger roll and pitch movements. Furthermore, the skew in the observed force tracings—the maximum is shifted to late support in the front legs and to early support in the hind leg—results in a further overall improvement, reducing roll and pitch movements. Clearly, the calculations rest on several simplifying assumptions. Nevertheless, it is shown that pursuing the goal of static stability may be detrimental. Like the human walkers, quadrupeds may gain speed and even stability with respect to body motions daring to roll over two feet from time to time. They rely on the fact that the retracting leg is placed in time to absorb the fall in time.

Similar observations are available for hexapods (Ting et al. 1994). Cockroaches use an alternating tripod as the basic pattern of foot placement. With speed, the percentage of double support decreases from 30% at 10 cm/s to 10% at 60 cm/s. The static stability margin given in percentage of the maximum stability margin that would be possible due to the configuration of the tripod drops from about 50% at 10 cm/s to 0% at 50 cm/s, the latter indicating static instability (Figure 5.24b, c). Furthermore, in individual strides roll, pitch, and yaw movements of the body where induced by eccentric forces. The cockroaches did not seem to minimize such movements but instead preferred compensation from step to step. Superficially, such a strategy seems to be rather dangerous because the height of the COG is very low compared to the animal's length and width and the dimensions of the tripod. However, compared to the stride period, the animals have sufficient time. I cite (Ting et al. 1994): "Therefore, the fraction of a stride period to fall to the ground is not necessarily lower in insects. The American cockroach *Periplaneta americana* actually requires a greater fraction to fall (1.1, slightly over a stride period) than does a trotting dog (0.96) when they are compared at stride frequencies normalized to maximum values." In fact, cockroaches are able to run with aerial phases and bipedally at highest speeds, exploring lift forces to raise the body (Full and Tu 1991). In the turtles described above, the ratio between falling time and stride period is much smaller, but their gait pattern prevents them from touching the ground (not always as you may observe in a zoo).

We have seen that by using dynamic postural stability, the speed of locomotion can be increased. This measure is explored in various robots. A further step forward toward higher speed would be possible if leg placement is taken into account. Evidently, such a concept must take the dynamics of a system into account, including for higher speeds, the alternation of flight, and contact phases. Fortunately, the theory of dynamical systems provides us with some instruments to investigate and characterize stability.

5.6.3 Attractive Systems: Self-Stability

As mentioned above in a step-to-step approach, a system recovering from imbalance needs to generate the forces and moments in the next steps. The neuronal system may register the deviations and as a response alter the stepping pattern or the forces generated. Some systems are able to do this without neuronal correction because the mechanics already responds in a proper way. This is actually not restricted to a step-to-step approach and is, for example, also valid for the situation in which posture is corrected after a disturbance (see below).

Let us return to the bipedal runner described by a simple spring–mass system (Section 5.2.1). There we assumed symmetrical runs, that is, runs in which the dynamics after midstance is mirrored by the dynamics before midstance. Only the correct configuration of initial or touchdown conditions (velocity of the point mass at touchdown, angle of attack of the leg) combined with proper system properties (spring length and stiffness, mass) seems to guarantee such a symmetric behavior. In the numerical solutions of the differential equations, it is necessary to search after such solutions. Translated to the runner this implies that it is not trivial to find a steady-state solution. Furthermore, due to variation of properties and conditions (e.g., stiffness, angle of attack, and humpy ground), it seems to be necessary to correct permanently. Such a permanent neuronal control would require knowledge about properties of the system and its current state. Misjudgment would result in either acceleration or deceleration of the system.

It turns out that provided a suitable range of initial conditions combined with system properties is used, the system is able to return to the envisioned state of steady locomotion by itself (self-stability) without combining intervention. The runner explores attractive behavior inherent to the nonlinear system of differential equations, and the return to the envisioned goal is due to mechanical system properties. Confronted with an unexpected or misjudged bump, the runner may roughly keep all his system parameters and the angle of attack of his leg constant. The bump itself results in the fact that the kinetic energy of the runner at touchdown in less than the one necessary for a symmetrical take-off. The leg spring will be compressed less and the ground reaction force will be reduced. The next flight phase will thus be of reduced height. Stepping down the bump in the next flight phase will reverse this game. Now, the system lands with a higher speed and the reaction force increases. The system is conservative, that is, it neither wins nor looses energy. Tricky and difficult to describe is the fact that the energy is redirected, that is, the elastic inverted pendulum redirects horizontal and vertical kinetic energy.

For hybrid systems, that is, systems with a countable number of states (discrete) and intercalary continuous states with low dimensions, the return map offers a transparent window into the behavior of the system. In such a map, the future state is depicted as a function of its previous state. If the state remains the same, the system is stable. A disturbance will push the system away from its current state. A stable system will return after the deflection to a stable situation. Fixing mass, spring resting length, and stiffness of the system, adjustment of the angle of attack is left to obtain a steady state or better, a stable steady state. To characterize each state consisting of contact and flight phase, we select the apex height. There the potential energy has its maximum value, and the vertical component of the kinetic energy is zero. For a given angle of attack, the system starting from different flight heights (potential energy) and speeds (kinetic energy) will respond with different forces and excursions, and only selected initial conditions will result in a steady state. In a graph in which the apex height for the next rebound y_{i+1} is plotted in dependence of the previous apex height h_i, the function $F(h_i)$ describing its dependence

$$y_{i+1} = f(y_i) \tag{5.122}$$

must cross the diagonal $y_{i+1} = y_i$ at the point denoted by y_F for steady-state solutions. Only if such a crossing occurs is a steady-state solution is possible. In our system the dependency $f(y_i)$ can be obtained by numerical integration of the equations given in Section 5.2. Analytical approximations are given in Geyer et al. (2005) and Ghigliazza et al. (2005). The graph also offers insight into whether such a solution is stable. Linearization in the immediate environment of the crossing implies that

$$y_{i+1} - y_F = \lambda\,(h_i - y_F) \tag{5.123}$$

so for $|\lambda|<1$, the difference diminishes with each step successively approaching the fixed point y_F, and the system is asymptotically stable. For $|\lambda|>1$, the distance increases, and the system is unstable; for $\lambda = 1$, the distance remains the same, and the system is neutrally stable; and for $\lambda = -1$, it is oscillating between two conditions. The smaller the $|\lambda|$, the faster the system will approach its fixed point. For the spring–mass system (e.g., mass $m = 80$ kg; spring length $l = 1$m; leg stiffness $k = 20\frac{kN}{m}$; speed $v = 5$ m/s; Figure 5.25 a, b), the return plots show that stable conditions can be reached for angles of attack α_0 measured between the horizontal and the spring (leg) axis between about 67° and 69°. The graph also visualizes that disturbances in the ground height of some centimeters are easily compensated. Lack of energy limits recovery from rather low heights, but from higher heights the system can recover using a large number of steps as long as it starts below the second upper crossing.

The fact that human runners approximately use self-stable configurations (Figure 5.25c) supports the relevance of stability. It should be noted that stability comes free of cost. If we assume that active stabilization requires energy, it may even save energy. This is also true if the strategy helps to avoid expensive equipment such as precise sensors. Translating into three dimensions seems to be straight forward. In contrast to a previous investigation (Seipel and Holmes 2005) that did not find stable fix-points for a 3D-spring-mass model, recent investigations (Peuker and Seyfarth, pers. com.) revealed self-stable 3D-motion. As left and right symmetry is assumed with respect to the movement strategy (touchdown angles), it seems to be crucial to use the local coordinate system. Besides modeling studies more experimental probing is necessary (after Daley and Biewener 2006).

During compliant bipedal walking (Geyer et al. 2006), the apex occurs around midstance (single support phase; cf. Section 5.2.4). Its position at step i is defined by its horizontal position with respect to the foothold Δx_i and its vertical position y_i. At the apex, the vertical speed is 0 and including its position the state of the system is characterized by the horizontal velocity $\Delta \dot{x}_i$. The total energy of the system

$$E_{tot} = \frac{m}{2}\Delta\dot{x}_i^2 + mgy_i + \frac{k}{2}(l_0 - l_i)^2 \tag{5.124}$$

with $l_i = \sqrt{\Delta x_i^2 + y_i^2}$. For a given speed or system energy, we can generalize the condition for stability of a fixed point lined out above using eigenvalues $\lambda_{1,2}$ of the Jacobian DR_{ab}. With $x_{i+1} = F(x_i)$ and $x_i = (\Delta x_i, y_i)$

$$DR_{ab} = \left.\begin{pmatrix} \dfrac{\partial F_1}{\partial \Delta x_i} & \dfrac{\partial F_1}{\partial y_i} \\ \dfrac{\partial F_2}{\partial \Delta x_i} & \dfrac{\partial F_2}{\partial y_i} \end{pmatrix}\right|_{(\Delta x_i, y_i)=(x_F, y_F)} \tag{5.125}$$

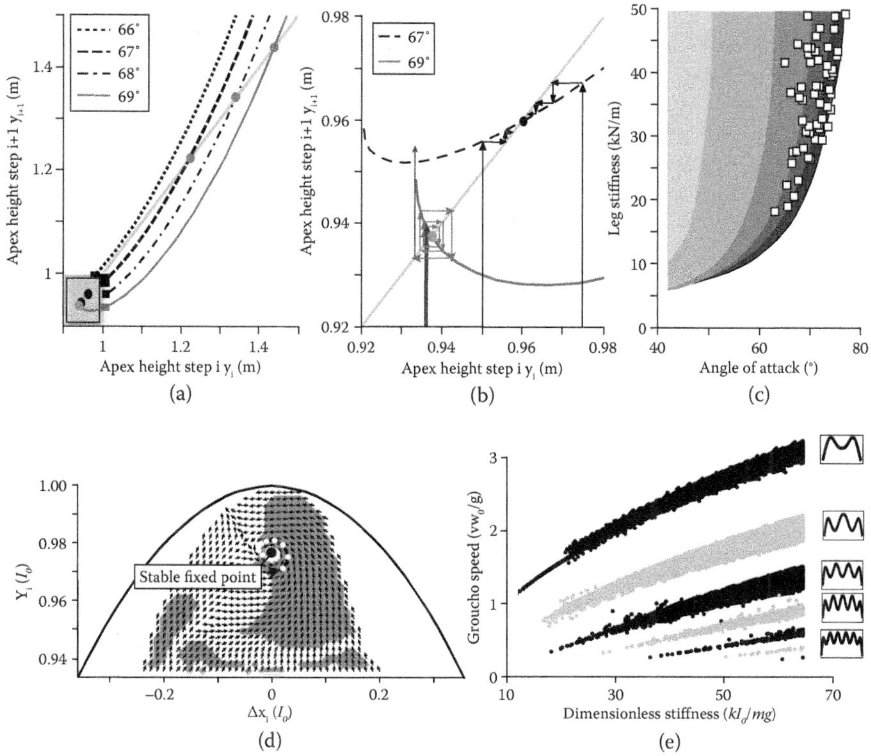

Figure 5.25. (a–c) Stability of a spring–mass runner. (a) $y_{i+1} = f(y_i)$ for angles of attack of 66°, 67°, 68°, and 69°. The light-shaded dots mark instable fixed points, the black dots in the lower corner mark stable fixed points. The shaded square is magnified in (b) omitting the characteristics for 66° and 68°. (b) Return plots in the vicinity of a stable (upper right corner) and an unstable (lower left corner) fixed point. The system starts with a value indicated by the arrow starting at the abscissa. The resulting apex height is used as an initial value for the next bounce (horizontal arrow) from where the next vertical arrow gives the next apex height, etc. (c) Mapping numerically the number of successful trials (shaded) reveals for different stiffness values and angles of attack reveals that the human runners (squares) run in an area in which the simulations resulted in at least six successful runs. The gray values of the shading are proportional to the number of successful steps. The white area on the right results in overrunning without a second successful step. (After Seyfarth, A., et al., A movement criterion for running. *J. Biomechanics*, 35, 649–655, Copyright 2002 a, b: fig. 3, p. 653, c: fig. 2, p. 652. With permission from Elsevier.) (d, e) Stable walking. (d) Apex return map for walking $\left(\alpha_0 = 69°; \frac{kl_0}{mg} = 17.8\right)$. The parabola denotes the circle where $\frac{1}{l_0}\sqrt{\Delta x_i^2 + y_i^2} = 1$. The arrows mark the direction of change in subsequent states $(\Delta x_i, y_i) \rightarrow (\Delta x_{i+1}, y_{i+1})$. Gray shading: areas of local stability $(|\lambda_{1,2}|<1)$. These areas do no not represent basins of attraction, instead, they just indicate local stability in the vicinity of a fixed point. (Geyer 2005, pers. comm.) (e) Stable walking modes. The stable running modes (not depicted) occur after a gap for higher speeds. On the right: typical force patterns. Asymmetric modes are not included. Depicted are solutions with at least 100 stable steps and an initial contact height of 0.9 leg length. (Andrada, pers. comm.)

$$\lambda_{1,2} = \frac{DR_{11} + DR_{22}}{2} \pm \sqrt{\left(\frac{DR_{11} + DR_{22}}{2}\right)^2 + DR_{12}DR_{21} - DR_{11}DR_{22}}. \tag{5.126}$$

For stability, it is required that in the neighborhood of the fixed point the eigenvalues are within the unit cycle $|\lambda_{1,2}| \leq 1$.

Mapping the development of the apex position $F(\Delta x_i, y_i)$ from step to step (return map; Figure 5.25d) within the directional field in a two-dimensional plot visualizes the attractive environment of a fixed point. Even though disturbances may drive the system away from its envisioned goal the system will return to a stable pattern.

During walking a number of modes can be observed which differ with respect to symmetry and or the pattern of the ground-reaction force (Figure 5.25e). For normal walking, the latter is double humped, however also mode with more humps can be observed in the simulations. Configurations for stable locomotion occur within distinct domains. Humans seem to prefer the use of stable domains during locomotion. Moreover, several tricks have been developed to enhance the stability domains. During running retraction of the swing leg before touchdown enhances stability (Seyfarth et al. 2003). In a retracting leg, the amount of retraction depends on the time of the swing phase. A bump in the ground shortens the swing phase and the leg hits the ground at a flatter angle of attack. Certainly the adjustment is not introduced by neuronal sensing of the ground but by employing a clever motor program which improves the mechanical response of a system in the event of unknown ground properties.

We have seen that stable locomotion can be observed for the spring mass template and the experimental data available support the idea that animals explore self-stability. The passive walkers described above (Section 5.3.2) represent self-stable modes of locomotion based on stiff-legged compound-pendulums. The concepts are finding their way into robotics. The next chapter will point toward a rather stimulating line of research and will show that the issue of self-stable locomotion is not reserved to large vertebrates but has been identified and investigated in depth for insects.

5.6.4 Stability of Insect Locomotion

Cockroaches swiftly run across irregular ground with remarkable ease. After strong lateral disturbance with chemical explosives generating about 15 body weight forces the animals moving at a speed of about 30 cm/s needed about 30 ms to recover and to return to their envisioned path (Jindrich and Full 2002). A step after the disturbance movement was similar to the undisturbed case. In a series of papers, Holmes and coworkers investigated employing analytical and numerical modeling the stability of running of cockroaches in the frontal plane. Here an outline of basic ideas is presented. According to experimental evidence the pattern of the ground reaction force and correspondingly the induced displacement of the center of mass could be explained by the action of a single spring–mass system [or spring-loaded inverted pendulum] replacing the simultaneously acting legs (Section 5.2.4). This corresponding reduction of degrees of freedom, which must be revised with respect to rotation (see below), strongly facilitates treatment. The model can be described by four nondimensional parameters: inertia, stiffness, attachment distance, and orientation angle. Attachment distance d is the distance at which the "hip" joint of the virtual leg is fixed onto the long axis of the elliptical body with respect to the COG. In a feed-forward manner, the foot is always placed at the same angle with respect to the body axis. Its state has been described by four quantities: the magnitude v and direction with respect to the body axis δ of the heading velocity, the heading or yaw angle θ, and its rotational speed $\dot{\theta}$. Using polar coordinates θ and r_G, ψ for position (distance, angle) of the COG with respect to the foot, the Lagrangian is given as

$$L = \frac{m}{2}\left(\dot{r}_G^{\,2} + r_G^{\,2}\dot{\psi}^2\right) + \frac{I}{2}\dot{\theta}^2 - V(l) \tag{5.127}$$

with the leg length l (n even, left; n odd, right)

$$l = \sqrt{r_G^{\,2} + d^2 + 2r_G d\sin\left(\psi - (-1)^n\theta\right)} \tag{5.128}$$

$V(l)$: potential function, for example, for an elastic spring $V(l) = \frac{C}{2}(l - l_0)^2$ with l_0 being the leg length at touchdown.

For fixed COP $d \equiv d_0$; for a moving attachment $d = (\psi-(-1)^n\theta)d_1$, the legs are assumed to be massless, with m mass of the body and I inertia of the body.

Because the foot does not induce torque with respect to the ground, the angular momentum around the foot is conserved

$$L_{foot} = mr_G^2\dot{\psi} \pm I\dot{\theta} = \text{const.} \tag{5.129}$$

and only 2 degrees of freedom remain.

The system is conservative during each stride; however, the angular momentum L_{foot} is traded from leg to leg at touchdown and take-off during each step.

Using a return map, the conditions for stable steady-state movements can be determined

$$\left(v_{n+1}, \delta_{n+1}, \theta_{n+1}, \dot{\theta}_{n+1}\right) = \mathbf{F}\left(v_n, \delta_n, \theta_n, \dot{\theta}_n\right) \tag{5.130}$$

For stability the absolute eigenvalues $|\lambda_j|$ of the 4 × 4 Jacobian matrix must be <1. Due to energy conservation and rotational symmetry, the eigenvalues with respect to speed v and rotation θ are equal to 1, that is, with respect to these variables the system is neutrally stable. Thus, alterations induced by disturbances are preserved. Altogether, the system fulfilling the condition that some eigenvalues are 1 and others <1 may be called marginally or partially asymptotically stable (Holmes et al. 2006). Below a critical speed v_c no stable gait exists. At critical speed a bifurcation is observed with two emerging branches: one branch with small oscillations that are stable and another branch with large oscillations that are typically unstable. For fixed leg attachment caudally to the COG ($d = d_0 < 0$), the system is stable with respect to the heading δ and rotational speed $\dot{\theta}$. It becomes unstable for attachments in front of the center of mass. Correspondingly, the system is stable if the hip moves backward during stance $d_1 < 0$. This result including the neutral stability of speed and orientation implies that changes in direction are readily possible, for example, by a change of d. Stability results in a loss of momentum and rotational kinetic energy at each step, which is converted into translational kinetic energy. The seemingly detrimental lateral oscillations observed in running cockroaches are necessary for a stable gait. The eigenvalues (Figure 5.26b) show a minimum in the range of speeds the animal prefers. Furthermore, the absolute ratio between inertia and the location of the virtual hip joint dimensions are close to realistic values in the region where fast recovery occurs. By changing the dimensionless parameters, the system alters its behavior. Superior stability could be obtained when adjusting leg length, stiffness, and touchdown angle with speed.

It should be noted that stability also occurs for a system in which reaction forces and foot positions are prescribed (e.g., Kubow and Full 1999; Schmitt and Holmes 2000). However, the foot forces must rotate with the animal, otherwise the resulting gate is unstable.

Between the behavior of the model described and the animal's behavior, there remain significant deviations. Schmitt et al. (2002) stated: "Due to the collapse of the stance support tripod to a single virtual leg, however, moments and yaw variations are significantly underestimated. The animal evidently has greater flexibility in generating large moments without large net forces, than a model with only a single effective leg along which forces are supposed to act." Introducing an appropriate hip-spring improved the description of the torques, but it created strong deviations with respect to force. This can be corrected by implementing six legs distributed along the body axis. Furthermore, the model tested with prescribed length changes of the springy legs and prescribed movements of the hips was stable with respect to speed of locomotion (work done due the prescribed changes) and lead to a stability channel that makes constant stepping frequencies unreasonable. The latter would require the crossing of unstable regions. These results are confirmed in models that mimic the leg geometry in detail. The linear springs are replaced by rotational springs whose torque-free rotational angles were prescribed to match the reaction forces measured for given fixed foot position at touchdown. Investigations of disturbances (Figure 5.26d) and the system recovered within one stride without any reflex implemented. This could not be achieved by the most recent neuromuscular models (Kukillaya et al. 2009). This is interpreted as a lack of appropriate feedback, for example, by exteroception not included in the model. The impressive neuromechanical model developed has similar geometry as the spring-leg model, but it includes antagonistic muscle pairs (Hill-type mode), including action potential–based activation dynamics. It is driven by a network of neural pattern generators and entails reflexes.

Insects use self-stable strategies for locomotion in the transversal plane. This is possible by the cooperative action of the legs involved. Some of the results found their way into the construction of legged machines. Theoretical investigations on hexapods also pointed out that force generation by muscles must not be detrimental for self-stability. Nevertheless, it might be worthwhile to treat this issue in more depth. We work out how muscle as a tissue with mechanical properties not very familiar to engineers may hamper or foster self-stability. Again, we use the chance to enhance our methodical repertoire.

5.6.5 Muscle–Skeletal Stability

Treating the legs as a spring is motivated by the principle behavior of the system. Scientists using this approach are well aware of the fact that this neglects important properties of the muscle-system, especially the properties of the muscle itself. The neural driven muscle–skeletal system generates spring-like behavior, but its reaction to

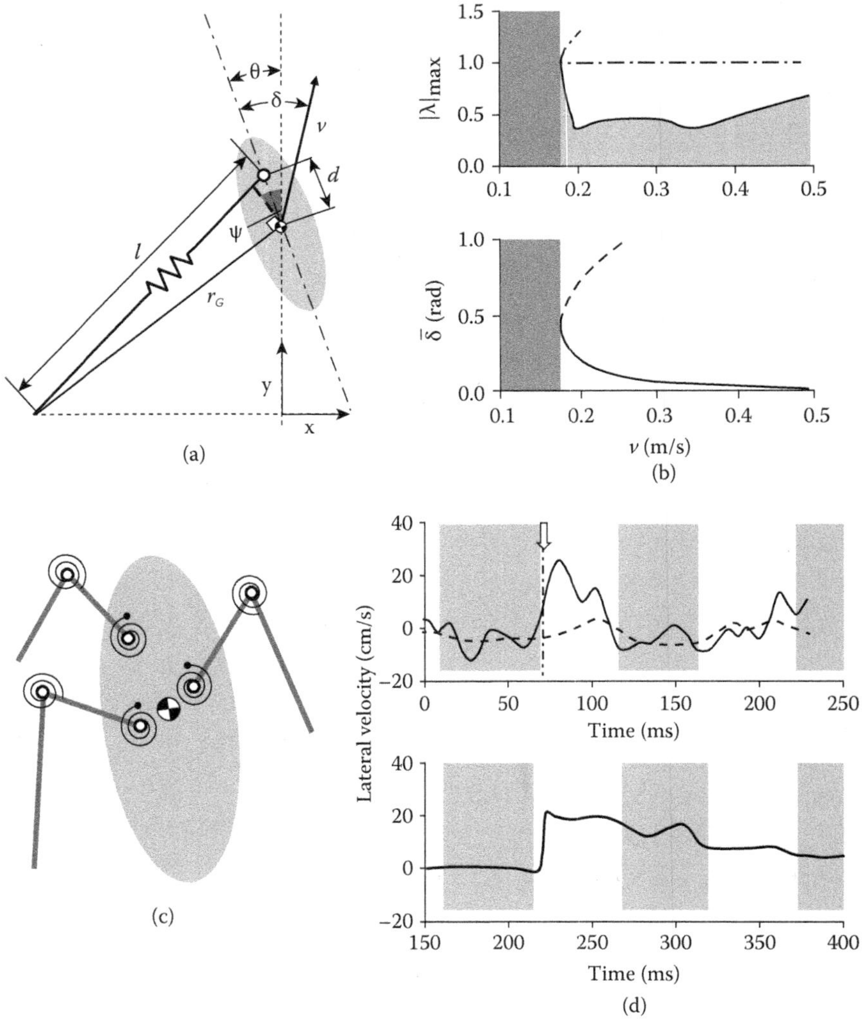

Figure 5.26. Stability of locomotion in the frontal plane. (a) Definitions and Lagrange parameters. (b) Analysis of a configuration with parameters observed in freely moving cockroaches including a hip peg moving during a step in dependency of the speed of locomotion v ($\dot{\theta} = 0$ at touchdown). In the shaded area below a critical speed no stable locomotion is possible. From above: Beyond the critical speed (above shaded region) a bifurcation occurs with a stable and an instable branch. In the stable branch, the absolute eigenvalues of δ and $\dot{\theta}$ are below 1 (solid line and shaded region in upper figure). Correspondingly, after a disturbance frontal rotation $\dot{\theta}$ diminishes and the body orients into the direction of the speed vector ($\delta = 0$). However, the eigenvalues with respect to v and θ are 1 (dash-dotted line in the upper figure) indicating neutral stability. Following a disturbance, the speed vector alters with respect to direction and magnitude. The average heading $\bar{\delta}$ diminishes with speed (lower figure; with kind permission from Schmitt, J., et al., Springer Science+Business Media: *Biological Cybernetics*, Dynamics and stability of legged locomotion in the horizontal plane: A test case using insects, 86, 2002, 345, fig. 3; Schmitt, J., and Holmes, P., *Biol. Cybernetics*, 83, 501–515, 2000). (c, d) Reaction of a model with major springy joints. (c) Model. (d) Above: lateral velocity—time series of a cockroach measured in the body frame and laterally perturbed by an explosive (normal line. After Jindrich, D.L., and Full, R.J., *J. Exp. Biol.*, 205, 2803–2823, 2002.) The dashed line marks for the same stepping pattern the average time course (S.E. < 2 cm/s). Below: the same parameter after a lateral disturbance (24 cm/s) of the model. (With kind permission from Kukillaya et al. Springer Science+Business Media, 2007, fig. 15; p. 391).

disturbances is “spring-like” only under special conditions. Assuming a constant activation for a muscle–skeletal system, it is impossible to generate a steady-state walking or running pattern. The mechanical nonlinear properties of muscles result in a nonconservative system, that is, during eccentric contractions muscles are dissipating energy, whereas during concentric phases they can generate positive work. Therefore, for steady state locomotion-coordinated activation patterns of muscles are necessary. In a situation stable at the principal level of the spring–mass (etc.) world, failure of the muscle–skeletal system to respond in a stabilizing way would make principal stabilization obsolete.

We are very well acquainted with the fact that a muscle is able to hold a load at a certain equilibrium length ($\mathbf{X}_s = (x_s, v_s) = (l_s, 0)$). What properties are essential that this state is stable? A disturbance moves the muscle away from such an equilibrium or fixed point to a state $\mathbf{X}$. If the system returns back to $\mathbf{X}_s$, that is, the distance $\Delta\mathbf{X} = \mathbf{X} - \mathbf{X}_s$ diminishes, then the system is stable. For a system described by a system of, in our case nonlinear, differential equations

$$\dot{\mathbf{X}} = F(\mathbf{X}) \text{ with } \mathbf{X} = \mathbf{X}_s + \Delta\mathbf{X} \tag{5.131}$$

Taylor expansion at the fixed point $\mathbf{X}_s$ results in

$$\Delta\dot{\mathbf{X}} = F'(\mathbf{X}_s)\Delta\mathbf{X} + \ldots = \mathbf{J}\Delta\mathbf{X} + o\left(\|\Delta\mathbf{X}\|\right) \tag{5.132}$$

with J being the Jacobian of the system and $o\left(\|\Delta\mathbf{X}\|\right)$ entails the high-order terms

$$\mathbf{J} = \left(\frac{\partial F_i}{\partial x_k}(\mathbf{X}_s)\right)_{i,k=1,\ldots,n} \tag{5.133}$$

In the case of exponential stability (e.g., linear system) and the absence of degenerated eigenvalues, the response of the system can be described as

$$\Delta\mathbf{X}(t) = \sum_{i=1}^{n} c_i e^{\lambda_i t}\mathbf{v}_i \tag{5.134}$$

with the eigenvalues λ_i of the Jacobian and the eigenvectors v_i. Lyapunov proved that if the real part of the eigenvalues of the Jacobian is all negative, the system is asymptotically stable: the more negative the eigenvalues, the faster the return to the equilibrium.

For our muscle example, in which the load acts along the line of action of the muscle force F_m, the latter must balance the load. In a standard, simplifying approach the muscle force can be described as a product of activation E, the force–length relation f_l, and the force–velocity relation f_v:

$$F_m = E \cdot f_l \cdot f_v \tag{5.135}$$

The system of first-order equations describing the motion can be written as

$$\dot{l}_m = v_m \tag{5.136}$$

$$\dot{v}_m = \frac{1}{m} E \cdot f_l(l_m) \cdot f_v(v_m) - g \tag{5.137}$$

Here the activation is assumed to be constant and the system is autonomous. It should be noted that this approach does not guaranty stability for nonautonomous systems (see below). The Jacobian of this system at the equilibrium length, where the muscle force equals the load, is

$$\mathbf{J} = \begin{bmatrix} 0 & 1 \\ a_1 & a_2 \end{bmatrix} \tag{5.138}$$

with

$$a_1 = \frac{1}{m} E \cdot \frac{\partial f_l}{\partial l_m} \cdot f_v \text{ and } a_2 = \frac{1}{m} E \cdot f_l \cdot \frac{\partial f_v}{\partial v_m}.$$

Stability of the equilibrium requires that the real component of the eigenvalues

$$\lambda_{1,2} = \frac{a_2}{2} \pm \sqrt{\frac{a_2^2}{4} + a_1} \tag{5.139}$$

must be negative. This results in the requirement that both a_1 and a_2 must be negative. As by definition, the terms m, E, f_l, and f_v are positive, the partial derivatives $\frac{\partial f_l}{\partial l_m}$ and $\frac{\partial f_v}{\partial l_m}$ must be negative to obtain stability. Thus, working at the ascending limb of the force–length curve as well as the typical shape of the force–velocity curve with increasing (decreasing) forces during eccentric (concentric) loading are basic requirements for self-stability of the system (Wagner and Blickhan 1999). Without any change of activation (e.g., assumed to be constant) the disturbed muscle should return to its equilibrium. We see that basic muscle properties are not just a biologic nuisance but that they have important positive consequences on the system level. It is clear that taking the natural environment of a muscle embedded into a muscle–skeletal system into account, some corrections or enhancements are necessary.

From Equation (5.139), we realize that the eigenvalues $\lambda_{1,2} = \mu + i \cdot \nu \in C$ are complex for $a_2^2 < 4a_1$. For complex eigenvalues, the vector of the solution in the phase plane spirals either away or toward the fixed point in dependence of the sign of the real part μ (Figure 5.27e). We realize that imaginary terms appear for steep slopes of the force–length curve combined with shallow slopes of the force–length curve. This is still the case if we more realistically include the leverage of a simple muscle–skeletal system. For such a system, again, it depends on the slope of the force–velocity curve now translated by the gear ratio (including morphologic details such as fiber length) whether oscillations occur or whether the system approaches the equilibrium critically damped (Rode et al. 2008). Within muscle–skeletal systems, slow muscles with force–velocity curves with high curvature seem to be adapted to posture where critical damping in the presence of disturbances is essential, whereas fast muscles with low curvature combined with short fibers are useful in a functional context where oscillations are desired. The judgment of such an adaptation requires the consideration of the gear ratio defined by morphology.

Let us take a second look at the latter to make the issue of gear ratio more transparent (Section 5.7). For a muscle embedded in a skeletal system, external loads F_L and velocities v_L are transformed by the leverage of the system (gear ratio). Correspondingly, the task of a muscle within a muscle–skeletal system must be considered in the context of this leverage. The effective mechanical advantage *EMA*, the ratio of the muscle lever r_m, and the lever of the load r_L (Biewener 2005) define the gear ratio $G = EMA$ between the muscle and the environment:

$$G = \frac{r_m}{r_L} \quad (5.140)$$

A higher gear ratio reduces muscle force for given external load

$$F_L = G \cdot F_m \quad (5.141)$$

and it increases

$$v_L = \frac{1}{G} \cdot v_M \quad (5.142)$$

muscle velocity for given external velocity. It should be noted that this gear ratio is also influenced by muscle geometry.

This modification affects stability analysis. The structure of the equations remains the same, but the coefficients of the Jacobian change. We consider stability of the position of the load x_L. The muscle force must now be rewritten considering the constant gear ratio

$$F_m = E \cdot f_{x_L} \cdot f_{v_L} \text{ and}$$

$$a_1 = \frac{1}{m} G \cdot E \cdot \frac{\partial f_{x_L}}{\partial x_L} \cdot f_{v_L}, \text{ and } a_2 = \frac{1}{m} G \cdot E \cdot f_l \cdot \frac{\partial f_{v_L}}{\partial v_L} \quad (5.143)$$

The gear ratio affects both the influence of the force–length curve as well as the force–velocity curve, both of which are essential for stabilization by muscles. A muscle with the task to hold a load for a given activation must generate as a goal a static equilibrium in which the moment generated by the weight or load $F_L = mg$ must be balanced by muscle force. For $v_m = 0$ the normalized force–velocity curve $f_v = 1$. At its optimum length, the muscle is able to generate its maximum (controllable) force. Above that point, no equilibrium is possible, whereas equilibria can be obtained below at any site of the force–length curve, that is, it is required that (Figure 5.27).

$$F_L < G \cdot F_m \text{ or } F_{rel} = \frac{G \cdot F_m}{F_L} > 1 \quad (5.144)$$

Here we assume the gear ratio is constant, which is reasonable for the example outlined above. For example, it is valid when considering the stability of the lower arm while a waitress carries a vine on her tablet (Wagner et al. 2007; Giesl and Wagner 2008). In nature, geometry is frequently rather nasty. Think about the gear ratio of our extensor muscles (Musculus quadriceps) supporting us during a knee bend. Here the gear ratio changes a lot during knee extension, that is, it now becomes strongly dependent on x_L.

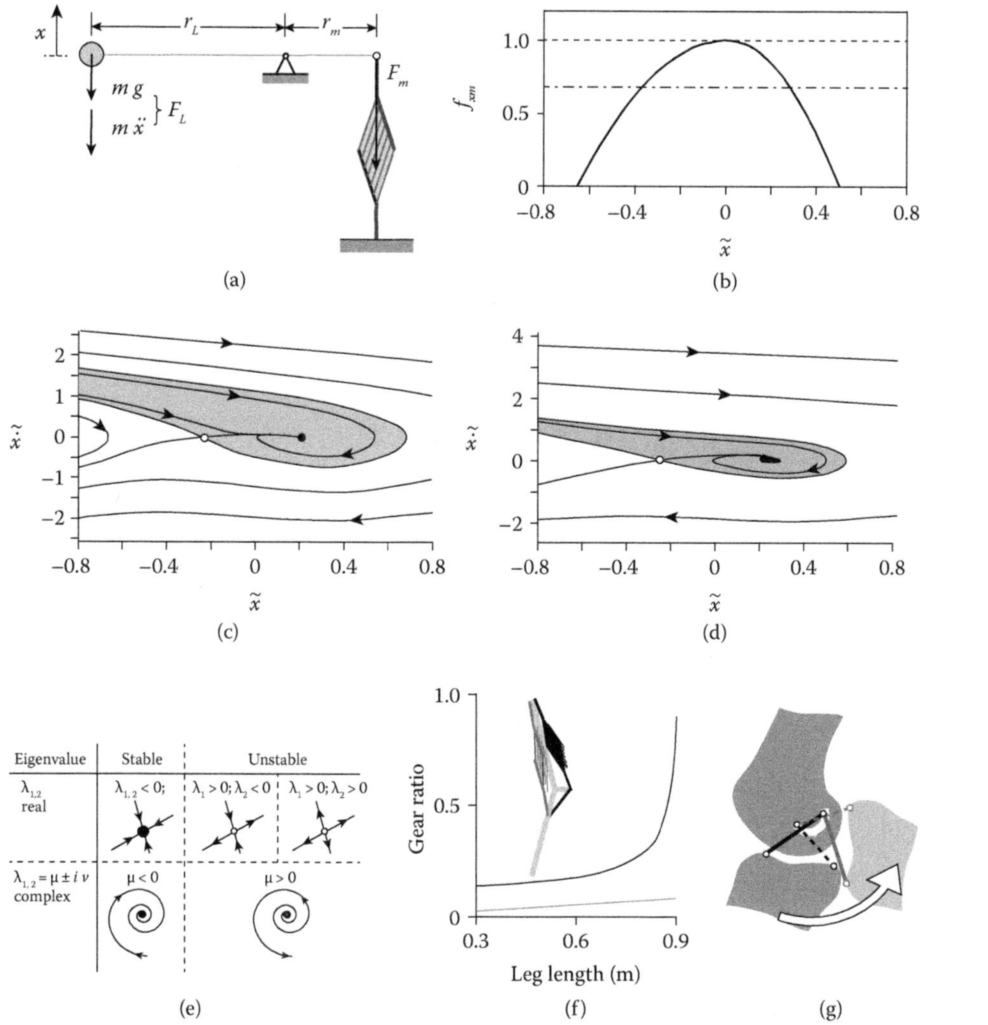

Figure 5.27. Stability of a simple muscle–skeletal system. (a) Muscle lifting a load via a lever defining a gear ratio. (b) Force–length curve. A basic requirement is that the static load does not exceed the length dependent capacity of the muscle. The dash-dotted line marks the point of operation for the simulations below. (c and d) Phase diagrams for muscles with different intrinsic velocities. With two fixed points: open circle: instable saddle point; closed circle: stable fixed point which is damped for the slow muscle (c, $v^*_{max} = 1.7$; *curvature* = 0.25) and oscillates for the fast muscle (d, $v^*_{max} = 4.1$; *curvature* = 0.4). Both static stabilization and dynamic oscillation may be explored in a proper design of the muscle–skeletal system. Note: A negative $\tilde{x}$ implies a long muscle, that is, the fixed saddle point is located on the descending slope of the muscle whereas the stable fixed point is located on the ascending slope of the muscle (cf. b). Shaded: Area of attraction in which the system stabilizes. $\tilde{x} = xG/2w_1$, where w_1 represents the width of the ascending limb of the force–length curve; $\tilde{\dot{x}} = \dot{x}\sqrt{2w_1 g/G}$. (Rode, pers. comm.; Rode, C., et al., Effect of muscle properties on postural stability of simple musculoskeletal system, *8th World Congress on Computational Mechanics (WCCM8)—5th European Congress on Computational Methods in Applied Sciences and Engineering (ECCOMAS 2008)*, Venice, Italy, 2008.) (e) Depending on the eigenvalues the fixed points are stable or unstable with different topologies of the environment within the phase space (trough, saddlepoint, peak, spiral). (f) Nasty geometries: The gear ratio of the knee extensors (dark muscle) strongly changes with leg extension (Sust, pers. comm.). Keeping the knee bent as most animals (and most robots) helps to avoid this critical situation. (g) The four-bar mechanism consisting of the cruciate ligaments (bold and stippled lines) and adjacent bones (femur and tibia, shaded) combined with a suitable surface of the bone has the effect that the center of rotation (crossing of the ligaments) wanders dorsally (white line) with leg extension (large arrow). This reduces the slope of the gear ratio and helps the extensors to stabilization the knee. (After Wagner, H., and Blickhan, R., *Biol. Cybernetics*, 89, 71–79, 2003; Wank, pers. comm.)

The function has actually a singularity once the leg is extended (Figure 5.27f).

$$F_m = E \cdot f_m(x_L, v_L) \text{ and } G(x_L)$$

$$a_1 = \frac{1}{m} E \cdot \frac{\partial\left(G(x_L) \cdot f_m(x_L, v_L)\right)}{\partial x_L} \text{ and}$$
$$a_2 = \frac{1}{m} E \cdot G(x_L) \cdot \frac{\partial f_m(x_L, v_L)}{\partial v_L} \quad (5.145)$$

A steep force–length curve becomes now necessary to compensate for the detrimental effects of the force–length curve (Wagner and Blickhan 1999). Because of this influence, the complicated geometry of the human joint with its surface determined by the four-bar linkage system, including the cruciate ligaments, results in wandering joint axes softening the singular gear ratio and thus facilitating stabilization of the almost extended knee joint (Figure 5.27g). Measures such as an antagonistic muscle (Wagner and Blickhan 2003) and considerable parallel elasticity are measures to support stability. However, such measures do not come free of metabolic cost and restrain motor control.

In all these considerations, we assumed activation is constant, that is, the differential equations were not autonomous. When considering an equation with explicit time dependency $E = E(t)$, the condition of negative eigenvalues of the Jacobian does not guaranty stability (see example in Markus and Yamabe 1960), and stability may be observed for trajectories in which positive eigenvalues occur. To investigate the stability of certain solutions $x_s(t)$ with $\dot{x}_s(t) = F(x_s, t)$, it is now necessary to investigate whether solutions in the neighborhood $x(t) = x_s(t) + \Delta X(t)$ are stable. The Floquet multipliers are measures to estimate stability for periodic functions in which $x_s(t) = x_s(t + T)$, as in our case, enforced by a periodic activation function (description textbooks on mechanics entailing nonlinear dynamics, e.g., Greiner 2002). The equation

$$\Delta\dot{\mathbf{X}}(t) = \mathbf{F}(\mathbf{x}_s + \Delta\mathbf{X}, t) - \mathbf{F}(\mathbf{x}_s, t) = \mathbf{G}(\Delta\mathbf{X}, t) \quad (5.146)$$

can be linearized using Taylor expansion in the environment of $x_s(t)$

$$\Delta\dot{\mathbf{X}}(t) = \mathbf{J}(t)\,\Delta\mathbf{X}(t) \quad (5.147)$$

with

$$\mathbf{J}(t) = \left.\frac{\partial \mathbf{G}}{\partial \Delta\mathbf{X}}\right|_{\Delta\mathbf{X}=0} = \left.\frac{\partial \mathbf{F}}{\partial \mathbf{x}}\right|_{\mathbf{x}_s(t)} \quad (5.148)$$

and $J(t) = J(t + T)$. For $x_s(t)$ being periodic in time J is periodic at the point of expansion $x_s(t)$.

The general solutions of a system of linear differential equations can be expressed within a fundamental system of linear independent basis solutions. A fundamental system consists of eigenvectors of the differential equation that, in turn, are solutions of the system of the following system of equations:

$$(\mathbf{J} - \lambda_i \mathbf{I})\mathbf{x} = 0 \quad (5.149)$$

with λ_i being the eigenvalues of the Jacobian.

In our simple 2D case, we obtain the two eigenvectors $v_1(t)$, $v_2(t)$ that we may choose to be for $t = 0$

$$\mathbf{v}_1(0) = \begin{pmatrix} 1 \\ 0 \end{pmatrix} \text{ and } \mathbf{v}_2(0) = \begin{pmatrix} 0 \\ 1 \end{pmatrix} \quad (5.150)$$

The superposition of the general solution can be written as

$$\Delta\mathbf{X}(t) = \mathbf{v}(t)\mathbf{c} \quad (5.151)$$

with the fundamental matrix $\mathbf{v} = (\mathbf{v}_1 \quad \mathbf{v}_2)$, the coefficients $\mathbf{c} = \begin{pmatrix} c_1 \\ c_2 \end{pmatrix}$, and $v(0) = \mathbf{I}$.

The solution $\Delta X(t)$ will, in general, not be periodical. However, due to the periodicity of the Jacobian, $\Delta X(t + T)$ is a solution of Equation 5.147, and the fundamental matrix $v(t + T)$ must emerge from an elementary matrix transformation

$$\mathbf{v}(t+T) = \mathbf{v}(t)\mathbf{C}$$
$$\mathbf{C} = \mathbf{v}(T) \quad (5.152)$$

From $\dot{\mathbf{v}}(t) = \mathbf{J}(t)\,\mathbf{v}(t)$ and $v(0) = 0$

$$\int_0^T J(t)v(t)dt = \mathbf{C} - \mathbf{I} \tag{5.153}$$

that is, matrix C can be calculated by integration.

The development of the solutions for large times can be judged by repeated application. Using the eigenvalues ζ_i of the fundamental matrix and its description in terms of its eigenvectors **u**,

$$v(T)\mathbf{u} = \zeta_i(T)\mathbf{u} \tag{5.154}$$

the repeated application to future period yields

$$\zeta_i(nT) = v(nT)\mathbf{u} = v(T)^n\mathbf{u} = \zeta_i^n(T)\mathbf{u}. \tag{5.155}$$

For $n \to \infty$

$$\zeta_i(T) = e^{\sigma_i T} \tag{5.156}$$

with the Floquet multipliers ζ_i and the Floquet exponents σ_i. The system is asymptotic stable if the absolute value of all multipliers $|\zeta_i| < 1$ or the real part of all exponent $\mathbb{R}(\sigma_i) < 0$.

For the knee bends described above, a sinusoidal reference trajectory x_r, for example, can be prescribed. The now necessary time-dependent activation can be calculated by reassembling Equations 5.136 and 5.137. This specifies the system of equation for the reference trajectory (Equation 5.146) and the Jacobian **J**. This is the basis for calculation of the fundamental matrix. The eigenvalues of the fundamental matrix as a function of the period v(T) are the Floquet multipliers. In fact, such oscillations with muscle parameters obtained previously in a muscle diagnostic session yield self-stable movements (i.e., with negative real values of the Floquet exponents; Wagner et al. 2007; Eriten and Dankowicz 2009). They can be observed even for cases where the eigenvalues of the Jacobian are positive and critically depend on geometry. Antagonistic arrangements with cocontraction support stabilization as well as—not surprisingly—parallel elasticity.

Unfortunately, so far the calculation of the Floquet multipliers depends on the selected reference trajectory and the numerical routines. General statements based on analytical calculations are not possible. In the calculations, it is necessary to avoid discontinuities (e.g., in the muscle properties while crossing from concentric to eccentric contractions).

We have seen how muscle properties such as the force–velocity curve can contribute to stabilization, and how this stabilization is influenced by gearing. In general, the investigation of stability is still rather cumbersome, especially because in most cases muscle activity has complex time dependencies and muscle–skeletal systems have high degrees of freedom. Nevertheless, in robotics the question whether a drive inherently contributes to stability may be relevant. We have also seen that the environment in which solutions are attracted toward the fixed point is limited. Depending on the task, small or large areas of attraction may be desirable. Unfortunately, besides using numerical sweeps, identifying basins of attraction is still rather cumbersome.

5.6.6 Basins of Attraction

So far, we only asked the question whether a system may be stable for a given point of operation. We did not specify the basin of attraction, that is, the set of initial conditions for which the system approaches an attractor. It does not help very much in terms of stability if this basin of attraction is diminutive. In this case, small disturbances would drive the system out of the region where attraction is achieved. In contrast, the large and deep basins of stability may not be desirable because such a strategy unloads the neural system with respect to the control of the specific posture or movement but also makes alterations rather difficult: it prevents maneuverability. Think about the difference between a jet fighter and a jumbo jet that indicates a strong relation between mechanics facilitating either fast maneuvers or stable motions, respectively.

For some cases, we have an idea about the range from numerical studies. We mentioned the strategy of retraction for the spring-mass model that enhances the range of attraction. This method can be complemented by intelligent feed-forward strategies of stiffness adjustment. This already shows that the intelligent design found in animals is able to adapt the basin of attraction dependent on demands. In muscle–skeletal systems, cocontraction of antagonists represents another method to enlarge the basin of attraction. We all know that we use energetically demanding cocontraction when we learn new movements. A fit, novice tennis player will be out of breath after some minutes, whereas the elderly teacher plays the game with greatest of ease. Cocontractions enhance the basin of attraction and thus prevent injuries; for example, it may prevent falling. However, it is not only expensive but also makes it difficult to move smoothly and makes it impossible to reach extreme movement goals such as a long jump (Section 5.7.7) or to play a fast sequence with the piano.

This relevant area of research is in early stages. For simple muscle–skeletal arrangements such as the bisegmental system of the human elbow, estimates of the basin of attraction are available (Figure 5.28). These calculations rest on the availability of a Lyapunov function V in the open neighborhood of a fixed point. Such a positive definite potential marks a well with a minimum (0) at the fixed point, that is, the derivatives along the solutions of the system must be negative. The difficulty is to find an appropriate Lyapunov function for a given system. By using linearization of the solution in the neighborhood of the fixed point local functions can be constructed. These can be extended by interpolating local solutions with radial basis functions (Giesl and Wagner 2006; Giesl and Wagner 2007; Giesl 2008).

Investigation of the contributions of mechanical and neuronal factors to stability and maneuverability largely depends on our ability to model the system and to predict alterations with changing parameters. They will largely enhance our understanding of the organization and operation muscle–skeletal systems.

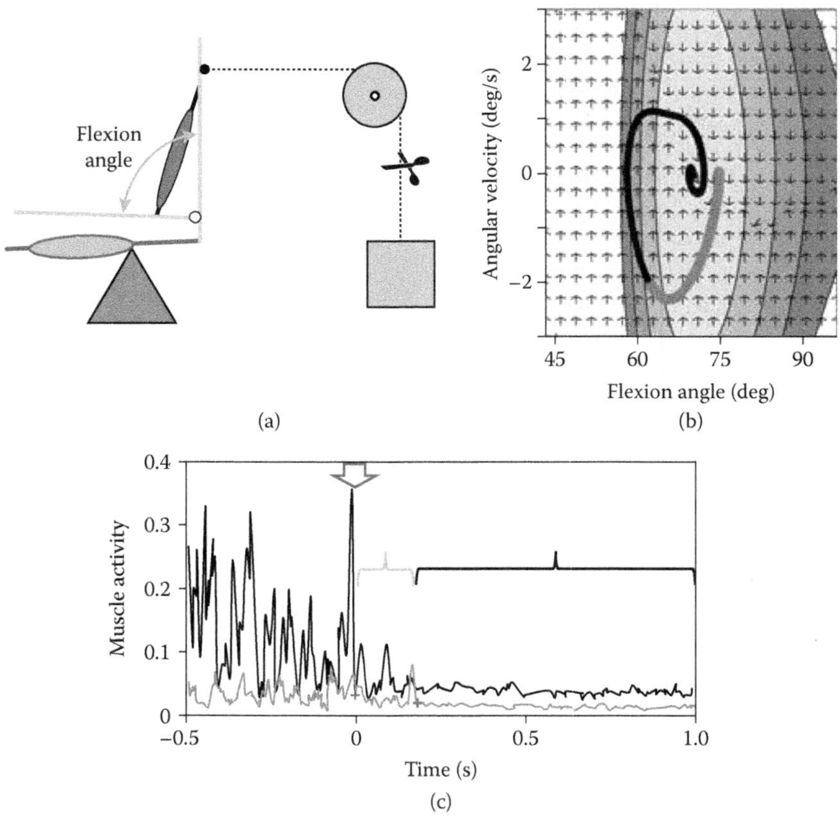

Figure 5.28. Quick release of a loaded elbow and its basin of attraction. (a) The arm holds a load that is suddenly released. The subject has the task to fix the position of the arm. (b) Phase graph (line) with underlying Lyapunov function (gray shading) describing a basin of attraction. In this case the oscillations of the arm occur within the basin. (c) Activity of the M. brachialis (black, cf. a) and its antagonist the M. triceps (gray). After release (arrow above) it takes less than 0.2 s (gray bracket and gray line in the phase plot (b) until the activity of both muscles ceases (black bracket and black line (b). Correspondingly, equilibrium is achieved in the second phase without altering muscle activity. (With kind permission from Giesl, P., and Wagner, H., Springer Science+Business Media: *Journal of Mathematical Biology*, Lyapunov function and the basin of attraction for a single-joint muscle–skeletal model, 54, 2007, 463, b, c: fig. 5.)

So far, we have focused on stationary situations such as walking at a given speed or holding a tray. The fitness of an animal also depends on its abilities in nonstationary situations. Here, stable performance may also be an issue. However, in the following, we ignore this issue and use jumping as an example to investigate how the goal of jumping height or jumping distance can be maximized. We take the chance to address basic biomechanical questions such as the interaction between elastic tissues and muscle, the use of biarticular muscles, the issue of accelerated distributed masses, and procedures of forward simulation.

5.7 JUMPING: MUSCLES AND CATAPULTS

5.7.1 Acceleration and Take-Off Velocity

We have already termed running "saltatory locomotion" because it consists of a (rather long) series of jumps. But jumping also represents an important mode of locomotion in legged animals. Think about the attack of a tiger or the flight jump of a frog or a flea. The jumper quickly overcomes distances and hurdles. A jump represents a single event in which the legged systems—but also fish on dry land—try to generate maximum take-off speed. It represents a "simple" movement with a clearly defined goal and as such is suited to learn about the organization and possibilities of legged systems.

Let us first approach the problem from a rather general perspective. Depending on the initial condition, we might consider jumps with a run-up where the latter provides the jumper with kinetic energy that must be redirected to obtain long distances, large heights, or both or those without a run-up in which the goal must be achieved starting from zero speed. In the latter case, all energy must be developed by the driving tissues and elements available during take-off. Correspondingly, the power that these tissues and elements are able to produce is crucial for the jumper. We return to this below and first address general issues.

The goal is quite simple, and the treatment is further simplified by limiting the consideration to vertical jumps. Starting at $t_0 = 0$ from zero speed $v_0 = 0$ the jumper wants to attain maximum take-off speed v_{TO} (and yes, a high take-off position). We know the jumper should maximize impulse p, that is,

$$p = mv_{TO} = \int_{t_0=0}^{t_{TO}} F(t)\,dt \qquad (5.157)$$

where m is the jumpers mass, g is the gravitational acceleration, and F is the net force with the net force F (F_R−mg for vertical jumps, F_R ground reaction force). The indices 0 and TO indicate initiation and take-off, respectively. Maximizing force and time lead to a high take-off velocity. The force is limited due to the actuator. It seems to be a good idea to enhance the jump's duration, that is, $t_{TO} - t_0$. However, this time is limited and related to speed and leg length. Note that take-off can take place before leg extension, but the jumper will certainly disconnect when leg length is reached.

Let us assume a jumper generates an average acceleration of

$$\frac{F_R - mg}{m} = \left(\frac{F_R}{mg} - 1\right)g = \lambda\, g \qquad (5.158)$$

Via the relation between potential and kinetic energy, this acceleration determines take-off speed and the height h of the ballistic phase

$$\lambda\, g\Delta y = \frac{v_{TO}^{\;2}}{2} = g\, h \qquad (5.159)$$

This results in

$$\lambda = \frac{h}{\Delta y} \qquad (5.160)$$

For $\lambda = 2$ equivalent to an average reaction force of three times body weight, the flight height is double the acceleration distance. A human jumper starting with bent knees and extending the leg by 0.5 m to a final length of 0.9 m would be able to propel his center of mass across 1.8 m. Obviously, the time to take-off increases with longer legs and decreases with increasing acceleration, because

$$t_{TO} = \sqrt{\frac{2\Delta y}{\lambda g}} \qquad (5.161)$$

For our human jumper, this yields convenient 0.23 s. For a locust jumping 10 times its leg length of say 5 cm, this time reduces to 32 ms. A tiny flea extends its legs by about 0.5 mm to reach a jumping height of 50 mm and has a mere 1 ms to produce its jump. We turn to this below.

Now, it is important that the same impulse generating the same take-off speed can make different use of leg length depending on the time distribution of the force. For different distributions of accelerations with the same average, the time course of the velocity differs, leading to different displacements as the integral of the velocity differs.

To obtain maximum speed at take-off under such a condition, it is better to accelerate late (Figure 5.29). The time of take-off (think of flight reactions) is shorter for early accelerations (biomechanical principles, Hochmuth 1974).

After obtaining a very basic understanding at which lines optimization takes place, starting with insect catapults, we turn to biological examples illustrating how elements of the motion system can be used to foster the goal of spectacular jumps.

5.7.2 Catapults: Delivering Energy Fast

We would easily go along with the notion that frogs are specialized jumpers (see below). However, the most spectacular jumpers compared to size seem to be insects. We already mentioned that for these animals, acceleration distances and times are rather limited. The peak forces estimated are about 18 times the body weight for the locust (Bennet-Clark 1975) and as calculated from accelerations the flea mentioned exerts average forces of up to about 130 times the body weight (Bennet-Clark and Lucey 1967; Bennet-Clark 1975). So with our estimates above of $\lambda = 10$ for the locust and $\lambda = 100$ for the flea, we are not far off reality. From this, the specific power produced by the locust (index *loc*) $P_{loc} = \sqrt{\frac{\lambda_{loc}^{3}}{\lambda_{hum}^{3}} \frac{\Delta y_{loc}}{\Delta y_{hum}}} P_{hum}$ is about 1.4 times the specific power of a human jumper (index *hum*). In a flea, the factor is about 4.5. The gear ratio does not alter power. Instead of investing into substantially more muscle power, which would result in increased cost, the animals invest in catapults (Figure 5.30). Like an archer, they slowly store energy in elastic tissues in a preparation phase and release it to provide energy for the jump. Thus, the movement is not any more limited by the intrinsic velocity of the muscle. If you are not convinced that the trick helps try to manually throw an arrow. Conveniently—but not always—a catch mechanism releases the antagonist from holding against it, and a release helps to unfold the power. Using such a strategy, you must be certain to be out of reach of a predator after that jump because you will need time to recharge the catapult. It should be noted that storage might be distributed in the skeleton and the tendon, and other strategies can provide jumps of similar magnitude (e.g., Burrows and Wolf 2002).

Without addressing scaling, we realize that being small does require and allow special solutions. Also, not discussed so far is the issue of landing. Once being airborne, a suitable take-off must be mirrored by a suitable landing. This is no problem for a flea, but for a dog it is. The length for deceleration is limited, and your system must be able to bear the unavoidable deceleration amplitudes. So, we expect some shift while we proceed in the following toward vertebrates. This shift is related to the delicate interaction between elastic tissues storing energy and muscles providing the power.

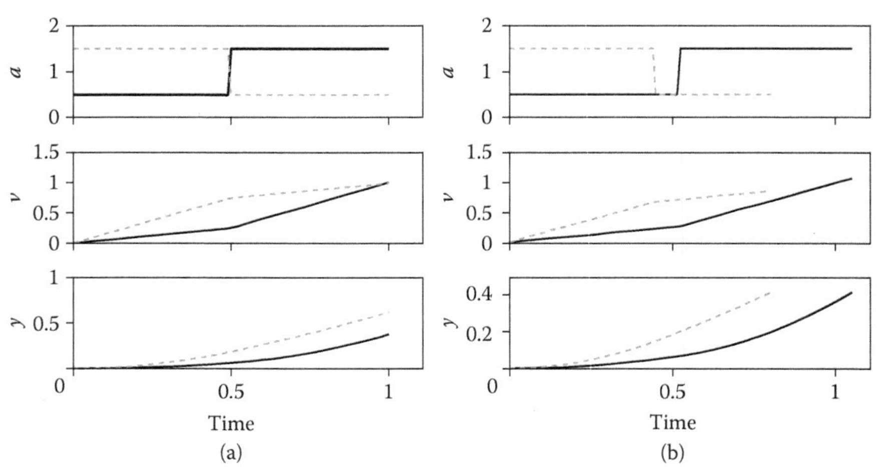

Figure 5.29. Influence of the time course of acceleration period on the final velocity and required displacement. (a) For the same impulse, that is, final velocity a late acceleration needs less distance. (b) Using the same final distance, a late acceleration can be used to achieve a larger final speed. y: displacement; v: velocity; a: acceleration.

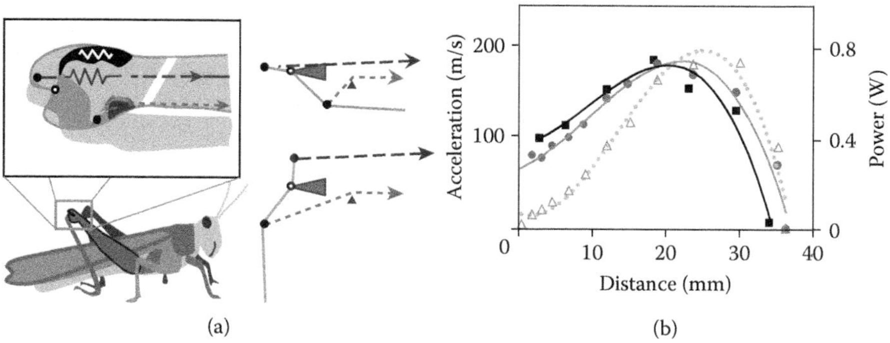

Figure 5.30. Jump of a locust. (a) Left: The third legs of locusts are not only long and powerful but also entail remarkable specializations. In the femur tibia joint (enlarged), the cuticular semilunar process (dorsal in black with white spring) allows for energy storage, functioning like a bending beam. Furthermore, the apodeme (black spring) connecting the powerful extensor to the tibia is used as a storage element. The less powerful flexor (ventral dotted line of action) connects via an apodeme to the tibia, possessing a latch that clicks into a hump (dark gray) on the femur. An accessory muscle (stippled) releases the latch. Right: Besides the latch joint geometry provides the flexor with large lever arms. (b) Model calculation (squares, black line) assuming isometric muscle action and the parallel release of strain energy in the apodeme and the semilunar process (quantified material properties) and accounting for the geometry result in rather similar patterns of acceleration of the center of mass as observed experimentally (filled circles, gray dotted line; from high-speed filming). The enhanced performance at the end of the jump might be due to contraction of the extensor. Triangles and dotted line: experimental power. (After Bennet-Clark, H.C., *J. Exp. Biol.*, 63, 53–83, 1975, a: fig. 1, p. 58, b: fig. 16; p. 79; Heitler, W.J., *J. Comp. Physiol. A Neuroethol. Sens. Neural Behav. Physiol.*, 89, 93–104, 1974; Scott, J., *Adv. Physiol. Educ.*, 29, 21–26.)

5.7.3 Matching and Amplifying Muscle Tendon Properties

Taking the caveats of limited leg length into account, jumping depends on the ability of the actuator to develop power. If the extension during take-off is not decoupled (see above), the muscle tendon complex must be able to produce the necessary forces and velocities. Studies on frog jumps have indeed demonstrated that the properties of the M. semimembranosus are well adapted to its task (Lutz and Rome 1994). The comparison between muscle experiment and kinematics revealed that the muscle operates in the plateau region of its force–length curve and close to its power maximum (Section 5.6.5 and see below). Such a remarkable matching requires the coadaptation of strength related to body mass, shortening velocity, gearing, and acceleration length. Lutz and Rome (1994) based the crucial estimates of muscle length and velocity on external kinematics. From the observed joint angles, the lengths of the investigated muscle and especially the lengths of its sarcomeres were estimated morphologically from specimens frozen under different angles.

A later study (Figure 5.31) used sonomicrometry to directly estimate fiber lengths of the M. plantaris. Sonomicrometry uses the traveling time of an ultrasonic pulse in the tissue to determine tissue length (Roberts and Marsh 2003). This direct measurement has the advantage to be able to distinguish between the length of the muscle tendon complex and the length of the active element. Hence, assuming a lumped parameter model the contribution of the series elastic element to power development can be identified. As mentioned, the geometrical gearing ratio defined as $G = \frac{r_m}{r_L}$ (Section 5.6.5) translates the velocity of the load directly to whole muscle speed. However, a compliant tendon introduces an additional degree of freedom, and the geometrical gearing is not sufficient to estimate the velocity of the muscle fibers. The measurements show a continuous increase in the velocity of the center of mass. In contrast, the muscle fibers show two maxima, with a higher value at the beginning of the jump where the animal reaches only about 10% of its final speed. A muscle-lever-mass model including a series elastic element and a variable gearing ratio has been used for interpretation. The gear ratio continuously decreases during a jump (similar to when a car driver shifts gears with increasing speed). Furthermore, in the initial phase of the jump the muscle shortens almost by half of

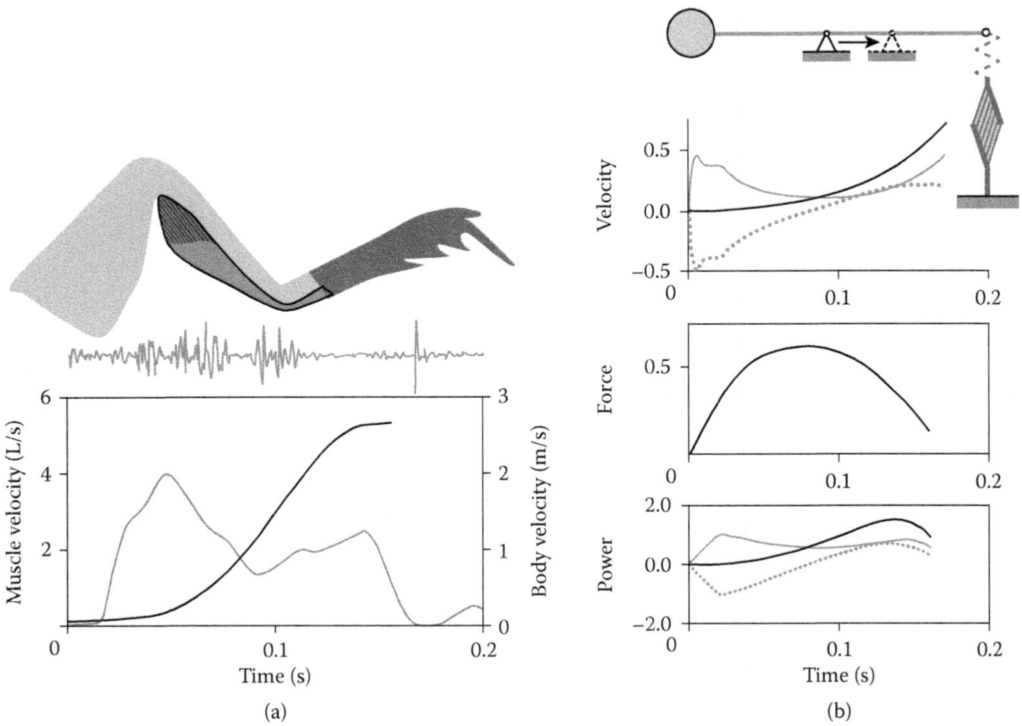

Figure 5.31. Influence of series elasticity on the function of the frog plantaris during jumping. (a) Above: M. plantaris in a frog's leg. The length of the fibers were detected using sonomicrometry. Underneath: Activity of a selected muscle during a jump. Below: Whereas the velocity of the center of mass (black line) increases continuously, the muscle fiber velocity (gray) displays two peaks. (b) A simple model entailing a mass, a variable gearing and a muscle (dark hatched)-tendon (dotted) units demonstrates the effect of the series elastic element, the latter including the contribution of the aponeurosis. Whereas the velocity of the total unit (black line) is increasing continuously, the muscle fiber (dark gray line) first lengthens the tendon and at the end accelerates the mass (dotted line). Below: The energy stored in the tendon is released in the second half of the jump and contributes significantly to the total power of the unit (black line). The muscle model includes a force–velocity and a force–length characteristic. (After Azizi, E., and Roberts, T.J., *Proc. Roy. Soc. B Biol. Sci.*, 277, 1523–1530, 2010; Roberts, T.J., and Marsh, R.L., *J. Exp. Biol.*, 206, 2567–2580, 2003, a: fig. 4a, p. 2575, b: fig. 7d, p. 2577.)

its shortening distance against the lengthening series elastic element, while the animal hardly moves. The combined action of both mechanisms has the effect of shifting the force and power (the local product of reaction force and velocity) into the later phase of the jump. The energy stored in the tendon in the first half of the propulsion is released in the second half, thereby substantially supporting the jump. The power provided by this arrangement is clearly above the power that can be delivered by the muscle. More significantly, the change in gearing supports the mechanism of storing energy in the tendon and beneficial generation of acceleration in the later phase of the jump (Figure 5.1). Because part of the shortening is covered by shortening of the tendon, the required contraction speed of the fiber is lower, thereby enhancing the power development in the late stages of the jump.

Above the action of the series elastic element, the muscle starts shortening from the descending slope of the force–length curve (Azizi and Roberts 2010). This is normally avoided in distal muscles involved in reactive movements, where muscle instability would be enforced (cf. Section 5.6.5). Here in purely concentric operating environment, it helps to shape the time dependency of the reaction force, shifting higher accelerations into late phases of the jump (Figure 5.29). Rules must be judged before the background of the task.

We have seen that optimizing the jump requires a delicate adjustment of properties and movement strategies. The more compliant constitutes will be introduced in robotics, the more it will be worthwhile to have a glimpse at biology to see how evolution solved the problem. We gained basic insight by using simple models. However, as soon as systems become more complex we need instruments to deal with the complexity. In general, engineers are well aware of procedures available for complex dynamic simulations. Applying them to complex muscle skeletal systems still represents a formidable task. Considering the vast degrees of freedom of biological but also of technical legged systems to learn, for example, how to optimize the use of given resources can turn out to be difficult. Jumping offers the advantage of an obvious optimization goal.

5.7.4 Optimizing Complex Muscle–Skeletal Systems

Due to its clear-cut general goal and common occurrence in sports diagnostics, the vertical human jump has been frequently used to develop and probe complex muscle–skeletal simulations. Most scientists educated in physics and engineering are attracted by the possibility to use a full-blown muscle–skeletal system implemented in a computer to mimic the performance of the natural example. For many, this has been a reasonable goal in itself. Despite increasingly available commercial tools and the acceleration of computer power, this is still not a trivial enterprise.

To arouse curiosity, let us first just list some outcomes derived from some of those simulations. We start with the simplest simulation, where the properties of knee- (and ankle-) extensors are lumped into the joint torques. Alexander (1995) simulated jumpers modeling a two- (or three-) segmented leg topped by a body. Due to the symmetric action of the legs and the planar modeling, posture was not a problem. For the two-segmented leg, the system has a single degree of freedom but five segments with mass and inertia. For the three-segmented leg, the indetermination is resolved by prescribing similar angles to both joints. The muscle is assumed fully activated and the torque developed depends on force–length and force–velocity properties and on the moment arm with respect to the knee. The tendon, the series elastic element, is modeled as a linear spring. Alexander (1995) also explored the use of catapults (see above). Based on these ingredients, Alexander (1995) could show that a counter movement jumper reaches higher jumps than a squat jumper—if he implements series elasticity. Nevertheless, only a fraction of the stored energy becomes effective in increasing jump height. It seems that the main effect is a different point of operation of the muscle on the force–velocity curve, that is, the shortening speed of the muscle fiber is reduced, enhancing force development. Reducing leg mass also helps to increase jumping height. During contact, the jumper has to accelerate his legs (especially foot and shank) less than the trunk. But the transmitted impulse must push the whole system, including the legs. This effect of the "Sprungmasse" was been first worked out by Hoerler (1974). He assumed the jumper consisted of two masses: one mass at the ground and the other mass driven by the actuators. Moreover, some work is wasted for rotating the legs. Because the legs do not swing, mass distribution does not matter. Jumping animals do not need light feet. Long legs certainly help (see above). Three segments help, too; they improve net mechanical advantage (gearing, Section 5.4).

This is a rather rich pool of insights obtained by simplifying and using comparison. The more detailed and thus accurate models did confirm (or predict) the relevance of series elastic elements (Anderson and Pandy 1993; van Soest and Bobbert, 1993; Bobbert 2001). They also confirmed the interpretation that the difference in performance in a countermovement jump is due to an improved operation of human muscles. It should be noted that at this point, none of the models considered the effect of stretch enhancement, which would require inclusion of the history of movement (e.g., Rode et al. 2009). Biarticular muscles seem to enhance jumping height (Pandy and Zajac 1991; Pereira et al. 2008). However, the significance of their biarticular action strongly depends on details in geometry (van Soest et al. 1993; Jacobs et al. 1996). We will turn to this issue below. With respect to mass and inertia, the corresponding forces become important toward the end of the jump. Ignoring inertia makes a difference (Pandy and Zajac 1991). A rather pioneering find was that the control of joint moments is much more difficult than the control of suitable muscle activation (van Soest and Bobbert 1993). This finding has partly triggered research on the impact of muscle properties on stability of movement (Section 5.6.5). Let us now have a closer look on detailed muscle–skeletal models such as pioneered by Hatze (1976). The first task is certainly to establish the equations of motion of the linkage system. The matrix algorithm used to describe the action of a double pendulum (see Equations 5.51 and 5.52) can be transferred to a four-segmented inverted pendulum, the model of the jumper consisting of a trunk, and a pair of thighs, shanks, and feet (e.g., after Pandy et al. 1990).

$$A(\varphi)\ddot{\varphi} = \mathbf{B}(\varphi)\dot{\varphi}^2 + \mathbf{C}(\varphi) + \mathbf{D}\,\mathbf{G}(\varphi)\,\mathbf{F}^T(\varphi,\dot{\varphi}) + \mathbf{T}(\varphi,\dot{\varphi}) \quad (5.162)$$

Four angles defining the vector $\boldsymbol{\varphi}$ are sufficient to describe the motion. The system mass-matrix $\boldsymbol{A}$ is now a 4 × 4 matrix. **B** is now a 4 × 1 vector entailing the Coriolis and centrifugal terms; vector **C** contains the gravitational terms; and **T** contains the external forces such as a torsion-spring-damper, taking care for interaction of the foot with the floor or the actually measured ground reaction force. In Pandy et al. (1990), the latter is implemented via a constraint enforcing the correct motion of the system's center of mass. The addend taking care for the inner (muscle-) forces is a little more complicated. $\mathbf{F}^T$ is a vector essentially also depending on the angles and their derivatives describing the muscle forces (see below). ***G*** defines the joint moments generated by the muscle, and **D** translates the joint moments into moments with respect to the segments. The explicit formulations of the matrices are given in Pandy et al. (1990). More recent algorithms (Spagele et al. 1999a, 1999b; Stelzer and von Stryk 2006) use the Jacobian matrix to describe translation and rotation terms or alternatively, numerically much more efficient factorizations of the mass matrix and its inverse (see below).

We treat here the forward simulation, that is, the calculation of movement from forces and muscle activation. It should be noted that by neglecting the muscle term, we can use the same rearranged equations to calculate joint torques and via optimization (see below) muscle forces if we know the system properties and the motion (inverse dynamics). Combinations of these approaches are possible, such as prescribing certain angles in a forward dynamics model as long as the designer avoids conflicting conditions.

The next step involves the formulation of the muscle properties. The notation used here deviates from the formulation used by Pandy et al. (1990). For each muscle, they are in general given as a product of the force displacement f_{xm}, force velocity f_{vm}, and activation E behavior (Section 5.6.5),

$$F_m = E \cdot f_{xm} \cdot f_{vm} \,. \quad (5.163)$$

the latter being described in the form of bilinear first-order differential equations describing the release and uptake of Ca^{2+}, the recruitment process depending on muscle stimulation, or both. Whereas f_{xm} and f_{vm} are normalized, being 1 at the maximum for f_{xm} and at the isometric condition for f_{vm} the activation E incorporates the strength of the muscle investigated. Therefore, E is factorized using a stimulation dependent $u_0(t)$ activation term u ($0 \leq u(t) \leq 1$) and a factor *A* giving the muscles strength that, in turn, is proportional to the area of its physiological cross section.

$$F_m = u \cdot A \cdot f_{xm} \cdot f_{vm} \text{ with} \quad (5.164)$$

$$\dot{u} = \frac{1}{\tau}\left(u_0(t) - u\right) \quad (5.165)$$

and different time constants for activation and deactivation. Depending on the formula used for approximating the different characteristics this results in 3 + 1 + 2 + 4 = 10 parameters to describe the dynamical properties of a single muscle.

More details such as the pinnation angle (Pandy et al. 1990) are possible. As outlined above, the investigation of the series elastic element was part of the investigations. This adds another degree of freedom to the system and depending on the description (e.g., linear, exponential, zero length), some more parameters. However, it strongly influences the routine. Without the series elastic element, the muscle forces accelerate the system resulting in altered angles and angular velocity that are fed back and are used to describe the new point of operation of the muscle. With a tendon in series, the length of the muscle now depends on the lengthening of the tendon determined by the muscles force. Solving Equation 5.164 for f_{vm} and inverting this equation results in an explicit nonlinear differential equation that can be solved by numerical integration.

To approach realistic behavior besides the active moments, it is necessary to include the passive properties of the joints. These are frequently formulated by exponential elastic functions (Fung 1993). Below we see that viscoelastic properties may be of advantage. It is helpful for the numerical analysis to use smooth force–length and force–displacement curves and to ensure that the velocity–force characteristic (the inverted notation is on purpose) is bijective.

From this, a system of first-order differential equations is established to describe the state space of the system, that is, the system's position and velocity in dependence of the control, which can be integrated using appropriate solvers.

$$\dot{\mathbf{x}} = \begin{pmatrix} \dot{\mathbf{x}}_p \\ \dot{\mathbf{x}}_v \\ \dot{\mathbf{x}}_u \end{pmatrix} = \begin{pmatrix} \mathbf{x}_v \\ A^{-1}\left(\mathbf{x}_p\right)\left(g\left(\mathbf{x}_p, \mathbf{x}_v, \mathbf{x}_u, t\right)\right) \\ \Gamma\left(\mathbf{u}_0, \mathbf{x}_u, \mathbf{x}_v\right) \end{pmatrix} \tag{5.166}$$

In this state space notation $\boldsymbol{\varphi}$, $\dot{\boldsymbol{\varphi}}$, and $\mathbf{u}$ are replaced by $\mathbf{x}_p$, $\mathbf{x}_v$, and $\mathbf{x}_u$. In general, this system must be solved (Figure 5.32) by considering constraints such as the described range of the stimulation and the positive muscle forces.

All this forms the background of an optimization problem. The first problem already arises with respect to the initial condition. Suppose the initial posture is prescribed, then mass and the passive loading of the joints determines the net joint moments, but the question arises how to distribute initial excitation among the synergists? A quadratic objective function to be minimized may be used (Pandy et al. 1990) to solve this problem,

$$J = \sum_i \left(\frac{F_i}{F_{i,\,iso}} \right)^2 \tag{5.167}$$

where F_i is the force of the ith muscle and $F_{i,\,iso}$ its isometric maximal force proportional to its physiological cross section.

Having satisfied the initial condition, the goal for the jump must be specified. This goal can be obtaining maximum height at take-off or minimizing differences between modeled movements and observed movements (Spagele et al. 1999a). The problem is now to find the stimulation pattern $\mathbf{u}_0(t)$ fulfilling this goal. Because the dimension of the optimization problem (number of muscles, time) is high and the integration time for the model may take much time even on modern computers, shooting should be guided by applying available optimization tools such as sequential quadratic programing (e.g., NPSOL or in the MATLAB libraries; compare also the optimization problems dealing with the segmented leg, Section 5.4.1).

Modern tools such as LMS Virtual Lab. (Leuven, Belgium), SIMPACK (Gilching, Germany), SPACAR (University of Twente, Enschede, The Netherlands), ADAMS (MSC, Santa Ana, California), SimMechanics (The MathWorks, Natick, Massachusetts), SIMM (Musculographics, Inc., Santa Rosa, California) in combination with SD/Fast (Symbolic Dynamics, Inc., Mountain View, California) can help to formulate the multibody equations or may even circumvent this. The software generates the equations based on data supplied by the user (or sometimes the company), defining the geometry and inertia of the system, the properties of the tissues, and the initial conditions. Anthropometric data are available in several collections (e.g., NASA 1978; Winters and Woo 1990). In investigations on animals, it may be necessary to obtain own estimates. The segmental center of mass in an arbitrary Cartesian system of coordinates is

$$\mathbf{r}_{com} = \frac{1}{m} \int \mathbf{r} \cdot \rho(\mathbf{r})\, dV \tag{5.168}$$

with $\mathbf{r}$ being the vector from the origin to the volume element, ρ being the tissue density, and V being the volume, and it can be approximated

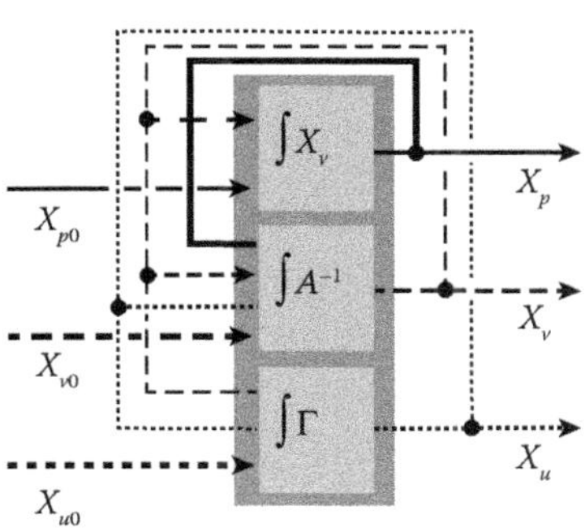

Figure 5.32. Forward modeling solving Equation 5.166. Arrows at the left: initial conditions; gray shaded: integration of equations; arrows at the right: resulting output.

using MRI and approximate densities of the tissues or by simply suspending the object from various directions. Similarly, the moment of inertia with respect to any defined axis

$$I = \int r^2 \cdot \rho(r)\, dV \tag{5.169}$$

where **r** vector from rotational axis to the volume element, can be calculated from MRI data. Calculations based on MRI data have the advantage that the images can be used to calculate the location of the principle axes of the inertia tensor I and the corresponding principal moments I_i:

$$\mathbf{I} = \begin{bmatrix} I_{xx} & 0 & 0 \\ 0 & I_{yy} & 0 \\ 0 & 0 & I_{zz} \end{bmatrix} \tag{5.170}$$

The symmetric ($I_{ij} = I_{ji}$; i.e., six independent elements) tensor with respect to any coordinate system through the center of mass can be determined by

$$\mathbf{I'} = \begin{bmatrix} I_{x'x'} & I_{x'y'} & I_{x'z'} \\ I_{y'x'} & I_{y'y'} & I_{y'z'} \\ I_{z'x'} & I_{z'y'} & I_{z'z'} \end{bmatrix} = \mathbf{RIR}^T \tag{5.171}$$

By using the rotation matrix **R** and its transposed $\mathbf{R}^T$ (e.g., Zatsiorsky 2002).

The scalar value $I_{CM,e}$, valid while rotating around any fixed axis through the center of mass defined by the unit vector ***e***, can be calculated by

$$I_{CM,e} = \mathbf{e}^T \mathbf{I'} \mathbf{e} \tag{5.172}$$

If the axis is displaced from the center of mass by r_{CM}, the parallel axis theorem can be used

$$I = I_{CM,e} + m \cdot r_{CM}^2 \tag{5.173}$$

with the segment mass m. Sufficient (≥6) independent trials are necessary if pendulum experiments are used to determine the moments of inertia from cadaver segments by using the properties of a physical pendulum

$$I = \frac{mg\, r_{CM}}{\omega^2} \tag{5.174}$$

or by using a torsion pendulum with the torsional spring stiffness D

$$I = \frac{D}{\omega^2} \tag{5.175}$$

where ω represents the angular frequency of the pendulum.

In general, it is necessary to calculate the moments with respect to a coordinate system centered at the center of mass. If we do not know the principal axes, it will be necessary to find the principal moments that can be determined as the eigenvalues of I′ that, in turn, can be used to determine the transforming rotation matrix and thus the direction of the principal axes.

Passive joint properties are rarely investigated in animals. Kargo et al. (2002) give ranges of motion for a frog species (*Rana pipiens*). The joints range of motion and the instantaneous center of rotations are determined from photographs while rotating the segments. The latter can strongly influence the gearing ratio (Section 5.6.5). Three segment markers allowed to define coordinate systems and to reconstruct the desired angles (e.g., Vaughan et al. 1996). A different experimental approach uses directly the transmission from joint rotation $d\varphi$ to muscle shortening $ds(\varphi)$ to calculate the angle dependent lever $r(\varphi)$:

$$r(\varphi) = \frac{ds(\varphi)}{d\varphi} \tag{5.176}$$

It should be noted that most joints are not adequately described by giving the range of motion. Consider, for example, your hip. Muscle stiffness largely determines the joint stiffness. You can test this by comparing your mobility with flexed and with an extended knee. You may be able to lift your knee to the chin if your knee is flexed, but (in most cases) not if it is extended. The biarticular muscle connecting both joints obviously has a strong influence on joint mobility. Joint stiffness increases gradually with deflection depending on the arrangement of the segmental chain. Furthermore, viscoelastic properties are not only a nuisance but also essential to stabilize the system (see below). We have now several reasons that make it difficult to define joint mobility by just giving an angle. Nevertheless, Kargo et al. (2002) could show that even the simple range of motion has strong influences on the behavior and performance of the system before considering the work of the actuators.

Tables are available with muscle properties with limited precision for humans (e.g., Winters and Woo 1990). It is possible to determine individual properties of muscle groups in action (e.g., Wagner et al. 2005). This is done by comparing simulations of the actions of a joint or a limb under maximal exertion with measured reaction forces, movements, and muscle activity. In animals such an activity on command is not possible, but in many cases experiments on isolated muscles are possible. In a few cases, groups using forward simulations carry out such experiments at least on selected muscles (e.g., Ahn et al. 2006; Guschlbauer et al. 2007; Siebert et al. 2010).

To summarize, this rough description of the simulation process documents its feasibility, but it also reveals that it represents a considerable enterprise requiring large amounts of data. Some of the results are insensitive to minor deviations; however, inaccuracies quickly add up. Despite the fast development of computer power, such calculations take ample calculation time and even professional tools are quickly overloaded. The early and pioneering jumping models of Hatze (1976) entailed detailed muscle models, but they resulted in strong deviations with respect to principal dynamics. It is always recommended to keep models simple. In case you prefer complex modeling, be sure that the model is able to provide you with an answer to your question. The next section leads us back to a simpler question related (not only) to jumping: it addresses the coordinating action of biarticular muscles.

5.7.5 The Role of Biarticular Muscles (A Tribute to van Ingen Schenau)

Biarticular muscles couple joint movements. The various effects of such a coupling are described by van Schenau (1989). The simplest action is symbolized by the pantograph mechanism (e.g., Fischer and Blickhan 2006) as long as the corresponding pantograph segment can work under tension as has been observed during locomotion of many species (Figure 5.33a).

Such a pantograph muscle can generate tension under different joint angles without changing length, that is, contraction velocity is decreased. The biarticular M. gastrocnemius may be able to extend the ankle in phases in which the monoarticular muscle (e.g., M. plantaris) is running out of power as long as the moment of the knee flexors is sufficient to overcome the opposing moment of the M. gastrocnemius at the knee. In such a situation, indeed a part of the power of the knee extensors that is lost at the knee due to the antagonistic operation of the M. gastrocnemius is not lost or absorbed by M. gastrocnemius lengthening, but instead is transferred to the ankle joint. That means the extending knee supports extension at

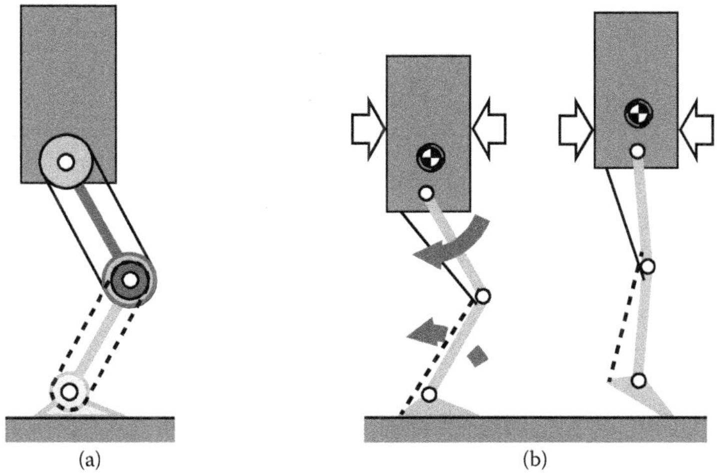

Figure 5.33. Biarticular muscles. (a) Biarticular muscles may act as pantographs synchronizing adjacent joints, generating axial leg extension, and transferring power. Deviations (levers, monoarticular muscles) brake these symmetrical actions. (b) Due to the lower mechanical advantage at the knee compared to the adjacent joints and the buttresses introduced by a heavy trunk and the ground biarticular knee flexors extend the knee (Lombard's paradox).

the ankle joint. van Schenau (1989) pointed out that this coupling can be used to control force direction. Joints generate force components at the leg tip tangential to the leg axis. To generate an axial force in a trisegmented leg, it is necessary to generate a counteracting moment at the second joint. The coactivated muscle helps to do this.

The net power generated in a leg is the result of muscle power. Power generated at a proximal joint can be transmitted by the pantograph to the distal joint. The pantograph muscle does not generate power: it maintains its length and the power absorbed at the proximal joint reappears at the distal joint. If we look more precisely at anatomical conditions in humans, we realize that levers of the biarticular muscles at the adjacent joints are different. Thus, an isometric biarticular muscle now transmits power, but it also redistributes moments and angles (gear ratio). Due to the different gear ratio at the joints, the effect with respect to the antagonists and the inertia, especially of the trunk, is differential. For example, the lever of the hamstrings at the hip is larger than the lever at the knee. Activation of the hamstring generates a larger moment at the hip than at the knee. Due to the fact that the foot is fixed and the inertia of the trunk is high, the extension of the hip induces a "paradox" extension of the knee (Lombard's paradoxon; Figure 5.33b). Also, the M. gastrocnemius can support knee extension in the vicinity of an almost extended joint. The lever at the knee is smaller than at the ankle. The net reaction force at the ankle joint points in front to the center of rotation of the knee and overrides the flexion induced by the muscle. In conclusion, provided the biarticular muscles are activated at the right time, they can contribute to extend a joint where they seem to act as flexors. The asymmetry is essential for such an effect. Human jumpers, that is, only those who have learned to jump, are able to explore this effect (Ertelt 2008). It remains to be shown whether the geometry in the legs of specialized animals adapts to enhance this effect.

Biarticular muscles connect the action of adjacent joints, and the result can be surprising depending on the distribution of levers and inertia. Pantographs are frequently used in robotics. I am not aware of a purposeful use of Lombard's paradoxon in engineering. To achieve high take of velocities, sequential action of adjacent joints is advantageous.

5.7.6 The Proximo-Distal Sequence: A Note

Complex muscle–skeletal models may hide many answers in the formidable degree of freedom of the system. In contrast, principles worked out by simple considerations may not surface in such a model as system properties and constraints may prevent them from being expressed. One of those principles is the proximo-distal sequence of motions. This principle that has been identified for many throwing motions in sport seems to be also roughly valid for the jump. For example, *Galago senegalensis*, well known for its formidable jumps, extends its leg in a proximo-distal sequence: (Aerts 1998) hip, knee, ankle, and midfoot (Figure 5.34). If we consider the angular velocities, they are also higher at the distal joints than at the hip. The absolute speed of the hip is certainly higher at the hip located close to the center of mass than on the midfoot located close to the toes and thus the floor until take-off. However, it is obvious that the relative speed of the joints with respect to center of mass increases in a proximodistal sequence, reaching its maximum at take-off at the toe. The jumper has the task to generate the momentum by extending his appendixes. The distal segments still accelerate the body when the proximal segments are already strongly extended and are not able to contribute to leg lengthening. Alternatively, the almost extended joints offer a reasonable buttress to the extension of the distal joint that can now release stored energy, thereby overcoming the limits due to intrinsic muscle speed. The toe must be fast with respect to the COG, otherwise it is not possible to generate an "ideal" force pattern with its maximum toward the end.

The multisegmented leg offers the advantage of a large working range. It can be flexed, for example, to facilitate protraction and it can be extended to provide a long acceleration distance. Using the whip effect, rotation can couple to extension and flexion. This may be less efficient in technical telescopes. Another efficient strategy to provide sufficient jumping power is to redirect energy accumulated during run-up. Now the legs must be able to bear high forces.

5.7.7 Jumping Far: Redirecting Energy

A run-up helps to improve human jumping, and this should be true for most animals. The reason

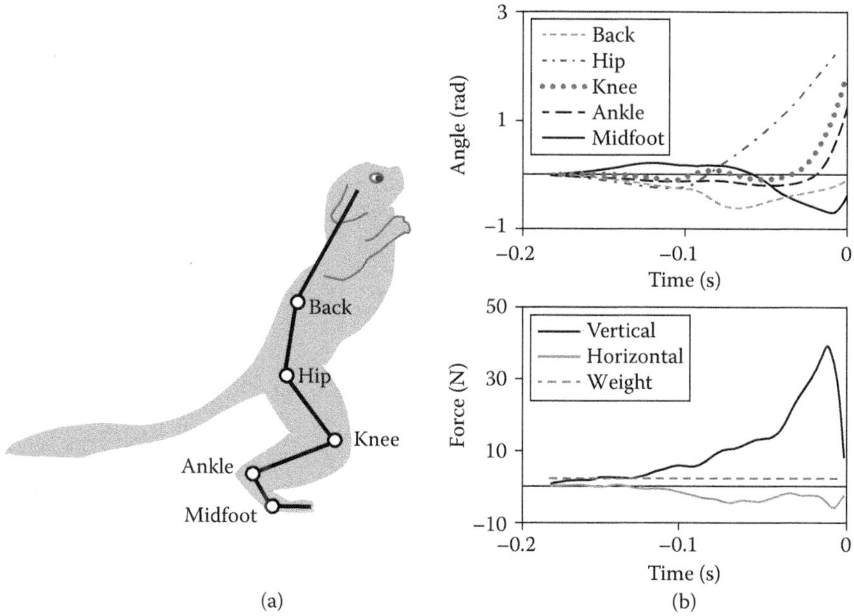

Figure 5.34. Jumping bushbaby. (a) Segment model. (b) Above: Change of angles during a typical jump. Neither the body nor the midfoot contributes to extension. In the other joint a proximo-distal sequence of extension is observed. Below: The coordination generates a peak reaction force at the end of the jump. (After Aerts, P., *Philos. Trans. Roy. Soc. Lond. Ser. B Biol. Sci.*, 353, 1607–1620, 1998, a: fig. 1, p. 1608, b: fig. 3, p. 1611. With permission of the Royal Society.)

is that we are not forced to generate energy for the jump by simply extending our legs, but we can exploit the kinetic energy from the run-up. In the run, we have a sufficient acceleration distance (even though running speed is limited for all animals). That this energy resource is considerable and how it can be used can be easily visualized by considering a human pole vault. There the elastic pole with adapted compliance and a skilled jumper allows to perfectly convert the kinetic energy of the run into jumping height. A run-up at 10 m/s results in a height of 5 m:

$$h = \frac{v^2}{2g} \tag{5.177}$$

Apparently, animals do not use poles, and humans are not able to redirect energy with similar efficiency by using their legs. Furthermore, no systematic studies are known that investigate this possibility in the animal kingdom. In running and hopping bipeds, this should be a suitable strategy in general. In fact, investigations on rock wallabies have shown that these animals use a moderate run-up and redirect the kinetic energy during a jump very much like humans do. And, in addition, they generate extraordinary jumping power (Figure 5.35a) in their muscle–skeletal system (McGowan et al. 2005). In quadrupeds, redirecting may play a role for the decelerating front legs and may be frequently hidden in investigations where the accelerating hind limbs dominate the focus of interest.

In the wallaby, actually only a small fraction of the kinetic energy of the approaching animal (~20%) was used to increase jumping height. The rest of the increase in the order of the total run-up energy was generated by the extending limb. The wallabies still had to fly a distance of about 2 m to land on the elevated post, and they did not like to lose too much speed. However, what about the human jumper? Let us assume the runner would use a springy leg to redirect all horizontal kinetic energy into vertical kinetic energy (Section 5.2.3). The energy of the run-up must be stored in the spring and then released in the redirected direction:

$$\frac{m}{2}v^2 = \frac{k}{2}\Delta l^2. \tag{5.178}$$

For a linear spring in which force F is proportional to compression Δl, this results in a maximum specific force or acceleration of

$$a = \frac{F}{m} = \frac{v^2}{\Delta l} \qquad (5.179)$$

The human jumper, having available a leg compression of about 0.5 m and running up at 10 m/s, would have to generate a peak acceleration of 20 g. However, in part, the energy is redirected by the pivoting spring. But even if we assume that half of the energy is redirected, we would still have to

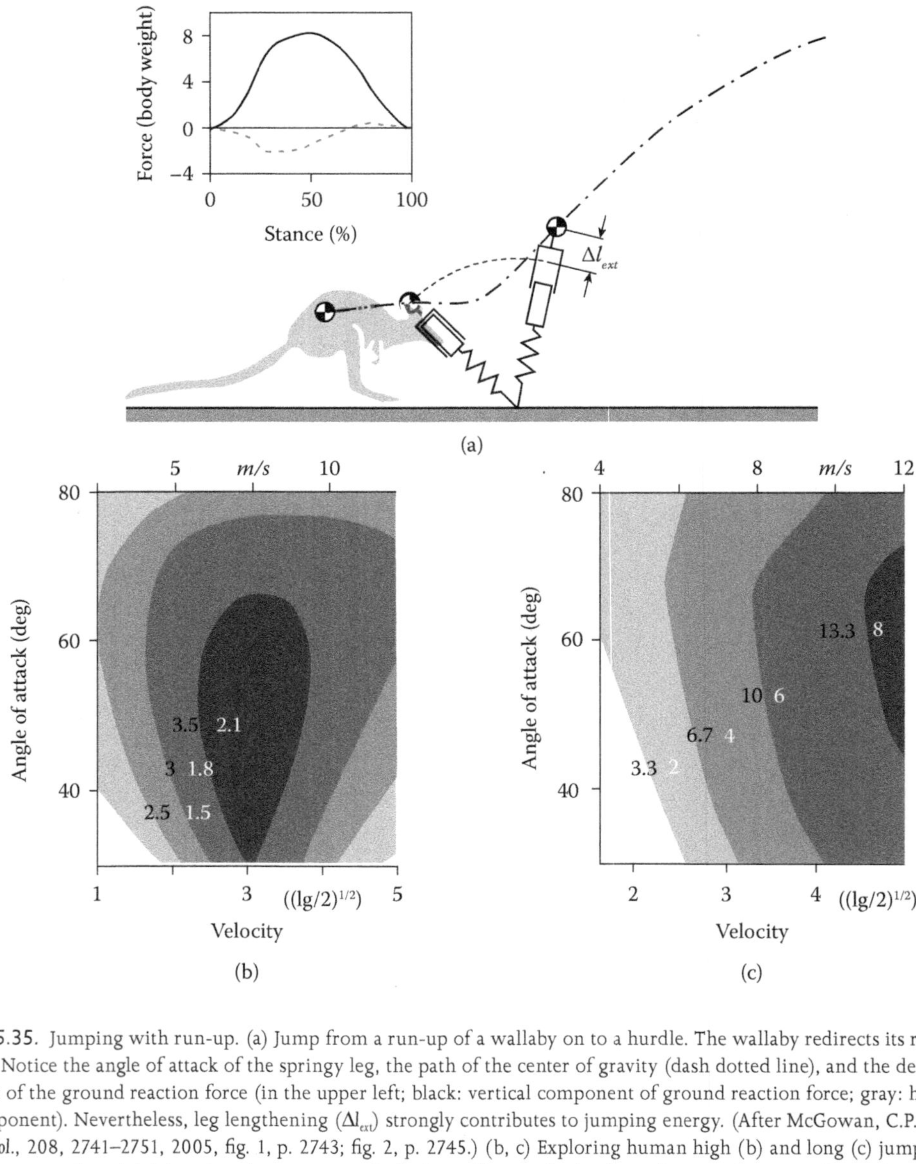

Figure 5.35. Jumping with run-up. (a) Jump from a run-up of a wallaby on to a hurdle. The wallaby redirects its run-up energy. Notice the angle of attack of the springy leg, the path of the center of gravity (dash dotted line), and the decelerating part of the ground reaction force (in the upper left; black: vertical component of ground reaction force; gray: horizontal component). Nevertheless, leg lengthening (Δl_{ext}) strongly contributes to jumping energy. (After McGowan, C.P., et al., J. Exp. Biol., 208, 2741–2751, 2005, fig. 1, p. 2743; fig. 2, p. 2745.) (b, c) Exploring human high (b) and long (c) jump with a two-segmented model driven by a knee extensor including a force–velocity dependency and a series elastic element. (a) Contours of jumping height (black: [l/2]; white: [m]) in dependence of run-up speed and angle of attack (between leg axis and the horizontal). Highest jumps are achieved at intermediate run-up speeds. (After Alexander, R.M., Philos. Trans. Roy. Soc. Lond. Ser. B Biol. Sci., 329, 7, 1990., fig. 4. With permission of the Royal Society.) (c) Contours of jumping distance in dependence of run-up speed and angle of attack. The optimal angle of attack of about 70° remains almost constant. Distance increases with ruin-up speed. (After Alexander, R.M., Philos. Trans. Roy. Soc. Lond. Ser. B Biol. Sci., 329, 8, 1990, fig. 5. With permission of the Royal Society.) Length of both segments is assumed as l = 0.6 m. These results are rather robust with respect to modeling details. (Continued)

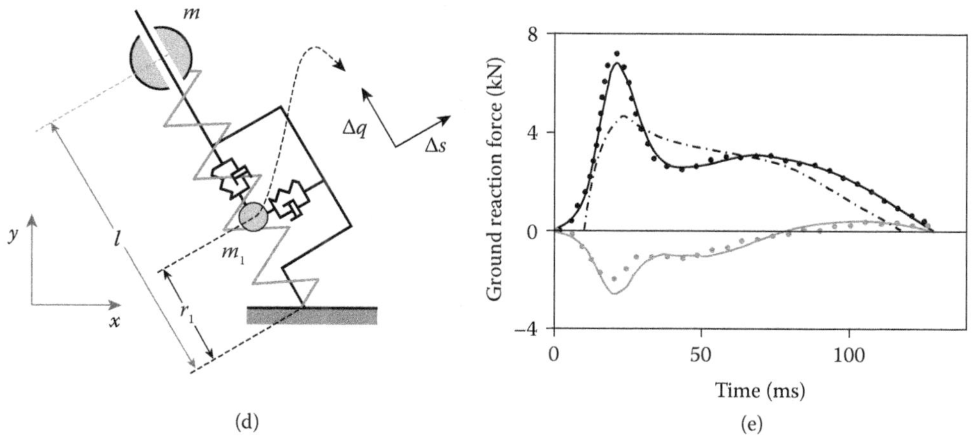

Figure 5.35. (Continued) Jumping with run-up. (d) Model for a springy jumper entailing a suspended wobbling mass. (e) The vertical (black) and horizontal (gray) experimentally observed (dotted line) forces are well described by the model (continuous lines). The muscle model (dash-dotted line, after Seyfarth, A., et al., *J. Exp. Biol.*, 203, 741–750, 2000) used in (c) is not able to describe the pattern of the ground reaction force. (After Seyfarth, A., et al., Dynamics of the long jump, *Journal of Biomechanics*, 32, 1262, Copyright 1999, fig. 4. With permission from Elsevier.)

cope with a maximum of 10 *g*. By using a pole, the human jumper is able to enhance distance (see above) to about 3 m, resulting in a moderate acceleration of about 3 *g*, assuming complete storage. It is obvious that a human runner without a pole will not be able to redirect his maximum run-up energy into a vertical jump. We also see in such a consideration that the resulting vector of the change of momentum is directed 45° backward to approach direction. Ignoring gravity, which introduces an asymmetry into the process of redirection, this must be the direction of the average force.

Animals such as a wallaby that is able to generate accelerations of 8 *g* and a bushbaby with even higher values (~12 *g*) and possessing long hind legs should have a much higher potential to redirect the approach energy.

In a long jump, the issue of redirecting energy is similar. Treating the flight as an oblique throw demonstrates that for a given absolute value of the speed v, a release angle of $\alpha = 45°$, and a little bit less if we consider the fact that the COG is y_0 above level at take-off, leads to a maximum distance x_d:

$$x_d = \frac{v^2 \sin(2\alpha)}{2g}\left(1+\sqrt{1+\frac{2y_0 g}{v^2 \sin^2(\alpha)}}\right) \quad (5.180)$$

In this case, redirection must be less brute and the change in momentum is directed toward about 67.5° and a purely elastic, that is, symmetric, process must cover this angle at midstance. The changes in momentum must be generated during contact during which the body may travel 1.5 m. Let us assume the jumper slows down and has an average speed of the center of mass of 0.75 × 12 m/s. Then, he would spend 0.17 s in contact with the board. To redirect the velocity, the jumper would need an average specific force of 5.5 mg or, considering a sinusoidal time course, peak forces of almost 9 mg. The problem is less accentuated than in the vertical jump; however, it is still there.

By roughly considering human-like knee extensors and exploring different angle of attack strategies, take-off angles can be observed that are closer to reality (Figure 5.35b, c; Alexander 1990; Seyfarth et al. 1999). The highest jump can only be obtained at a moderate run-up speed and an angle of attack of about 45°. The jump distance increases with speed and has its maximum for angles of attack of about 70° and take-off angles between 10° and 20° at high speeds. The example points out that in the case of the long jump, the force amplitude becomes critical and it is also necessary to build up the force in a rather short time. Again, the problem is reduced if the jumper has long legs and can distribute the acceleration across a longer time.

We have seen that redirecting energy can be rather efficient. Now part of the acceleration distance is located outside the animal's muscle–skeletal system. Future robots will be able to use such a strategy. The fact that real legs come with mass and an exact speed match between foot and ground is the exception implies that natural and robot legs must accommodate impacts. In the following, we learn how this is handled and turned into an advantage in human jumpers.

5.7.8 Distal Masses and Wobbling Masses

So far, we neglected in our considerations the leg's mass. Especially, a human jumper has high distal masses in contrast to most cursorial species. But there, too, the question arises whether we can neglect the masses. At preferred running speed, the modifications in the joint torques that can be contributed to the segment inertia amount to only about 5% of those necessary at midstance. However, the force tracings of jumpers display initial peaks at impact that are high compared to the midstance forces. Muscle properties are not sufficient to describe the time courses. The initial peaks are largely due to the deceleration of the distal limb. The jumper approaches at a high speed and places his legs to quickly build up high forces. He cannot "give," at touchdown he needs the whole contact time for the redirection process involving high redirecting forces. Thus, the jumper faces the problem that after the swing, the leg tip approaches the ground at about running speed and is stopped at an instant (consider the spikes of the shoes!), whereas the center of mass is allowed to travel further the foot stops instantly. This alone results in a considerable impact. In fact, to avoid a huge impact it is necessary that only a fraction of the foot (mass ~1 kg) is decelerated within a distance of about 1 cm, whereas other parts are allowed to travel further or deflect. The mechanism results in a delayed or smoothened contribution of the further proximal parts. Despite of this local deflection, the reaction forces still must be transmitted via the stiffening muscle–skeletal system toward the COG of the jumper. The bone as a rather stiff material only interrupted by cartilage layers swiftly transmits a part of the induced shock proximally where it even reaches the head. Clearly, the zigzag arrangement of the leg and the bending of the spine help to reduce shock transmission because part of the energy is consumed in joint torsion. Another trick is to suspend visco elastically weights to the skeleton that take up energy. In very high and slender buildings, such passive or active auxiliary dampers are used to reduce wind- or earthquake-induced sway. The muscle–skeletal system delivers abundant soft tissue; there is no need to add weights. In the simulations, a weight suspended to the springy, slightly extending telescope leg (Figure 5.35d) is able to describe the time course of the impact (Seyfarth et al. 1999).

Let us have a brief glimpse onto the equations that are now denoted in polar coordinates (see Equation 5.18):

$$\ddot{l} = \dot{\alpha}^2 l - \frac{k}{m}\left(l - l_0(\alpha)\right) - g\sin\alpha \quad (5.181)$$

$$\ddot{\alpha} = -\frac{1}{ml^2}\left(r_1\, F_s - \Delta s\, F_q\right) - \frac{1}{l}\left(2\dot{l}\,\dot{\alpha} + g\,\cos\alpha\right) \quad (5.182)$$

where l is distance between upper mass m and the foot point (point of rotation); α is the angle between leg and the horizontal plane; k is stiffness of the leg spring; l_0 is the resting length of the spring that increases linearly with the angle of rotation to consider leg extension; g is gravitational acceleration; r_1 is the site of the axial suspension of the leg mass m_1, where the mass is assumed to be mounted on the spring axis; and $F_{q,s}$ is the radial and tangential component of force generated by corresponding swing mass deflection (Δq, Δs). Using these parameters, it is nice to explicitly see the contribution of the centrifugal effects to the radial acceleration (first term in Equation 5.181 on the right) and the Coriolis-acceleration (term preceding the gravitational term in Equation 5.184) (Equation 5.182). Suspended to the stiff (weightless) axes guiding the spring, the distal mass does not affect the axial acceleration of the distal mass. This decoupled approach is certainly a simplification. The force term in the bracket simply marks the cross-product for the torque generated by the oscillations of the swing mass with respect to the foot-hold. The movements of the swing mass are a little bit more complicated. The small linear accelerations receive contributions from angular, centrifugal, and Coriolis-accelerations:

$$\Delta\ddot{q} = \Delta s\ddot{\alpha} + r_1\dot{\alpha}^2 + 2\Delta\dot{s}\dot{\alpha} + \left(\frac{F_q}{m_1}\right) - g\sin\alpha \quad (5.183)$$

$$\Delta\ddot{s} = -r_1\,\ddot{\alpha} + \Delta s\dot{\alpha}^2 - 2\Delta\dot{q}\dot{\alpha} + \left(\frac{F_s}{m_1}\right) - g\cos\alpha \quad (5.184)$$

Nonlinear viscoelastic functions are used to describe the force exerted by the swing mass entailing the distal masses (skeleton and wobbling masses) decelerated in the impact:

$$F(\Delta q, \Delta\dot{q}) = -\left(c_q\,\mathrm{sig}(\Delta q) + d_q\Delta\dot{q}\right)\left|\Delta q\right|^{v_q} \quad (5.185)$$

$$F(\Delta s, \Delta\dot{s}) = -\left(c_s\,\mathrm{sig}(\Delta s) + d_s\Delta\dot{s}\right)\left|\Delta s\right|^{v_s} \quad (5.186)$$

where sig is the signum function. In the simulation, the constants for stiffness c, damping d, and the exponent v describing the degree of nonlinearity are assumed to be the same ($c_q = c_s$, $d_q = d_s$, and $v_q = v_s$).

The parameters are adapted to fit the experimental observations. This revealed the distal mass as a fraction (0.27) of the rest mass m of the jumper (for an average person, this would be more than leg weight) connected at 0.25 of the initial leg length from the contact point to the leg axis. The initial velocity of the swing mass is obtained as being 5.3 m/s, a value that is higher than the reported speed of the foot at touch-down of about 4 m/s, but less than the running speed of 8.2 m/s. Furthermore, the distal mass moves at an angle of about 30° toward the floor. This is in agreement with observations. The leg moves downward and with respect to the center of mass backward. It retracts and thus accommodates part of the relative speed in running direction, reducing the horizontal impact but enhances the vertical impact. The impact forces are almost critically damped. Thus, the energy is lost and cannot be recovered. However, this passive impact significantly contributes to total momentum (Figure 5.1). Even though its effect may be overestimated because we ignore asymmetry introduced by the muscle's force–velocity curve, models avoiding the impact by starting with initial conditions with the foot on the ground or lacking a wobbling mass (e.g., Gruber et al. 1998) are not able to describe the dynamics correctly independent of model complexity. The development of torques using a maximally activated eccentrically and subsequently concentrically loaded muscle is asymmetric with higher forces in the eccentric phase. However, this does not replicate magnitude and the time course of the reaction force, especially if the critical muscle activation is taken into account. As described for the frog jump, the series elastic element can be used to store and release energy and, more importantly, to put the contractile element beyond the amortization period into an eccentric loading condition enhancing muscle force (Seyfarth et al. 2000).

Talking about optimization, does it help to slap the foot toward the ground in a gripping motion as jumpers do to enhance jumping distance? It certainly does. Imagine a world without gravity (even though in that world a runner would not be able to run-up). The mass of the object is accelerated to a high speed. It is assumed not having any vertical component. Now, the system consisting of two segments connected by a hinge suddenly explosively opens. The momentum remains the same and at first nothing happens as the center of mass preserves its velocity. Now, the one part moving downward has an inelastic impact with the ground proportional to its vertical speed. This inelastic impact alters the speed of the total system. The work done by extending during the flight pays off! Legged systems may use inelastic impacts of appendages to redirect during locomotion; more importantly, they certainly do this to damage objects. In that case the object absorbs the energy and does the major part of the damping. Our karatekas are by far not the most effective users of this principle. In the mantis shrimp (*Odontodactylus scyllarus*, ~0.3 m long) the shell-smashing, catapult-driven (10,000 g) appendage moving about 0.02 m reaches a tip speed of about 20 m/s. The object's devastation (e.g., the latter may be an aquarium wall) is even supported by cavitation effects (Patek et al. 2004; Patek and Caldwell 2005).

Consideration of wobbling masses has increasingly moved into the focus of scientists. Certainly, they play a large role in impacts as described for the human long jump or for landing situations. However, because impacts are a concomitant effect of legged locomotion (see an estimate for human running in Günther et al. 2003), they play a rather general role. Critical damping, that is, the avoidance of disturbing oscillations, imposes a restraint on the mechanical properties of the soft tissue involved, especially the musculature. Inactive musculature should not generate major tension to reduce the cost to move the muscle–skeletal system. In contrast, the impact should have a minor effect on shape: you would not like to lose your muscle. To be effective as a shock-damping system, the viscoelastic properties must be adapted to this purpose. The active muscle with its intrinsic adjustable elasticity (force–length curve) and damping (force–velocity curve) can help in this situation. However, this again increases cost. Evolution may have found some compromise. In fact, it is argued that dampening oscillations in the distal leg by activating a muscle may significantly contribute to the cost of locomotion (Wakeling et al. 2003). Taking the concept of controllable wobbling masses into account may help in the future to reduce shock waves in agile robots, thereby enhancing life time. In contrast, organs of different densities within the animal's body certainly interact. The idea that a visceral piston could be used to drive respiration (e.g., Bramble 1989) could not be substantiated in more rigorous studies (Young et al. 1992; Simons 1999).

Besides the accelerations induced by the impact at touchdown, accelerations generated by muscle forces or motors also can drive segmented systems into oscillations. This is addressed in the next section. Here we use the example of a spider's jump where we would intuitively assume that the open hydraulic system provides for sufficient damping.

5.7.9 Segmental Oscillations: A Spider's Jump

We have seen (Section 5.2) that vibrational modes of the system can be explored to facilitate locomotion. In addition, we have seen in the preceding section that the control of local oscillations represents an important aspect in the tuning of a motion system that responds to an impact. Here, I use a spider's jump to elucidate the fact that the suppression of segmental oscillations is not only an issue in robotics.

Spiders have an open circulatory system. They pump the hemolymph (blood) from the heart located in the abdomen (opisthosoma) via arteries into the body, the legs, and other appendages where it is released into the open space between the tissues encapsulated by the cuticular shell. Spiders use this fluid to hydraulically extend the legs at major joints. At these joints the axes of rotation are located at the outer rim of the leg tube, and muscles are only able to flex the joint. The pump to generate the pressure is located in the prosoma, a segment entailing the brain, the stomach, and the "head," from where the eight legs originate and radiate outward. It consists of muscles connecting the dorsal and the ventral cuticular sheets and the ventral attachments of the legs (coxae). Contractions of these muscles generate a pressure and shift the fluid to the extending legs. Spectacular jumps based on this mechanism are known for jumping spiders (Parry and Brown 1959a, 1959b). Large spiders are much more limited with respect to their ability to generate jumps (see below), but some can still generate spectacular jumps (Figure 5.36a, b).

Numerical models are well suited to describe the basic features of such a jump (Zentner et al. 2000, Zentner 2003). It is assumed that the hind legs are the only legs accelerating the trunk, and they start from a posture typical for prepared jumps (Weihmann et al. 2010). The legs are approximated as a chain of three tubes connected by dorsal hinge joints closed by a flexible membranous bellow and operate in the sagittal plane. The pressure-driven chain is fixed to the floor via a hinge joint and drives a body whose orientation is assumed to be fixed (no angular acceleration). The pressure extends the leg, that is, the leg segments and the body are accelerated, resulting in a ground reaction force. Differently from muscle forces (cf. section 5.5.4), now the properties of the fluid must be considered. The pressure does work dW due to the torque M rotating the joint by $d\varphi$ that, in turn, can be expressed by the static force F_p equal to the local pressure p times the cross section A acting at the center of this surface with the lever arm of r with respect to the joint:

$$dW = M\,d\varphi = r\,F_p\,d\varphi = r\,A\,p\,d\varphi = dV\,p \quad (5.187)$$

Thus, the work is the product of local pressure and displaced volume dV (Parry and Brown 1959a, 1959b; Blickhan and Barth 1985). The torque due to the forces F_M generated by the momentum of the hemolymph is of much lower order (<5 × 10^{-5} the static pressure force F_p):

$$F_M = \frac{d(m_F v_F)}{dt} = \frac{dm_F}{dt} v_F + m_F \frac{dv_F}{dt} = \frac{\rho_F}{A}\left(\dot{V}^2 + V\ddot{V}\right) \quad (5.188)$$

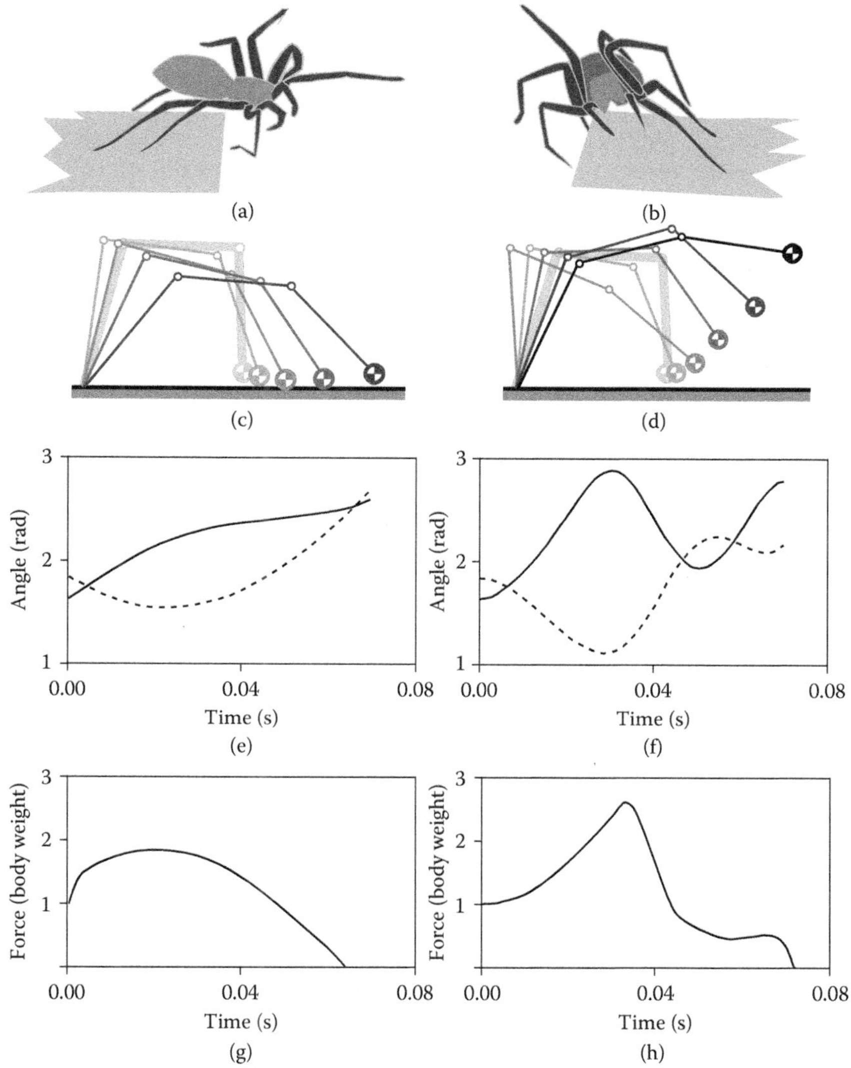

Figure 5.36. (a, b) Hunting spiders (~3 g) are able to jump. (a) Forward jump. (b) The radial leg symmetry allows jumps in different directions (here to the side), even backward (not shown). (After Weihmann, T., et al., *J. Comp. Physiol. A Neuroethol. Sens. Neural Behav. Physiol.*, 196, 424, 2010, fig. 3.) (c through h) Modeling the jump of tarantulas. Left column (c, e, g): small spider (1.8 g; v_{TO} = 0.3 m/s); Right column (d, f, h): large spider (8.7 g; v_{TO} = 1.1 m/s). Both jump with a pressure of 65 kPa. (c, d) Stick figures of hydraulic extension of the hind limb. Whereas the small spider (c) only shows a countermovement in the tibia metatarsus joint (distal) the large spider (d) shows large oscillations in both major joints. This is more obvious in the time dependencies of both angles (e, f); normal line: proximal femur–tibia joint angle; dashed line: distal tibia–metatarsus joint angle. (g, h) The oscillations influence the (vertical) ground reaction forces). In fact, for larger spiders they disrupt the ground reaction force. Oscillations can be reduced by additional damping reducing take-off speed. (Zentner, pers. comm.)

where m_F is mass of the fluid, v_F is velocity of the fluid, and ρ_F is density of the fluid.

Significant is the pressure drop due to viscosity (μ, dynamic viscosity). The viscosity of the hemolymph is close to the values of water. Because the segmental tube (r_{seg}) is largely filled with flexor muscles, the channels (r_{ch}) providing space for the supply are rather narrow ($\frac{r_{ch}}{r_{seg}} \approx \frac{1}{3}$). The pressure drop Δp along the segment with length l_{seg} can be described using the Hagen–Poiseuille equation:

$$\Delta p = \frac{8\mu\, l_{seg}}{\pi\, r_{ch}^4}\dot{V} \qquad (5.189)$$

As in the forward simulations described above, it is necessary to take the nonlinear passive joint stiffness at the joint into account defining the range of motion. In a system where inherent damping is obvious, we might be inclined to neglect the passive viscosity of the joint. Whereas this is indeed sufficient for small individuals, it is not sufficient for large, adult animals exceeding a body mass of 3 g. The simulations show that without considering a strong viscous damping at the joint, the three segments start to oscillate (see Figure 5.36 for an example of differently sized tarantulas). For hunting spiders of a 10% up-scaled size, the segmental oscillations result in early detachment of the leg (if the claws do not prevent this). In our treatment of the stability of the trisegmented leg in Section 5.4.4, segment inertia has been ignored. Segment masses impose an additional stability problem. Sufficient inherent damping is necessary to provide a smooth harmonic leg extension. This damping can be introduced by viscoelastic passive properties of the joint or by muscle activity. In spiders, this certainly reduces performance and (among other issues) imposes a limit to performance of large animals.

Any linkage system driven by muscles, hydraulics, or actuators should face the problem of segmental oscillations. Dampening these oscillations will at least cost energy and as in the case of large spiders, it will reduce performance. To my knowledge, systematic investigations are not available. We close at that point, our series of issues related to the generation of high take-off speeds during jumping. The following section addresses the issue well known for human jumpers but of general relevance how lacking interaction with the ground the jumper is able to adjust his or her landing position. Because landing from an aerial phase is not restricted to jumping, this is of general significance and gives us the chance to stress the significance of inertial properties.

5.7.10 Flying without Wings: Controlling Attitude without Fluid Forces

Jumpers, and also runners, spend some time in the air without any contact with ground. This "flight phase" is generally simplified as being ballistic. This is justified for the description of the trajectory of the center of mass as long as aerodynamic forces can be neglected (cf. Chapters 3 and 4). In many cases, and especially for animals of intermediate and small size, animals explore aerodynamic forces to enhance or stabilize flight phases. Another issue is the movement of the appendages relative to the center of mass. This is not a minor issue. In saltatory gaits, the flight phase offers the window to prepare for the landing, that is, for the initial conditions for the contact phase. We know that the flight phase must offer sufficient time to reposition appendices. This can be critical at high speeds of locomotion. From a different perspective, the contact prepares for suitable flight, that is, the requirements impose a constraint on to the contact phase and vice versa. Systematic and thorough studies on this interrelation are to my knowledge not available. The situation is of similar importance for jumps. Even though there is much more time available, the repositioning of the appendages during flight is critical to prepare for landing. Furthermore, especially for systems not built to explore aerodynamic forces, controlling posture can become a major issue.

Neglecting aerodynamic forces during the flight phase, the center of mass follows a ballistic curve, and the angular momentum is conserved. However, the position of the appendices with respect to the center of mass can be altered. Even for the case of zero angular momentum at take-off, this allows for a repositioning of trunk and appendices. The vector of the total angular momentum L can be taken as the sum of segmental angular moments L_i:

$$\mathbf{L} = \sum \mathbf{L}_i \tag{5.190}$$

Because the total moment is conserved during flight at any instant, a change of the angular moment of one segment must be accompanied by a similar change of the remaining segments. The simplest example is the scissor movement of legs against the body in sailing style of the human jump that affects only the angular momentum perpendicular to the sagittal plane. More subtly, indicating possibilities and limits, is the hitch-kick style (like running in the air) in which one leg is retracted in an extended position and protracted in a flexed position, resulting in a stepwise erection of the body.

The most spectacular example is the falling cat that has intrigued the scientific community at the turn of the nineteenth century (Figure 5.37).

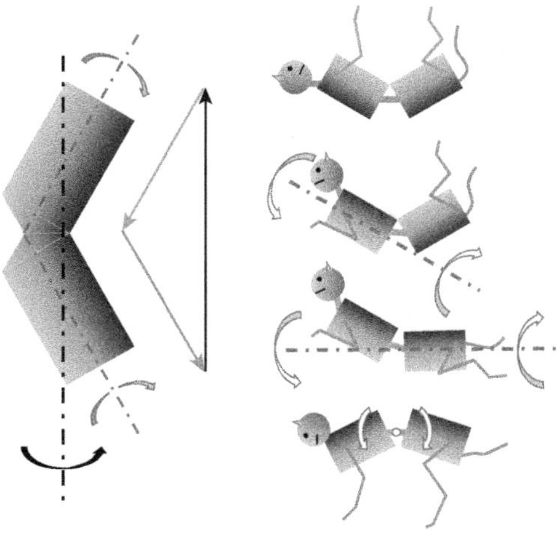

Figure 5.37. Repositioning at zero angular momentum—the falling cat. (a) Two co-rotating (gray dash-dotted lines and arrows) oblique cones counter-rotate together (black dash-dotted line, black arrow) conserving total momentum vectors on the right (after Frohlich 1979, fig. 11, p. 589, with permission from American Association of Physics Teachers). (b) A sequence of counter rotations exploiting different moments of inertia and corresponding differences in angular velocity allow the cat to turn when released from upside down. (After Hay, J.G., **The Biomechanics of Sports Techniques**, Prentice Hall, NJ, 1993, fig. 6–44, p. 164.) The real animal will explore a mixture of both mechanisms.

A cat thrown into the air does nothing until it reaches the apex and then turns on its feet. If you change the experiment by just holding it upside down on its legs and release (in a sufficient height above the ground), the animal again turns to land on its feet, despite the fact that the total angular momentum vanishes. Two explanations can be put forward that might be intertwined in reality. One explanation is the "hula technique." Think of generating a hula-movement while you are in the air. Then, each point in the cross sections in the transverse plane performs circular movements around the longitudinal axis. The angular momentum that can be contributed to this movement must be balanced by an opposite rotation that is, in this case, a twist of the segments. A hula movement in the air results in a twist of body and legs. A different explanation uses a sequence of movements. Again, the first and important step is to attain a C-shape. The front rotates axially at low inertia against the rear, which is angled with respect to this axis of rotation and has high inertia. Now, the cat faces the floor. Subsequently, the body parts switch their roles and the animal twists around the rear with the front angled with respect to the twisting axis, thus largely remaining in position. The moment of inertia of the cat's tail is low; thus, the tail can only serve for adjustments. This is much different in animals with a heavy tail. Geckos use their tail for righting after falling from a wall (Jusufi et al. 2008) and arboreal lizards that autotomized their tail overturn posterior (backward rotation; Gillis et al. 2009). This backward somersault is surprising as the center of mass shifts forward after losing the tail. The authors observed that the lizards with tails touch with their tail the ground after take-off, generating a counteracting torque. Biology can be surprising.

With respect to the total system, the relative movement of appendages represents a deformation and thus a time-dependent moment of inertia of the system. Changes with respect to inertia induce changes in the vector of the rotational velocity:

$$\mathbf{L} = \mathbf{I}(t)\omega(t) \tag{5.191}$$

Controlled deformation while in the air can induce movements that can be used to control the attitude of the animal during flight.

It should be noted that although angular momentum is conserved in the deforming structure, rotational energy is not. For example, a gymnast jumping a somersault and doubling his rotational speed by halving inertia doubles his rotational energy. The necessary energy must be supplied by muscles deforming the body of the gymnasts.

Again, a special issue is stability. Let us consider for simplicity the movement of a rectangular solid with three different moments of inertia with respect to their principal axes ($I_{xx} > I_{yy} > I_{zz}$). Rotations around their major [ω_x 0 0] or minor [0 0 ω_z] axes are stable because infinitesimal changes of the orientation of the rotation axis would require altered angular momentum. However, rotations around the intermediate axis [0 ω_y 0] are instable because in the vicinity of this rotational axis, rotations are possible with the same angular momentum, resulting in a tumbling motion. This problem is enforced if two or more principal moments are similar. In deforming bodies (or moving chains of multibody systems), tumbling can also occur and introduces rotation among undesired rotational axes. Selection of suitable motions to avoid tumbling is not a trivial issue in such cases (see example for gymnasts;

Yeadon and Mikulcik 1996). Note that the angular momentum with respect to altered coordinates (index B) can be expressed as

$$\mathbf{L}^B = \mathbf{R}_B\mathbf{L} = \mathbf{R}_B\mathbf{I}\,\omega = \mathbf{R}_B\mathbf{I}\left[\mathbf{R}_B^{-1}\mathbf{R}_B\right]\omega$$
$$= \left[\mathbf{R}_B\mathbf{I}\,\mathbf{R}_B^{-1}\right]\left[\mathbf{R}_B\,\omega\right] = \left[\mathbf{R}_B\mathbf{I}\,\mathbf{R}_B^{-1}\right]\omega^B = \mathbf{I}^B\,\omega^B \tag{5.192}$$

where R_B represents the rotation matrix transforming the original system of coordinates into the new directions and $\mathbf{R}_B^{-1} = \mathbf{R}_B^T$ (cf. Equation 5.171).

Mass distribution has consequences for animals. We have seen the consequences of pendulum frequencies, impact loading, and redirection in the air. Especially, distally a large fraction of the mass is just that of the actuator itself. Now, placement of an actuator must allow segment actuation. However, it is obvious that many aspects interplay, resulting in distributions complying with different demands depending on the scope and priorities in the behavior of the animals.

Our treatment so far is not exhaustive, and many important aspects remain to be addressed. We already mentioned the issue of landing, and in the introduction the vastly different modes of locomotion not addressed such as the creeping of a snail, the brachiation of gibbons, or the climbing of a gecko. We use the last section to point toward developments in robotics related to the treatment above.

5.8 RELEVANCE TO ROBOTICS

The emergence of robotics as a field is closely related to the development of computational power. Nevertheless, from the earliest stages to the present, engineers take nature as an example. Frequently, the motivation has been to build "artificial men or animals." Think about Leonardo's knight robot serving for L. Sforza, Duke of Milano (ca. 1495; Rosheim 2006). In modern times, engineers still are annoyed by the gap between the elegance and performance of animals compared to that of mimicking machines. For biologists, robots can serve as testbeds, allowing for rejection or validation of theories about function. This chapter does not go into technical or theoretical depth, but by describing some examples that may represent a gadget for the purpose of demonstration or a full-blown autonomous machine, it should give a glimpse into the abundant variety of realizations. As in the evolutionary introduction, we here have a look beyond legged locomotion and we start with creeping machines to highlight the broad and widely distributed efforts in engineering to unearth stimulating ideas from nature.

5.8.1 Artificial Creepers

There are both vertebrate and invertebrate species that cover terrestrial ground by creeping. Here we stick to phylogeny and start with constructions derived from invertebrate paradigms.

A recent robot takes an example on the movement of nematodes (Bao Kha et al. 2009; Cohen and Boyle 2011) and their flipping and undulating motions. The pressure-filled tube has not been copied. However, the longitudinal elasticity and the segmented electro-active polymer muscles combined with essentials of the neuronal organization were successfully used to mimic the motion pattern on a robot 2000 times the size of its animal model. Some artificial worms with possible applications in the medical area use peristaltic motions mimicking annelids (Section 5.1.2). The elongated systems are subdivided into segments surrounded with compliant walls. Each segment is able to undergo cycles in which the segments thicken and shorten and then extend (Figure 5.38). Due to friction, the extended segments serve as anchors. While the wave of thickening segments travels backward, the extended segments are pushed forward. Change in shape of the flexible hull has been, for example, induced by servomotors (Figure 5.38c) translating the movement of cranks onto flexible leaf springs (Omori et al. 2008). This allows by differential actuation of both motors turning maneuvers. Other designs use shape memory alloy to induce contraction and a counteracting expanding silicon shell in each segment (Menciassi et al. 2004), or they use simply inflatable balloons in which the sequence of central filling and emptying is determined by orifices between the segments (Glozman et al. 2010). Although many solutions are able to progress, they differ in terms of controllability and interaction with the environment, such as the pressure against tube walls, friction, and flexibility. Pressure against a tube wall sometimes facilitates

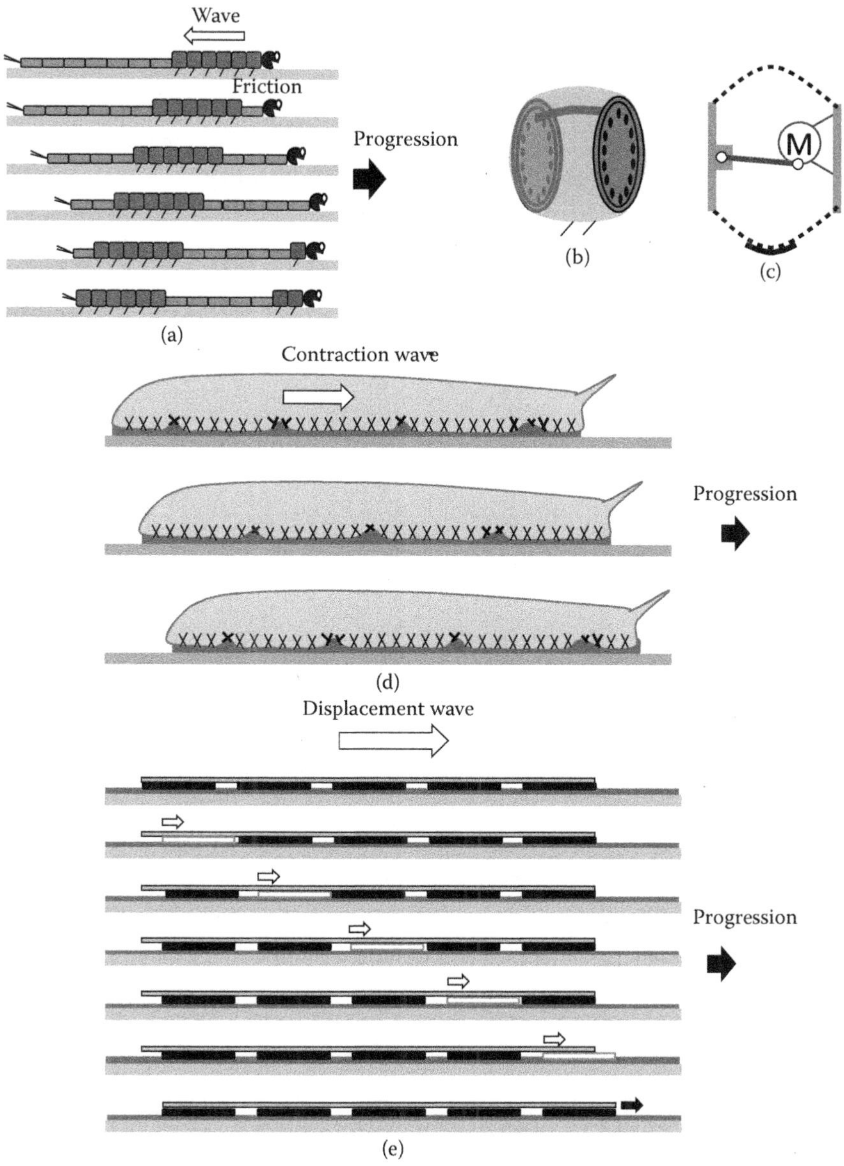

Figure 5.38. Creeping creatures. (a) Annelid creeping by means of a caudad traveling wave of shortening and thickening segments. (b) Scheme of an annelid segment. Segmental elongation (by contraction of circular muscles) and shortening (by longitudinal muscles spanning the segment). Setae (bottom) provide for directed friction. (c) Mechanism of the robot segment. (After Omori, H., et al., Locomotion and turning patterns of a peristaltic crawling earthworm robot composed of flexible units, *IROS 2008. IEEE/RSJ International Conference on Intelligent Robots and Systems*, pp. 1632, 2008, fig. 7. With permission from Institute of Electrical and Electronics Engineers.) The plates (gray at left and right side) are placed within a flexible tube (dashed lines) expanding for segment shortening. Rubber plates (bottom: black) deliver friction. Length change is induced by a motor with a crank. (d) In snails the contraction wave of the oblique pedal muscles (Denny 1981) travels anteriad allowing the foot within the slightly (here strongly exaggerated) lifted intersegmental wave to be pulled forward with respect to the static sections. Dark shaded: mucus between ground and foot. (e) In the robot (Chan et al. 2005, 2007) most plates remain in place (black plates), whereas one segment (open shaded) is shifted forward (arrow) at a time. After a circle, the gliding rail (open frame above the segments) is shifted (filled arrow) into the new position indicating progression.

anchoring, but in delicate tubes such as intestines or arteries, pressure should be limited.

Some worms move one segment at a time and explore the difference between the low sliding friction of the moving segment compared to the higher static friction experienced by the numerous stationary segments (e.g., Zimmermann and Zeidis 2007; Zimmermann et al. 2009). Obviously, such a solution does not depend on segment thickening. Closely related are realizations based on the inch-worm principle. In this case, "attachments" are restricted to the leading and trailing edge. The trailing attachment that might be clamped to the ground provides support, whereas the body extends and the leading edge seeks for a new attachment. Thereafter, the contact at the trailing edge can be detached and the body shortened to attach again at the trailing edge.

In addition to a—most frequently—anteriorly traveling muscular wave within the single extended foot of a snail, these animals add mucus. Having dealt with the peristaltic movement observed in annelids and related machines, we are familiar with the backward traveling wave and the friction allowing for propulsion. Although sticky slime certainly helps to enhance friction necessary for propulsion, it seems to be nuisance for the "interwave" segments in which the muscle is slightly lifted and moves forward (Figure 5.38d). Yes, due the shearing induced by forward movement, viscosity of the mucus reduces strongly, but still substantial work has to be done against sliding friction (Lai et al. 2010). The advantage: A robot using this principle can attach to the ceiling and the load is distributed over a large surface. So far, robots mimicking a snail still lack their own mucus production and the wonderful flexible foot of the animals, but the slimy robots are under way (Chan et al. 2007).

In early versions of snake-like robots, that is, slender terrestrial robots progressing in an undulating motion, a large number of wheeled segments where connected by joints. The wheels ensured differences in longitudinal and transversal frictions. The motors actuated the yaw of the joint, whereas the roll and pitch were produced passively. This "active chord mechanism" allowed the production of undulatory waves and propulsion (see Hirose and Morishima 1990). The snake paradigm is rather popular. Recent developments relying on "skin" friction and an actuation allowing for 3D motions enables snake robots to explore a variety of gaits, including side winding, to crawl on stairs, to move within and around pipes, and to swim (Mori and Hirose 2006; Hatton and Choset 2010a, 2010b). The task for the control to manage such systems with its high degrees of freedom still remains challenging.

5.8.2 Legged Vehicles

Legged machines have been built at least since the Renaissance (Rosheim 2006). In such systems, the scientists tried to copy shape and some basic functions by using tools for engineering such as levers, cogs, transmissions, and gears. The idea that animals can serve as an elegant example still prevails today. However, today we are much more curious about the way nature achieves elegant and versatile systems.

Hexapods were built in a variety of sizes. Truck-sized machines such as The Ohio State University's Adaptive Suspension Vehicle (Song and Waldron, 1989; Pugh et al. 1990) machine took advantage of static stability of such a system (Section 5.6). In fact, robotics enforced the development of concepts on the relationship between gait and stability. The vehicle was able to crawl, moving one leg at a time, but also like a quick cockroach explored an alternating tripod (speed ca. 1 m/s). The adaptive legs allowed crossing rough ground; slopes up to 37°; and by combinations of sagittal and lateral motion, the vehicle was able to turn on the spot. Control of the two-segmented leg was simplified by introducing a pantograph mechanism (Section 5.7.5). Because the vehicles possessed low compliance, adaptation to ground had to be delegated to the controls. Different from their tiny animal paragons, but adequate for large systems, a sprawled posture and work of the legs against each other were avoided. A quite different route was chosen by the constructors of RHex (Saranli et al. 2001). They explicitly relied on stability and adaptability of a compliant system. They prescribed their much smaller machine (0.5-m body length) with springy legs that by rotation (open loop) mimicked a tripod gait. The machine is fast (maximum speed up to ~2.5 m/s). Due to its inherent compliance, it is self-stable (Section 5.6) and extremely robust when covering rough ground. It is able to climb stairs on rough slopes of up to 45°. Like in wheeled vehicles, turning is induced by differential speed at both sides. It is able to run upside down and to return on its belly.

This largely passive machine so far outperforms many others relying on sophisticated control such as systems mimicking geometry and degrees of freedom of stick insects (e.g., Weidemann et al. 1994; Roennau et al. 2010). At high speed, the robot bounces and its gait becomes dynamic.

Surprisingly agile is the quadruped BigDog (0.91 m long and 0.76 m high; Boston Dynamics, Waltham, Massachusetts). This vehicle has hydraulically operated legs with three segments (Figure 5.40a). Unfortunately, the engineers stick to the Aristotelian counter-orientation of the legs (Section 5.4). Nevertheless, the machine can walk and climb up a slope of 35°. It trots amazingly smooth across rough terrain (up to ~2 m/s) and is able to bound and to jump across a ditch. Leg compliance is implemented passively and by control. Balance is maintained by leg placement and simultaneously the attitude of the body is controlled (posture). A lateral push of the machine results in a large, quite natural sideward step. This machine is a true successor of the early bouncing robots in which springy telescope legs were used (Raibert 2000; Raibert et al. 2008).

So far, in our biomechanical treatment we did not cover the issue of climbing. The ability to climb as listed above for some quadruped robots can be largely improved in specialized machines. Because static and gliding friction cannot guarantee foothold at steep slopes, the key is in this case a suitable attachment. Engineers for many years presented solutions using hooks, grippers, and suction cups (Nagakubo and Hirose 1994; Yoneda et al. 2001; Maempel et al. 2008; See also *Annual Proceedings of the Climbing and Walking Robots Network*). Provided that there is a chance, gripping can allow for rather effective locomotion in robots mimicking rats (Maempel et al. 2008) or even walking and brachiating apes (Zhiguo et al. 2010). But it does not help on surfaces without grips where suction cups may come into play. To be able to cope with rough surfaces where suction cups will leak, some small machines explore the uneconomic aerodynamic suction (Longo and Muscato 2008). Unraveling the secrets of attachments in arthropods and vertebrates (Autumn et al. 2000, 2006; Kesel et al. 2003; Spolenak et al. 2005; Federle and Clemente 2008; Barnes et al. 2009a; Rind et al. 2011) boosted the development of a new generation of machines. These machines use spiny feet, claws, and especially highly structured hairpads to vertically climb on smooth as well on rough surfaces (Figure 5.39a, b). They usually move one foot at a time, and the kinematics must support attachment as well as detachment of the feet. To overcome adhesion, the toes are peeled off (Goldman et al. 2006; Sangbae et al. 2008; Haynes et al. 2009; Boesel et al. 2010).

Instead of climbing across obstacles, such obstacles of several meters could be surmounted by jumping. Dealing with quadrupedal locomotion, we already mentioned the springy telescope invented by Marc Raibert (rev. in Raibert 2000). Controlling the angle of attack of the leg and the

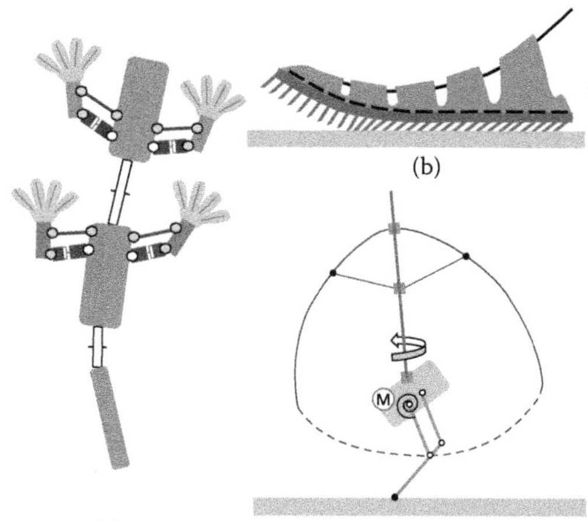

Figure 5.39. Bioinspired climbers and jumpers. (a, b) Stickybot, a robot mimicking climbing of a gecko. (After Sangbae, K., et al., Whole body adehsion: Hierarchical, directional and distributed control of adhesive forces for a climbing robot, *IEEE International Conference on Robotics and Automation*, Roma, 2007; Sangbae, K., et al., *IEEE Trans. Robotics*, 24, 1, 2008, fig, 1, with permission from Institute of Electrical and Electronics Engineers.) (a) Stickybot while climbing on a wall. The flexible feed attach to the wall with dry adhesive. They are attached to pantograph legs, the lower arm including a springy link (brackets). This arm is actuated. The body segments are also linked elastically. (b) Schematic longitudinal section of a toe. (After Sangbae, K., et al., *IEEE Trans. Robotics*, 24, 3, 2008, fig. 5, with permission from Institute of Electrical and Electronics Engineers.) Oblique plastic hairs allow for attachment. Dorsally, a push–pull wire allows peeling the toe starting from its tip as has been observed in nature. (c) EPFL jumping robot mimicking a locust (Kovac et al. 2010). A motor bends the pantograph leg and winds up a spring that can be unloaded for the jump. The mechanism is placed close to the bottom of a wire cage providing for a compliant landing and the stability of a tumbler. To direct the jump, yaw and pitch of the mechanism within the cage can be adjusted before take-off.

legs thrust generated by an air spring, the robot was able to dynamically balance and to change speed. Kangaroo-like arrangements were also under way, demonstrating the feasibility of segmented legs (Zeglin 1991). As in animals, being small can be an advantage with respect to jumping. In fact, small lightweight robots (Armour et al. 2007) show spectacular directed jumps mimicking the mechanism discovered in locusts (Section 5.7). Surrounding the mechanism with compliant wire cage provides for protection during landing and for recovery (Kovac et al. 2010). Not only take-off but also landing must be mastered for a successful jump (Figure 5.39c).

Much effort has been put forward into the development of bipedal walkers and runners. Humanoid machines are envisioned to become human companions, for example, as toys or in health care; artificial birds and even more so the rather similar artificial dinosaurs attract a wide audience. Nevertheless, especially to combine stable posture at slow speed and to be able to run at high speed represents still a rather challenging goal. Currently hundreds of prototypes have been developed, and this list will be outdated before the ink is dry. Some of them are commercially available. This is especially true for small robots (<60 cm) that are used as testbeds for intelligent algorithms by engineers. They have remarkable success, for example, at regulated competitions such as the RoboCup (www.robocup.org) where fast, real-time reactions within a changing environment are demanded. Some of the smaller prototypes are able to jump a backward somersault or show remarkable stability while practicing martial arts. Taller humanoids still largely represent platforms for research and development. The major challenge is the control of above 20 degrees of freedom. The most popular humanoid is ASIMO (Honda, Japan). The chimp-sized (1.3 m; 48 kg; maximum speed 2.5 m/s) robot is able to walk forward, backward, and sideward and to climb stairs. Furthermore, it is able to fall into a slow run and, for example, carry a tray while walking (Sakagami et al. 2002). The HRP-4 (KAWADA Industries, Tokyo, Japan) is a little bit taller (1.51 m; 39 kg) and able to walk (about 0.5 m/s) with almost extended knees and constant waist height and to turn rather dynamically (Kaneko et al. 2011). In the strongly controlled machines leaving only little space for natural dynamics, walking with almost extended knees does represent a significant measure to reduce cost of locomotion. Lola, a humanoid (1.80 m; 55 kg; maximum speed 0.9 m/s; Technical University of Munich, Munich, Germany; Figure 5.40b) has learned to walk fast and across a bumpy ground (Buschmann et al. 2011). Besides of the ambitioned control issue, one key is a carefully designed, lightweight structure and a proper mass distribution. By now, most machines clearly left the static stability phase of early days in which the center of mass was carefully shifted across the polygon of support provided by large feet (Section 5.6.1). Many robots explore ZMP control (Section 5.6.2). Nevertheless, in general, posture is maintained rather rigidly, leaving only little space for passive dynamics and self-stabilizing mechanisms. At this stage, PETMAN (Boston Dynamics), a dynamically heel-toe walking biped, inherited dynamic stability from its hopping and bouncing precursors and comes closest to this goal (Boston Dynamics 2011). The human-sized robot marches at high speed (up to 2 m/s). It is able to crouch, to kneel down, and to get up again, and even performs pushups. Unfortunately, details are not published. In many cases, the machines are less spectacular and far from being truly autonomous robots. However, most of them help to refine constructions and algorithms, and above all to improve our understanding of the mechanics and control of locomotion and their respective interaction (Wisse et al. 2004; Iida et al. 2009; Seyfarth et al. 2009; Koepl and Hurst 2011).

This short review may give the reader an idea of the wide range of current developments in the area of bioinspired terrestrial robots. Taking resort to natural examples has been and will be a successful strategy. By building such machines, we are able to test our understanding of the biological systems (e.g., Long 2012). In the robots listed above, there is a trend toward system compliance. The compliance may be attributed to corresponding materials (e.g., RHex) or largely to control (e.g., BigDog). Our understanding on how such properties lead to "modes" of locomotion will improve and inspire engineers. It seems that the tuning of material properties to the task without losing versatility is one of the key issues where we still can learn from nature. Hereby, details such as geometry and arrangement will reveal their physical meaning. We only touched the issue of the muscular drive. As such, a drive incorporates compliance and damping

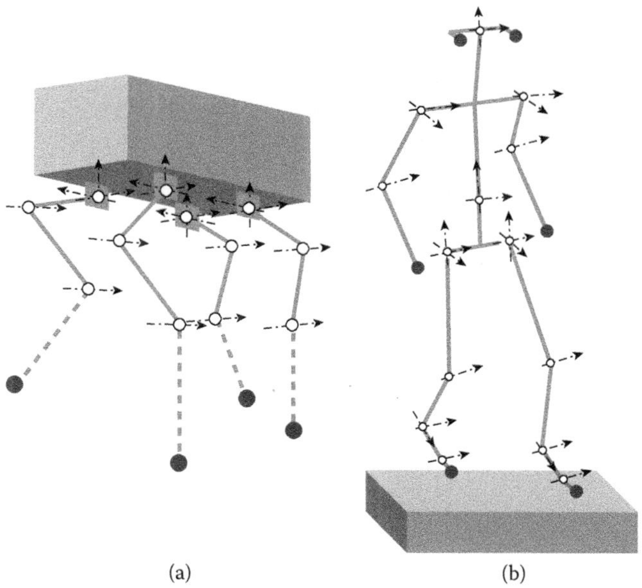

Figure. 5.40. Degrees of freedom of a popular quadruped and a biped. (a) BigDog. (After Raibert, M., et al., BigDog, the rough terrain quadruped robot, *17th World Congress of the International Federation of Automatic Control (IFAC)*, Seoul, pp. 10822–10825, 2008.) Each leg is attached to the trunk with three rotational degrees of freedom. Knee and ankle are implemented as hinge joints. The distal segment (dashed) entails a telescope spring. (b) Lola. (After Buschmann, T., et al., Experiments in fast biped walking, *2011 IEEE International Conference on Mechatronics (ICM)*, pp. 863–868, 2011.) The legs of the humanoid posses a spherical hip joint and a hinge joint at the knee. The ankle comprises two hinge joints. Rolling of the feet is facilitated by another hinge joint at the toes. The trunk can be rotated in two directions with respect to the pelvis, similarly the head as well as the arms at the shoulder with respect to the trunk. The elbow is implemented as a hinge joint.

identifying technical drives with similar properties will be an important area of future research. We are just beginning to reveal how the intelligent organization of muscle–skeletal systems facilitates control. This includes the adjusted implementation of materials such as the muscular tissue. It is highly probable that shock absorption by wobbling tissues that come for free in animals will also find its way in the construction of fast machines. And for large machines, we may admit that protecting the lightweight segments with a viscoelastic damping coat may be suitable. We will probably not copy the multipurpose muscle tissue, but use several translations to mimic the different functions. Animal joints may also serve as examples. We are able to make decent copies for joint replacement. However, there may be much more than the simple degree of freedom and load bearing to be worthwhile to transfer into robotics. Biomechanists developing forward models are aware of the significance to implement proper passive joint properties. It may well be that in future plastic skeletons and rubber drives with embedded flexible electronics may dominate machines, and technical runners may look and perform more and more like their animal model. The direction of development will strongly depend on future tasks devoted to such machines. Currently, tasks such as exploration, transport, and support seem to be within reach. With engineers becoming more familiar with the concepts, their fantasy may open realms so far unexplored by their animal precursors and spur technical evolution.

SYMBOLS

a:	coefficient (e.g., Fourier), element (matrix)
b:	constant
d:	distance between center of mass and hip (cockroach), (jumping) distance
$\mathbf{e}$:	unit vectors
e:	coefficient of restitution; e_g: variant
f:	frequency, function (e.g., muscle, normalized to 1)
g:	gravitational acceleration
h:	height, distance between line of action and joint, hip height, apex height
m:	mass, leg number
n:	positive integer (e.g., number of legs)
k:	stiffness
l:	length of leg, of mathematical pendulum, of leg segments
p:	impulse, momentum, scaling exponent, pressure
q:	$\frac{t_c}{T_n}$, displacement of suspended mass
$\mathbf{r}$:	position vector
r:	storage ratio
s:	stride length, displacement of suspended mass
sig:	signum function
t:	time
$\mathbf{t}$:	tangential vector
u:	velocity, eigenvector, activationvector
v:	velocity
$\mathbf{v}$:	velocity, eigenvector
w:	displacement (bar), eigenvalues
y:	vertical displacement
x:	horizontal displacement, variable
A:	area
C:	constant
CR:	compression ratio
CT:	cost of transport
D:	torsional spring stiffness
DF:	duty factor
E:	energy, Young's modulus, excitation (muscle)
ER:	energy recovery
F:	function
$\mathbf{F}$:	force
G:	global function, gear ratio
H:	sum of heights
$\mathbf{H}$:	Hessian
I:	moment of inertia, identity matrix
J:	second moment of inertia, optimization function
$\mathbf{J}$:	Jacobi matrix
L:	Lagrange function
$\mathbf{L}$:	angular momentum
M:	moment, torque
P:	power
Q:	shearing force
R:	rotation matrix
S:	distances (static stability)
T:	(stride-) period; time dependent term of bar deflection
V:	potential, volume
W:	work
X:	space dependent term of bar deflection; real number
$\mathbf{X}$:	state
α:	angle of the spring between horizontal and axis
β:	glancing angle
γ:	slope angle, scalar factor
δ:	angle between heading and velocity
φ:	angle of the spring between vertical and axis, angle of swing leg, angles of segments with respect to the vertical
λ:	dimensionless length (indicating the bending mode of the bar), coefficient in cost function, eigenvalue, acceleration ratio
μ:	real part (of complex number), dynamic viscosity
ν:	imaginary part (of complex number), exponent (power function)
ψ:	inclination angle of rotation axis (static stability); angle center of gravity (cockroach)
ρ:	density
σ:	Floquet exponent
θ:	orientation of body with respect to y-axis; yaw angle
ω:	cycle frequency
ζ:	eigenvalue (Floquet multiplier)
φ:	phase angle
Λ:	coefficient in collision models

Operators

$\dot{}$:	first time derivative
$\ddot{}$:	second time derivative
$'$:	first position derivative
$''$:	second position derivative
(4):	fourth derivative
$\circ$:	scalar product
$\times$:	vector cross product
d:	differential
det:	determinant
fract:	fractional part

∂: partial differential
Δ: virtual differential
Δ: difference
∇: nabla vector

Indices, Suffixes

0: refers to $t = 0$, fundamental mode
1,2,3: segment numbers, cost functions, margins
12, etc: between segments 1 and 2
a: aerial or flight phase
c: contact
ch: channel
ds: double support (distance)
ext: external
gravity: gravitational acceleration
gyr: gyration
h: homogeneous
hum: human
hor: horizontal
i: 1,...,n
iso: isometric
inh: inhomogeneous
k: 1,...,n
kin: kinetic (energy)
l: leg, length (muscle)
leg: leg
lead: leading
loc: locust
L: load
m: muscle ...
max: maximum
n: natural (frequency); leg number, step number
p: pressure
pot: potential (energy)
rot: rotational
s: (stable) goal (of state)
seg: segment
spring: generated by the spring
step: referring to a step (in contrast to a stride)
ten: tendon
trail: trailing
trans: translational
tot: total (energy)
u: unloaded (leg length)
v: velocity, force velocity
x: component in x
CM: center of mass
COP: center of pressure
F: force, fluid
$\hat{F}$: Froude related number (facilitates description of collisions)
Fr: Froude (speed)
H: hip
P: index to S distance of the centers of supports/stride length
R: index to S longitudinal leg tip translation during support/stride length, reaction force
T: transposed (matrix)
TD: touchdown
TO: take-off
bold: vectors, tensors, matrices

REFERENCES

Books

Alexander, R. M. 1992. *Exploring Biomechanics: Animals in Motion*. W. H. Freeman, New York.

Alexander, R. M. (1988). Why Mammals Gallop. *American Zoologist* 28, 237–245.

Alexander, R. M. 2006. *Principles of Animal Locomotion*. Princeton University Press, New York.

Bekker, M. G. 1956. *Theory of Land Locomotion: The Mechanics of Vehicle Mobility*. University of Michigan Press, Ann Arbor, MI.

Biewener, A. A. 1989. *Scaling body support in mammals:* limb posture and muscle mechanics. *Science* 245, 45–48.

Biewener, A. A. 1992. *Biomechanics: Structures and Systems. A Practical Approach*. Oxford University Press, New York.

Biewener, A. A. 2003. *Animal Locomotion*. Oxford University Press, New York.

Borelli, G. A. 1680. *De Motu Animalium*. Angeli Bernabo, Rome.

Cartmill, M. 1985. Climbing. In *Functional Vertebrate Morphology* (eds. M. Hildebrand, D. M. Bramble, K. F. Liem and D. B. Wake), pp. 73–89. Harvard University Press, Cambridge, MA.

Edwards, J. L. 1985. Terrestrial locomotion without appendages. In *Functional Vertebrate Morphology* (eds. M. Hildebrand, D. M. Bramble, K. F. Liem and D. B. Wake), pp. 159–172. Harvard University Press, Cambridge, MA.

Ertelt, T. 2008. *Kraftmorphologie der menschlichen Beinbewegung—Elektrmyographische und kinematische Einflüsse frequenzbedingter Schlittensprünge*. Kovac, Hamburg.

Full, R. J. 1989. Mechanics and energetics of locomotion: From bipeds to polypeds. In *Energy Transformations in Cells and Organisms* (eds. W. Wieser and E. Gnaiger), pp. 175–182. Thieme, Stuttgart.

Fung, Y. C. 1993. *Biomechanics*. Springer, New York.

Giesl, P. and Wagner, H. 2006. On the determination of the basin of attraction for stationary and periodic movements. In *Fast Motions in Biomechanics and Robotics*, vol. 340 (eds. M. Diehl and K. Mombaur), pp. 147–166. Springer, Berlin.

Giesl, P. and Wagner, H. 2008. Mathematical stability analysis in biomechanical applications. In *Mathematical Biology Research Trends* (ed. L. B. Wilson), pp. 261–274. Nova Science, Hauppauge, NY.

Gordon, M. S., and E. C. Olson. 1995. *Invasions of the Land: The Transitions of Organisms from Aquatic to Terrestrial Life*. Columbia University Press, New York.

Greiner, W. 2002. *Classical Mechanics: Systems of Particles and Hamiltonian Dynamics*. Springer, Heidelberg, Germany.

Hay, J. G. 1993. *The Biomechanics of Sports Techniques*. Prentice Hall, Upper Saddle River, NJ.

Hildebrand, M., D. M. Bramble, K. F. Liem, and D. B. Wake. 1985 *Functional Vertebrate Morphology*. Harvard University Press, Cambridge, MA.

Hochmuth, G. 1974. *Biomechanik sportlicher Bewegungen*. Sportverlag, Berlin.

Inman, V. T., Ralston, H. J., Todd, F. 1981 *Human walking*. Williams & Wilkins, Baltimore, pp. 154.

Little, C. 1990. *The Terrestrial Invasion: An Ecophysiological Approach to the Origins of Land Animals*. Cambridge University Press, Cambridge, UK.

Long, J. H. 2012. *Darwin's Devices. What Evolving Robots Can Teach Us About the History of Life and the Future of Technology*. Basic Books, New York.

Margeria, R. 1976. *Biomechanics and Energetics of Muscular Exercise*. Clarendon Press, Oxford, UK.

Maus, M., Rummel, J. and Seyfarth, A. (2008). Stable upright walking and running using a simple pendulum based control scheme. In *Advances in Mobile Robotics* (eds. L. Marques, A. de Almeida, M. O. Tokhi and G. S. Virk), pp. 623–629. World Scientific, Coimbra, Portugal.

McMahon, T. A. 1984. *Muscles, Reflexes and Locomotion*. Princeton University Press, Princeton, NJ.

NASA. 1978. *Anthropometric Source Book*. NASA Scientific and Technical Information Office, Springfield, VA.

Nussbaum, M. C. 1986. *Aristotle's De Motu Animalium*. Princeton University Press, Princeton, NJ.

Raibert, M. 2000. *Legged Robots That Balance*. MIT Press, Cambridge, MA.

Rosheim, M. 2006. *Leonardo's Lost Robots*. Springer, Heidelberg, Germany.

Shubin, N. 2008. *Your Inner Fish: A Journey into the 3.5-Billion-Year History of the Human Body*. Vintage, Random House, New York. 256 pp.

Snodgras, R. E. 1935. *Principles of Insect Morphology*. McGraw-Hill, New York (reprint 1993, Cornell University Press).

Song, S.-M., and K. J. Waldron. 1989. *Machines That Walk: The Adaptive Suspension Vehicle*. MIT Press, Cambridge, MA.

Vaughan, C. L., B. L. Davis, and J. C. O'Connor. 1996. *Dynamics of Human Gait*. Human Kinetics, Champaign, IL.

Vukobratovic, M., B. Borovac, D. Surla, and D. Stikic. 1990. *Biped Locomotion—Dynamics, Stability, Control, and Application*. Springer, Heidelberg, Germany.

Winter, D. A. 2009. *Biomechanics and Control of Human Movement*. Wiley, New York.

Winters, J. M., and S. L. Y. Woo. 1990. *Multiple Muscle System*. Springer, New York.

Zatsiorsky, V. M. 2002. *Kinetics of Human Locomotion.* Human Kinetics, Champaign, IL.

Zimmermann, K., I. Zeidis, and C. Behen. 2009. *Mechanics of Terrestrial Locomotion: With a Focus on Non-pedal Motions Systems.* Springer, Heidelberg, Germany.

Reviews

Alexander, R. M. (2005). Models and the scaling of energy costs for locomotion. *J Exp Biol* 208, 1645–1652.

Biewener, A. A. (2005). Biomechanical consequences of scaling. *J Exp Biol* 208, 1665–1676.

Blickhan, R., Seyfarth, A., Geyer, H., Grimmer, S., Wagner, H. and Guenther, M. (2007). Intelligence by mechanics. *Philos Trans A Math Phys Eng Sci* 365, 199–220.

Dickinson, M. H., Farley, C. T., Full, R. J., Koehl, M. A. R., Kram, R. and Lehman, S. (2000). How animals move: An integrative view. *Science* 288, 100–106.

Fischer, M. S. and Blickhan, R. (2006). The tri-segmented limbs of therian mammals: Kinematics, dynamics, and self-stabilization—A review. *J Exp Zool A Comp Exp Biol* 305, 935–952.

Kukillaya, R., Proctor, J. and Holmes, P. (2009). Neuromechanical models for insect locomotion: Stability, maneuverability, and proprioceptive feedback. *Chaos* 19, 026107.

Long, J. A. and Gordon, M. S. (2004). The greatest step in vertebrate history: A paleobiological review of the fish-tetrapod transition. *Physiol Biochem Zool* 77, 700–719.

Saibene, F. and Minetti, A. E. (2003). Biomechanical and physiological aspects of legged locomotion in humans. *Eur J Appl Physiol* 88, 297–316.

van Ingen Schenau, G. J. (1989). On the action of bi-articular muscles, a review. *Neth J Zool* 40, 521–543.

Vukobratovic, M. and Borovac, B. (2004). Zero-moment point—Thirty five years of its life. *Int J Humanoid Rob* 1, 157–173.

Witte, H., Hoffmann, H., Hackert, R., Schilling, C., Fischer, M. S. and Preuschoft, H. (2004). Biomimetic robotics should be based on functional morphology. *J Anat* 204, 331–342.

Woledge, R. C., Curtin, N. A. and Homsher, E. (1985). Energetic aspects of muscle contraction. *Monogr Physiol Soc* 41, 1–357.

Research Papers

Aerts, P. (1998). Vertical jumping in *Galago senegalensis*: The quest for an obligate mechanical power amplifier. *Philos Trans Roy Soc Lond Ser B Biol Sci* 353, 1607–1620.

Ahlborn, B. K., Blake, R. W. and Megill, W. M. (2006). Frequency tuning in animal locomotion. *Zoology (Jena)* 109, 43–53.

Ahn, A. N., Meijer, K. and Full, R. J. (2006). In situ muscle power differs without varying in vitro mechanical properties in two insect leg muscles innervated by the same motor neuron. *J Exp Biol* 209, 3370–3382.

Aleshinsky, S. Y. (1986a). An energy "sources" and "fractions" approach to the mechanical energy expenditure problem–II. Movement of the multi-link chain model. *J Biomech* 19, 295–300.

Aleshinsky, S. Y. (1986b). An energy "sources" and "fractions" approach to the mechanical energy expenditure problem–IV. Criticism of the concept of 'energy transfers within and between links'. *J Biomech* 347, 307–309.

Alexander, R. M. (1988). Why mammals gallop. *American Zoologist* 28, 237–245.

Alexander, R. M. (1990). Optimum take-off techniques for high and long jumps. *Philos Trans Roy Soc Lond Ser B Biol Sci* 329, 3–10.

Alexander, R. M. (1991). Energy-saving mechanisms in walking and running. *J Exp Biol* 160, 55–69.

Alexander, R. M. (1992). A model of bipedal locomotion on compliant legs. *Philos Trans Roy Soc Lond Ser B Biol Sci* 338, 189–198.

Alexander, R. M. (1995). Leg design and jumping technique for humans, other vertebrates and insects. *Philos Trans Roy Soc Lond Ser B Biol Sci* 347, 235–248.

Anderson, F. C. and Pandy, M. G. (1993). Storage and utilization of elastic strain energy during jumping. *J Biomech* 26, 1413–1427.

Armour, R., Paskins, K., Bowyer, A., Vincent, J., Megill, W. and Bomphrey, R. (2007). Jumping robots: A Biomimetic Solution to Locomotion Across Rough Terrain. *Bioinspiration & Biomimetics* 2, S65–S82.

Armstrong, S. (1990). Fog, wind and heat: Life in the Namib desert. *New Scientist* 127, 46–50.

Ashley-Ross, M. A., Perlman, B. M., Gibb, A. C. and Long, J. H. (2014). Jumping sans legs: Does elastic energy storage by the vertebral column power terrestrial jumps in bony fishes? *Zoology*: 117, 7–18.

Autumn, K., Hsieh, S. T., Dudek, D. M., Chen, J., Chitaphan, C. and Full, R. J. (2006). Dynamics of geckos running vertically. *J Exp Biol* 209, 260–272.

Autumn, K., Liang, Y. A., Hsieh, S. T., Zesch, W., Chan, W. P., Kenny, T. W., Fearing, R. and Full, R. J. (2000). Adhesive force of a single gecko foot-hair. *Nature* 405, 681–685.

Azizi, E. and Roberts, T. J. (2010). Muscle performance during frog jumping: Influence of elasticity on muscle operating lengths. *Proc Roy Soc B Bio Sci* 277, 1523–1530.

Bao Kha, N., Jordan, H. B., Abbas, A. D.-S. and Netta, C. (2009). A C. elegans-inspired microrobot with polymeric actuators and online vision. In *Proceedings of the 2009 International Conference on Robotics and Biomimetics*. Guilin, China: IEEE Press.

Barnes, W. J. P., Scholz, I. and Federle, W. (2009a). Structural correlates of adhesion and friction in tree frog toe pads. *Comp Biochem Physiol A Mol Integr Physiol* 153A, S123.

Barnes, W. J. P., Scholz, I., Smith, J. M. and Baumgartner, W. (2009b). Ultrastructure and physical properties of an adhesive surface, the toe pad epithelium of the tree frog, *Litoria caerulea* White. *J Exp Biol* 212, 155–162.

Bejan, A. and Marden, J. H. (2006). Unifying constructal theory for scale effects in running, swimming and flying. *J Exp Biol* 209, 238–248.

Bennet-Clark, H. C. (1975). The energetics of the jump of the locust *Schistocerca gregaria*. *J Exp Biol* 63, 53–83.

Bennet-Clark, H. C. and Lucey, E. C. (1967). The jump of the flea: A study of the energetics and a model of the mechanism. *J Exp Biol* 47, 59–67.

Biewener, A. A., Farley, C. T., Roberts, T. J. and Temaner, M. (2004). Muscle mechanical advantage of human walking and running: Implications for energy cost. *J Appl Physiol* 97, 2266–2274.

Biknevicius, A. R. and Reilly, S. M. (2006). Correlation of symmetrical gaits and whole body mechanics: Debunking myths in locomotor biodynamics. *J Exp Zool A Comp Exp Bio* 305, 923–934.

Blickhan, R. (1989). The spring-mass model for running and hopping. *J Biomech* 22, 1217–1227.

Blickhan, R. and Barth, F. G. (1985). Strains in the exoskeleton of spiders. *J Comp Physiol A Neuroethol Sen Neural Behav Physiol*, 157, 115–147.

Blickhan, R. and Full, R. (1992). Mechanical work in terrestrial locomotion. In *Biomechanics: Structures and Systems. A Practical Approach* (ed. A. A. Biewener), pp. 75–96. New York: Oxford University Press.

Blickhan, R. and Full, R. J. (1993). Similarity in multilegged locomotion: Bouncing like a monopode. *J Comp Physiol A Neuroethol Sen Neural Behav Physiol* 173, 509–517.

Bobbert, M. F. (2001). Dependence of human squat jump performance on the series elastic compliance of the triceps surae: A simulation study. *J Exp Biol* 204, 533–542.

Boesel, L. F., Greiner, C., Arzt, E. and del Campo, A. (2010). Gecko-inspired surfaces: A path to strong and reversible dry adhesives. *Adv Mat* 22, 2125–2137.

Boston Dynamics. (2011). Retrieved from http://www.bostondynamics.com/robot_petman.html

Bramble, D. M. (1989). Axial-appendicular dynamics and the integration of breathing and gait in mammal. *American Zoologist* 29, 171–186.

Browning, R. C. and Kram, R. (2009). Pound for pound: Working out how obesity influences the energetics of walking. *J Appl Physiol* 106, 1755–1756.

Bullimore, S. R. and Burn, J. F. (2004). Distorting limb design for dynamically similar locomotion. *Proc Roy Soc B Bio Sci* 271, 285–289.

Bullimore, S. R. and Burn, J. F. (2006). Consequences of forward translation of the point of force application for the mechanics of running. *Journal of Theoretical Biology* 238, 211–219.

Burrows, M. and Wolf, H. (2002). Jumping and kicking in the false stick insect *Prosarthria teretrirostris*: Kinematics and motor control. *J Exp Biol* 205, 1519–1530.

Buschmann, T., Favot, V., Lohmeier, S., Schwienbacher, M. and Ulbrich, H. (2011). Experiments in fast biped walking. In *2011 IEEE International Conference on Mechatronics (ICM)*, IEEE, Istanbul, Turkey, pp. 863–868.

Carroll, S. (2016). *The Big Picture. On the Origins of Life, Meaning and the Universe Itself*. Dutton, New York, 480 pp.

Cartmill, M. (1985). Climbing. In *Functional Vertebrate Morphology* (eds. M. Hildebrand, D. M. Bramble, K. F. Liem and D. B. Wake), pp. 73–89. Cambridge, MA: Harvard University Press.

Cavagna, G. A. (1975). Force platforms as ergometers. *J Appl Physiol* 39, 174–179.

Cavagna, G. A., Heglund, N. C. and Taylor, C. R. (1977). Mechanical work in terrestrial locomotion: Two basic mechanisms for minimizing energy expenditure. *Am J Physiol* 233, R243–R261.

Cavagna, G. A. and Kaneko, M. (1977). Mechanical work and efficiency in level walking and running. *J Physiol* 268, 467–481.

Cavagna, G. A., Saibene, F. P. and Margaria, R. (1963). External work in walking. *J Appl Physiol* 18, 1–9.

Chan, B., Balmforth, N. J. and Hosoi, A. E. (2005). Building a better snail: Lubrication and adhesive locomotion. *Phys Fluids* 17, 113101.

Chan, B., Ji, S., Koveal, C. and Hosoi, A. E. (2007). Mechanical devices for snail-like locomotion. *J Int Mat Sys Struct* 18, 111–116.

Chapman, A. E. and Caldwell, G. E. (1983). Kinetic limitations of maximal sprinting speed. *J Biomech* 16, 79–83.

Clack, J. A. (2006). The emergence of early tetrapods. *Palaeogeogr Palaeoclimatol Palaeoecol* 232, 167–189.

Cohen, N. and Boyle, J. H. (2011). Swimming at low Reynolds number: A beginners guide to undulatory locomotion. *Cont Phys* 51, 103–123.

Crofts, S. B. and Summers, A. P. (2011). Biomechanics: Swimming in the Sahara. *Nature* 472, 177–178.

Daley, M. A. and Biewener, A. A. (2006). Running over rough terrain reveals limb control for intrinsic stability. *Proc Natl Acad Sci USA* 103, 15681–15686.

Denny, M. W. (1981). A quantitative model for the adhesive locomotion of the terrestrial slug, *J Exp Biol* 91, 195–217.

Dowling, J. J., Durkin, J. L. and Andrews, D. M. (2006). The uncertainty of the pendulum method for the determination of the moment of inertia. *Med Eng Phys* 28, 837–841.

Eriten, M. and Dankowicz, H. (2009). A rigorous dynamical-systems-based analysis of the self-stabilizing influence of muscles. *J Biomech Eng* 131, 011011.

Farley, C. T., Blickhan, R., Saito, J. and Taylor, C. R. (1991). Hopping frequency in humans: A test of how springs set stride frequency in bouncing gaits. *J Appl Physiol* 71, 2127–2132.

Federle, W. and Clemente, C. J. (2008). Pushing versus pulling: Division of labour between tarsal attachment pads in cockroaches. *Proc Roy Soc B Bio Sci* **275**, 1329–1336.

Fischer, M. S. and Witte, H. (2007). Legs evolved only at the end! *Philos Transact A Math Phys Eng Sci* 365, 185–98.

Frohlich, C. (1979). Do springboard divers violate angular momentum conservation? *American Journal of Physics* 47, 583–592.

Frolich, L. M. and Biewener, A. A. (1992). Kinematic and electromyographic analysis of the functional-role of the body axis during terrestrial and aquatic locomotion in the Salamander *Ambystoma tigrinum*. *J Exp Biol* 162, 107–130.

Full, R., Earls, K., Wong, M. and Caldwell, R. (1993). Locomotion like a wheel? *Nature* 365, 495–495.

Full, R. J., Blickhan, R. and Ting, L. H. (1991). Leg design in hexapedal runners. *J Exp Biol* 158, 369–390.

Full, R. J. and Koehl, M. A. R. (1993). Drag and lift on running insects. *J Exp Biol* 176, 89–101.

Full, R. J. and Tu, M. S. (1991). Mechanics of a rapid running insect: Two-, four- and six-legged locomotion. *J Exp Biol* 156, 215–231.

Gal, J. M. (1993a). Mammalian spinal biomechanics. I. Static and dynamic mechanical properties of intact intervertebral joints. *J Exp Biol* 174, 247–280.

Gal, J. M. (1993b). Mammalian spinal biomechanics. II. Intervertebral lesion experiments and mechanisms of bending resistance. *J Exp Biol* 174, 281–297.

Garcia, E. and de Santos, P. G. (2005). An improved energy stability margin for walking machines subject to dynamic effects. *Robotica* 23, 13–20.

Garcia, M., Chatterjee, A. and Ruina, A. (2000). Efficiency, speed, and scaling of two-dimensional passive-dynamic walking. *Dynam Stab Sys* 15, 75–99.

Garcia, M., Chatterjee, A., Ruina, A. and Coleman, M. (1998). The simplest walking model: Stability, complexity, and scaling. *J Biomech Eng* 120, 281–288.

Geyer, H. (2005). *Simple models of legged locomotion based on compliant leg behavior. PhD Dissertation.* Fakulty of Social and Behavioral Science, Friedrich-Schiller-University, Jena.

Geyer, H., Seyfarth, A. and Blickhan, R. (2005). Spring-mass running: Simple approximate solution and application to gait stability. *J Theor Biol* 232, 315–328.

Geyer, H., Seyfarth, A. and Blickhan, R. (2006). Compliant leg behaviour explains basic dynamics of walking and running. *Proc Roy Soc B Bio Sci* 273, 2861–2867.

Ghigliazza, R. M., Altendorfer, R., Holmes, P. and Koditschek, D. (2005). A simply stabilized running model. *SIAM Review* 47, 519–549.

Giesl, P. (2008). Construction of a local and global Lyapunov function for discrete dynamical systems using radial basis functions. *J Approx Theo* 153, 184–211.

Giesl, P. and Wagner, H. (2007). Lyapunov function and the basin of attraction for a single-joint muscle-skeletal model. *J Math Biol* 54, 453–464.

Gillis, G. B., Bonvini, L. A. and Irschick, D. J. (2009). Losing stability: Tail loss and jumping in the arboreal lizard *Anolis carolinensis*. *J Exp Biol* 212, 604–609.

Glozman, D., Hassidov, N., Senesh, M. and Shoham, M. (2010). A self-propelled inflatable earthworm-like endoscope actuated by single supply line. *IEEE Trans Biomed Eng* 57, 1264–1272.

Goldman, D. I., Chen, T. S., Dudek, D. M. and Full, R. J. (2006). Dynamics of rapid vertical climbing in cockroaches reveals a template. *J Exp Biol* 209, 2990–3000.

Gordon, M. S., Ng, W. W. and Yip, A. Y. (1978). Aspects of the physiology of terrestrial life in amphibious fishes. III. The Chinese mudskipper *Periophthalmus cantonensis*. *J Exp Biol* 72, 57–75.

Greven, H. and Schüttler, L. (2001). How to crawl and dehydrate on moss. *Zoologischer Anzeiger* 240, 341–344.

Gruber, K., Ruder, H., Denoth, J. and Schneider, K. (1998). A comparative study of impact dynamics: wobbling mass model versus rigid body models. *J Biomech* 31, 439–444.

Gunther, M. and Blickhan, R. (2002). Joint stiffness of the ankle and the knee in running. *J Biomech* 35, 1459–1474.

Gunther, M., Keppler, V., Seyfarth, A. and Blickhan, R. (2004). Human leg design: Optimal axial alignment under constraints. *J Math Biol* 48, 623–646.

Günther, M., Sholukha, V. A., Kessler, D., Wank, V. and Blickhan, R. (2003). Dealing with skin motion and wobbling masses in inverse dynamics. *J Mech Med Biol* 3, 309–335.

Günther, M. and Weihmann, T. (2012). Climbing in hexapods: A plain model for heavy slopes. *J Theor Biol* 293, 82–86.

Guschlbauer, C., Scharstein, H. and Buschges, A. (2007). The extensor tibiae muscle of the stick insect: Biomechanical properties of an insect walking leg muscle. *J Exp Biol* 210, 1092–1108.

Hackert, R., Schilling, N. and Fischer, M. S. (2006). Mechanical self-stabilization, a working hypothesis for the study of the evolution of body proportions in terrestrial mammals? *C R Palevol* 5, 541–549.

Hatton, R. L. and Choset, H. (2010a). Generation of gaits for snake robots: Annealed chain fitting and key frame wave extraction. *Auton Robots* 28, 271–281.

Hatton, R. L. and Choset, H. (2010b). Sidewinding on slopes. In *2010 IEEE International Conference on Robotics and Automation (ICRA)*, pp. 691–696.

Hatze, H. (1976). The complete optimization of a human motion. *Math Biosci* 28, 99–135.

Haussler, K. K., Bertram, J. E., Gellman, K. and Hermanson, J. W. (2001). Segmental in vivo vertebral kinematics at the walk, trot and canter: A preliminary study. *Equ Veter J Suppl*, 33, 160–164.

Haynes, G. C., Khripin, A., Lynch, G., Amory, J., Saunders, A., Rizzi, A. A. and Koditschek, D. E. (2009). Rapid pole climbing with a quadrupedal robot. In *IEEE International Conference on Robotics and Automation*, pp. 2767–2772.

Heglund, N. C., Taylor, C. R. and Mcmahon, T. A. (1974). Scaling stride frequency and gait to animal size: Mice to horses. *Science* 186, 1112–1113.

Heglund, N. C., Willems, P. A., Penta, M. and Cavagna, G. A. (1995). Energy-saving gait mechanics with head-supported loads. *Nature* 375, 52–54.

Heitler, W. J. (1974). The locust jump. *J Comp Physiol A Neuroethol Sen Neural Behav Physiol* 89, 93–104.

Hildebrand, M. (1965). Symmetrical gaits of horses. *Science* 150, 701–708.

Hildebrand, M. (1977). Analysis of asymmetrical gaits. *J Mammal* 58, 131–156.

Hirose, S. and Morishima, A. (1990). Design and control of a mobile robot with an articulated body. *Int J Rob Res* 9, 99–114.

Hoerler, E. (1974). Hochsprungmodell. *Jugend und Sport* 30, 271–273.

Hogan, N. and Sternad, D. (2013). Dynamic primitives in the control of locomotion. *Front Comput Neurosci* 7, 1–16

Holmes, P., Full, R. J., Koditschek, D. and Guckenheimer, J. (2006). The dynamics of legged locomotion: Models, analyses, and challenges. *SIAM Review* 48, 207–304.

Huat, O. J., Ghista, D. N., Beng, N. K. and John, T. C. C. (2004). Optimal stride frequency computed from the double-compound pendulum of the leg, and verified experimentally as the preferred stride frequency of jogging. *Inter J Comput App Tech* 21, 46–51.

Iida, F., Minekawa, Y., Rummel, J. and Seyfarth, A. (2009). Toward a human-like biped robot with compliant legs. *Rob Auton Syst* 57, 139–144.

Ijspeert, A. J., Crespi, A., Ryczko, D. and Cabelguen, J. M. (2007). From swimming to walking with a salamander robot driven by a spinal cord model. *Science* 315, 1416–1420.

Jacobs, R., Bobbert, M. F. and van Ingen Schenau, G. J. (1996). Mechanical output from individual muscles during explosive leg extensions: The role of biarticular muscles. *J Biomech* 29, 513–523.

Jayes, A. S. and Alexander, R. M. (1980). The gaits of chelonians: Walking techniques for very low speeds. *J Zool* 191, 353–378.

Jayne, B. (1986). Kinematics of terrestrial snake locomotion. *Copeia* 22, 915–927.

Jindrich, D. L. and Full, R. J. (2002). Dynamic stabilization of rapid hexapedal locomotion. *J Exp Biol* 205, 2803–2823.

Jusufi, A., Goldman, D. I., Revzen, S. and Full, R. J. (2008). Active tails enhance arboreal acrobatics in geckos. *Proc Natl Acad Sci USA* 105, 4215–4219.

Kaneko, K., Kanehiro, F., Morisawa, M., Akachi, K., Miyamori, G., Hayashi, A., et al. (2011). Humanoid robot HRP-4—Humanoid robotics platform with lightweight and slim body. In *2011 IEEE/RSJ International Conference on Intelligent Robots and Systems (IROS)*, pp. 4400–4407.

Kargo, W. J., Nelson, F. and Rome, L. C. (2002). Jumping in frogs: Assessing the design of the skeletal system by anatomically realistic modeling and forward dynamic simulation. *J Exp Biol* 205, 1683–1702.

Kesel, A. B., Martin, A. and Seidl, T. (2003). Adhesion measurements on the attachment devices of the jumping spider *Evarcha arcuata*. *J Exp Biol* 206, 2733–2738.

Kesel, A. B., Martin, A. and Seidl, T. (2004). Getting a grip on spider attachment: An AFM approach to microstructure adhesion in arthropods. *Smart Mater Struct* 13, 512–518.

Knuesel, H., Geyer, H. and Seyfarth, A. (2005). Influence of swing leg movement on running stability. *Hum Movement Sci* 24, 532–543.

Koepl, D. and Hurst, J. W. (2011). Force control for plan spring-mass running. In *IEEE International Conference on Intelligent Robots and Systems (IROS)*.

Kovac, M., Schlegel, M., Zufferey, J.-C. and Floreano, D. (2010). Steerable miniature jumping robot. *Auton Robots* 28, 295–306.

Kubow, T. M. and Full, R. J. (1999). The role of the mechanical system in control: A hypothesis of self-stabilization in hexapedal runners. *Philos Trans Roy Soc Lond Ser B Biol Sci* 354, 849–861.

Kukillaya, R., Proctor, J. and Holmes, P. (2009). Neuromechanical models for insect locomotion: Stability, maneuverability, and proprioceptive feedback. *Chaos* 19, 026107.

Kuo, A. D., Donelan, J. M. and Ruina, A. (2005). Energetic consequences of walking like an inverted pendulum: Step-to-step transitions. *Exerc Sport Sci Rev* 33, 88–97.

Kuznetsov, A. N. (1995). Energetical profit of the third segment in parasagittal legs. *J Theo Biol* 172, 95–105.

LaBarbera, M. (1983). Why the wheels won't go. *Am Nat* 121, 395–408.

Lai, J. H., del Alamo, J. C., RodrÃguez-RodrÃguez, J. and Lasheras, J. C. (2010). The mechanics of the adhesive locomotion of terrestrial gastropods. *J Exp Biol* 213, 3920–3933.

Lauga, E., Lee, S. Y., Bush, J. W. M. and Hosoi, A. E. (2008). Crawling beneath the free surface: Water snail locomotion. *Phys Fluids* 20.

Lee, D. V. and Meek, S. G. (2005). Directionally compliant legs influence the intrinsic pitch behaviour of a trotting quadruped. *Proc Roy Soc B Bio Sci* 272, 567–572.

Lissmann, H. W. (1950). Rectilinear locomotion in a snake (*boa occidentalis*). *J Exp Biol* 26, 368–379.

Longo, D. and Muscato, G. (2008). Adhesion techniques for climbing robots: State of the art and experimental considerations. In *Advances in Mobile Robotics: Proceedings of the Eleventh International Conference on Climbing and Walking Robots and the Support Technologies for Mobile Machines* (eds. L.Marques, A.de Almeida and M. O. Tokhi), pp. 6–28. Coimbra, Portugal: World Scientific.

Lutz, G. J. and Rome, L. C. (1994). Built for jumping: The design of the frog muscular system. *Science* 263, 370–372.

Maempel, J., Andrada, E., Witte, H., Trommer, C., Karguth, A., Fischer, M., Voigt, D. and Gorb, S. N. (2008). INSPIRAT—Towards a biologically inspired climbing robot for inspection of linear structures. In *Advances in Mobile Robotics: Proceedings of the Eleventh International Conference on Climbing and Walking Robots and the Support Technologies for Mobile Machines* (eds. L.Marques, A.de Almeida and M. O.Tokhi), pp. 206–2013. Coimbra, Portugal: World Scientific.

Maladen, R. D, Ding,Y., Li, C. and Goldman, D. I. (2009). Undulatory swimming in sand: Subsurface locomotion of the sandfish lizard. *Science* 325: 314–318.

Markus, L. and Yamabe, H. (1960). Global stability criteria for differential systems. *Osaka Math J* 12, 305–317.

Maus, H. M., Lipfert, S. W., Gross, M., Rummel, J. and Seyfarth, A. (2010). Upright human gait did not provide a major mechanical challenge for our ancestors. *Nat Commun* 1, 70.

McGeer, T. (1990). Passive dynamic walking. *Int J Rob Res* 9, 62–82.

McGhee, R. B. and Frank, A. A. (1968). On the stability properties of quadruped creeping gaits. *Math Biosci* 3, 331–351.

McGowan, C. P., Baudinette, R. V., Usherwood, J. R. and Biewener, A. A. (2005). The mechanics of jumping versus steady hopping in yellow-footed rock wallabies. *J Exp Biol* 208, 2741–2751.

McMahon, T. A., Valiant, G. and Frederick, E. C. (1987). Groucho running. *J Appl Physiol* 62, 2326–2337.

McMahon, T. A. and Cheng, G. C. (1990). The mechanics of running: How does stiffness couple with speed? *J Biomech* 23 Suppl 1, 65–78.

Menciassi, A., Gorini, S., Pernorio, G. and Dario, P. (2004). A SMA actuated artificial earthworm. In *Proceedings. ICRA '04. IEEE International Conference on Robotics and Automation*, vol. 4, pp. 3282–3287.

Minetti, A. E. and Alexander, R. M. (1997). A theory of metabolic costs for bipedal gaits. *J Theo Biol* 186, 467–476.

Mochon, S. and McMahon, T. A. (1980). Ballistic walking: An improved model. *Math Biosci* 52, 241–260.

Mori, M. and Hirose, S. (2006). Locomotion of 3D snake-like robots—Shifting and rolling control of active cord mechanism ACM-R3. *J Robot Mechatron* 18, 521–528.

Mrozowski, J., Awrejcewicz, J. and Bamberski, P. (2007). Analysis of stability of the human gait. *J Theo Appl Mech* 45, 91–98.

Nagakubo, A. and Hirose, S. (1994). Walking and running of the quadruped wall-climbing robot. In *IEEE International Conference on Robotics and Automation*, pp. 1005–1012.

Nagy, P. V., Desa, S. and Whittaker, W. L. (1994). Energy-based stability measures for reliable locomotion of statically stable walkers: Theory and application. *Int J Rob Res* 13, 272–287.

Nyakatura, J. A. (2016). Learning to move on land. *Science* 353: 120–121.

Omori, H., Hayakawa, T. and Nakamura, T. (2008). Locomotion and turning patterns of a peristaltic crawling earthworm robot composed of flexible units. In *IROS 2008. IEEE/RSJ International Conference on Intelligent Robots and Systems*, 2008, pp. 1630–1635.

O'Reilly, J. C., Summers, A. P. and Ritter, D. A. (2000). The evolution of the functional role of trunk muscles during locomotion in adult amphibians. *Am Zool* 40, 123–135.

Ortega, J. D. and Farley, C. T. (2007). Individual limb work does not explain the greater metabolic cost of walking in elderly adults. *J Appl Physiol* 102, 2266–2273.

Pandy, M. G. and Zajac, F. E. (1991). Optimal muscular coordination strategies for jumping. *J Biomech* 24, 1–10.

Pandy, M. G., Zajac, F. E., Sim, E. and Levine, W. S. (1990). An optimal control model for maximum-height human jumping. *J Biomech* 23, 1185–1198.

Parry, D. A. and Brown, R. H. J. (1959a). The hydraulic mechanism of the spider leg. *J Exp Biol* 36, 423–433.

Parry, D. A. and Brown, R. H. J. (1959b). The jumping mechanism of salticid spiders. *J Exp Biol* 36, 654–662.

Patek, S. N. and Caldwell, R. L. (2005). Extreme impact and cavitation forces of a biological hammer: Strike forces of the peacock mantis shrimp *Odontodactylus scyllarus*. *J Exp Biol* 208, 3655–3664.

Patek, S. N., Korff, W. L. and Caldwell, R. L. (2004). Biomechanics: Deadly strike mechanism of a mantis shrimp. *Nature* 428, 819–820.

Pereira, R., Machado, M., Dos Santos, M. M., Pereira, L. N. and Sampaio-Jorge, F. (2008). Muscle activation sequence compromises vertical jump performance. *Serbian J Sports Sci* 2, 85–90.

Prilutsky, B. I., Herzog, W. and Leonard, T. (1996). Transfer of mechanical energy between ankle and knee joints by gastrocnemius and plantaris muscles during cat locomotion. *J Biomech* 29, 391–403.

Pugh, D. R., Ribble, E. A., Vohnout, V. J., Bihari, T. E., Walliser, T. M., Patterson, M. R. and Waldron, K. J. (1990). Technical description of the adaptive suspension vehicle. *Int J Rob Res* 9, 24–42.

Putnam, C. A. (1993). Sequential motions of body segments in striking and throwing skills: descriptions and explanations. *J Biomech* 26 Suppl 1, 125–135.

Raibert, M., Blankepoor, K., Nelson, G., Playter, R. and Team, B. D. (2008). Bigdog, the rough terrain quadruped robot. In *17th World Congress of the International Federation of Automatic Control (IFAC)*, Seoul, pp. 10822–10825.

Rechenberg, I. (2009). Im Gespräch: Ingo Rechenberg über eine neu entdeckte Art rollende Spinnen für den Mars. Fragen: Anna Loll. *Frankfurter Allgemeine Zeitung*, p. 7. January 10, 2009.

Rind, F. C., Birkett, C. L., Duncan, B.-J. A. and Ranken, A. J. (2011). Tarantulas cling to smooth vertical surfaces by secreting silk from their feet. *J Exp Biol* 214, 1874–1879.

Roberts, T. J. and Marsh, R. L. (2003). Probing the limits to muscle-powered accelerations: Lessons from jumping bullfrogs. *J Exp Biol* 206, 2567–2580.

Rode, C., Siebert, T. and Blickhan, R. (2009). Titin-induced force enhancement and force depression: A "sticky-spring" mechanism in muscle contractions? *J Theo Biol* 259, 350–360.

Rode, C., Siebert, T., Till, O. and Blickhan, R. (2008). Effect of muscle properties on postural stability of simple musculoskeletal system. In *8th World Congress on Computational Mechanics (WCCM8)—5th European Congress on Computational Methods in Applied Sciences and Engineering (ECCOMAS 2008)*. Venice, Italy.

Roennau, A., Kerscher, T. and Dillmann, R. (2010). Design and kinematics of a biologically-inspired leg for a six-legged walking machine. In *2010 3rd IEEE RAS and EMBS International Conference on Biomedical Robotics and Biomechatronics (BioRob)*, pp. 626–631.

Ruina, A., Bertram, J. E. and Srinivasan, M. (2005). A collisional model of the energetic cost of support work qualitatively explains leg sequencing in walking and galloping, pseudo-elastic leg behavior in running and the walk-to-run transition. *J Theo Biol* 237, 170–192.

Sakagami, Y., Watanabe, R., Aoyama, C., Matsunaga, S., Higaki, N. and Fujimura, K. (2002). The intelligent ASIMO: System overview and integration. In *IEEE/RSJ International Conference on Intelligent Robots and Systems*, 2002. vol. 3, pp. 2478–2483.

Sangbae, K., Spenko, M., Trujillo, S., Heyneman, B., Mattoli, V. and Cutkosky, M. R. (2007). Whole body adehsion: Hierarchical, directional and distributed control of adhesive forces for a climbing robot. In *IEEE International Conference on Robotics and Automation*, Roma.

Sangbae, K., Spenko, M., Trujillo, S., Heyneman, B., Santos, D. and Cutkosky, M. R. (2008). Smooth vertical surface climbing with directional adhesion. *IEEE Trans Robot* 24, 1–10.

Saranli, U., Buehler, M. and Koditschek, D. E. (2001). RHex: A simple and highly mobile hexapod robot. *Int J Rob Res* 20, 616–631.

Sasaki, K., Neptune, R. R. and Kautz, S. A. (2009). The relationships between muscle, external, internal and joint mechanical work during normal walking. *J Exp Biol* 212, 738–744.

Schmitt, J., Garcia, M., Razo, R. C., Holmes, P. and Full, R. J. (2002). Dynamics and stability of legged locomotion in the horizontal plane: A test case using insects. *Biol Cybernet* 86, 343–353.

Schmitt, J. and Holmes, P. (2000). Mechanical models for insect locomotion: Dynamics and stability in the horizontal plane I. Theory. *Biol Cybernet* 83, 501–515.

Scott, J. (2005). The locust jump: An integrated laboratory investigation. *Adv Phys Educ* 29, 21–26.

Secor, S. M., Jayne, B. C. and Bennett, A. F. (1992). Locomotor performance and energetic cost of sidewinding by the snake *Crotalus cerastes*. *J Exp Biol* 163, 1–14.

Seipel, J. E. and Holmes, P. (2005). Running in three dimensions: Analysis of a point-mass sprung-leg model. *Int J Rob Res* 24, 657–674.

Selles, R. W., Bussmann, J. B., Wagenaar, R. C. and Stam, H. J. (2001). Comparing predictive validity of four ballistic swing phase models of human walking. *J Biomech* 34, 1171–1177.

Seok, S., Wang, A., Chuah, M. Y., Hyun, D. J., Lee, J., Otten, D. M., Lang, J. H. and Kim, S. (2015). Design Principles for Energy-Efficient Legged Locomotion and Implementation on the MIT Cheetah Robot. *IEEE/ASME Trans Mechatron* 20, 1117–1129.

Seyfarth, A., Blickhan, R. and Van Leeuwen, J. L. (2000). Optimum take-off techniques and muscle design for long jump. *J Exp Biol* 203, 741–750.

Seyfarth, A., Friedrichs, A., Wank, V. and Blickhan, R. (1999). Dynamics of the long jump. *J Biomech* 32, 1259–1267.

Seyfarth, A., Geyer, H., Gunther, M. and Blickhan, R. (2002). A movement criterion for running. *J Biomech* 35, 649–655.

Seyfarth, A., Geyer, H. and Herr, H. (2003). Swing-leg retraction: A simple control model for stable running. *J Exp Biol* 206, 2547–2555.

Seyfarth, A., Gunther, M. and Blickhan, R. (2001). Stable operation of an elastic three-segment leg. *Biol Cybernet* 84, 365–382.

Seyfarth, A., Iida, F., Tausch, R., Stelzer, M., von Stryk, O. and Karguth, A. (2009). Towards bipedal jogging as a natural result of optimizing walking speed for passively compliant three-segmented legs. *Int J Rob Res* 28, 257–265.

Siebert, T., Weihmann, T., Rode, C. and Blickhan, R. (2010). *Cupiennius salei*: Biomechanical properties of the tibia-metatarsus joint and its flexing muscles. *J Comp Physiol B* 180, 199–209.

Simon, M. A., Woods, W. A. J., Serenbrenik, Y. V., Simon, S. M., Griethuijsen, L. I. V., Socha, J. J., Wah-Keat, L. and Trimmer, B. (2010). Visceral locomotory pistoning in crawling caterpillars. *Curr Biol* 20, 1458–1463.

Simons, R. S. (1999). Running, breathing and visceral motion in the domestic rabbit (*Oryctolagus cuniculus*): Testing visceral displacement hypotheses. *J Exp Biol* 202, 563–577.

Song, S.-M. and Waldron, K. J. (1987). An analytical approach for gait study and its applications on wave gaits. *Int J Rob Res* 6, 60–71.

Spagele, T., Kistner, A. and Gollhofer, A. (1999a). Modelling, simulation and optimisation of a human vertical jump. *J Biomech* 32, 521–530.

Spagele, T., Kistner, A. and Gollhofer, A. (1999b). A multi-phase optimal control technique for the simulation of a human vertical jump. *J Biomech* 32, 87–91.

Spolenak, R., Gorb, S. and Arzt, E. (2005). Adhesion design maps for bio-inspired attachment systems. *Acta Biomater* 1, 5–13.

Srinivasan, M. (2010). Fifteen observations on the structure of energy-minimizing gaits in many simple biped models. *J Roy Soc Int* 8, 74–98.

Srinivasan, M. and Holmes, P. (2008). How well can spring-mass-like telescoping leg models fit multipedal sagittal-plane locomotion data? *J Theo Biol* 255, 1–7.

Stelzer, M. and von Stryk, O. (2006). Efficient forward dynamics simulation and optimization of human body dynamics. *ZAMM—J Appl Math Mechanics/Z Angew Math Mech* 86, 828–840.

Ting, L. H., Blickhan, R. and Full, R. J. (1994). Dynamic and static stability in hexapedal runners. *J Exp Biol* 197, 251–269.

Usherwood, J. R. and Bertram, J. E. A. (2003). Understanding brachiation: Insight from a collisional perspective. *J Exp Biol* 206, 1631–1642.

Usherwood, J. R., Williams, S. B. and Wilson, A. M. (2007). Mechanics of dog walking compared with a passive, stiff-limbed, 4-bar linkage model, and their collisional implications. *J Exp Biol* 210, 533–540.

van den Doel, K. and Pai, D. K. (1996). Performance measures for robot manipulators: A unified approach. *Int J Rob Res* 15, 92–111.

van Soest, A. J. and Bobbert, M. F. (1993). The contribution of muscle properties in the control of explosive movements. *Biol Cybernet* 69, 195–204.

van Soest, A. J., Schwab, A. L., Bobbert, M. F. and van Ingen Schenau, G. J. (1993). The influence of the biarticularity of the gastrocnemius muscle on vertical-jumping achievement. *J Biomech* 26, 1–8.

Verdaasdonk, B. W., Koopman, H. F. and van der Helm, F. C. (2009). Energy efficient walking with central pattern generators: From passive dynamic walking to biologically inspired control. *Biol Cybernet* 101, 49–61.

Vukobratovic, M. and Borovac, B. (2004). Zero-moment point - thirty five years of its life. *Int J Hum Robot* 1, 157–173.

Wagner, H. and Blickhan, R. (1999). Stabilizing function of skeletal muscles: An analytical investigation. *J Theo Bio* 199, 163–179.

Wagner, H. and Blickhan, R. (2003). Stabilizing function of antagonistic neuromusculoskeletal systems: An analytical investigation. *Biol Cybernet* 89, 71–79.

Wagner, H., Siebert, T., Ellerby, D. J., Marsh, R. L. and Blickhan, R. (2005). ISOFIT: a model-based method to measure muscle–tendon properties simultaneously. *Biomech Model Mechanobiol* 4, 10–19.

Wagner, H., Giesl, P. and Blickhan, R. (2007). Musculoskeletal stabilization of the elbow—Complex or real. *J Mech Med Biol* 7, 275–296.

Wakeling, J. M., Liphardt, A. M. and Nigg, B. M. (2003). Muscle activity reduces soft-tissue resonance at heel-strike during walking. *J Biomech* 36, 1761–1769.

Weidemann, H. J., Pfeiffer, F. and Eltze, J. (1994). The six-legged TUM walking robot. In *Proceedings of the IEEE/RSJ/GI International Conference on Intelligent Robots and Systems '94. 'Advanced Robotic Systems and the Real World', IROS '94*, vol. 2, pp. 1026–1033.

Weihmann, T., Karner, M., Full, R. J. and Blickhan, R. (2010). Jumping kinematics in the wandering spider *Cupiennius salei*. *J Comp Physiol A Neuroethol Sen Neural Behav Physiol* 196, 421–438.

Winter, D. A. and Robertson, D. G. (1978). Joint torque and energy patterns in normal gait. *Biological Cybernetics* 29, 137–142.

Wisse, M., Schwab, A. L. and van der Helm, F. C. T. (2004). Passive dynamic walking model with upper body. *Robotica* 22, 681–688.

Witte, H., Hoffmann, H., Hackert, R., Schilling, C., Fischer, M. S. and Preuschoft, H. (2004). Biomimetic robotics should be based on functional morphology. *J Anat* 204, 331–342.

Woledge, R. C., Curtin, N. A. and Homsher, E. (1985). Energetic aspects of muscle contraction. *Monogr Physiol Soc* 41, 1–357.

Yeadon, M. R. and Mikulcik, E. C. (1996). The control of non-twisting somersaults using configuration changes. *J Biomech* 29, 1341–1348.

Yoneda, K., Ota, Y., Ito, F. and Hirose, S. (2001). Quadruped walking robot with reduced degrees of freedom. *J Robot Mechatron* 13, 190–197.

Young, I. S., Alexander, R., Woakes, A. J., Butler, P. J. and Anderson, L. (1992). The synchronization of ventilation and locomotion in horses (*Equus caballus*). *J Exp Biol* 166, 19–31.

Zeglin, G. J. (1991). Uniroo: A one legged dynamic hopping robot. In *Department of Mechanical Engineering*, Bachelor of Science, Massachusetts Institute of Technology, Cambridge, MA.

Zentner, L., Petkun, S. and Blickhan, R. (2000). From the spider leg to a hydraulik device. *Technische Mechanik*, 20, 21–29.

Zhiguo, L., Aoyama, T., Sekiyama, K., Hasegawa, Y. and Fukuda, T. (2010). Walk-to-brachiate transfer of multi-locomotion robot with error recovery. In *2010 IEEE/RSJ International Conference on Intelligent Robots and Systems (IROS)*, pp. 166–171.

Zimmermann, K. and Zeidis, I. (2007). Worm-like locomotion as a problem of nonlinear dynamics. *J Theo Appl Mech* 45, 179–187.

INDEX

A

D

E

F

G

H

I

J

K

L

M

N

O

P

Q

R

S

T

U

V

W

Z